THIS BOOK IS THE PROPERTY OF:

STATE ____________
PROVINCE ____________
COUNTY ____________
PARISH ____________
SCHOOL DISTRICT ____________
OTHER ____________

Book No. ________
Enter information in spaces to the left as instructed

ISSUED TO	*Year Used*	*CONDITION*	
		ISSUED	*RETURNED*

PUPILS to whom this textbook is issued must not write on any page or mark any part of it in any way, consumable textbooks excepted.

1. Teachers should see that the pupil's name is clearly written in ink in the spaces above in every book issued.
2. The following terms should be used in recording the condition of the book: New; Good; Fair; Poor; Bad.

SOUTH-WESTERN

Pre-Algebra

Claudia R. Carter
Mississippi School for Mathematics and Science
Columbus, Mississippi

Barbara O. Eatherly
North Lamar High School
Paris, Texas

Anita M. Johnston
Napoleon High School
Napoleon, Michigan

MA01AA
SOUTH-WESTERN PUBLISHING CO.
CINCINNATI, OH DALLAS, TX LIVERMORE, CA

Senior Acquisitions Editor: Eve Lewis
Production Editor: Mark Cheatham
Editorial Production Manager: Carol Sturzenberger
Production Artist: Sophia Renieris
Marketing Manager: Gregory Getter

Cover photo by Bray Ficken

ISBN: 0-538-61375-0

Library of Congress Catalog Card Number: 90-63342

2 3 4 5 6 7 8 9 Ki 8 7 6 5 4 3 2

Printed in the United States of America

Authors

CLAUDIA R. CARTER is a mathematics teacher at the Mississippi School for Mathematics and Science, Columbus, Mississippi. She has taught all secondary levels of mathematics and is an active advisor for Mu Alpha Theta. Mrs. Carter is a Woodrow Wilson National Fellowship Foundation Master Teacher, a consultant for IBM as a demonstrator of the Mathematics Exploration Toolkit, and president-elect of the Mississippi Council of Teachers of Mathematics. She earned her M.A. degree at the University of New Orleans. She was the recipient of the Presidential Award for Excellence in Mathematics Teaching in 1989.

BARBARA O. EATHERLY is the mathematics department chairperson at North Lamar High School, Paris, Texas. She has taught mathematics study skills, as well as junior and senior high school mathematics. A supertrainer for the Texas Staff Development Project, Mrs. Eatherly received the General Telephone and Electronics Growth Initiatives for Teachers grant. She received her M.S. degree from East Texas State University. She is a member of her district's curriculum revision committee and the Texas Council of Teachers of Mathematics.

ANITA M. JOHNSTON is the mathematics department head at Napoleon High School, Napoleon, Michigan. In addition to teaching all levels of mathematics, she has coached her school's team in the Mathematics Academic Games since 1977. She has coauthored a monograph for the Michigan Council of Teachers of Mathematics and is a popular speaker. She earned her M.A. degree at Michigan State University. She was the recipient of the Presidential Award for Excellence in Mathematics Teaching in 1989. She is currently the satellite teacher for Tune in Math & Science (TIMS).

Consultants

Karen O'Neal
Klein High School Spring, Texas

Susanne Westegaard
Montgomery-Lonsdale Public School Montgomery, Minnesota

James J. Zwick
Guion Creek Middle School Indianapolis, Indiana

CONTENTS

CHAPTER 3 Fractions and Decimals 74

CHAPTER 4 Data Analysis 112

CHAPTER 5 Working with Variables 150

CHAPTER 6 Solving Equations and Inequalities 192

CHAPTER 7 Problem Solving 238

CHAPTER 8 Ratios, Proportions, and Percents 276

CHAPTER 11 Relations and Functions 398

CHAPTER 12 Graphing Relations and Functions 424

PHOTO ACKNOWLEDGEMENTS

All photos Image Bank Chicago unless indicated by (*).

Cover photo by Bray Ficken.

CHAPTER 1 p. 2: Charles Place; p. 5: Stephen Marks; p. 6: Sobel/Klonsky; p. 9: Jake Rajs, 1987; p. 11: Janeart Ltd.; p. 12: Mel DiGiacomo; p. 15: John Kelly; p. 16: G+J Images; p. 19: Bernard Van Berg; p. 21: G.V. Faint; p. 23: Guido Alberto Rossi; p. 25: Don Carroll; p. 27: S.V.+B Productions; p. 29: Stephen Wilkes; p. 31: Jeff Smith; p. 33: Marc Romanelli; p. 35: Mel DiGiacomo.

CHAPTER 2 p. 40: Patti McConville; p. 43: Stephen Wilkes; p. 45: Garry Gay; p. 47: Patti McConville; p. 48: Dan Witt; p. 51: Robert Kristofik; p. 53: Weinberg-Clark; p. 55: Guido Alberto Rossi; p. 56: Luis Castaneda; p. 59: John Kelly; p. 61: Whitney Lane; p. 63: Robbi Newman; p. 65: Michael Melford; p. 66: Louis H. Jawitz; p. 69: Flip Chalfant.

CHAPTER 3 p. 74: Michael Melford; p. 77: Weinberg-Clark; p. 79: Nicholas Foster; p. 81: Murray Alcosser; p. 82: John P. Kelly; p. 85: Curtis Willocks; p. 86: Sobel/Klonsky; p. 88: Al Satterwhite; p. 89: Andy Caulfield; p. 93: Andy Caulfield; p. 95: John P. Kelly; p. 97: Lou Jones; p. 99: R. Pan, Stockphotos, Inc.*; p. 101: G.V. Faint; p. 102: John P. Kelly; p. 105: Bill Varie, 1987; p. 106: Swedowsky.

CHAPTER 4 p. 111: Cliff Feulner; p. 114: Al Satterwhite; p. 116: Hans Neler; p. 119: Leo Mason; p. 120: Gregory Heisler; p. 123: Michael Melford; p. 129: David Perry Lahrbuck; p. 131: Garry Gay; p. 133: Santi Visalli; p. 135: Weinberg-Clark; p. 137: Guido Alberto Rossi; p. 138: Brett Frommer; p. 140: Don Klumpp; p. 142: Bernard Roussel; p. 144: Stephen Marks.

CHAPTER 5 p. 150: Patrick Eden; p. 153: Marvin E. Newman; p. 154: Dominique Sarroute; p. 157: Dan Witt; p. 159: Ron Colby; p. 161: Larry Pierce, Sam Zarember; p. 165: Grafton Marshall Smith; p. 169: Janeart Ltd.; p. 171: Elyse Lewin; p. 173: Jean Pierre Pieuchot; p. 175: Lou Jones; p. 177: Mel DiGiacomo; p. 179: Steve Niedorf; p. 181: Kevin Forest; p. 183: Eric Schweikardt; p. 185: Jake Rajs; p. 187: Luis Castaneda.

CHAPTER 6 p. 192: Barrie Rokeach; p. 195: Sobel/Klonsky; p. 197: J. Coolidge, 1988; p. 199: Murray Alcosser; p. 200: Garry Gay; p. 203: Larry Dale Gordon; p. 205: Maria Taglienti; p. 207: Karl Hentz; p. 209: W. Scmitz; p. 211: Brett Froomer; p. 213: Tom Mareschal; p. 215: Maria Taglienti; p. 217: Rocha Filho Rubens; p. 219: Obremski, Alex Stewart; p. 220: Mel DiGiacomo; p. 222: Stephen Hunt; p. 225: Nicholas Foster; p. 227: Amanda Clement, Stockphotos, Inc.*; p. 229: John Kelly; p. 231: Williamson/Richards.

CHAPTER 7 p. 238: Alain Choisnet; p. 241: Michael Salas; p. 242: T. Kawaerski; p. 244: Murray Alcosser; p. 244: Michael Tcherevkoff; p. 248: Gabriel M. Covian; p. 250: Lou Jones; p. 253: Dag Sundberg; p. 256: Alvis Upitis; p. 257: David De Lossy; p. 261: Barrie Rokeach; p. 263: Garry Gay; p. 265: Izumi Ohta; p. 266: Mel DiGiacomo; p. 269: Steven Hunt.

CHAPTER 8 p. 276: Dan Witt; p. 278: Nancy Brown; p. 279: Mw. A. Logan; p. 282: Steve Dunwell; p. 283: W. Von dem Bussche; p. 287: Al Hamdan; p. 290: Jeff Spielman, Stockphotos, Inc.*; p. 291: Maria Taglienti; p. 292: Robert E. Burdick, Stockphotos, Inc.*; p. 294: Michael Quackenbush; p. 297: ZAO Productions; p. 299: NASA, Don Carroll, 1987; p. 303; Steve Hunt; p. 304: Alvis Upitis; p. 307: G. V. Faint, Ocean Images, Inc.; p. 310: Stephen Marks; p. 311: Gary S. Chapman.

CHAPTER 9 p. 318: David J. Maenza; p. 321: Andy Caulfield; p. 322: Garry Gay; p. 325: Marks Productions; p. 327: Joseph Brignolo; p. 329: Murray Alcosser; p. 331: Bill Hickey; p. 333: Mark Romanelli; p. 334: Williamson/Edwards; p. 337: G.V. Faint; p. 339: Walter Bibikow, 1988; p. 341: Barrie Rokeach; p. 345: Michael Melford; p. 346: Jay Freis; p. 349: Chris Alan Wilton; p. 351: M. DeCamp; p. 353: David Jeffrey; p. 355: Al Satterwhite.

CHAPTER 10 p. 360: Pete Turner; p. 363: Lou Jones, Frank Whitney; p. 367: Bill Varie; p. 368: Ira Block; p. 372: Steve Dunwell; p. 373: Terje Rakke; p. 375: Steve Dunwell; p. 377: G.V. Faint; p. 379: G + J Images, John P. Kelly; p. 383: ZAO Grimberg; p. 385; Barrie Rokeach; p. 387: Gregory Heisler; p. 388: Murray Alcosser; p. 391: Steve Krongard, 1990; p. 393: Guido Alberto Rossi.

CHAPTER 11 p. 398: Guido Alberto Rossi; p. 401: Luis Veigo; p. 403: J. Carmichael; p. 405: Edward Bower, 1989; p. 409: DW Productions; p. 410: Peter Ginter; p. 413: A.M. Rosario; p. 414: Michael Tcherevkoff; p. 417: Antonio Rosario.

CHAPTER 12 p. 424: Benn Mitchell; p. 427: Alvis Upitis, Pamela J. Zilly; p. 429: Lou Jones; p. 431: H. Peterhofen, Kenneth Redding; p. 433: Joe Devenney; p. 435: Andre Gallant; p. 436: Romilly Lockyer; p. 439: Obremski, Alex Stewart; p. 442: David J. Maenza; p. 443: Alfred Gescheidt.

CHAPTER 13 p. 450: Santi Visalli; p. 452: Domonique Sarroute; p. 454: Obremski, Alex Stewart; p. 457: Martie Pie, Chris Hackett; p. 458, Pete Turner; p. 459: Sobel/Klonsky; p. 461: Margaret W. Peterson; p. 463: John Kelly; p. 465: Larry Dale Gordon, Patrick J. LaCroix; p. 467: Michael Melford, 1989, Ken Cooper; p. 469: Walter Bibikow, 1989; p. 471: Murray Alcosser; p. 473: Terje Rakke; p. 475: Lynn N. Stone; p. 477: Steve Dunwell; p. 479: John P. Kelly.

CHAPTER 14 p. 484: Francisco Ontanon; p. 487: Pete Turner; p. 489: L. Dennis; p. 491: Michael Rochipp; p. 492: Gregory Heisler, Steve Dunwell, 1984; p. 495: Roger Miller; p. 496: Joe Azzara; p. 497: Grafton Marshall Smith; p. 499: Co Rentmeester; p. 500: Nicholas Foster; p. 501: Brett Froomer; p. 503: Michael Melford; p. 505: Geoffrey Gove; p. 507: Al Satterwhite; p. 508: Kay Chernush; p. 511: Philip A. Harrington.

PREFACE

To the Student

How would you answer the question, "What is mathematics?"

We hope that as you read this text and join your teacher in problem-solving, you will understand that mathematics is the study of patterns. As you study new skills and explore the patterns that lead to a conclusion, make up your own rule in your own words. Then apply your rule to real-life situations. To survive in the world, you need computational skills. Equally important, you need to be able to estimate whether the answer the calculator or computer gave you was reasonable.

We hope that you will gain a vision of mathematics as a way to explain the patterns in the world around you. Someone might write a poem to explain the world. You would write an equation. We would expect you to find many uses of mathematics. You might see mathematics in the pattern of your zip code, in the reporting of cold temperatures with negative integers, or in the discounted price of an item. But we want you to begin to create mathematical models to explain relationships. For example, is there an equation that would relate the length of your foot and your shoe size or the temperature outside and the number of chirps of a cricket? Begin to look at your world through the eyes of mathematics and see what mathematical models you can discover.

We have taught for many years. We know that we cannot force you to actively listen, to predict and problem-solve, or to learn and use the language of mathematics. Only you can do this for yourself. It will mean a commitment to think, to question the old ways of doing things, to study and do homework, and to live with the confusion that occurs when new learning takes place. But we need you to learn to think mathematically. The technology that runs our society operates with the language of mathematics. As we increase our use of technology, it is obvious we need an increasing number of citizens literate in algebra.

INTRODUCTION TO PRE-ALGEBRA

To be successful in any endeavor, you need to have goals and plans for how to reach those goals. Take a few minutes to learn how to use this book to help you learn pre-algebra and prepare for further mathematics courses.

GETTING STARTED

You may find it helpful to keep a notebook that contains important information. Your teacher may add or change what you need to keep in this notebook, but, to begin with, start with these ideas.

Vocabulary lists For each unfamiliar word, definition, or property, write the correct definition as given in the text. Also include an example, if that will help. Then rewrite the definition and/or example in your own words.

Notes and questions Depending upon your teacher's method of teaching, this section should contain notes on how to complete exercises and/or examples. In this section, you should write any questions, examples, or exercises that need further explanation. You may want to mark questions with an * or ? in the margin. Then, when you get a satisfactory explanation or are able to answer your own questions, you can circle the symbol (* or ?) to show that you no longer need to be concerned about that problem.

Assignments Write the complete assignment **everyday**. This may prevent "panic" phone calls to classmates, asking, "Do you know the math assignment?"

Formulas or properties You may wish to make this a separate section in the last half of your notebook. By having all this information in one place, you can quickly find what you need. Be sure to include any needed examples or explanations for each item.

Homework Your teacher will tell you whether you should keep your homework in your notebook or whether you should turn it in each day.

USING THIS BOOK

Chapter Openers

Read each introduction and answer the "Ask Yourself" questions as best you can. Study the photograph. How do you think this photograph relates to the material you will be studying in this chapter? Make any appropriate notes in your notebook. Mark with an * or a ? any questions in "Ask Yourself" that you cannot answer at this time. Do not worry if you cannot answer all questions now. You will be able to answer them by the time you finish studying the chapter.

Chapter Summary

Read the chapter summary and vocabulary lists found at the end of the chapter. **Yes, read this now!** Knowing what you will be learning may help you understand various concepts, and the need to study them, as you come to them within the chapter.

Warm-Up Activity

This is a fun activity to begin each lesson. Sometimes these activities provide a puzzle or challenge that expands on previously learned materials. Sometimes these activities will help you develop additional problem-solving strategies.

Introduction

Algebra is related to many real-world situations. These introductions, whether short or long, provide a real-life setting for each lesson.

Skills Development

Here is where you will find the important part of the lesson. Study each example and solution. Be sure to note any questions in your notebook.

Exercises

Here is where you can show what you have learned. Remember, there is no such thing as a dumb question—just be sure you have really tried to do all of the exercises before you ask questions. Also remember that working with a partner might help. In most jobs, you will not be expected to solve every problem on your own. By working with other students, you may be able to help each other. Also, do not be afraid to try to use whatever technology is available. Calculators and computers can take much of the drudgery out of "doing math." Be sure to check with your teacher before using such technology. Sometimes there is a reason that you need to "plod through" the arithmetic or algebra of a particular problem or set of exercises.

Chapter Review
This section reviews concepts and key words as well as provides practice exercises. Be sure you understand this material before going on to the chapter test.

Chapter Test
This is a practice over the material covered in the chapter. Again, be sure you know how to do each exercise. Ask questions!

TAKING TESTS

There are several things you can do to help get the best score possible when taking tests.

Be prepared. This is a key to successful test-taking. Try these ideas:

- Review the vocabulary.
- Read the chapter summary.
- Reread any examples that you had difficulty with.
- Be sure all *'s and ?'s in your notebook are circled (⊛ and ⓘ).
- Have all materials you will need—paper, pencil, calculator, graph paper, etc.

Relax. Being nervous is normal. If you are prepared, that nervous feeling will probably leave once you start the test.

Work neatly. Be sure all of your work is readable. If you complete several steps in a problem, circle the answer.

Read directions. Often, silly errors can be avoided by simply following directions.

Reread each word problem. Be sure you have answered the question that was asked. For example, you know that each bus holds 30 students. You are to find how many buses are needed to transport 45 students. By using algebra and arithmetic, your answer is $1\frac{1}{2}$. But, the number of buses needed is 2, not $1\frac{1}{2}$!

AFTER THE TEST

Review any exercises you did not successfully complete. Ask questions about material you still do not understand. You may need to know these concepts in the very next chapter!

Successful Strategies For Problem-Solving

You solve problems every day. When you make a decision as to which outfit to wear or which way to get to school, you are solving problems. When you go to the store, you solve problems as you decide what to buy, how to pay for it, and how to check that you have been charged the correct price and received the correct change.

KINDS OF PROBLEMS

You will have many opportunities to solve problems in this book. Some of these will be the traditional word problems—by the end of the course you will learn to solve many different kinds of word problems. But you will also learn to solve many other kinds of problems.

Applications

Some problems will relate directly to real-life situations that you, your friends, your parents, or others solve in everyday life. Many of these relate directly to other subjects that you may study in school. To alert you to these problems, the symbol at the left is used.

How do you think each problem below is related to this general type of problem? Do not solve the problems.

1. Jenny knows that her plant for the science fair grows 2 centimeters per week. The science fair is in 3 weeks. If the plant is 5 centimeters tall now, how tall will it be at the science fair?
2. Your mother knows that there is $150 in the checking account. If she writes a check for $48, how much will be left in the account?

Technology

Some problems may involve complicated or repeated mathematical computations. Often you can use technology to help solve these problems. Using a calculator or a computer to solve problems allows you to concentrate on what the problem is and how to solve it, rather than concentrating on the arithmetic needed to solve the problem. The symbols at the left alert you when the use of technology might be of help. You will also find more information on calculators and computers in the Reference Section at the end of this textbook.

Would you use a calculator or a computer to solve the following problem? Explain your choice. Write an original problem that you would use a calculator or a computer to solve.

3. Julio found the following pattern while trying to solve a problem:

$A = 2 \times 3$
$B = 2 \times 3 \times 4$
$C = 2 \times 3 \times 4 \times 5$
$D = 2 \times 3 \times 4 \times 5 \times 6$

$J = ?$

Logic Problems

Logic problems often cannot be solved by just using arithmetic. They require many different strategies to be solved. You will learn more about these strategies later in this section. Logic problems include word puzzles, mathematical puzzles, geometric puzzles, brain teasers, and optical illusions. Many of these are fun to solve.

Two examples of logic problems are given below.

4. Karri is taller than Lau. Mimi is shorter than Karri. William is taller than Lau. Who is tallest?

5. The same digit is used for A in this number: AA10A. What is the value of A if the number is divisible by 5?

Explorations

Sometimes you will be asked to find answers and look for a pattern. These types of explorations allow you to expand your problem-solving powers by using what you know to generalize or find a general rule for a pattern. When doing these kinds of problems, it is important to really try to first find the pattern and then write the general rule. But it is equally as important to test whether your rule works.

Following is a problem where you must first find the pattern and then generalize the pattern. Notice how a table is used to help solve the problem.

6. Count the number of diagonals. Copy and complete the table. How many diagonals does a figure with 10 sides have?

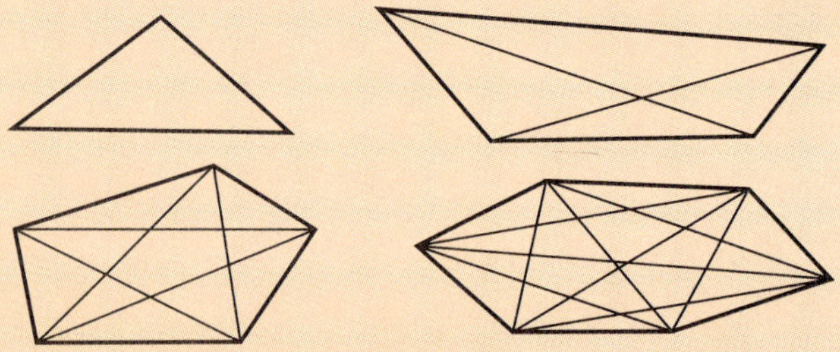

Number of sides	Number of diagonals
3	0
4	
5	
6	
10	

Writing Mathematics

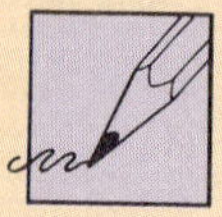

One way to learn to solve word problems is to write your own. At the end of some lessons, you may be asked to use certain information to write your own word problems. You may be asked to do some research and write a report about your findings. The symbol at the left lets you know when you are expected to write an original problem or report. A sample problem follows.

7. Suppose you go to a store with $50. Write a word problem, assuming you buy one or more of the items shown on the signs below.

Rings $12 each	Hats $18 each	Gloves $9 pair	Scarves $8 each

SOLVING PROBLEMS

When you solve problems, you need to have a plan. In the first chapters of this book, you will be solving problems that require mainly arithmetic computation. To solve these problems, you need to follow these steps:

Read. Decide what you are to find and which numbers will help you find the solution.

Plan. Decide whether you can use mental math, estimation, paper-and-pencil, or technology to solve the problem. Next, decide how you can find the answer. Some lessons may give you a special way to find the solution, such as use a chart or table or work backwards. Others will just use computation.

Compute. Use your plan to do the computation.

Answer. Reread the question and write your solution. Be sure your solution answers the question.

Check. Use estimation to check that your computation is correct. Then check that your answer is reasonable. If a problem asked how many students are needed to do a certain job, an answer of $2\frac{1}{2}$ would not be reasonable. You cannot have $\frac{1}{2}$ a person.

If you follow this plan, you should be successful in solving most of the problems in the first few chapters of this textbook. Chapter 7 expands on this plan to give you more problem-solving power.

CHOOSE A METHOD

When solving a problem, you must decide on which method will work best to solve the problem: mental math, estimation, paper-and-pencil, technology, or other methods. Each method can be used for particular kinds of problems. For instance,

Mental Math An exact answer is needed, and the numbers are small enough to compute mentally.

Estimation An approximate answer is good enough. The numbers usually can be rounded to make mental computation easy. This does not give an exact answer.

Paper-and-Pencil You need a record of your computations or you need to see your work on paper. These answers might be exact answers or estimates.

Technology Exact answers are needed, but you do not need a record of your work. Calculators and computers work well for complicated arithmetic procedures. Computers work best when the same procedures have to be repeated or expanded upon many times.

Other Methods These might involve a combination of any of the above, or making a graph, using a chart, extending a pattern, and so on.

Think about each of these problems. Tell which method you would use to solve each problem and explain why you would use that method. Do not solve the problems

8. You are at a restaurant. The bill is $30. How much should you leave for a tip?

9. You worked 8 hours for $6 per hour. How much did you earn?

10. Your mother earned $28,456.34 and your father earned $26,454.99. They paid $11,231 in taxes. How much did they have left after taxes?

11. You work as an accountant for a business. You have to prepare the paychecks. You know how much each person earns per hour, how many hours each person works, and what amounts are to be deducted from each person's paycheck. What method would you use?

PROBLEM-SOLVING STRATEGIES

You will learn many different problem-solving strategies throughout this textbook. Below is a list of some of the strategies that you will be learning.

- Act out the problem.
- Find a counterexample.
- Find a pattern.
- Make a generalization.
- Make a list.
- Make a model.
- Make or use a diagram.
- Make or use a graph.
- Make or use a table or chart.
- Make or use a tree diagram.
- Predict the answer and test your prediction.
- Solve a simpler, related problem.
- Use a formula or a mathematical model.
- Use a specific example.
- Use logical reasoning.
- Work backwards.
- Write an equation.
- Write an inequality.

For each problem, choose an appropriate strategy and explain why you chose that strategy. List any other strategies that you think might work—explain why. Do not solve the problem.

12. Zella gave Barnard 3 apples. Then she gave Gaspar 2 apples and Waldo 12 apples. Finally, she gave Marta 7 apples. Zella had 3 apples left. With how many apples did Zella start?

13. A farmer has a field that is 100 meters long and 100 meters wide. Suppose the farmer places a post every 10 meters. How many posts will be needed to go around the entire field?

14. Yvonne painted the outside of a large cube. The cube was made up of 8 smaller cubes. How many sides of the smaller cubes are painted? How many sides of the smaller cubes are not painted?

15. Rita said that she was thinking of two numbers whose sum is 64 and whose product is 880. What two numbers was she thinking of?

REFERENCE SECTION

At the end of your textbook, you will find a Reference Section. Take a few minutes now to find the type of information that is contained in the Reference Section. See if you can find each of the following items and decide why it is included in the Reference Section.

- Table of Contents
- Index
- Tables of Weights and Measures
- Mathematical Tables
- Science Information

The information you need to solve a problem may not be readily available. You often must use reference materials to solve a problem. Throughout the year, you will be asked to look up information found in the Reference Section in order to solve a problem.

There are many other sources of information that you should learn to use in order to solve problems. Use your learning center, library, or other source to find information to solve the problems below.

16. Write a report on one planet in the solar system. Which planet would you choose? Why? What type of mathematical information would you include in your report? What sources could you use?

17. You are to write your own mathematics problems that relate to the type of work you hope to do sometime in the future. What career will you choose to report on? Where can you find information on that career?

18. Your mathematics assignment uses the current interest rates for various loans and savings accounts. Where can you go to find this information? Why is it important to know where to find this information?

DIFFERENT STYLES

Compare how you thought you would solve the problems above with the methods suggested by other classmates. Answer these questions.

- Did you agree on how the application problems related to real life?
- Did you agree on whether to use a calculator or computer to solve Julio's problem?
- For Exercises 8–11, did you agree on the best method?
- For Exercises 12–15, did you agree on which problem-solving strategy to use?
- For Exercises 16–18, did you agree on which source of information to use?

For most of the questions above, your answer was probably "No." We all learn in different ways. Therefore, we also solve problems in different ways. Often, there is not just one right way to solve a problem. There may be one way that is easiest or fastest, but for many real-life problems many different methods and strategies can be used.

As you work through problems in this textbook, be willing to discuss with other students and your teacher your method for solving the problem. Often, you may find that other students can help you become a better problem-solver by showing you alternative methods or strategies. However, no matter what strategy or method is used, you should always write down the information that you used to solve a problem. Then you can go back and review your work at any time.

SUCCESS

Remember, whether completing a group of exercises or solving word problems, you can succeed. Do not be afraid to ask questions. There is a good possibility that other students have the same questions!

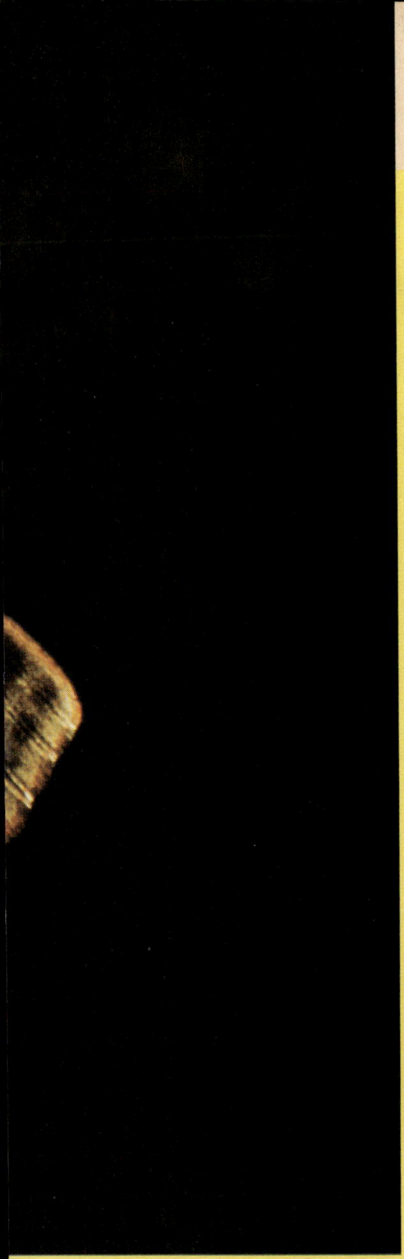

Patterns are found in our lives everyday—in traffic, in nature, and in everyday mathematics.

Patterns in traffic can be a help as well as a hindrance. So that traffic will flow as smoothly as possible, engineers try to locate highways and interchanges by collecting data about the traffic. Computers sort the data, and resulting patterns help make intelligent decisions for future traffic flow.

Patterns in nature can be observed in both plants and animals along with the universe. The butterfly shows how these patterns can be beautiful.

Patterns are also found in numbers. In order to learn about the properties of numbers, and to develop additional skills for solving problems, you need to study patterns about numbers.

In this chapter, you will study patterns associated with our number system.

Ask Yourself

- *What is unique about our number system?*
- *What are prime numbers?*
- *What are some short cuts in computation?*
- *What is meant by an exponent?*

CHAPTER 1

PATTERNS AND PROPERTIES OF NUMBERS

SECTION 1–1

PATTERNS IN EVERYDAY LIFE

WARM-UP ACTIVITY

The numbers 1, 2, 3, 4, 5, and 6 are used to fill the circles so that each side of the triangle adds up to 9.

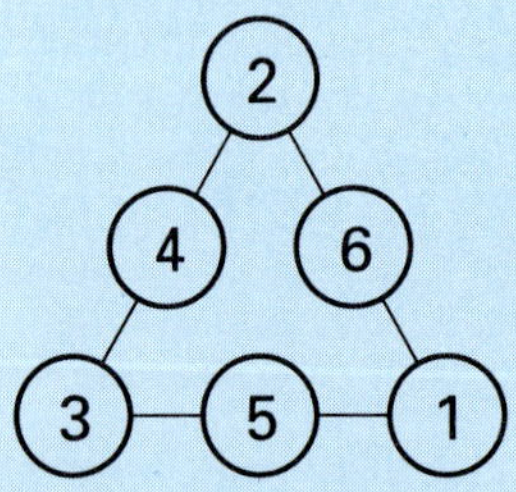

• Use the numbers 1 to 6 so that each side adds to 10.

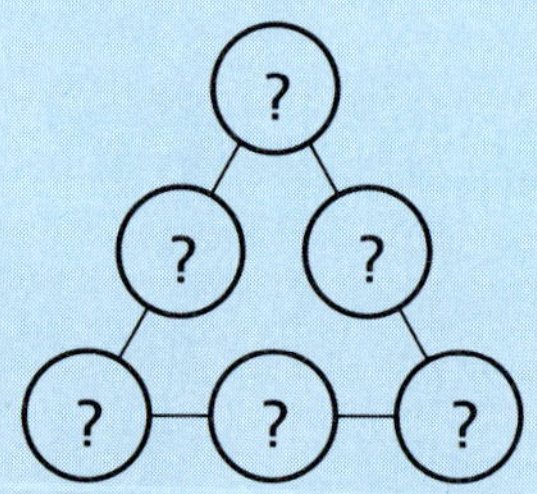

• Use the numbers 1 to 6 again so that each side adds to 11.

INTRODUCTION

People use numbers everyday for identification, ordering, or ranking. Sometimes the use of patterns is hidden. For example, when making a long-distance telephone call or using a ZIP code, the position of the numbers is very important.

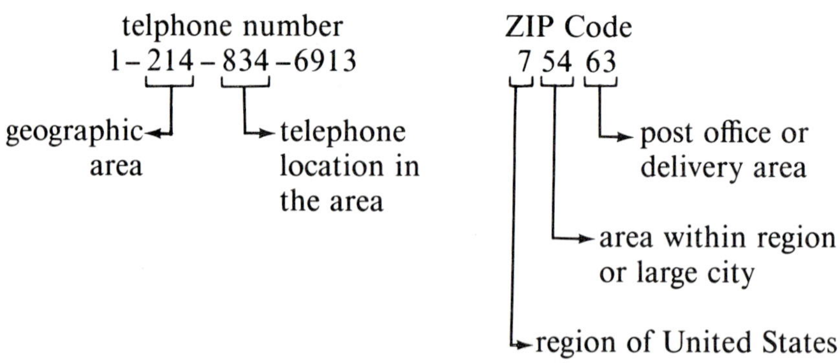

SKILLS DEVELOPMENT

Telephone numbers and ZIP codes are used for identification. Numbers are also used for ordering and ranking. Page numbers bring order to a book and help you find information. **Ranking** means assigning places based on an outcome. For example, first place in the 100-meter dash goes to the person who finishes in the least amount of time.

Example 1

Tickets were sold for the school play. Prizes were given to the top three sellers. Chris sold 4 tickets, Jack sold 25 tickets, Jana sold 19 tickets, Betty sold 5 tickets, and Jon sold 2 tickets. Find the top three winners.

Solution Making a list that shows the number of tickets sold from highest to lowest is a good problem-solving strategy for this problem. Jack won first place, Jana second, and Betty third.

Tickets	Sold By
25	Jack
19	Jana
15	Betty
4	Chris
2	Jon

Example 2

Who bought their ticket first, Jim or Jean?

Jean Jim

Solution Since the number 153,774 comes before 153,775 when counting, then Jean bought her ticket before Jim.

Numbers can also be organized for special kinds of ranking.

Example 3

Four students were top sellers of fruit in a fund-raising sale. A runoff sale was held to determine winners of prizes for the highest daily sales and a grand prize winner of total sales.

On Monday, Wednesday, and Friday, Bonnie sold \$30.50, \$27.50, \$64.75, respectively. On those days Doris sold \$38.50, \$49.00, \$45.50. On Monday, Bob had no sales. On Wednesday and Friday, he had \$67.00 and \$62.00. On Monday, Wednesday, and Friday, Judy sold \$20.25, \$36.75, and \$65.00.

a. Who were the daily prize winners?

b. Who won the Grand Prize?

Solutions A problem-solving strategy here is to put the information in a chart or table.

Fund-Raising Fruit Sale

	Monday	Wednesday	Friday	Total Sales
Bonnie	\$30.50	\$27.50	\$64.75	\$122.75
Doris	38.50	49.00	45.50	133.00
Bob	____	67.00	62.00	129.00
Judy	20.25	36.75	65.00	122.00

a. Compare amounts in each column for daily prize winners. Doris, Bob, and Judy are the daily prize winners.

b. Use your calculator. Find the total of the row with the highest amount. So, Doris is the Grand Prize winner.

EXERCISES

1. Each day after basketball practice C.J. and Sue each shoot 25 free throws. Below are the results. Who made the most free throws?

	Monday	Wednesday	Friday
C.J.	16	18	15
Sue	20	19	9

In mathematics, ranking is based on numerical value. However, in language it is often based on alphabetical order. The eight cities on the right are ranked by their high temperature on a certain day.

Temperature (degrees F)	City
90°	Bangkok
88°	Singapore
66°	Sydney
61°	Athens
54°	Berlin
41°	Copenhagen
36°	Oslo
22°	New York City

2. Rank these cities and their temperature using alphabetical order.

3. What is the advantage of each type of ranking?

Many items require a serial number.

4. Make a list of such items: bicycles, cars, radios, etc.

5. For two items on your list, obtain a sample of the type of serial number used.

6. Give one or more reasons for making and keeping such a list.

During a July expedition, a group of scientists caught and tagged 15 polar bears in Fairbanks, Alaska. The first animal tagged was numbered PB-10792, the second was numbered PB-10793, and so on.

7. What was the number of the fifteenth bear tagged?

8. Several months later a polar bear wearing tag number PB-10814 was seen outside Fairbanks. Was this one of the bears tagged during the July expedition?

Mrs. Harvey drives to work three days a week. Rush hours are from 8A.M.–9A.M., 11:30A.M.–1P.M., and 4:30P.M.–7P.M.. During rush hours it is a one-hour drive. At other times, 35 minutes.

9. Mrs. Harvey leaves for work at 7:20A.M. and starts for home at 5:15P.M.. About how long does she spend driving to and from work each week?

10. How much of Mrs. Harvey's time is delayed by traffic?

SECTION 1–2

PLACE VALUE

WARM-UP ACTIVITY

Work independently.

Materials
magazines, Sunday newspaper

Use magazines or a section of the Sunday newspaper. Make a list of ways numbers are used. Compare your list with those of other students in the class.

INTRODUCTION

In early times, various symbols were used to write numbers. The name or symbol used to represent a number is called a **numeral.**

Roman	I	V	X	L	C	D	M
Hindu-Arabic	1	5	10	50	100	500	1,000

The number system you use today is the **decimal number system.** It is sometimes called the Hindu-Arabic system of numeration. There are only 10 number symbols:

1, 2, 3, 4, 5, 6, 7, 8, 9, 0

Any number can be written using these ten number symbols, which are called **digits.**

SKILLS DEVELOPMENT

Each digit of a number has a certain **place value.** The position of a digit in a number tells *which place* the digit is in—the units place, tens place, hundreds place, and so on. For example, in the number 2578, the digit 5 is in the hundreds place.

hundreds place → 5; tens place → 7

thoudsands place → 2578 ← ones place

SKILLS DEVELOPMENT

Each digit of a number has a certain **face value.** The actual digit (0, 1, 2, 3, 4, 5, 6, 7, 8, 9) tells *how many* ones, tens, hundreds, thousands, and so on, there are.

- In the number 5164, the digit 5 tells you that there are five thousands.
- In the number 3982, the digit 3 tells you that there are three thousands.

Numbers can be written in different forms. **Decimal form** is the most common. 6731 is written in decimal form.

Expanded form describes the meaning of a number.

Decimal form *Expanded form*

$$6731 = 6000 + 700 + 30 + 1$$
$$= 6 \times 1000 + 7 \times 100 + 3 \times 10 + 1$$

Language form describes the number in words. To express or write numbers in words, your knowledge of place value and face value is used. $6731 = 6000 + 700 + 30 + 1$

The numeral 6731 is read as *six thousand, seven hundred thirty-one.* The numeral 7,845,163 is read as *seven million, eight hundred forty-five thousand, one hundred sixty-three.*

Example 1

Write the word name for each.

a. 17,033 b. 102,300

Solutions a. Seventeen thousand, thirty-three.
b. One hundred two thousand, three hundred.

When writing a word name for a number, use commas as they would appear in the decimal form—to separate periods of *three* digits.

Place value and face value are also used to decide which numbers are greater or smaller than other numbers..

Remember:
$<$ means *is less than*
$>$ means *is greater than*

Example 2

Replace ○ with $<$ or $>$.

a. 481 ○ 439 b. $3 \times 100 + 2 \times 10 + 8$ ○ 314

Solutions

a. $481 > 439$ since the hundreds digits are the same, but for the tens digits $8 > 3$.

b. $3 \times 100 + 2 \times 10 + 8 = 328$
Since $328 > 314$ then, $3 \times 100 + 2 \times 10 + 8 > 314$

EXERCISES

Match the expanded form with the decimal form on the right.

1. $4 \times 100 + 5 \times 10 + 1$	**A.** 4501
2. $4 \times 1{,}000 + 5 \times 10 + 1$	**B.** 40,510
3. $4 \times 10{,}000 + 5 \times 100 + 1 \times 10$	**C.** 40,051
4. $4 \times 1{,}000 + 5 \times 100 + 1$	**D.** 451
5. $4 \times 10{,}000 + 5 \times 10 + 1$	**E.** 4051

Write each number in decimal form.

6. three hundred forty-five

7. two thousand, one hundred sixteen

8. ninety-five thousand, forty

9. one million, one hundred thousand, one

Write the numbers in the following in decimal form.

10. The Tibetan yak ranges at a higher altitude than any other mammal. It was seen in the Himalayas at an altitude of six thousand, ninety-six meters.

11. Footprints of the Abominable Snowman have been reported at an altitude of seven thousand, ten meters in the Himalayas.

Which number is greater? Replace ○ with $<$ or $>$.

12. 409 ○ 490 **13.** 639 ○ 640 **14.** 2361 ○ 2631

15. 9009 ○ 9100 **16.** 1999 ○ 2000 **17.** 24,301 ○ 24,299

What is the place value and face value of the digit 7? The first one has been done for you.

18. 3761 hundreds place; 700 **19.** 3070 **20.** 2798

21. 79,436 **22.** 733,242,936 **23.** 7,913,248 **24.** 721,311

Compare. Write T for true or F for false.

25. $4 \times 100 + 3 \times 10 < 9 \times 100 + 9 \times 10 + 9$

26. $6 \times 10{,}000 + 4 > 9 \times 100 + 9 \times 10 + 9$

Write each dollar amount in words.

27. \$27 **28.** \$48 **29.** \$348 **30.** \$1005

Numbers used to describe the solar system are large. Write each number in decimal form.

31. The diameter of the sun is one million, three hundred ninety-two thousand kilometers.

32. Jupiter is seven hundred seventy-seven million, nine hundred fifty thousand kilometers from earth.

Follow the directions and answer the questions.

33. Use your calculator. Press 5, 6, 2, and 3. What is the value of the 6?

34. Now, press [×]. Then press [2], then [=]. What is the place value of the 6?

35. Now, press [×]. Then press [8], then [=]. What is the place value of the 6?

36. Refer to the Roman/Hindu-Arabic numeral chart at the beginning of this section. Use what you know to write and add values of Roman numerals. For example, the current year, your birthday, your birth year, social security.

SECTION 1–3

ROUNDING AND ESTIMATING

WARM-UP ACTIVITY

The average heart beats 70 times per minute when relaxed and makes 180 beats per minute during strenuous exercise.

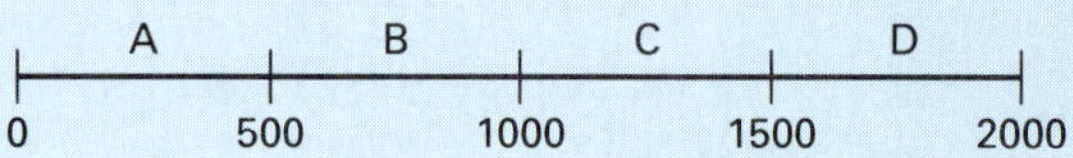

Estimate the answer. Do not calculate. Write the letter for which each number of heartbeats would fall.

1. when relaxed for: **a)** 5 minutes **b)** 12 minutes **c)** $\frac{1}{4}$ hour

2. during strenuous work for: **a)** 5 minutes **b)** 9 minutes

INTRODUCTION

Approximate numbers are found by rounding. Approximate means *about.* For example,

- About 250,000,000 people live in the United States.
- The Amazon River is about 4000 miles long.
- About 42,000 people read the sports section in our local daily newspaper.
- A typical stadium will hold about 50,000 people.

SKILLS DEVELOPMENT

Use your understanding of place value to **round to the nearest** ten.

Look at the ones digits. All are less than 5.		Look at the ones digits. All are greater than or equal to 5.	
481, 482, 483, 484	Round to 480.	485, 486, 487, 488, 489	Round to 490.

The skills for rounding to the nearest apply to greater numbers.

Example 1

a. Round 4783 to the nearest 100.
b. Round 15,504 to the nearest 1000.

Solutions

a. $4783 \approx 4800$ → To round to the nearest 100, look at the tens digit.

b. $15{,}504 \approx 16{,}000$ → To round to the nearest 1000, look at the hundreds digit.

The symbol ≈ may be used to show rounded numbers. It means *is approximately equal to.*

Skills with rounding can help you estimate to check whether the answer to a problem is reasonable.

Example 2

The number of visitors at each exhibit of the zoo were 16,183 at the African Pavilion and 9864 at the Reptile Pavilion. Which pavilion had more visitors? How many more?

Solution Subtract. Check by estimating.

		Check:	
African Pavilion	16,183	Rounds to	16,000
Reptile Pavilion	− 9,864	Rounds to	−10,000
	6,319	about	6,000

The answer is reasonable.

There were 6319 more visitors at the African Pavilion.

Rounding helps you estimate sums and differences.

Example 3

Estimate the sum and the difference to the nearest thousand.

a. 7231 + 958 + 1601 b. 9542 − 8841

Solutions Round each number to the nearest thousand.

a.
$$\begin{array}{rr} 7{,}231 & 7{,}000 \\ 958 & 1{,}000 \\ \underline{+1{,}601} & \underline{+2{,}000} \\ & 10{,}000 \end{array}$$

b.
$$\begin{array}{rr} 9{,}542 & 10{,}000 \\ \underline{-8{,}841} & \underline{-9{,}000} \\ & 1{,}000 \end{array}$$

The sum is about 10,000. The difference is about 1000.

Rounding can also help you estimate products and quotients.

Example 4

Estimate products and quotients.

a. 851 × 79 b. 3152 ÷ 48

Solutions Round each number. Then multiply or divide.

a.
$$\begin{array}{rr} 851 & 900 \\ \underline{\times\ 79} & \underline{\times\ 80} \\ & 72{,}000 \end{array}$$

b.
$$48\overline{)3152} \quad 50\overline{)3000}^{\,60}$$

The quotient is about 60.

The product is about 72,000.

EXERCISES

Estimate the answer.

1. 873 − 148 **2.** 72 × 61 **3.** 4896 ÷ 72

Each answer is estimated. Which is reasonable? Choose A, B, or C.

		A	B	C
4.	423 + 385	500	800	1000
5.	295 − 48	50	150	250
6.	15 × 58	120	800	1200
7.	4985 ÷ 69	70	90	700

Round each number to the accuracy indicated.

8. The water tank held 2154 L of water. (nearest 10L)

9. The record for high kicks is 8491. (nearest 100)

Estimate the answer.

10. One week a farmer sold 7231 ripe mangoes, 958 green mangoes, and 601 melons. About how many pieces of fruit did he sell in all?

11. A supermarket gave refunds for 9542 aluminum cans and 8841 bottles. How many more cans than bottles were returned?

12. During this year Eliza scored 6983 points. She scored 4213 points last year. How many more points did she score this year?

Use your calculator.

	Numbers	Round to nearest	"Check It"
13.	3,649 438 1,752 1,955	hundreds	7,800
14.	7,652 38,407 49,500 150,498	thousands	246,000
15.	63,462 145,651 195,338 99,000	ten thousands	510,000

a. Clear your calculator.
b. Round the numbers in each section.
c. Enter the rounded number into your calculator.
d. Push [+] after each rounded entry.
e. The display on your calculator should equal the "Check It" number.
f. Clear. Then start again.

16. Look at the photograph. Estimate large items of things found at a stadium, such as hot dogs, adults, etc. Make problems of your own.

FACTORS, MULTIPLES, AND EXPONENTS

WARM-UP ACTIVITY

Use 24 centimeter cubes. Arrange as many rectangles with the cubes as possible. The following arrangement shows 4 rows of 6 cubes.

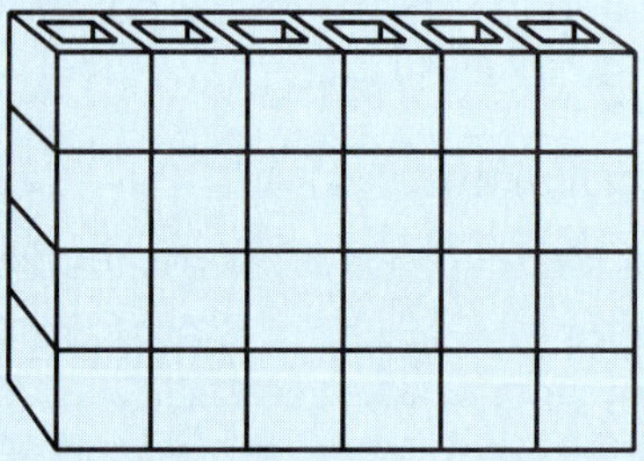

INTRODUCTION

The different arrangements of the 24 cubes may suggest the idea of **factoring.** The number 24 can be expressed as factors in different ways.

The set of all the factors of 24 is {1, 2, 3, 4, 6, 8, 12, 24}. The brackets show the members of a set.

SKILLS DEVELOPMENT

A factor of a number is always less than or equal to the number. Use your calculator to find factors.

Example 1

Write the set of all the factors of 40.

Solution Writing pairs of factors helps you check that all factors are included.

$1 \times 40 = 40$ → The number 1 is a factor of every number.
$2 \times 20 = 40$
$4 \times 10 = 40$
$5 \times 8 = 40$
$8 \times 5 = 40$ → It is not necessary to continue since these factors are already recorded.

The set of all factors of 40 is {1, 2, 4, 5, 8, 10, 20, 40}.

To test whether a number is a factor you need only check whether the number divides evenly.

Example 2

Verify each statement.

a. 4 is a factor of 124.
b. 4 is not a factor of 125.

Solutions a. $4\overline{)124}$ with quotient 31 → No remainder. 4 divides evenly.

So, 4 is a factor of 124.

b. $4\overline{)125}$ with quotient 31 r1 → Remainder.
124
1

So, 4 is not a factor of 125.

A number divides evenly if it is a factor. Since 4 is a factor of 124, then 124 is **divisible** by 4.

The numbers 2, 4, 6, 8, and so on are **multiples** of 2. To find the multiples of 2, multiply 2 by 1, 2, 3, 4, and so on.

Example 3

Find the multiples of 2.

Solution $1 \times 2 \rightarrow 2$ $\quad 2 \times 2 \rightarrow 4$ $\quad 3 \times 2 \rightarrow 6$ $\quad 4 \times 2 \rightarrow 8$ $\quad 5 \times 2 \rightarrow 10$ . . .

A multiple of 2 has as one of its factors the number 2. Each multiple of 2 is divisible by 2.

The symbol . . . means *and so on.*

The multiples of 2 are 2, 4, 6, 8, 10, . . .

A multiple of a number is always greater than or equal to the number.

A number can be a multiple of more than one number.

- 8 is a multiple of 2 since 8 is divisible by 2.
- 8 is a multiple of 4 since 8 is divisible by 4.

Symbols express numbers in a more compact way. The sum of a group of the same number can be expressed more quickly by using multiplication.

$4 + 4 + 4 \rightarrow 3 \times 4 = 12$ $\qquad 3 + 3 + 3 + 3 + 3 \rightarrow 5 \times 3 = 15$

There is also a shorter way to express the same number as a factor.

$2 \times 2 \times 2 \times 2 \times 2 = 32$ $\qquad$ base $\rightarrow 2^5 \leftarrow$ exponent

2^5 means 2 is used as a factor 5 times.

2^5 is said to be in **exponential form.** Each part has a special meaning. 5 is the **exponent.** It tells how many equal factors there are. 2 is the **base.** It shows what all the equal factors are. The result is said to be a **power** of 2.

2^5 is read 2 to the fifth power.
4^3 is read 4 to the third power.
5^4 is read 5 to the fourth power.

Example 4

Find the value of 3^3.

Solution $3^3 = 3 \times 3 \times 3 = 27$

EXERCISES

1. Write the number 36 as a product of two factors in as many ways as you can.

2. Write the set of factors of 36.

Rewrite these numbers in exponential form.

3. 16 **4.** 25 **5.** 27 **6.** 81

Name the base and the exponent.

7. 1^2 **8.** 2^3 **9.** 3^7 **10.** 7^3

Write three multiples of each number.

11. 2 **12.** 3 **13.** 5 **14.** 6

15. 7 **16.** 8 **17.** 9 **18.** 10

Write each of the following in exponential form.

19. $5 \times 5 \times 5$ **20.** 9×9 **21.** $4 \times 4 \times 4 \times 4$

22. $10 \times 10 \times 10 \times 10$

Write each number as a product of two factors in as many ways as you can. Then write the set of factors for each number.

23. 100 **24.** 24 **25.** 45 **26.** 18

Does each number have both 3 and 5 as factors?

27. 36 **28.** 15 **29.** 105 **30.** 35

31. 21 **32.** 330 **33.** 103 **34.** 30

Find the value.

35. 4^2 **36.** 2^4 **37.** 1^3 **38.** 9^2

39. 3^3 **40.** 10^2 **41.** 10^3 **42.** 1^{100}

SECTION 1–5

PRIME NUMBERS

WARM-UP ACTIVITY

Copy and complete the chart. Use the divisibility tests in the table below. Check your work with a calculator.

Divisible By

	Number	2	3	4	5	6	8	9	10
1.	49,863	×	√	×					
2.	28,020								
3.	31,415								
4.	29,687								
5.	92,145								
6.	68,360								

☑ if number divides evenly. ☒ if a remainder

Number Divisible by	Divisibility Tests
2	if the number is even
3	sum of the digits divisible by 3
4	last two digits divisible by 4
5	if the number ends in 0 or 5
6	the number is even and divisible by 3
8	the number is even and divisible by 4
9	the sum of digits divisible by 9
10	the last digit must be zero

SKILLS DEVELOPMENT

This chart shows that the numbers 2, 3, 5, and 7 have only two different factors. These numbers are called **prime numbers.**

The numbers 4, 6, 8, 9, and 10 have more than two different factors. These numbers are called **composite numbers.**

Number	Factors	Number of Different Factors
1	1	1
2	1,2	2
3	1,3	2
4	1,2,4	3
5	1,5	2
6	1,2,3,6	4
7	1,7	2
8	1,2,4,8	4
9	1,3,9	3
10	1,2,5,10	4

Any whole number other than 1 is either a prime number or a composite number. The number 1 has only itself as a factor, $1 = 1 \times 1$. So, it is neither a prime nor a composite number.

To find the prime factors of a number,

- write two factors of the number, then
- continue to factor any composite numbers.

Organize your search for prime factors by using a **factor tree.**

Example 1

Write 80 as a product of prime factors.

Solution

$$80$$

$$4 \times 20$$

$$2 \times 2 \times 2 \times 10$$

$$2 \times 2 \times 2 \times 2 \times 5$$

So, $80 = 2 \times 2 \times 2 \times 2 \times 5$ is the prime factorization.

To find the prime factors the starting point may be different, but the prime factors you obtain will always be the same.

Example 2

Find the prime factorization of 144.

Solutions

144

12	×	12	→ Find factor pairs.
3 × 4	×	3 × 4	→ Continue factoring.
3 × 2 × 2	×	3 × 2 × 2	→ These are prime.
3 × 3	×	2 × 2 × 2 × 2	→ Rearrange factors.
3^2	×	2^4	→ $(3^2)(2^4)$

So, the prime factorization in exponential form is $(3^2)(2^4)$

EXERCISES

Which numbers are prime? Composite? Give reasons for your answer.

1. 2 **2.** 7 **3.** 10 **4.** 19 **5.** 5

6. 9 **7.** 21 **8.** 41 **9.** 12 **10.** 11

Find the missing factors.

11. ____ × 10 = 30 **12.** 8 × ____ = 56 **13.** ____ × 18 = 72

14. ____ × 12 = 192 **15.** 14 × ____ = 84 **16.** 19 × ____ = 209

Express each number as a product of its prime factors.

17. 12 **18.** 50 **19.** 75

20. 48 **21.** 100 **22.** 200

All the prime factors of each number are shown. What is the number?

23. 2, 3, 5 **24.** 2, 2, 3, 3 **25.** 2, 2, 3, 5, 5

26. 2, 3, 5, 5 **27.** 2, 3, 3, 5, 7 **28.** 2, 2, 2, 3, 13

Write the prime factors of each number.

29. 4 **30.** 16 **31.** 25 **32.** 27 **33.** 49

34. Two prime factors of a number are 2 and 3. What is another factor of the number?

35. The prime factors of a number are 3, 5, and 7. What is another factor of the number?

Write the prime numbers between

36. 2^2 and 3^2 **37.** 3^2 and 4^2 **38.** 4^2 and 5^2

39. 5^2 and 6^2 **40.** 6^2 and 7^2 **41.** 7^2 and 8^2

42. Use your answers in the above questions to predict the number of prime numbers between 9^2 and 10^2.

43. Write the prime numbers between 9^2 and 10^2. How many did you find?

44. How close was your prediction in Question 42 to your answer in 43?

The digits 2, 3, and 4 can be written to make six different numbers.

45. List the six numbers. **46.** Which ones are NOT prime?

The chart gives examples of another way of classifying numbers. Find the sum of its factors, excluding the number itself:

perfect—the sum equals the number

deficient—the sum is less than the number

abundant—the sum is greater than the number

Number	6	5	12
Factors Excluding the Number Itself	1, 2, 3	1	1, 2, 3, 4, 6
Sum of Factors	6	1	16
Name	perfect	deficient	abundant

Classify each number as perfect, deficient, or abundant.

47. 16 **48.** 36 **49.** 54 **50.** 28

SECTION 1–6

GREATEST COMMON FACTOR AND LEAST COMMON MULTIPLE

WARM-UP ACTIVITY

Algebra is a special area of mathematics that makes use of language and patterns in numbers. One such pattern is that every even number greater than two is the sum of two prime numbers.

List five even numbers greater than two and express them as the sum of two prime numbers.

INTRODUCTION

Did you realize that when you change gears on a bicycle you need to pedal faster? Gears are related in a special way. Use your skills with multiples to predict information about gears.

Refer to the gears in the diagram. After one complete turn of gear A, gear B has made less than a complete turn.

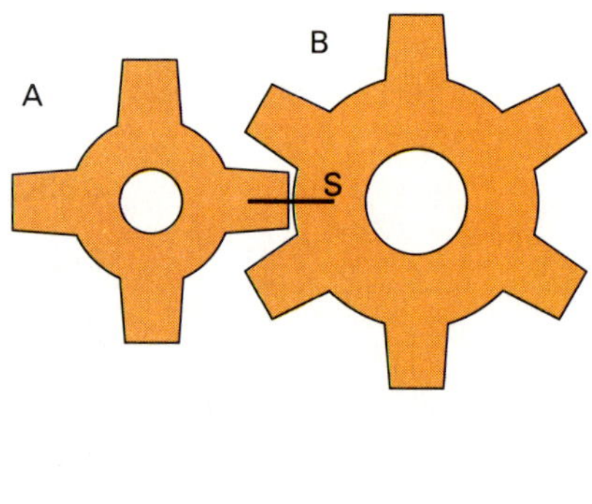

How many times will gear A and gear B turn so that the marks are at the same starting point S?

Gear A has 4 teeth. The pattern is 4, 8, 12, 16, 20, . . .
Gear B has 6 teeth. The pattern is 6, 12, 18, 24, 30 . . .

Compare the patterns of gears A and B to the multiples of 4 and 6. In order to answer the question about gears A and B, find the **least common multiple.**

It is helpful to know how to find both the greatest common factor and least common multiple of numbers.

SKILLS DEVELOPMENT

One is a common factor of any two numbers.

To understand the meaning of a **greatest common factor**, list all the factors of 12 and 18.

The factors of 12 are	(1)	(2)	(3)	4	(6)	12	
The factors of 18 are	(1)	(2)	(3)		(6)	9	18
The common factors are	1,	2,	3,		6		

So, the greatest common factor (GCF) of 12 and 18 is 6. A common factor of two numbers divides both numbers evenly.

You can use the product of prime factors to find the greatest common factor of 12 and 18.

$12 = (2) \times 2 \times (3)$
$18 = (2) \times 3 \times (3)$

Greatest Common Factor $= 2 \times 3$, or 6

The greatest common factor of 12 and 18 is 2×3 or 6.

In order to answer the question about gears A and B in the Introduction, find the **least common multiple.**

Number of turns	1	2	3	4	5 . . .
Gear A (LCM)	4	8	12	16	20 . . .
Gear B (LCM)	6	12	18	24	30 . . .

After 3 turns of gear A and 2 turns of gear B, the marks will be together at point S. The least common multiple is 12.

Use your knowledge of prime factors to find the least common multiple of two numbers.

Example 1

Find the least common multiple of 18 and 24.

Solution

18 = 2 × 3 × 3

24 = 2 × 2 × 2 × 3

First, write all the common prime factors.

LCM = 2 × 3 × 2 × 2 × 3

= $2^3 \times 3^2$

Then, write all the remaining factors. Use exponents when possible.

So, the least common multiple of 18 and 24 is $2^3 \times 3^2$ or 72.

Thus, 72 is divisible by 18 and 24.

Example 2

Find the greatest common factor (GCF) and least common multiple (LCM) of 12, 18, and 24 using prime factors.

Solution To find the least common multiple, write the GCF.

12	18	24
4 × 3	6 × 3	6 × 4
2 × 2 × 3	2 × 3 × 3	2 × 3 × 2 × 2

The GCF is the product of the factors common to all three numbers and equals 2 × 3.

12 = 2 × 2 × 3 → Write the GCF, 2 × 3.

18 = 2 × 3 × 3 → Then multiply the remaining factors common to two numbers, 2.

24 = 2 × 2 × 2 × 3 Then multiply by the

= 2 × 3 × 2 × 2 × 3 remaining factors, 3 and 2.

= $2^3 \times 3^2$ → Combine terms.

The LCM of 12, 18, and 24 is $2^3 \times 3^2$ or 72.

Parentheses show which calculations to do first.

Example 1

Calculate.

$26 + 63 \div (15 - 8) \times 2^2 \div 3 - 3$ → Record the original expression as the first step.

Solution

$26 + 63 \div (15 - 8) \times 2^2 \div 3 - 3$ → Do the calculations in parentheses first.

$26 + 63 \div 7 \times 2^2 \div 3 - 3$ → Evaluate expressions involving exponents next.

$26 + 63 \div 7 \times 4 \div 3 - 3$ → Multiplication and division are done next in the order they appear.

$26 + 9 \times 4 \div 3 - 3$

$26 + 36 \div 3 - 3$

$26 + 12 - 3$ → Then addition and subtraction are done in the order they appear.

$38 - 3$

35

So, $26 + 63 \div (15 - 8) \times 2^2 \div 3 - 3 = 35$.

Example 2

Find the value of $\frac{5 + 3 \times 5}{4 + 24 \div 4}$.

Solution $\frac{5 + 3 \times 5}{4 + 24 \div 4} = \frac{5 + 15}{4 + 6}$

$= \frac{20}{10}$

$= 2$

Steps for order of operations are used with the calculator as well as the computer.

Example 3

Calculate $16 - 16 \div 4$ using
a. pencil and paper **b.** calculator **c.** computer

Solutions

a. $16 - 16 \div 4$ → Pencil and paper: Divide first.
$16 - 4 = 12$ → Subtract.
So, $16 - 16 \div 4 = 12$.

b. $16 - 16 \div 4$ → Calculator: Press [1] [6] [÷] [4] [=]
$16 - 4$ → Press [1] [6] [−] [4]
12 → Press [=]
So, using a calculator, $16 - 16 \div 4 = 12$.

c. $16 - 16 \div 4$ → Calculate using BASIC.
10 PRINT 16 − 16/4 [ENTER] → Input.
Then, RUN 12 → The system performs order of operations.
So, using a computer, $16 - 16 \div 4 = 12$.

In BASIC
+ is +
− is −
× is *
÷ is /

EXERCISES

Find the value of each expression.

1. $5 \times 3 + 2$ **2.** $5 \times 3 - 4$ **3.** $18 \div 3^2 + 4$

4. $10 + 2 \times 4$ **5.** $10 - 2 \times 4$ **6.** $15 - 8 \div 4$

7. Why are there different answers for each?

a. $25 \div (5 \div 5)$ **b.** $25 \div 5 \div 5$

8. Why are the answers the same for a and b?

a. $6 \times (5 \times 2)$ **b.** $6 \times 5 \times 2$

Find the value of each expression.
Which has the greatest value?

9. $8 \times 4 \div 2$ **10.** $3 \times 4 \times 5$ **11.** $8 \div 4 \times 2$

Find the value of each expression.

Which has the least value?

12. $18 - 6 - 4$ **13.** $5 \times 6 \div 30$ **14.** $16 + 4 - 20$

Calculate.

15. $\frac{36 \div (5 + 1)}{2 + (8 \div 8)}$ **16.** $\frac{16 \div 4 + 14}{8 \div 2 - 1}$ **17.** $\frac{4 + 5 \times 4}{8 - 3 \times 2}$

18. $\frac{3 \times 4 - 2}{12 \div 3 + 1}$ **19.** $\frac{3 \times (1 + 3)}{(8 + 12) \div 5}$ **20.** $\frac{19 - 3 \times 5}{5 - 3 \div 3}$

Replace ○ with <, >, or =.

21. $8 \times 6 \div 3 \bigcirc 3 + 3 \div 3$ **22.** $8 - 25 \div 5 \bigcirc 16 \div 4 + 3$

23. $2^3 + 3 \times 4 \bigcirc 4^2 + 3^2$ **24.** $5^2 - 4^2 + 2 \bigcirc 6^2 - 5^2$

25. $\frac{12 + 3}{4 + 1} \bigcirc \frac{28 - 4}{4 \times 3}$ **26.** $\frac{16 + 16}{4^2 \times 2} \bigcirc \frac{7 \times 8 - 6}{8 \times 6 + 2}$

Place parentheses to find the value.

27. $36 \div 4 + 2$; value of 6 **28.** $16 - 6 \div 4$; value of 0

29. $8 + 45 \div 9 - 6$; value of 23 **30.** $27 + 15 \div 3$; value of 14

Calculate.

31. $\frac{(6 + 7) \times 8}{14 - (8 - 2)}$ **32.** $\frac{36 - (3 \times 6)}{12 - (4 + 2)}$ **33.** $\frac{(14 + 2) \div 2^2}{8 \div (4 \div 2)}$

34. To take a flight, you must

- leave the ground, or *depart.*
- land at a destination, or *arrive.*
- leave the airplane, or ☐☐☐☐☐☐☐.

To find the correct word, write the answers in order from least to greatest.

[N] $2 \times (4 + 2)$

[A] $(9 \times 3) - (8 \times 2)$

[E] $(26 + 13) \div 3$

[L] $2 \times 4 + 2$

[P] $28 - 3 \times 7$

[E] $24 - 3 \times (5 + 1)$

[D] $\frac{8 + 32 \div 4}{4 \times 4}$

SECTION 1–8

PROPERTIES OF NUMBERS

WARM-UP ACTIVITY

Explore some language analogies to the mathematical properties. Read each pair of sentences. Tell if the meaning is kept the same.

1. a. They ate pie and cheese.
b. They ate cheese and pie.

2. a. The dog chased the cat.
b. The cat chased the dog.

3. a. Al met Sue and Mike.
b. Al and Sue met Mike.

4. a. The coats and hats are red.
b. The coats are red and the hats are red.

INTRODUCTION

In this section we will look at some of the important patterns in mathematics—properties of numbers. We will use properties to help make calculations easier. Because the patterns are important, we will look for them when exploring different sets of numbers and when trying various operations.

SKILLS DEVELOPMENT

Natural numbers are the set of counting numbers. This is written as N = {1, 2, 3, 4, . . .}. Natural numbers can be shown on a **number line.** Each number is called a **coordinate** of the point.

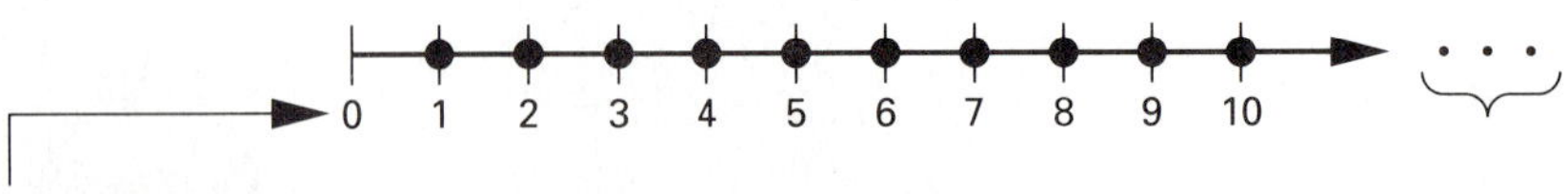

The zero was discovered much later. The natural numbers including zero are called the **whole numbers.**

When adding any two whole numbers, the answer is a whole number.

$2 + 6 = 8 \qquad 3 + 9 = 12 \qquad 146 + 241 = 387$

Also when multiplying any two whole numbers, the answer is a whole number.

$2 \times 4 = 8 \qquad 3 \times 7 = 21 \qquad 13 \times 22 = 286$

The set of whole numbers is **closed** for addition and multiplication. This means that the sum and product of two whole numbers is always a whole number.

Other special words are used to describe the properties of whole numbers.

Commutative property: The order in which two whole numbers are added or multiplied does not change the answer.

Example 1

Add or multiply.

a. 4 + 5

b. 39 × 41

Solutions

a. $\underbrace{4 + 5}_{9} = \underbrace{5 + 4}_{9}$

b. $\underbrace{39 \times 41}_{1,599} = \underbrace{41 \times 39}_{1,599}$

So, changing the order of the addends or multiplicands does not change the answer.

Associative property: In addition or multiplication, grouping the addends or factors in different ways does not change the answer.

Example 2

Add or multiply.

a. $2 + (3 + 4)$

b. $12 \times (6 \times 3)$

Solutions

a. $2 + (3 + 4) = (2 + 3) + 4$

$$2 + 7 = 5 + 4$$

$$9 = 9$$

So, grouping the addends differently does not change the answer.

b. $12 \times (6 \times 3) = (12 \times 6) \times 3$

$$12 \times 18 = 72 \times 3$$

$$216 = 216$$

So, grouping the factors differently does not change the answer.

Distributive property: The operation of multiplication distributes over the operation of addition.

Example 3

Add or multiply.

a. $4 \times (3 + 5)$

b. $4 \times 3 + 4 \times 5$

Solution

a. $4 \times (3 + 5)$

$$4 \times 8$$

$$32$$

b. $4 \times 3 + 4 \times 5$

$$12 + 20$$

$$32$$

So, $4 \times (3 + 5) = 4 \times 3 + 4 \times 5$.

The number 1 is an important number in the set of natural and whole numbers.

$$15 \times 1 = 15 \qquad 46 \times 1 = 46 \qquad 1 \times 38 = 38$$

The property of 1 for multipliction states that the product of 1 and any number always equals the original number. Because when you multiply by 1 the number is unchanged, one is called the **multiplicative identity.**

Zero is also an important number in the set of whole numbers.

$$15 + 0 = 15 \qquad 15 \times 0 = 0$$

The **addition property of 0** states that the sum of zero and any number is that number. Because when you add a 0 the number is unchanged, zero is called the **additive identity.**

The **multiplication property of 0** states that the product of 0 and any number is always 0.

You cannot divide by 0.

EXERCISES

The words *commutative, associative,* and *distributive* are properties of the set of whole numbers. Which property is shown in each of the following?

1. $4 + (2 + 3) = (4 + 2) + 3$

2. $491 + 365 = 365 + 491$

3. $4 \times (3 + 5) = 4 \times 3 + 4 \times 5$

4. $(3 \times 6) \times 2 = 3 \times (6 \times 2)$

5. Karena went to 6 football games. At each game she paid \$2.00 for a ticket and \$2.00 for food. How much money did she spend at the six games? Find the answer two different ways.

The set of whole numbers is closed for addition and multiplication. Is the set of whole numbers closed for

6. division

7. subtraction

True or False?

8. The sum of an odd number and an even number is an even number.

9. The product of an even number and an odd number is an odd number.

10. The sum of two odd numbers is an odd number.

11. The product of two natural numbers can be 0.

12. Every number has 0 as a factor.

13. Think of sentences similar to those in the Warm-Up Activity that illustrate the properties of whole numbers.

CHAPTER 1 SUMMARY

In this chapter you looked at the numbers around you. Some were used for identification purposes and some were used for computation. The idea that the value of a digit depends on its place in the number is called place value. With your understanding of place value, you practiced your rounding skills. Being able to round numbers helped you to approximate and estimate the reasonableness of your answers.

You explored number patterns in a branch of mathematics called number theory. The factors of a prime number are only itself and one. Numbers with more than two factors were called composite numbers. Being able to identify prime numbers allowed you to write a number as a product of its prime factors. Using prime factors, you found the greatest common factor and least common multiple of numbers.

You learned that there are rules for order of operations. Many calculators and all computers follow those rules. You also studied a more formalized approach to number patterns. You learned to describe natural and whole numbers using set notation and the number line. You studied properties of numbers that helped you shortcut calculations and do problems mentally. Sometimes you could do problems faster in your head than with a calculator.

VOCABULARY

additive identity
approximate
associative property
base
closed
commutative property
composite numbers
decimal form
decimal numeral system
distributive property
divisible
estimate
expanded form
exponent
exponential form
face value
factor tree
factors
greatest common factor
Hindu-Arabic numeral
language form
least common multiple
multiples
multiplication identity
multiplication property of 0
natural numbers
number line
numeral
place value
power
prime numbers
quantity
ranking
Roman numeral
round
whole numbers

CHAPTER REVIEW

Lucky claims his favorite number is 37,773. Round that number to the indicated place value.

1. nearest hundred
2. nearest thousand

Stanley was planning a vacation to the mountains. He was looking at a brochure that carried information about hotels. He read this advertisement.

Hideaway Quiet Inn
342 Waterfront Blvd., Somewhere, CA 99605;
TELE: 1-818-555-3400; 1-800-555-1313.
258 rooms. Located off Center Freeway at Waterfront.

Regular Rates: S-$89; D-$99; XP-$10

3. What is the ZIP code of the Hideaway Quiet Inn?
4. What is the toll-free telephone number of the inn?
5. What is the cost for a double room for one night?
6. What place is the value of the 4 in the street number 342?
7. You pass 240 Waterfront and the next address you see is 460 Waterfront. What do you conclude?

Estimate each answer.

8. 847 + 953 + 601
9. 72 × 61
10. 752 ÷ 47

Complete the following.

11. Write the set of factors for 42.
12. Write the set of multiples of 4 between 25 and 36.
13. Write 60 as a product of prime factors. Leave answer in exponential form.

Identify the following numbers as prime or composite.

14. 6
15. 7
16. 1
17. 17

Solve.

18. Spider-Long-Legs lands on every 5th step when it climbs a staircase. Spider-Shorter-Legs lands on every 2nd step when it climbs. If the spiders climb the same staircase of 25 steps, what number steps will they both step on?
19. Find the greatest common factor and least common multiple of 28 and 42. Be sure to identify them properly.

Calculate.

20. 42 + (58 + 126)
21. 25 × (4 × 69)
22. 0 ÷ 78
23. 36 × 21 + 36 × 79
24. 48 × 115 − 48 × 15
25. 78 ÷ 1

CHAPTER 1 TEST

You wrote check number 378 for seventy-five dollars.

1. Write in amount of the check, using numerals.
2. What will the number of the next check you write be?

Bernard's lucky number is 12,345. Round his lucky number to the indicated place value.

3. nearest ten
4. nearest ten-thousand

Tell what property is shown.

5. $3 \times 5 + 7 \times 5 = (3 + 7) \times 5$
6. $12 + 20 = 20 + 12$
7. $(3 \times 2) \times 50 = 3 \times (2 \times 50)$
8. $2758 \times 1 = 2758$

Calculate.

9. $8 + (4 - 1)$
10. $(8 + 3) \times 4$
11. $\frac{18 \div 9 + 3}{(17 - 2) \div 3}$

Complete.

12. List multiples of 5 between 12 and 25.
13. List even whole numbers less than 7.
14. Write 56 as a product of its prime factors in exponential form.
15. Write the set of factors of 56.
16. A number is even. What are possible last digits of the number?

Find the greatest common factor and least common multiple for each pair of numbers.

17. 6, 9
18. 12, 28
19. 30, 72

In Exercises 20–21, find the correct answer and estimate to check your result.

20. The Western Football final had a paid attendance of 59,409 while the Eastern final had a record attendance of 54,530. How many more attended the Western final?
21. Deposits of \$1534 and \$894 were made into an account with a balance of \$12,458. What is the final balance? *Hint:* Deposits are added to your account. The final balance is the total amount of money you have.

SKILLS PREVIEW

Your work with integers will build on your basic skills. You need to practice these skills. Using the basic skills quickly and accurately will help you to check answers and estimate whether answers are reasonable.

1. 8×7
2. $7 - 2$
3. $28 \div 4$
4. $17 - 9$
5. $15 \div 3$
6. 8×4
7. $5 + 3$
8. $3 + 7$
9. $14 - 5$
10. 2×8
11. $35 \div 5$
12. 5×1
13. $9 \div 3$
14. 8×0
15. $9 + 6$
16. $8 + 4$
17. $42 \div 6$
18. 9×5
19. 5×4
20. $12 - 5$
21. 3×7
22. $6 + 4$
23. $16 - 7$
24. $32 \div 4$
25. 8×2
26. $30 \div 6$
27. 7×7
28. $9 + 9$
29. 7×6
30. $9 + 3$
31. 3×6
32. $24 \div 3$
33. $8 + 9$
34. $7 + 3$
35. $72 \div 8$
36. $14 - 8$
37. $12 - 7$
38. $7 + 6$
39. 5×8
40. 4×6
41. 7×2
42. $13 - 4$
43. $35 \div 7$
44. 9×6
45. $10 - 4$
46. $8 + 8$
47. $28 \div 4$
48. 6×6
49. $81 \div 9$
50. $13 - 5$
51. $63 \div 7$
52. $64 \div 8$
53. $25 \div 5$
54. $11 - 8$
55. $6 + 9$
56. 6×8
57. 3×9
58. $36 \div 4$
59. $13 - 7$
60. $40 \div 5$

The National Weather Service publishes a table for estimating wind chill. Part of that table is shown below.

Wind Chill Factor Chart

Speed (km/h)	Actual thermometer reading (°C)												
	10	5	0	−5	−10	−15	−20	−25	−30	−35	−40	−45	−50
8	9	3	−3	−8	−13	−18	−23	−28	−33	−38	−44	−50	−57
16	4	−2	−9	−15	−21	−26	−33	−38	−44	−50	−57	−64	−72
24	2	−6	−13	−20	−26	−33	−40	−45	−51	−58	−65	−73	−81
32	0	−8	−16	−22	−30	−37	−44	−49	−55	−63	−71	−79	−88
40	−1	−9	−18	−25	−32	−40	−47	−53	−59	−67	−76	−83	−93
48	−2	−11	−19	−27	−34	−41	−50	−55	−62	−70	−78	−87	−97
56	−3	−12	−20	−28	−36	−42	−52	−57	−64	−72	−81	−89	−99

Notice in the table that there are numbers that have a "−" in front of them. These numbers are called negative numbers and are a part of the set of numbers called integers. This chapter will show you how to work with integers. You will also learn ways we use integers in our daily lives.

Ask Yourself

- *Where else have you seen negative numbers?*
- *What ways are negative numbers used?*
- *Who might use negative numbers in their work?*

CHAPTER 2

WORKING WITH INTEGERS

SECTION 2–1

WHAT IS AN INTEGER?

WARM-UP ACTIVITY

Work in cooperative groups. You will need these materials.

6 coins or two-colored disks
a 12-inch ruler
a marker for each player

Place all markers at the 0 end of the ruler. The first player tosses all 6 coins. For each coin that lands heads up, move that player's marker toward 12. For each coin that lands tails up, move that player's marker toward 0. However, a marker cannot go back any farther than 0. Players take turns tossing all 6 coins until one player reaches the 12-inch mark. If no one reaches 12 in two turns, the player closest to 12 is the winner.

INTRODUCTION

In the activity above, suppose that 3 coins landed heads up and 3 coins landed heads down. We could show this by writing $+3$ and -3.

We say that $+3$ (**positive** three) and -3 (**negative** three) are **opposites.** These numbers are **integers.** Integers are the set of whole numbers and their opposites. This is written as

$$I = \{\ldots -3, -2, -1, 0, 1, 2, 3, \ldots\}.$$

We can use integers to show opposites.

4°C above zero	an increase of 12 mL
4°C below zero	a decrease of 12 mL

SKILLS DEVELOPMENT

Integers can be shown on a **number line.**

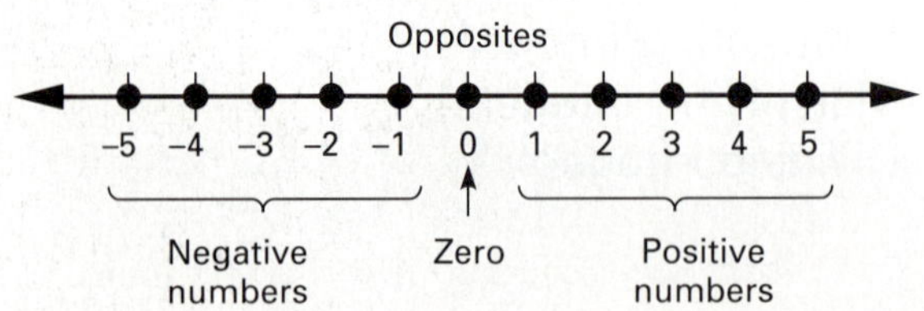

Zero is neither negative nor positive. The opposite of 0 is 0. Positive integers can be written without their sign (+). Negative integers must always be written with their sign (−).

Example 1

a. What is the opposite of 5?

b. What is the opposite of −17?

Solutions **a.** The opposite of 5 is −5.

b. The opposite of −17 is 17.

Think: The opposite of −17 is −(−17) or 17.

As with whole numbers, the farther to the right, the greater the number; and, the farther to the left, the less the number.

Example 2

Remember:
< means less than.
> means greater than.

Replace ○ with < or >.

a. 2 ○ −2 **b.** −23 ○ −20

Solutions **a.** Since 2 is to the right of −2 on a number line, then $2 > -2$.

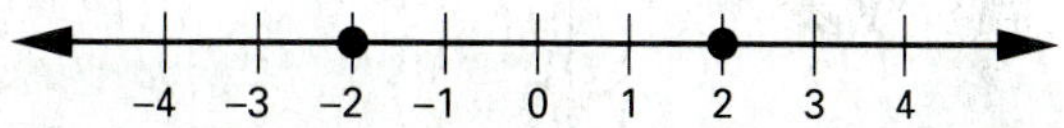

b. Since −23 is to the left of −20 on a number line, then $-23 < -20$.

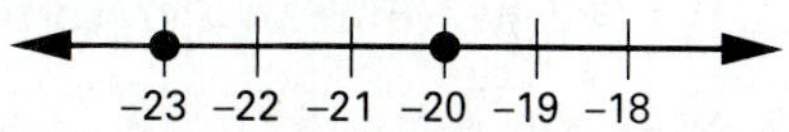

Integers are used in many different ways.

Example 3

Write integers to represent each numerical application.

a. One day a stock rose 2 points.

b. The football team lost 3 yards.

c. One patient lost 5 pounds in one month, lost 1 pound the following month, and gained no weight the next month.

Solutions **a.** +2 or 2 **b.** −3 **c.** −5, −1, +0

EXERCISES

Use an integer to show each of the following. Be prepared to explain your choice.

1. 6 steps forward

2. a profit of $12

3. a loss of 14 points

4. an increase of 8 m^2

5. 9 steps up

6. a loss of $6

Left and *right* are opposites. Use −2 m to mean 2 m to the left and +2 m to mean 2 m to the right. What do the following mean?

7. −3 m

8. +4 m

9. −12 m

10. +36 m

11. 0 m

12. −148 m

Above sea level and *below sea level* are opposite ideas used for elevation. Write an integer for each of the following.

13. Mount Everest is the highest point on earth, 8840 m above sea level.

14. A whale has been found at a depth of 1130 m below sea level.

15. The lowest road is along the shores of the Dead Sea at 390 m below sea level.

16. The highest highway, 5630 m above sea level, is found in Tibet.

Write the opposite of each integer.

17. −2 **18.** 1 **19.** −36 **20.** 18

Which integer in each pair is greater?

21. 3, 5 **22.** −2, 1 **23.** 0, 8

24. −6, −3 **25.** 5, −4 **26.** −3, 0

Replace ○ with < or >.

27. 3 ○ 6 **28.** 6 ○ −1 **29.** 0 ○ −2

30. 0 ○ 4 **31.** −12 ○ −13 **32.** −2 ○ 143

Draw a number line and label the approximate location of each number.

33. −3, −5, 0 **34.** −21, −15, −18 **35.** −10, 0, 20

Tides are affected both by the sun and the moon. The moon has the greater effect since it is closer to the earth. Write an integer for each of the following.

36. In Lake Superior the tide is hardly noticed and rises about 5 centimeters.

37. Often storms allow the tide to rise as much as 24 inches.

38. In the Red Sea, the tide dropped about 8 centimeters.

Draw a number line. Show positive and negative integers on it. Refer to the number line as you write a paragraph that explains how positive and negative integers illustrate the idea of opposites.

JUST FOR FUN

Change the Warm-Up Activity game by using two rulers set with their 0-points together. Now you can move to the left of 0 if there are more tails than heads. How does this change the game? Make up new rules. Decide how to tell who wins.

SECTION 2–2

ADDING INTEGERS

WARM-UP ACTIVITY

A good problem-solving strategy is to make an ordered list. Below is a list of some temperatures recorded on a day in January. Arrange these in order from least (coldest) to greatest (warmest). Look at a real thermometer if you need help. Then answer the questions.

4 5 0 10 −18 2 −6

What was the coldest temperature?

What was the warmest temperature?

Do you think these temperatures are given above in the order they occurred? Explain your answer.

INTRODUCTION

In Section 2–1 you learned different types of opposites. Since these positive and negative reactions happen in a mixed-up order, you must be able to add opposites to find a total.

As two teams play football, they are going to gain yardage (+) and lose yardage (−). To find their total yardage, the teams must be able to add integers.

SKILLS DEVELOPMENT

A number line can be used to show gains and losses in a football game. Let 0 be the line of scrimmage (starting point) for all plays. Gaining yardage is positive, so move to the right. Losing yardage is negative, so move to the left. The total yardage (net gain or loss) is the ending point on the number line.

Example 1

On the first play a team gained 5 yards. On the next play they lost 8 yards. What is the net (total) yardage for both plays?

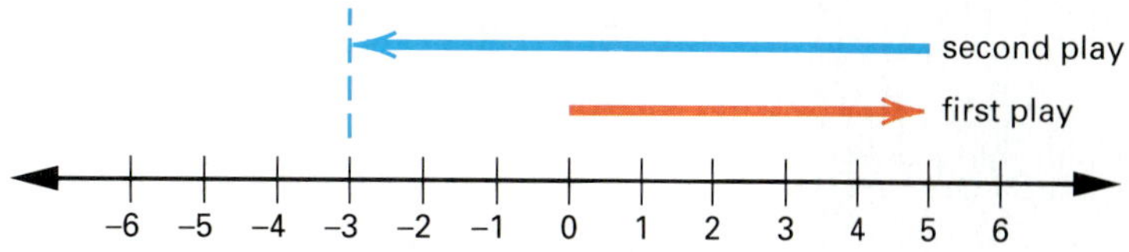

Solution

First play: 5 to the right. Second play: 8 to the left. End point is -3. There is a net loss of 3 yards. You can write: $5 + -8 = -3$.

Example 2

This time our team has a loss of 1 on the first play and a gain of 4 on the second play. What is the net yardage?

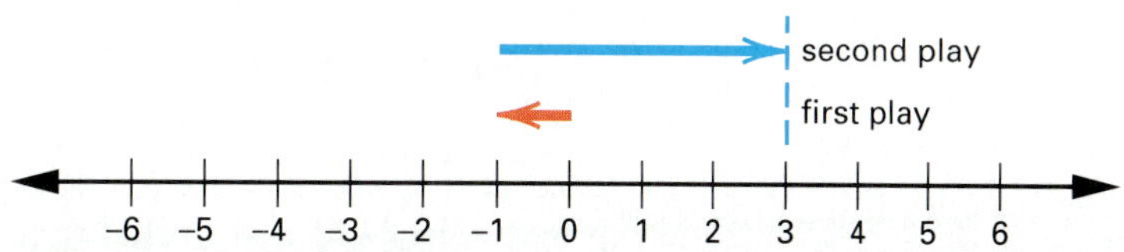

Solution

From 0, move 1 space to the left and then 4 spaces to the right. End on 3. There is a gain of 3 yards. You can write: $-1 + 4 = 3$.

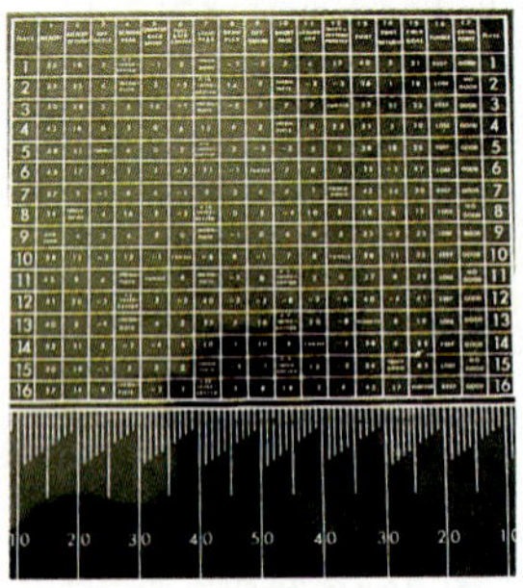

On a number line, $+7$ is at a distance of 7 units from 0. The **absolute value** of $+7$ (or 7), written $|+7|$, is 7. The absolute value of -5, written $|-5|$, is 5, because -5 is 5 units from 0. Absolute value can never be negative because it represents the distance a point is from 0 on a number line.

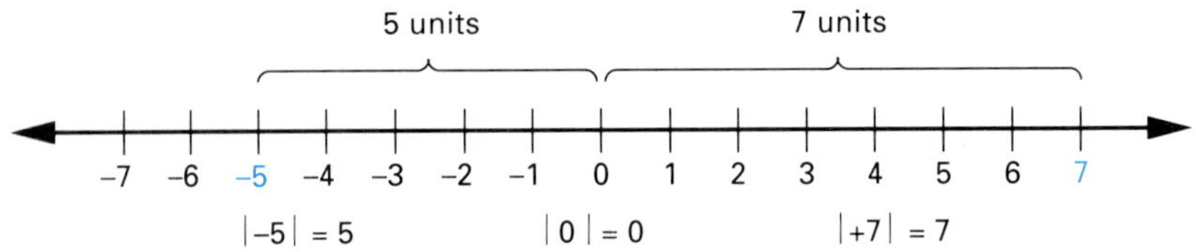

Example 3

What two integers have an absolute value of 4?

Solution

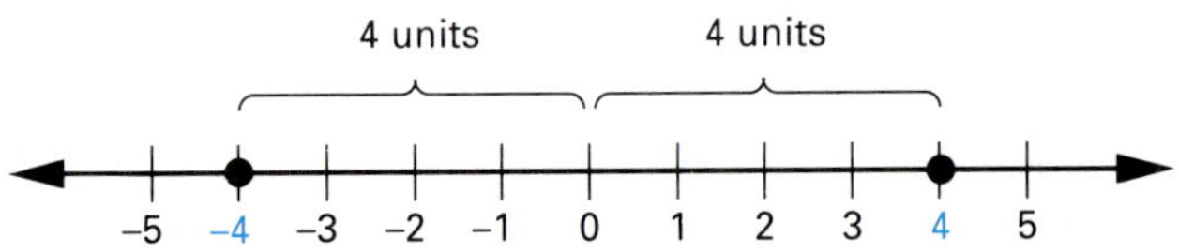

-4 and $+4$ (or 4) are both 4 units from 0. So, $|-4| = 4$ and $|+4|$ (or $|4|$) $= 4$.

Example 4

Find the absolute value of -15 and $+18$.

Solution

-15 is 15 units from 0. So, $|-15| = 15$ and $|+18| = 18$.

EXERCISES

Add these integers, using a number line to find the answer.

1. $7 + 2$ **2.** $7 + -2$ **3.** $-7 + 2$

4. $-7 + -2$ **5.** $-3 + -5$ **6.** $-1 + 1$

7. What is the sign of the answer when adding two positive integers?

8. What is the sign of the answer when adding two negative integers?

9. What is the sign of the answer when a negative is added to a positive?

10. What is the sum of a number and its opposite?

Eventually you will add integers without a number line. Use the patterns in Exercises 1–10 to find each sum.

11. $-6 + 3$ **12.** $-6 + -2$ **13.** $0 + -8$

14. $2 + -4$ **15.** $3 + -3$ **16.** $8 + -3$

17. $-3 + -4$ **18.** $-5 + 8$ **19.** $-8 + 9$

20. $-1 + 1$ **21.** $5 + -4$ **22.** $-2 + -4$

To add integers, work from left to right, adding two at a time.

$$-4 + 3 + -1 = (-4 + 3) + -1 = -1 + -1 = -2$$

Add these groups of integers.

23. $6 + -12 + 6$ **24.** $-8 + -3 + 12$ **25.** $-2 + -1 + 3 + 9$

Think critically and answer the following.

26. Find another way to get the correct answers in Exercises 23–25 instead of working left to right.

27. Try your method from above to find the sum: $-2 + 8 + -6 + -14$.

28. Did you get -14 for the answer to Exercise 27? If your method did not work, revise it, and try again.

Solve.

29. In a dart game, Carol gained 8 points, lost 3 points, and lost 6 points. George lost 11 points, gained 11 points, and lost 1 point. Whose score is greater?

30. In a football game, the Zees gained 2 yards, lost 5 yards, and then gained 38 yards. How far are they from the starting point?

31. At noon the temperature was only $-2°$. By 4 P.M., it had risen $5°$. What was the temperature at 4 P.M.?

32. Write a problem of your own. Explain how to find the answer.

SUBTRACTING INTEGERS

WARM-UP ACTIVITY

Making a drawing is a good problem-solving strategy. Make a drawing like the one at the right and use it to answer these questions.

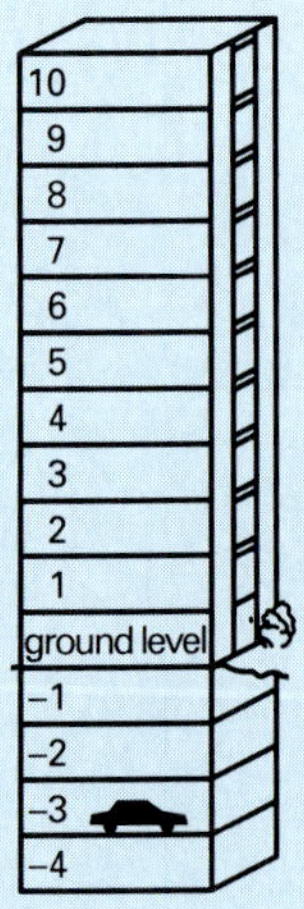

If you start on the floor labeled 1 and go up 3 stories, on what floor will you be?

If you start on the floor labeled garage level −2 and go up 3 stories, on what floor will you be?

Make up your own problems. Explain how to use your drawing to find the answer to each problem.

INTRODUCTION

Gina lives on the second floor of an apartment building. How many floors does she have to travel to visit her friend Toby on the ninth floor? Which direction does she go?

This problem can be written as $^{+}9 - {}^{+}2 = {}^{+}7$ or $9 - 2 = 7$.

She must travel 7 floors up.

Gina's family parks their car on the third level below ground. How many floors does Gina have to travel to get to the car? Which direction must she go?

This problem can be written as $^{-}3 - {}^{+}2 = {}^{-}5$ or $^{-}3 + {}^{-}2 = {}^{-}5$.

She must travel 5 floors down.

SKILLS DEVELOPMENT

You know that $8 - 2 = 6$ and you learned that $8 + -2 = 6$. Since both expressions have the same answer, $8 - 2$ must equal $8 + -2$.

$$3 - 1 = 2 \qquad 3 + -1 = 2$$

$$10 - 5 = 5 \qquad 10 + -5 = 5$$

$$17 - 8 = 9 \qquad 17 + -8 = 9$$

This shows that to subtract an integer, add its opposite.

The opposite of an integer is called its **additive inverse.** The sum of an integer and its additive inverse equals 0. This is called the **inverse property of addition** $(a + -a = 0)$.

$$4 + -4 = 0 \qquad 1 + -1 = 0 \qquad 21 + -21 = 0$$

To subtract integers: Rewrite the problem as addition.
Use the addition rules to find the answer.

Example 1

a. $3 - -2$ b. $3 - 4$

c. $-4 - 5$ d. $-1 - -4$

Solutions a. $3 - -2 = 3 + 2 = 5$ b. $3 - 4 = 3 + -4 = -1$

c. $-4 - 5 = -4 + -5 = -9$ d. $-1 - -4 = -1 + +4 = 3$

Note: $3 - -2$ may also be written $3 - (-2)$ or $3 - {}^{-}2$. In this book we will use the first and simplest format.

From your study of whole numbers, you know that addition and subtraction are inverse operations. For example, since

$$6 + 7 = 13, \text{ then } 13 - 7 = 6.$$

Addition and subtraction of integers are also inverse operations. If you know that $-3 + 5 = 2$, then you also know that $2 - 5 = -3$.

Example 2

Use inverse operations to find $-3 - -5$.

Solution Think $-5 + \text{what} = -3$?
The answer is 2, so $-3 - -5 = 2$.

EXERCISES

1. What is the sum of a number and its additive inverse?

2. What is the additive inverse of -5?

3. 16 and -16 are additive inverses. Why?

Rewrite as addition.

4. $1 - 8$

5. $-7 - 6$

6. $15 - -5$

7. $-3 - -12$

Calculate.

8. $5 - -3$

9. $15 - 12$

10. $-5 - -4$

11. $0 - -4$

12. $-14 - 11$

13. $79 - 81$

14. $35 - -19$

15. $-9 - -3$

Solve.

16. The temperature in the store is 18°C. In the cold storage room it is −11°C. How much colder is the cold storage room than the store?

17. The highest shade temperature recorded was in Libya in 1922. It was 136°F. The lowest temperature recorded was in Antarctica in 1960. There it was 260°F below the temperature given in Libya. What is the lowest temperature on record?

18. Write a description of your plan to find this sum. Then find the sum. $3 + -8 + -7 + 6$.

19. Which is greater? Then explain how you decided. $13 - -3$ or $-12 + 24$?

20. To subtract an integer, you ______.

The following involve both addition and subtraction. When you rewrite the problem as addition, change only the subtraction part.

Example: $2 - 3 + 4 = 2 + -3 + 4$
$= (2 + -3) + 4$
$= -1 + 4 = 3$

Calculate.

21. $-9 + 2 - -2$

22. $4 - -2 + 8$

23. $-4 + 9 - 9$

24. $6 - -4 - 1$

Write the related subtraction statement.

25. $7 + -3 = 4$

26. $-10 + 2 = -8$

27. $-11 + -4 = -15$

28. $9 + -14 = -5$

Use the idea of inverse operations to find the missing addends.

29. $3 - \square = 0$

30. $-6 - \square = 2$

31. $-4 - \square = -9$

32. $4 - \square = 10$

Solve.

33. The temperature of molten lava is 1730°C. The surface temperature of Venus is 480°C. Which is hotter? By how much?

34. An airplane was 12,843 meters high. It dropped 750 meters and then rose 780 meters. What is its new altitude?

35. On four successive plays a football team gained 6 yards, lost 10 yards, gained 14 yards, and lost 8 yards. What was the team's net gain or loss?

SECTION 2–4

MULTIPLYING INTEGERS

WARM-UP ACTIVITY

In a magic square the sum of each row and column is the same. That number for this square is −10. Find the missing integers.

5	?	−8	?
?	0	?	?
?	−4	−5	1
−7	?	4	?

Explain how you completed the magic square. Try making a magic square of your own. If you find that too easy, try making a magic square where the diagonals also have the same total as each row and column.

INTRODUCTION

Patterns are a part of nature. Patterns allow traffic planners to construct highways that can handle large amounts of cars, buses, and trucks. In real life, people have to look for and utilize patterns all the time.

In mathematics, patterns are also very important. In this section, you will learn to look for patterns. You will use these patterns to make conjectures (guesses) and generalizations (rules) for completing mathematical operations with integers.

SKILLS DEVELOPMENT

Skills and concepts you learn in one area of mathematics help you learn new skills. For example, in your work with whole numbers, you used a simpler way to find the sum of like numbers. You multiplied.

	Addition	Multiplication
Whole numbers:	$3 + 3 + 3 + 3 = 12$	$4 \times 3 = 12$
Integers:	$(+3) + (+3) + (+3) + (+3) = +12$	$(+4)(+3) = +12$
	$(-3) + (-3) + (-3) + (-3) = -12$	$(+4)(-3) = 12$

As in your earlier work, you can use a pattern to show the product of integers. From the pattern it appears that:

positive integer × negative integer = negative integer   positive integer × positive integer = positive integer

Study the following patterns for multiplying integers.

$(+2)(+3) =$		$(-3)(+2) =$
$(+1)(+3) =$	The products are	$(-3)(+1) =$
$(0)(+3) =$	decreasing or	$(-3)(0) =$
$(-1)(+3) =$	increasing by 3.	$(-3)(-1) =$
$(-2)(+3) =$		$(-3)(-2) =$

You should have observed the following pattern.

$(-2)(+3) = -6$ $(-3)(-2) = +6$

negative integer × positive integer = negative integer   negative integer × negative integer = positive integer

Example 1

Find these products.

a. 4(−2) **b.** (−3)(−15) **c.** −7(10)

Solutions

a. 4(−2) = −8 →

b. (−3)(−15) = 45 →

c. −7(10) = −70 →

> Think:
> A positive times a negative is negative.
> A negative times a negative is positive.
> A negative times a positive is negative.

Example 2

A plane descends 150 meters per minute. What is the change in altitude after 3 minutes?

Solution Descending is a negative direction, so descending 150 meters will be represented by −150. This occurs for 3 minutes, so multiply by 3.

3(−150) = −450 The change in altitude is −450 meters.

Example 3

A plane is descending 150 meters per minute. How much higher was it 3 minutes ago?

Solution Descending would be a negative direction, so descending 150 meters will be represented by −150. Three minutes ago will be represented by −3.

(−3)(−150) = 450 The plane was 450 meters higher.

EXERCISES

Study the patterns. Complete each equation.

1. (+4)(+2) = +8
(+4)(+1) = ___
(+4)(+_) = 0
(___)(−1) = −4
(+4)(−2) = ___

2. (+2)(+4) = +8
(___)(+4) = +4
(0)(+4) = ___
(___)(+4) = −4
(−2)(___) = −8

Use the patterns from Exercises 1 and 2 to find these products.

3. $(+4)(-3)$ **4.** $(-3)(+4)$

5. What do you notice about your answers to Exercises 3 and 4?

Copy and complete the following summary.

6. The product of a negative and a positive integer is ______.

7. The product of a positive and a positive integer is ______.

8. The product of a negative and a negative integer is ______.

Find each product.

9. $4(-1)$ **10.** $6(7)$ **11.** $-3(-4)$

12. $4(-3)$ **13.** $3(-4)$ **14.** $-4(-6)$

15. $-4(6)$ **16.** $-7(-7)$ **17.** $8(-9)$

To find the product of three or more numbers, you can multiply from left to right, in order.

Example:
$$\begin{aligned}(3)(-8)(-5) &= [(3)(-8)](-5)\\ &= (-24)(-5)\\ &= 120\end{aligned}$$

18. $(-5)(-3)(-1)$ **19.** $(4)(-3)(10)$

20. $(-6)(5)(-4)$ **21.** $(-4)(-3)(-20)$

22. What property allows you to solve $(3)(-8)(-5)$ by finding $(3)[(-8)(-5)]$? Why is this method easier?

A broken water tank is losing seven gallons of water each minute.

23. How much water is lost in 5 minutes?

24. How much more water was in the tank 3 minutes ago (negative)?

SECTION 2–5

DIVIDING INTEGERS

WARM-UP ACTIVITY

Use the integers shown and one of the operations, +, −, or × to write a true statement. The first one has been done for you.

a. −2, −3, −5 [Use addition. $(-2) + (-3) = -5$]

b. −2, 1, −3

c. −6, 2, −4

d. 5, 2, −3

e. −4, −6, 2

INTRODUCTION

Lydia worked during the summer helping a golf pro. One of her duties was to find the average score per round for each golfer during a tournament. Rosann scored −2, −4, 1, and 1 on her four rounds. To find Rosann's average, Lydia had to divide integers. In this lesson, you will develop your integer division skills.

SKILLS DEVELOPMENT

To learn how to subtract integers, you used the idea that subtraction was the **inverse operation for addition.** Multiplication and division are also inverse operations. That is, because $4 \times 5 = 20$, you know that

$$20 \div 5 = 4 \text{ or } \frac{20}{5} = 4.$$

Follow this pattern with integers.

Multiplication	Division	Multiplication	Division
$(+3)(+6) = +18$	$\frac{+18}{+6} = +3$	$(+3)(-6) = -18$	$\frac{-18}{-6} = +3$
$(-3)(+6) = -18$	$\frac{-18}{+6} = -3$	$(-3)(-6) = +18$	$\frac{+18}{-6} = -3$

Example 1

Use the multiplication in the first column to find the quotients in the second column.

a. $(+4)(+6) = +24$ $\quad \frac{+24}{+6} = —$

b. $(-4)(+6) = -24$ $\quad \frac{-24}{+6} = —$

c. $(+4)(-6) = -24$ $\quad \frac{-24}{-6} = —$

d. $(-4)(-6) = +24$ $\quad \frac{+24}{-6} = —$

Solutions a. 4, because $4(6) = 24$
b. -4, because $(-4)(6) = -24$
c. 4, because $4(-6) = -24$
d. -4, because $-4(-6) = 24$

Example 2

Study the patterns below. When are the quotients positive? When are the quotients negative?

a. $\frac{-32}{8} =$ b. $\frac{32}{-8} =$ c. $\frac{-32}{-8} =$ d. $\frac{32}{8} =$

Solutions a. -4, because $-4(8) = -32$
b. -4, because $-8(-4) = 32$
c. 4, because $4(-8) = -32$
d. 4, because $4(8) = 32$

Example 3

Reread the assignment Lydia was given at the INTRODUCTION of this section. Find Rosann's average score per round.

Solution Average $= \frac{-2 + -4 + 1 + 1}{4}$ → Find the sum of the scores.

$= \frac{-4}{4}$ → Divide by the number of scores.

$= -1$

Rosann's average score per round was -1.

EXERCISES

Find each quotient. You rewrite each question using multiplication to help you.

1. $\frac{+30}{+6}$ **2.** $\frac{-30}{+6}$ **3.** $\frac{+30}{-6}$ **4.** $\frac{-30}{-6}$

Copy and complete each sentence.

5. The quotient of two positive integers is ______.

6. The quotient of a negative and a positive integer is ______.

7. The quotient of two negative integers is ______.

For each of the following decide whether the quotient is positive or negative. Then write the quotient.

8. $\frac{-36}{+9}$ **9.** $\frac{+54}{-9}$ **10.** $\frac{+49}{+7}$

11. $-36 \div -12$ **12.** $15 \div -3$ **13.** $0 \div -10$

14. $64 \div -32$ **15.** $-72 \div 72$ **16.** $72 \div -9$

Find the average for each group of integers.

17. $-3, -4, 3, -6, 5$ **18.** $8, -12, 7, 5$

There are various trenches on the floor of the Pacific Ocean. Their depths are shown at the right.

Name of Trench	Depth
Mid-America Trench	−6669 m
Ryukyu Trench	−7507 m
New Hebrides Trench	−7570 m
Peru–Chile Trench	−8064 m
Aleutian Trench	−8100 m
Palau Trench	−8138 m
Japan Trench	−8412 m

19. Find the average depth of these seven trenches.

20. How much deeper is the New Hebrides Trench than the Mid-America Trench?

Use the Reference Section in the back of this book or another source. (If you use another source, cite the source and enclose a copy of the information with your work.)

21. Find the average distance of the nine planets in our solar system from the Sun.

22. Find the average diameter of these planets.

23. Use information you can find in the Reference Section to write two of your own problems. Then explain how to solve each.

Solve.

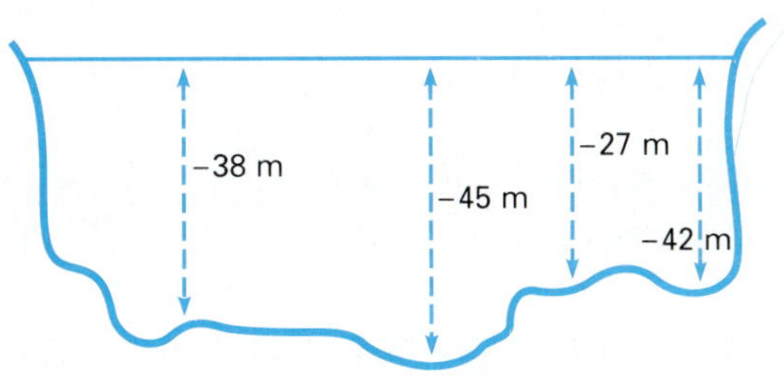

24. During a recent expedition, scientists made this profile of the bottom of a lake. Based on the results shown, what is the average depth of the lake?

25. The average depth of a lake is 11 feet. Is it safe to dive into the lake at any random point? Why or why not?

SECTION 2–6

ORDER OF OPERATIONS WITH INTEGERS

WARM-UP ACTIVITY

Why is a goose like an icicle? To solve the riddle, find the value of each variable. Copy the blanks and integers on a separate sheet of paper. Then place the variable letter in the space above its value. Some values appear twice. **Do not write in the textbook.**

$I = 7 \times -2$	$C = -12 \div -4$	$B = -3 \times 5$
$T = -3 \times 4$	$S = 4 \times -7$	$R = -26 \div 2$
$N = -5 + 10$	$O = -3 \times -2$	$E = -13 - -7$
$U = 7 + -10$	$W = 15 \div -3$	$G = 75 \div 5$
$A = -7 + 6$	$D = 8 \div -2$	

___ ___ ___ ___ ___ ___ ___   ___ ___
-15 -6 3 -1 -3 -28 -6 -14 -12

___ ___ ___ ___ ___   ___ ___ ___ ___
15 -13 6 -5 -28 -4 6 -5 5

INTRODUCTION

shut off the motor	get in the car	get out of the car
drive on the road	start the car	fasten your seat belt

The cards above list six different steps in traveling by car. What is the proper order in which these should be done in order to have a successful trip?

Just as with traveling by car and working with whole numbers, the order of operations also applies in your work with integers.

SKILLS DEVELOPMENT

Two students used their calculators to complete this problem: $-8 + 4 \times -2 \div -1$. Each found a different answer. The answers they found were: -8 and 0. Discuss how each answer was found. Review the order of operations for whole numbers, then find which answer is correct.

Example 1

Simplify $(-8)(-4) \div 2 - (-3)(-2)$.

Solution $(-8)(-4) \div 2 - (-3)(-2)$

$32 \div 2 - (-3)(-2)$ → Multiply and divide from left to right.

$16 - (-3)(-2)$

$16 - (6)$

$16 - 6$ → Add or subtract from left to right.

10

Example 2

a. Subtract the sum of -49 and 46 from 45.

b. Divide the product of -10 and -8 by -4.

Solution First, write the sentence using mathematical symbols. Then solve the problem.

a. $45 - (-49 + 46)$

$45 - -3$ → Work inside the () first.

$45 + 3$

48

b. $(-10)(-8) \div -4$

$80 \div -4$

-20

Example 3

Simplify $3(4)^2 - 2(-2)$.

Solution $3(4)^2 - 2(-2)$

$3(16) - 2(-2)$ → Evaluate powers first.

$48 - -4$ → Multiply first.

$48 + 4$ → Add or subtract.

52

EXERCISES

Explain each step in each of the following.

1. $\frac{(-22 + 2) \div -5}{-8 \div (8 \div 4)}$

 a. $\frac{-20 \div -5}{-8 \div 2}$

 b. $\frac{4}{-4}$

 c. -1

2. $16 - [3(6 - 3) - 12]$

 a. $16 - [3(3) - 12]$

 b. $16 - [9 - 12]$

 c. $16 - -3$

 d. 19

For each of the following, state which operation needs to be used first.

3. $48 \div -6 + 3$ **4.** $42 + 12 \div -2$ **5.** $(-8)(-5) \div 2$

Find the value of each of the following.

6. $48 \div -6 + 3$ **7.** $(42 + 12) \div -2$ **8.** $(-8)(-5) \div 2$

9. $42 + 12 \div -2$ **10.** $(-8)(-5) \div 2^2$ **11.** $-14 \div 2 - 6^2$

12. $-9 - (3)[2(2 - 3)]$ **13.** $(-4)[(-3)(-2) + 4]$ **14.** $\frac{-18 + 6}{(-3)(4)}$

15. $\frac{(-16 + 4) \div 2}{8 \div -8 + 4}$ **16.** $(-3)^2 + -8 \div -2$ **17.** $\frac{-8 + 2}{-3 + 2}$

Write the following using mathematical symbols, then use your calculator to simplify.

18. Add -38 to the sum of -43 and 85.

19. Multiply the sum of -24 and -14 by -25.

20. Subtract -36 from -28.

21. Divide the sum of -144, -48, and 96 by -12.

22. Subtract $(-3)^2$ from the sum of -8 and -2.

23. Add the product of -8 and -2 and the quotient of -24 and 6.

Read the following computer program. Remember that * means multiply and / means divide. What is the printout for each program?

24.

```
10 Let A = −5
20 Let B = 6
30 C = ((A*B)+A−B)/B
40 PRINT C
50 D = (B+(A+A+A))/(A+B)
60 PRINT D
70 END
```

25.

```
10 Let A = 6
20 Let B = −2
30 C = A*(B−(B*A))
40 PRINT C
50 D = C/A*(B−(B*A))
60 PRINT D
70 END
```

Use the computer programs above to answer each question.

26. If in Exercise 24, you change A to 5 and B to -2, what is the new printout?

27. If in Exercise 25, you change A to 5 and B to -2, what is the new printout?

28. In Exercise 24, can $B = 0$? Explain your answer.

29. In Exercise 25, what restrictions must be put on the values of A and B? Explain your answer.

Solve each problem.

30. A mountain climber started at 700 feet above sea level. She then climbed up 450 feet, down 280 feet, and up 700 feet. How high above sea level is she now?

31. A submarine started at sea level. It descended 600 feet then rose 400 feet. At what level was the submarine then?

32. The results of a game showed these scores: -28, 14, -17, 21, and -5. What was the average score?

SECTION 2–7

INTEGERS OUTSIDE THE CLASSROOM

WARM-UP ACTIVITY

Sue has lots of nickels, dimes, and quarters. Make a list of the different ways she can combine the coins to make 75 cents. You may wish to use a table like this one.

Dimes		Nickels		Quarters		Total Value
Number	Value	Number	Value	Number	Value	

INTRODUCTION

Throughout this chapter you have worked mathematics problems dealing with many activities. This section includes much of what you have done and introduces new situations. Among these new situations, you will use mathematics and reference skills to solve problems that arise outside of the mathematics classroom.

SKILLS DEVELOPMENT

An important skill in mathematics is using a table to summarize data. For example, scientists have recorded data to explore the combined effect of low temperatures and high winds. People in the Arctic or the Antarctic can get severe frostbite without knowing it. To a person traveling at a speed of 32 km/h, or if the wind speed is 32 km/h, then an actual air temperature of −20°C feels like a temperature of −44°C. In the chart, the row for 32 km/h and the column for −20°C intersect at −44°C, the **wind chill factor.**

Wind Chill Factor Chart

Speed (km/h)	Actual thermometer reading (°C)												
	10	5	0	−5	−10	−15	−20	−25	−30	−35	−40	−45	−50
8	9	3	−3	−8	−13	−18	−23	−28	−33	−38	−44	−50	−57
16	4	−2	−9	−15	−21	−26	−33	−38	−44	−50	−57	−64	−72
24	2	−6	−13	−20	−26	−33	−40	−45	−51	−58	−65	−73	−81
32	0	−8	−16	−22	−30	−37	−44	−49	−55	−63	−71	−79	−88
40	−1	−9	−18	−25	−32	−40	−47	−53	−59	−67	−76	−83	−93
48	−2	−11	−19	−27	−34	−41	−50	−55	−62	−70	−78	−87	−97
56	−3	−12	−20	−28	−36	−42	−52	−57	−64	−72	−81	−89	−99

Another important skill is being able to locate information and use various reference resources. As you know, there is a Reference Section at the back of your textbook.

Example 1

You are working on a science report about Saturn. You need to compare Saturn's distance from the sun to Earth's distance from the sun.

Solution Use the Reference Section in the back of this book, or an encyclopedia. Since you need information on more than one planet, it would probably be best to look up "Solar Systems."

After locating the proper information, you find that Saturn is 1,427,000,000 kilometers from the sun, while Earth is only 150,000,000 kilometers from the sun. Saturn is 1,277,000,000 kilometers further from the sun than the Earth.

Example 2

Your doctor has told you that you need to gain or lose a little weight. You are not sure the doctor is correct. How can you check this out?

Solution To verify this, you look up the table for recommended weights. Find your height and type of body frame. Then read what your recommended weight should be. Note that this information is given as a range, not just a single number.

EXERCISES

At Shackleton, Antarctica, the temperature was −25° C with no wind.

1. Find the wind chill factor if the wind was blowing at 8 km/h.

2. A wind of 32 km/h would cause the wind chill factor to drop by how many degrees?

Solve each problem.

3. At Mirny, Antarctica, the temperature was −30° C with no wind. Melanie drove her snowmobile at 32 km/h. What was the wind chill factor?

4. On Ellesmere Island in the Arctic the wind chill factor was −50° C. If the wind speed was 16 km/h, what was the actual air temperature?

5. Scientists had protective clothing made to withstand temperatures as low as −65° C. What maximum wind speed for an air temperature of −40° C could they withstand?

A vet recorded the changes in mass of an animal during the first 8 weeks of its life: −118 g, −89 g, −49 g, −10 g, 25 g, 80 g, 110 g, 115 g.

6. Find the net change in mass.

7. Find the average change in mass each week.

Use the Reference Section in the back of this textbook. A doctor notes each person's weight, height, and body build. Tell how much weight each person should gain or lose to be at the proper level for build and weight. Your answer will be a range rather than just one answer.

8. Jeff is 5 feet 6 inches tall, medium frame, and weighs 150 pounds.

9. Anna is 5 feet tall, small frame, and weighs 95 pounds.

10. Sun-Lee (female) is 5 feet 1 inch tall, medium frame, and weighs 98 pounds.

For a history project, you are doing a report on the changes in population in the states between 1800 and 1900. You need to incorporate the following information in your report.

11. Between 1800 and 1900, what was the change in population in Connecticut?

12. Between 1900 and 1800, what was the change in population in New Hampshire?

13. Write another problem that relates to this situation. Include an explanation of how to solve the problem and the answer.

You are a forest ranger in a national park. As a conservationist, you know that each bear in the park requires about 50 km^2 to survive comfortably. Use this information to solve each problem.

14. From your research, you find that there are 50 bears in approximately 1000 km^2. From this information, determine what possible problems are and recommend a solution.

You find, when examining the complete reports for the park, the following information.

Area Code	Number of Bears	Area (km^2)
A1	50	1000
A2	12	2500
B1	20	3000
B2	200	25,000
C1	50	2500
C2	100	4000

15. Determine how many bears need to be moved within the park. Make a list to show how you could rearrange the bear population to avoid overcrowding.

16. Using the information above, or the Wind Chill Factor Chart, make up three problems, outline how to solve them, and write out the solutions.

CHAPTER 2 SUMMARY

The purpose of this chapter has been to learn about integers. You learned to compare, add, subtract, multiply, and divide integers. When subtracting you rewrote the problem as addition to find the answer. Then you combined the operations to work integer problems using the correct order of operations. You read word expressions and rewrote them as mathematical expressions.

As you worked through the last sections, you learned to apply what you have learned to situations that you encounter outside of the mathematics classroom.

VOCABULARY

absolute value
additive inverse
integers
inverse property of addition
negative
number line
opposites
positive

CHAPTER REVIEW

Write an example of the following.

1. a definition of integers

2. the opposite of 37

3. the additive inverse of -12

4. a loss of 10 yards

Replace the ○ with > or <.

5. $-6 \bigcirc 0$

6. $14 \bigcirc -29$

7. $-52 \bigcirc -17$

Answer each of the following.

8. When adding $3 + -6$, is the answer positive or negative? Explain your answer.

9. When multiplying -3 and -5, is the answer positive or negative? Explain your answer.

Calculate.

10. $-6 + -3$

11. $-11 - 3$

12. $54 - -9$

13. $2 + -2$

14. $-22 + 12$

15. $-8 - -1$

Calculate.

16. $8 + -2 - -3$

17. $-3 + -6 + -8$

18. $2 - 3 + -38$

19. $(4)(-3)$

20. $(-7)(12)$

21. $(-11)(-5)$

22. $(10)(6)$

23. $\frac{15}{-3}$

24. $\frac{-48}{4}$

25. $(-35)(3)$

26. $\frac{18}{-18}$

27. $\frac{-28}{-7}$

28. $-48 \div 8 + 2$

29. $(-6)(-5) \div 2$

30. $12 \div -6 + 16 \div 4$

31. $(-2)^2 - (-3)^2$

32. $\frac{-24 - 6}{-6 - 4}$

33. $\frac{9 - 3 - 12 - 6}{-3}$

Solve each problem.

34. During a given month the high temperature in New Mexico was 68°F. The same month, the low temperature was −15°F. By how much do the extremes differ?

35. The stock of NAP, Inc., dropped 2 points each day for 3 days. What was the change in points after the 3 days?

36. The change in student enrollment in the High Grades Independent School District for the first half of the year is shown. Find the average monthly change of enrollment.

September	+26
October	−11
November	+3
December	0
January	+7

Write each as a mathematical expression, then calculate.

37. Subtract −2 from the sum of −16 and 3.

38. Divide the sum of −8 and 4 by −2.

39. Add the sum of 12 and −1 to −11.

40. Multiply the sum of −3 and −4 by the sum of 23 and −12.

Using the Wind Chill Factor Chart on page 67, determine which has the lower temperature, A or B.

41. a. Traveling on a snowmobile at 16km/h; temperature is −5°C.

b. Cross-country skiing at 8km/h; temperature is −15°C.

CHAPTER 2 TEST

Write a suitable integer for each of these.

1. a profit of 6 dollars

2. 5 steps down

Replace ○ with > or <.

3. 6 ○ −2

4. −3 ○ −1

5. −8 ○ 8

Answer the following.

6. When are sums negative?

7. When are products and quotients negative?

8. What are some key words that represent subtraction?

9. List the rules for the correct order of operations.

10. What are the steps used to find an average?

Calculate.

11. $-5 + 4$

12. $1 - 3$

13. $-5 - -6$

14. $5 + -2 - -4$

15. $(-4)(2)$

16. $-12 \div 4$

17. $16 \div -4$

18. $(-2)^3$

19. $-(2)^3$

Simplify.

20. $-3 + -12 \div -3$

21. $40 \div -4 - 2 + 2(-8 \div 2)$

22. $9 \div 3 + -3$

23. $-3 + (-2)^2 + (-3)^2$

Solve.

24. The tide was −36 cm from its normal position over a 4-hour period. How much did it change each hour?

25. During an experiment in their physical science class, the students recorded the following temperature changes: after 1 minute +3, after 2 minutes +1, after 3 minutes +1, after 4 minutes −2, and after 5 minutes −4. Find the total change in degrees after 5 minutes.

26. The temperatures recorded for a week were: −8°C, −6°C, −4°C, −3°C, 0°C, 3°C. Find the average daily temperature.

SKILLS PREVIEW

Write a fraction to tell how much of each figure is shaded.

1. 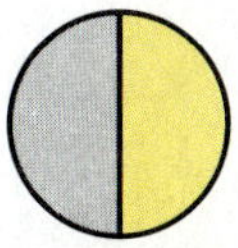

2. 

3. 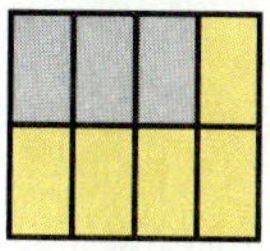

4. 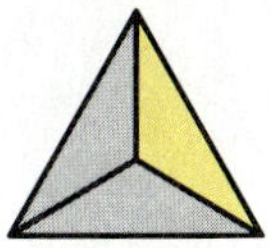

5. 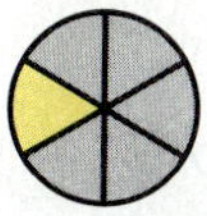

6. 

Write a decimal to tell how much of each figure is shaded.

7. 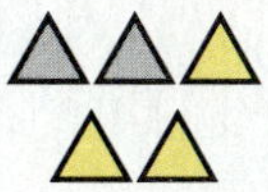

8. 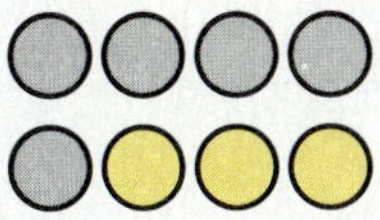

Write a decimal to tell how much of each figure is shaded.

9. 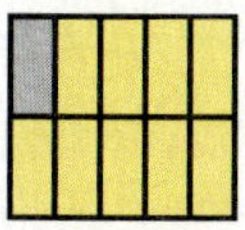

10. 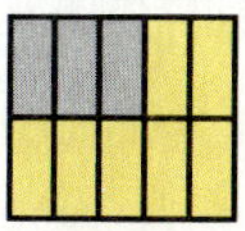

11. 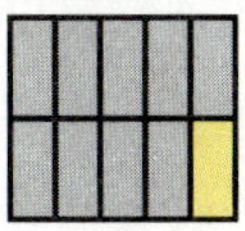

Use your calculator to find each answer.

12. $158 + 23$ **13.** $517 - 69$

14. $78 - 115$ **15.** $-26 - -81$

16. $142 + -300$ **17.** $701 - 36$

18. $(43)(11)$ **19.** $(-87)(4)$

20. $(-45)(-8)$ **21.** $\frac{-1600}{80}$

22. $\frac{-441}{-147}$ **23.** $\frac{625}{-25}$

Complete the following.

24. Draw a line segment. Show how to separate it into two equal parts.

25. Draw a line segment. Show how to separate it into four equal parts.

Explain the method you would use to solve each problem. Do not solve.

26. You have decided on a sandwich and milk for lunch. The total cost is \$3.50. If you give the cafeteria worker a \$5 bill, how much change should you receive?

27. For a birthday present, you received \$20. After buying 2 compact disks for \$9.95 each, how much money do you have left?

DMP SLD TXS
.4 5/8 3 20 1/8

To be successful in our society, you must be able to perform operations with fractions and decimals. You use fractions and decimals everyday. You can eat $\frac{1}{4}$ pound of meat, buy $6\frac{1}{4}$ yards of denim material, or run $\frac{1}{2}$ mile. Whenever you deal with money, you are dealing with decimals. Think of other situations where you have used fractions or decimals. Think of ways that your parents use them. Have you ever read about buying stocks and bonds? Fractions are used to show how much a stock costs.

In this chapter you will learn to compare, add, subtract, multiply, and divide fractions and decimals. Learning these skills will help you deal with fractions and decimals in real life. You will also use those same skills as you study algebra.

Ask Yourself

- *What do the fractions in this photograph represent?*
- *How have you used fractions today?*
- *How have you used decimals today?*
- *How have your parents or teachers used fractions and decimals this week?*
- *How are fractions and decimals used when working with algebraic ideas?*

CHAPTER 3
FRACTIONS AND DECIMALS

SECTION 3–1

DECIMALS AND FRACTIONS

WARM-UP ACTIVITY

A farmer had four children. They lived on a square farm. The farmer decided to keep a quarter of the farm for herself and to share the remaining land equally among her children. How would she divide this piece of land equally among her four children?

INTRODUCTION

Look at the chapter opening photograph. How does it show uses of fractions and decimals?

You go to the grocery store and buy groceries. Are you using fractions or decimals? Be careful. You probably used both. You could have purchased $1\frac{1}{2}$ pounds of chicken and gotten \$4.80 in change.

SKILLS DEVELOPMENT

Fractions describe the parts of a whole. You can write a fraction to show what part of a pizza is on plates. Suppose there are 3 slices of a pizza on plates. There are 8 slices in a whole pizza.

$\underline{3}$ ← number of parts on plates
8 ← number of equal parts in the whole

So, $\frac{3}{8}$ of the pizza is on plates.

numerator → $\frac{3}{8}$ means $3 \div 8$ or $8\overline{)3}$.
denominator → 8

Fractions with the special denominators of 10, 100, 1000, and so on, are often written as decimals.

$3\frac{426}{1000}$ can be written as the decimal 3.426. The number has 3 ones, 4 tenths, 2 hundredths, 6 thousandths. The decimal point separates the whole number part from the fractional part.

0	.	2		
3	.	2	1	
0	.	6	3	8

$\frac{2}{10}$ written in decimal form is 0.2.

$3\frac{21}{100}$ written in decimal form is 3.21.

$\frac{638}{1000}$ written in decimal form is 0.638.

The value of each digit in a decimal depends on its place. The 4 in the decimal numeral 3.426 means the number has 4 tenths or $\frac{4}{10}$. What does the 2 mean? The 6? The 3?

To read decimals, complete the middle column to find a pattern.

0.3	10	3 tenths
0.05		5 hundredths
0.936		936 thousandths
0.0421		421 ten-thousandths

How many places does the decimal number have? How many zeros does the number you wrote in the middle column have? Use this rule to read decimals aloud.

You can use patterns to change decimals to fractions. Look at the number of decimal places and the number of zeros in the denominator in the examples below. You say the names of these *fractions* in just the same words as you said the names of the *decimals* in the last example. Remembering the pattern helps you to change decimals to fractions and fractions to decimals.

$0.8 = \frac{8}{10}$ $\quad 0.05 = \frac{5}{100}$ $\quad 0.936 = \frac{936}{1000}$ $\quad 0.0421 = \frac{421}{10{,}000}$

Example 1

a. Write a fraction to show how much is shaded.

b. Identify the denominator and numerator.

c. Write in words what the fraction means.

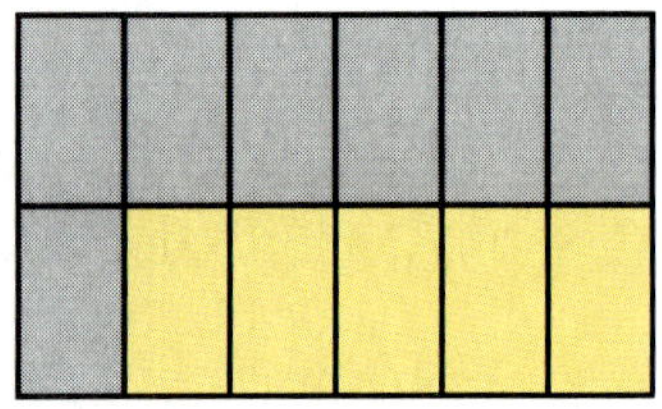

Solutions a. $\frac{7}{12}$ is shaded.

b. 7 is the numerator. 12 is the denominator.

c. $\frac{7}{12}$ means that 7 parts of the whole (12 parts) are shaded.

Example 2

Write each fraction as a decimal. Then write the name in words.

a. $\frac{3}{10}$ b. $\frac{41}{100}$ c. $1\frac{3}{1000}$

Solutions a. $\frac{3}{10} = 0.3$, 3 tenths

b. $\frac{41}{100} = 0.41$, 41 hundredths

c. $1\frac{3}{1000} = 1.003$, 1 and 3 thousandths

Example 3

Using a calculator to divide, write each fraction as a decimal.

a. $\frac{1}{8}$ b. $\frac{49}{500}$ c. $\frac{140}{350}$

Solutions a. $\frac{1}{8} = 0.125$

b. $\frac{49}{500} = 0.098$

c. $\frac{140}{350} = 0.4$

EXERCISES

Write each fraction in decimal form.

1. $\frac{5}{10}$ **2.** $\frac{65}{100}$ **3.** $\frac{385}{1000}$

4. $\frac{4}{100}$ **5.** $\frac{36}{1000}$ **6.** $\frac{8}{1000}$

7. $\frac{25}{10}$ **8.** $\frac{126}{100}$ **9.** $\frac{2865}{1000}$

Write each decimal as a fraction.

10. 0.23 **11.** 0.197 **12.** 0.003

13. 0.023 **14.** 1.97 **15.** 0.3

Express each answer as a fraction and as a decimal. Then write your answer in words.

16. What part is shaded?

17. What part is not shaded?

What is the place value of each underlined digit? The first one is done for you.

18. $4\underline{8}.56$ Place value is ones.

19. $4.\underline{8}3$ **20.** $\underline{6}5.7$ **21.** $0.5\underline{8}$

22. $94.00\underline{6}$ **23.** $\underline{2}05.801$ **24.** $1\underline{8}05$

Use your calculator. Write each amount as a fraction and then as a decimal.

25. You bought one-fourth pound of meat.

26. Your change was two and one-half dollars.

27. You need seven-eighths of a cup of sugar for your recipe.

28. You ran three-fourths of a mile today.

SECTION 3–2

COMPARING FRACTIONS AND DECIMALS

WARM-UP ACTIVITY

FRIENDLY NUMBERS

The numbers 284 and 220 are called friendly (or amicable) numbers. **Why?** The factors of the number 284, except for itself, are 1, 2, 4, 71, and 142. The factors add up to 220. Do the factors of 220, except for 220, add up to 284?

Use a calculator to test 1184 and 1210 for friendship.

INTRODUCTION

Many problems can be solved by comparing fractions or decimals. You can compare fractions if the denominators are the same. Then just compare the numerators. Fractions with the same or common denominators are called **like fractions**

$3 > 1$, so $\frac{3}{4} > \frac{1}{4}$ $\qquad$ $5 < 7$, so $\frac{5}{10} < \frac{7}{10}$

But we need to compare fractions with unlike denominators, too. Suppose Ellis has ridden $\frac{2}{5}$ mile and Forbes has ridden $\frac{3}{10}$ mile. Who is farther ahead, Ellis or Forbes? You need to decide whether $\frac{2}{5}$ or $\frac{3}{10}$ is larger. If you find a common denominator, you can compare the fractions. If you change the fractions to decimals (use a calculator), you can compare the fractions in decimal form by using place value.

SKILLS DEVELOPMENT

To find an equivalent fraction, multiply the given fraction by a suitable form of 1.

$\frac{2}{3} \times \frac{2}{2} = \frac{4}{6}$ $\qquad$ Multiply by a form of 1.

$\frac{2}{3} \times \frac{3}{3} = \frac{6}{9}$

$\frac{2}{3} \times \frac{5}{5} = \frac{10}{15}$ $\qquad$ $\frac{4}{6}, \frac{6}{9}, \frac{10}{15}$ are equivalent fractions.

Diagrams can be used to show two fractions that have the same value.

$\frac{2}{3}$ and $\frac{6}{9}$ are called **equivalent fractions.**

The fraction $\frac{5}{10}$ is not in lowest terms. We usually prefer to write $\frac{1}{2}$ instead. To write fractions in lowest terms, divide numerator and denominator by their common factors. This is like dividing by a form of 1. It does not change the value of the fractions.

$$\frac{4}{6} = \frac{2 \cdot 2}{3 \cdot 2} = \frac{2}{3} \cdot \frac{2}{2}$$

Since $\frac{2}{2}$ is just 1, we know $\frac{4}{6} = \frac{2}{3}$.

A fraction is in **lowest terms** if the numerator and denominator have no common factors.

Example 1

Write $\frac{10}{25}$ in lowest terms.

Solution $\frac{10}{25} = \frac{2 \cdot 5}{5 \cdot 5} = \frac{2}{5}$

Since 2 and 5 have no common factors, $\frac{2}{5}$ is in lowest terms.

Fractions in which the numerator is smaller than the denominator are called **proper fractions.** These are proper fractions: $\frac{1}{2}$, $\frac{2}{5}$, and $\frac{4}{7}$.

Fractions in which the numerator is greater than or equal to the denominator are called **improper fractions.** These are improper fractions: $\frac{5}{3}$, $\frac{3}{2}$, and $\frac{8}{7}$. Improper fractions can be written as **mixed numbers.** For example, $\frac{3}{2} = 1\frac{1}{2}$. The number $1\frac{1}{2}$ is called mixed because it has a fractional part and a whole part.

Example 2

Write $\frac{23}{4}$ as a mixed number.

Solution $\frac{23}{4}$ means $23 \div 4$, so divide.

$$4\overline{)23} \quad 5 \text{ R3}$$

$\frac{23}{4}$ means 5 wholes and a part left over, the remainder 3, so for $\frac{23}{4}$ write $5\frac{3}{4}$.

Some fractions can be written in decimal form exactly, for example, $\frac{1}{8} = 0.125$. But some fractions cannot be written as an exact decimal. Using a calculator, divide to find $\frac{2}{3}$.

$$\frac{2}{3} = 0.66666\ldots$$

Since the digits repeat, 0.666 . . . , $\frac{2}{3}$ is called a **repeating decimal.** You may also write $\frac{2}{3} = 0.\overline{6}$. The bar shows that the 6 repeats.

If a fraction can be written exactly as a decimal, the digits do not repeat. When we divide to find $\frac{1}{8}$ on a calculator, we get 0.125. The digits do not repeat. This is called a **terminating decimal.** Terminating means ending. The nonzero digits end.

Example 3

Let's return to the race in the Introduction. Who is farther ahead, Ellis or Forbes?

Solution Ellis has travelled $\frac{2}{5}$ mile, and Forbes has travelled $\frac{3}{10}$ mile. To compare the fractions, write an equivalent fraction for $\frac{2}{5}$ with denominator 10.

$$\frac{2}{5} = \frac{2}{5} \cdot \frac{2}{2} = \frac{4}{10}$$

Now compare $\frac{4}{10}$ and $\frac{3}{10}$. Since $\frac{4}{10} > \frac{3}{10}$, we know $\frac{2}{5} > \frac{3}{10}$. Ellis is ahead of Forbes.

EXERCISES

Match each fraction with its decimal equivalent.

1. $\frac{1}{10}$	**a.** 0.04
2. $\frac{1}{100}$	**b.** 0.1
3. $\frac{1}{25}$	**c.** 0.05
4. $\frac{1}{20}$	**d.** 0.01

An hour has 60 minutes. What part of an hour is each time interval shown?

5. 30 min

6. 15 min

7. 45 min

8. 12 min

9. 10 min

10. 75 min

Replace $\bigcirc$ by $>$, $<$, or $=$ to make each statement true.

11. $\frac{2}{3} \bigcirc \frac{5}{6}$

12. $\frac{3}{8} \bigcirc \frac{2}{3}$

13. $\frac{7}{5} \bigcirc \frac{4}{3}$

14. $\frac{3}{4} \bigcirc \frac{18}{24}$

15. $\frac{5}{12} \bigcirc \frac{3}{8}$

16. $\frac{7}{9} \bigcirc \frac{7}{12}$

17. $\frac{4}{5} \bigcirc \frac{5}{8}$

18. $\frac{1}{3} \bigcirc \frac{5}{15}$

19. $\frac{9}{20} \bigcirc \frac{8}{25}$

20. $\frac{1}{5} \bigcirc \frac{1}{4}$

Find the value of $\square$.

21. $2\frac{1}{4} = 1\frac{\square}{4}$

22. $3\frac{2}{3} = 2\frac{\square}{3}$

23. $4\frac{3}{4} = 3\frac{7}{\square}$

24. $2\frac{7}{10} = 1\frac{17}{\square}$

25. $3\frac{1}{8} = \square\frac{9}{8}$

26. $4\frac{4}{9} = \square\frac{13}{9}$

Arrange the following groups of numbers from least to greatest.

27. $\frac{3}{7}, \frac{2}{7}, \frac{4}{7}, \frac{5}{7}, \frac{6}{7}, \frac{1}{7}$

28. $\frac{1}{3}, \frac{1}{2}, \frac{1}{4}, \frac{1}{5}, \frac{1}{6}, \frac{1}{1}$

SECTION 3–3

RATIONAL NUMBERS

WARM-UP ACTIVITY

Copy the number line. Write each letter in its approximate position.

F = $1\frac{3}{4}$ R = $\frac{13}{5}$ O = $\frac{7}{8}$ T = $3\frac{5}{4}$

G = $\frac{1}{5}$ O = $\frac{24}{10}$ I = $2\frac{5}{3}$

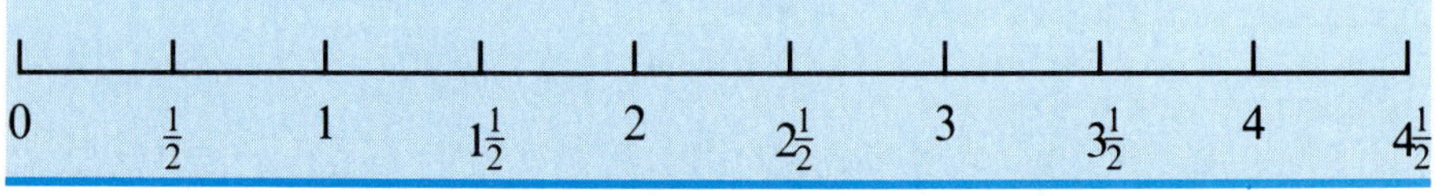

INTRODUCTION

When you add two integers, you obtain an integer.

$$-5 + -6 = -11$$

Because the sum is still an integer, integers are said to be closed with respect to addition.

When you subtract or multiply two integers, you obtain an integer.

$$-3 - -5 = +2 \qquad (-3)(+5) = -15$$

Integers are said to be closed with respect to subtraction and multiplication.

But you do not always obtain an integer when you divide.

$$\frac{7}{3} \rightarrow 3\overline{)7}^{\;2\text{ R}1} \rightarrow 2\frac{1}{3}$$

Because division sometimes results in remainders, integers are not closed with respect to division.

SKILLS DEVELOPMENT

To obtain a set of numbers that are closed with respect to division, a larger set, called **rational numbers,** were needed. Every rational number can be written as an integer divided by an integer. For example, $-\frac{8}{7}$, $\frac{1}{5}$, and $\frac{12}{3}$.

Of course, the denominator must not equal zero. (Remember, zero is an integer.) For the rational numbers to be closed under division, they must include the integers. How can you write an integer, say 5, in the form of a rational number? Many ways! For example, $\frac{5}{1}$, $\frac{10}{2}$, $\frac{-25}{-5}$ are all ways to write 5 as a rational number. Is zero a rational number? Yes, because $0 = \frac{0}{1}$ or $\frac{0}{-3}$.

You know decimals are one way of writing fractions, so decimals are rational numbers. For example, $0.78 = \frac{78}{100}$, which is an integer divided by an integer.

Rational numbers can be located on a number line. The number $\frac{1}{2}$ is midway between 0 and 1. You can also mark the number $-\frac{1}{2}$, midway between 0 and -1. A rational number such as $\frac{1}{2}$ has a decimal form of 0.5.

To compare rational numbers, think of the number line, just as you did for integers.

Example 1

Use <, >, or = to complete the following.

a. $-\frac{2}{3} \bigcirc -\frac{1}{3}$ **b.** $\frac{3}{4} \bigcirc \frac{1}{2}$ **c.** $-0.5 \bigcirc 0.25$

Solutions Use a number line.

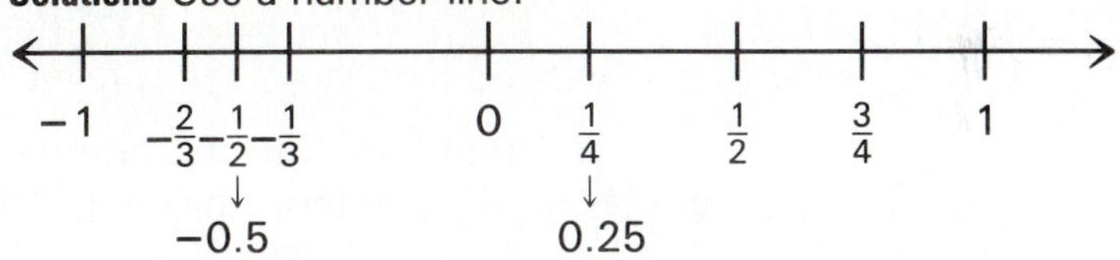

a. $-\frac{2}{3}$ is to the left of $-\frac{1}{3}$.
b. $\frac{3}{4}$ is to the right of $\frac{1}{2}$.
c. -0.5 is to the left of 0.25.

You round decimals using the same rules as for whole numbers. To round 4.83 to the nearest tenth, you would write 4.8. You look at the digit in the place to the right of the tenths place. Since the digit is less than 5, write 4.8. If you were rounding 3.76 to the nearest tenth, you would write 3.8, because 6 is greater than 5.

How would you round 6.45 to the nearest tenth? Look at the diagram below. The ten numbers in the 640s can be separated into two groups of equal size.

6.40 6.41 6.42 6.43 6.44	These numbers round down to 6.4.
6.45 6.46 6.47 6.48 6.49	These numbers round up to 6.5.

Example 2

Round 7.5232 to the nearest hundredth.

Solution Since you are rounding to the nearest hundredth, look at the thousandths place. That digit, 3, is less than 5, round down to 7.52.

You can round any rational number to the nearest integer. The open circles on the number lines mean that the point is not included.

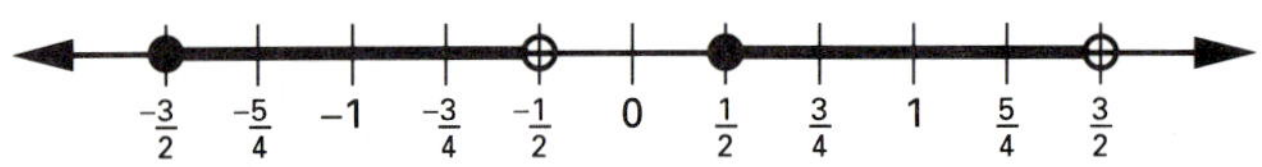

These rational numbers round to -1 as the nearest integer.

These rational numbers round to 1 as the nearest integer or whole number.

Example 3

Damietta had \$25. She went to three stores and spent \$7.82, \$6.72, and \$9.13. Does she have enough money for a \$4.00 movie ticket?

Solution A quick mental estimate sometimes is enough to answer this kind of question. For a quick estimate, round each amount: Damietta spent \$8, \$7, and \$9, or \$24 in all. No, Damietta does not have enough money to go to the movies.

EXERCISES

Write <, >, or = to complete the following.

1. $\frac{1}{4} \bigcirc -\frac{3}{4}$ **2.** $-\frac{1}{3} \bigcirc \frac{1}{3}$ **3.** $-\frac{3}{8} \bigcirc -\frac{7}{8}$

4. $\frac{2}{3} \bigcirc -\frac{1}{3}$ **5.** $\frac{1}{5} \bigcirc \frac{5}{4}$ **6.** $\frac{4}{5} \bigcirc -\frac{1}{2}$

7. $0.5 \bigcirc -0.5$ **8.** $-1.5 \bigcirc -1$ **9.** $1.25 \bigcirc 1.5$

10. $5.25 \bigcirc 6$ **11.** $-3 \bigcirc -3.25$ **12.** $2.75 \bigcirc -3.5$

Round each of the following to the indicated value.

13. 4.756, nearest hundredth **14.** −0.37865, nearest tenth

15. 5.055, nearest tenth **16.** −6.5153, nearest integer

17. $\frac{1}{3}$, nearest whole number **18.** $-1\frac{2}{5}$, nearest integer

19. $-\frac{1}{5}$, nearest integer **20.** $3\frac{5}{9}$, nearest whole number

The stock market uses rational numbers.

Stock	Sales (Hds)	High (yr)	Low (yr)	Close (today)	Net Change
QRT	669	174	$171\frac{3}{4}$	174	$-\frac{3}{4}$
Readi	10525	$23\frac{7}{8}$	$23\frac{1}{4}$	$23\frac{1}{2}$	$+\frac{1}{4}$
Style H	1027	$29\frac{5}{8}$	$29\frac{1}{8}$	$29\frac{1}{8}$	$+\frac{1}{8}$
name of company	number of shares (stocks) sold	highest price paid per share	lowest price paid per share	today's final price per share	change from yesterday's final price

21. The number of QRT stocks sold were 669 hundred. Write the number in decimal form. The amounts under high, low, close, and net change are in dollars. Hds under Sales stands for hundreds.

22. How many shares of Style H were sold?

23. How much, in dollars and cents, did Readi decline?

SECTION 3–4

ADDING AND SUBTRACTING DECIMALS

WARM-UP ACTIVITY

- Write a decimal equivalent for each common fraction.
- Find your answer in the chart.
- Cross out the letter above the answer.
- The remaining letters, when rearranged, will give you the answer to this riddle: What goes farther the slower it goes?

Money

1. $\frac{3}{4}$ 2. $\frac{7}{20}$ 3. $\frac{3}{8}$

4. $4\frac{1}{4}$ 5. $\frac{25}{10}$ 6. $\frac{9}{4}$

7. $\frac{4233}{1000}$ 8. $\frac{2}{3}$ 9. $\frac{7}{9}$

10. $\frac{9}{12}$ 11. $\frac{7}{15}$ 12. $\frac{3}{12}$

N	A	Y	C	I	M	K	H	P
2.15	$0.\overline{6}$	0.15	0.75	2.5	1.6	4.25	0.25	$0.\overline{7}$

R	O	U	S	T	E	W	D
0.35	0.23	0.375	2.25	$0.\overline{6}$	0.56	$0.4\overline{6}$	$4.2\overline{3}$

INTRODUCTION

Bobbie's time in the first race was 2.3 seconds better than Andy's. Andy's time was 1.2 seconds better than Carla's. By how many seconds was Bobbie's time better than Carla's?

SKILLS DEVELOPMENT

Adding decimals is like adding whole numbers. You must use place value notation to write the digits in the right columns. This is easy for decimals because you just line up the decimal points. Remember, subtraction is adding the opposite.

$$\begin{array}{r} 592 \\ +\ 1{,}096 \\ \hline 1{,}688 \end{array} \qquad \begin{array}{r} 298.54 \\ +\ 42.136 \\ \hline 340.676 \end{array} \qquad \begin{array}{r} 209 \\ -\ 83 \\ \hline 126 \end{array} \qquad \begin{array}{r} 493.8 \\ -\ 8.46 \\ \hline 485.34 \end{array}$$

Example

Roma walked 2.5 km on Monday, 3.7 km on Tuesday, and 3.4 km on Wednesday. How far did she walk in all?

To check whether your answer is reasonable, you can use your rounding skills.

Solution

Distance walked — Find the sum.

Think: Is my answer reasonable?

Distance walked		Find the sum.		
Monday	2.5 km	2.5	round to	3
Tuesday	3.7 km	3.7	round to	4
Wednesday	3.4 km	+ 3.4	round to	3
		9.6	estimate	10

The total distance Roma walked is 9.6 km. The answer is reasonable.

EXERCISES

Use rounding to choose the correct answer.

	a	b	c
1. 4.56 + 75.37	79.93	7.993	799.3
2. 13.4 + 7.59	209.9	20.99	2.099
3. 369.2 − −9.83	37.903	379.03	3.7903
4. 65.4 − −14.6	80.0	8.0	0.08
5. 26.8 − 8.1	34.9	18.7	1.87
6. 66.8 − 12.5	7.9	79.3	54.3
7. 126.9 + −32.8	94.1	49.1	159.7
8. 12.93 + −1.86	14.79	11.07	1.07

Find the value for each of the following.

9. 6.3 + 4.2

10. −13.9 − 4.8

11. 13.6 − −12.2

12. 9.63 + 4.18

13. 9.32 + −12.68

14. −11.296 − −9.213

15. 6.83 − 1.38

16. 3.21 + 21.73

17. 4.8 − −2.36

18. −12 + 6.93

Each cash register tape shows purchases. Estimate the total amount spent. Then find the actual amount spent.

19.	**20.**	**21.**
$4.84	$7.97	$0.15
0.13	.48	6.01
5.37	1.05	0.29
0.60	.26	3.27
2.59	3.09	8.43

Solve.

22. A three-sided garden has one side 4.5 m, another side 3.7 m. The total distance around the garden is 11.6 m. Find the length of the third side.

23. In preparing for a beach party, Trina and Sacha bought hot dogs and fixings. The hot dogs cost $15.36 and the fixings, $5.26. If $20.00 was collected, how much more do they need to collect?

24. Part of a cash register receipt was torn off. Find the missing value.

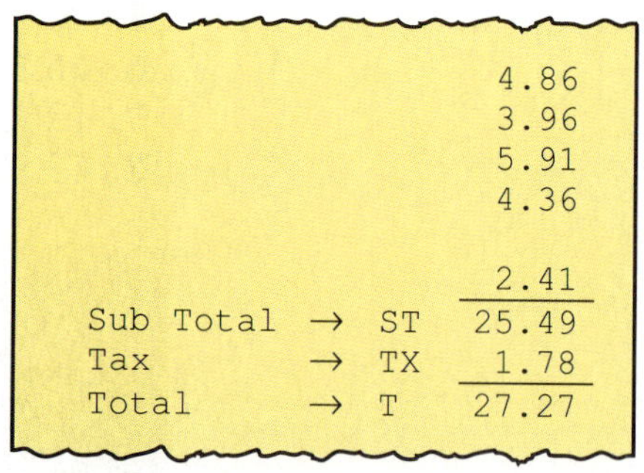

Calculate.

25. 4.7 + 2.8

26. 7.83 + (−9.6)

27. (−2.7) + 1.83

28. 5.5 − 4.6

29. 3.094 − (−6.88)

30. (−2.378) − 13.558

George has a paper route. He likes to collect from a few customers every day of the week except Sunday. One week he collected these amounts:

Monday	$ 7.85	Thursday	$17.75
Tuesday	$14.75	Friday	$21.50
Wednesday	$ 8.95	Saturday	$22.25

31. Find the total amount he collected.

32. His expenses for the week (that is, what he paid for the papers) were $67.75. Find how much money he made.

33. Use the data above to write another word problem. Then write a solution for your problem.

At a swimming meet the following times were posted.

100-Meter Freestyle		**200-Meter Freestyle**	
Alfred H.	1:30.5	Michelle S.	1:55.8
Jean S.	1:33.4	Johneen K.	2:03.1
Galois Z.	1:37.6	Sophie K.	2:04.0
400-Meter Freestyle		**1,500-Meter Freestyle**	
Clarence R.	5:40.8	Fabiola M.	20:59.9
Murray T.	5:50.7	Susan H.	21:00.7
Brian V.	6:09.5	Kittie K.	21:14.6

34. What is the difference between the times of the first and third place swimmers in each event?

35. Which race was the closest? Who were the swimmers?

36. Does it seem like the 200-meter freestyle takes twice as long to swim as the 100-meter freestyle?

37. Does the 400-meter freestyle take about four times as long as the 100-meter freestyle?

38. Use the data above to write two word problems. Exchange problems with other students and solve them.

SECTION 3–5

ADDING AND SUBTRACTING RATIONAL NUMBERS

WARM-UP ACTIVITY

$\frac{2}{5}$	$\frac{4}{6}$	$\frac{1}{3}$	$\frac{5}{10}$	$\frac{3}{4}$
$\frac{12}{6}$	$\frac{15}{20}$	$\frac{50}{100}$	$\frac{3}{10}$	$\frac{4}{20}$
$\frac{2}{6}$	$\frac{6}{12}$	$\frac{6}{8}$	$\frac{10}{15}$	$\frac{3}{9}$

Write each of the above fractions on a card. Working with a partner, sort the cards into groups. Each group of cards should show equivalent fractions.

INTRODUCTION

The pizza Scott and Denny ordered was cut into 8 slices. Scott ate 3 slices and Denny ate 2 slices. What fraction of the pizza was eaten? Scott ate $\frac{3}{8}$ of the pizza. Denny ate $\frac{2}{8}$ of the pizza.

SKILLS DEVELOPMENT

To add fractions that have the same denominator, just add the numerators.

$\frac{3}{8} + \frac{2}{8} = \frac{5}{8}$ Think: $3 + 2 = 5$

Scott and Denny ate $\frac{5}{8}$ of the pizza in all.

At the beginning of the trip, Dino's gas tank was $\frac{3}{4}$ full. By the end of the trip it was $\frac{1}{4}$ full. What fraction of a tank was used on the trip?

$\frac{3}{4} - \frac{1}{4} = \frac{3-1}{4} = \frac{2}{4}$ or $\frac{1}{2}$

Thus, $\frac{1}{2}$ of a tank of gas was used.

To add or subtract fractions that do not have the same denominator, first find equivalent fractions that have the same denominator.

Example 1

a. $\frac{1}{4} + \frac{1}{2}$

b. $\frac{5}{9} - \frac{1}{6}$

Solutions

a. $\frac{1}{4} + \frac{1}{2} = \frac{1}{4} + \frac{2}{4} = \frac{3}{4}$

Think: $\frac{1}{2} = \frac{2}{4}$
Add the numerators of like fractions.

b. $\frac{5}{9} - \frac{1}{6} = \frac{10}{18} - \frac{3}{18} = \frac{7}{18}$

18 is the lowest common denominator.
$\frac{5}{9} = \frac{10}{18}$ and $\frac{1}{6} = \frac{3}{18}$

To add or subtract mixed numbers, you may have to find equivalent fractions first. Always express answers in lowest terms.

Example 2

Add $6\frac{3}{8}$ and $2\frac{3}{4}$.

Solution

$$\begin{array}{rl} 6\frac{3}{8} = & 6\frac{3}{8} \\ +2\frac{3}{4} = & 2\frac{6}{8} \\ \hline & 8\frac{9}{8} \end{array}$$

→ Remember: a mixed number is in lowest terms when the fraction part is in lowest terms.

$8\frac{9}{8} = 8 + \frac{8}{8} + \frac{1}{8}$

$8 + 1 + \frac{1}{8} = 9\frac{1}{8}$

→ $\frac{9}{8}$ is greater than 1, $8\frac{9}{8}$ is not in lowest terms. Change $8\frac{9}{8}$ to $9\frac{1}{8}$.

Example 3

Subtract $1\frac{2}{3}$ from $3\frac{1}{6}$.

Solution

$$\begin{array}{rl} 3\frac{1}{6} = & 3\frac{1}{6} \\ -\ 1\frac{2}{3} = & 1\frac{4}{6} \\ \hline & 2\frac{7}{6} \\ & -\ 1\frac{4}{6} \\ \hline & 1\frac{3}{6} = 1\frac{1}{2} \end{array}$$

Think: $\frac{4}{6} > \frac{1}{6}$, rename $3\frac{1}{6}$.
$3\frac{1}{6} = 2 + 1 + \frac{1}{6} = 2 + \frac{7}{6}$
Subtract as before. Write answers in simplest form.

You can use the same ideas when adding and subtracting any rational numbers.

Example 4

a. Find $-\frac{2}{3} + -\frac{1}{5}$.

b. Find $-6\frac{1}{2} - -2\frac{3}{4}$.

Solutions a.

$-\frac{2}{3} = -\frac{10}{15}$ → Find equivalent fractions.

$+ -\frac{1}{5} = -\frac{3}{15}$ → Think about adding integers. $-10 + -3 = -13$

$-\frac{13}{15}$ → Write the answer.

b. $-6\frac{1}{2} - -2\frac{3}{4} = \quad -6\frac{1}{2} + 2\frac{3}{4}$. Why?

The signs are different, so we need to find the **difference.**

$-6\frac{1}{2} = -6\frac{2}{4} = -5\frac{6}{4}$

$+ 2\frac{3}{4} = \ 2\frac{3}{4} = \ 2\frac{3}{4}$

$-3\frac{3}{4}$

$-6\frac{1}{2}$ is farther from zero, the answer is negative.

EXERCISES

Which answer, **a** or **b**, is a reasonable estimate?

	a	b		a	b
1. $\frac{7}{9} + 2\frac{1}{9}$	1	3	**2.** $2\frac{1}{2} - 3\frac{1}{4}$	-1	1
3. $5\frac{13}{17} - 1$	5	6	**4.** $-\frac{2}{3} - -\frac{3}{4}$	0	-1

Add or subtract. Give each answer in lowest terms.

5. $\frac{1}{8} + \frac{3}{8}$ **6.** $\frac{3}{4} + \frac{1}{4}$ **7.** $\frac{5}{9} + \frac{4}{9}$

8. $\frac{3}{5} - \frac{1}{5}$ **9.** $\frac{9}{10} - \frac{7}{10}$ **10.** $\frac{5}{7} - \frac{3}{7}$

11. $\frac{2}{3} + \frac{1}{4}$ **12.** $\frac{7}{8} + \frac{1}{2}$ **13.** $\frac{9}{10} + \frac{3}{5}$

14. $\frac{5}{8} - \frac{1}{3}$ **15.** $\frac{3}{4} - \frac{2}{3}$ **16.** $\frac{5}{6} - \frac{1}{4}$

Solve each problem.

17. Cassandra took $\frac{3}{4}$ hour to walk to school. Gustave rides to school on his bicycle in $\frac{1}{3}$ hour. Who takes longer to go to school? How much longer?

18. The earth's crust is composed of elements. Many of these are extracted for industrial use. The chart shows the three most common elements. What fraction of the earth's crust is made up of these three elements?

Element	Oxygen	Silicon	Aluminum
Fraction of the earth's crust	$\frac{2}{5}$	$\frac{3}{10}$	$\frac{1}{25}$

19. One hour before the game started, $\frac{1}{8}$ of the stadium was filled. In the next 30 min, another $\frac{1}{4}$ of the stadium was filled. At 30 min to game time, what fraction of the stadium was filled?

Add or subtract. Watch the signs! Be sure all answers are in simplest form.

20. $\frac{3}{4} - -\frac{1}{4}$

21. $-\frac{2}{5} + -\frac{1}{2}$

22. $\frac{3}{5} - -\frac{1}{10}$

23. $-\frac{3}{4} + -\frac{1}{8}$

24. $-\frac{3}{4} - \frac{2}{3}$

25. $-\frac{1}{4} + -\frac{4}{5}$

26. $\frac{3}{5} - -\frac{2}{3}$

27. $-\frac{1}{8} - -\frac{3}{4}$

28. $-\frac{3}{5} + \frac{7}{10}$

Solve each problem. Be careful when adding or subtracting mixed numbers.

29. Juanita used $2\frac{3}{4}$ yards of material to make a sail for a large model boat. Raven used $1\frac{1}{4}$ yards less than Juanita. How much material did Raven use?

30. Bernie used $1\frac{2}{3}$ cups of milk and $\frac{3}{4}$ cup of other liquids to make a dish for dinner. How much liquid did he use in all?

31. Alexandra took $3\frac{1}{2}$ hours to drive to her grandparents' house. Traffic was worse on her way home. It took her $4\frac{1}{4}$ hours to drive home. How much longer did it take to drive home than to drive to her grandparents' house?

Add or subtract. Express all answers in lowest terms.

32. $-1\frac{1}{2} + 3\frac{2}{3}$

33. $-\frac{4}{5} - -2\frac{1}{2}$

34. $4\frac{5}{8} - 6\frac{7}{8}$

35. $-2\frac{3}{4} + -1\frac{1}{2}$

36. $1\frac{5}{8} + -2\frac{1}{4}$

37. $-5\frac{1}{2} - -5\frac{1}{2}$

SECTION 3–6

MULTIPLYING AND DIVIDING DECIMALS

WARM-UP ACTIVITY

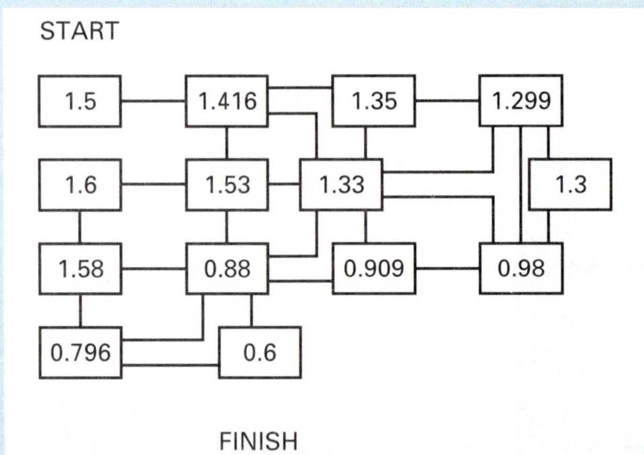

Mark a path that begins at 1.5 and ends at 0.6. Choose the path so that each number on the path is less than the next number on the path. How many boxes did you pass through to get to the finish?

INTRODUCTION

There are many real-life situations that require the use of decimals.

a. Suppose you know that it takes about 0.3 hour to read a section of a book. How could you find how long it would take to read the whole book if there are 12 sections?

b. Suppose you buy three videotapes for $14.95 each. The clerk says the total is $55.25. Will you actually pay that much for 3 videotapes?

c. Suppose you earn $4.65 per hour and worked 16.5 hours last week. How could you find how much you earned?

In each situation, you could add to find the answer. However, just as with whole numbers, multiplying would be much faster.

SKILLS DEVELOPMENT

Study these examples. Can you find the pattern that was used to place the decimal point in each answer?

$$\begin{array}{r}78\\ \times\ 4\\ \hline 312\end{array}\quad \begin{array}{r}78\\ \times\ 0.4\\ \hline 31.2\end{array}\quad \begin{array}{r}78\\ \times\ 0.04\\ \hline 3.12\end{array}\quad \begin{array}{r}7.8\\ \times\ 4\\ \hline 31.2\end{array}\quad \begin{array}{r}7.8\\ \times\ 0.4\\ \hline 3.12\end{array}\quad \begin{array}{r}7.8\\ \times\ 0.04\\ \hline 0.312\end{array}\quad \begin{array}{r}0.78\\ \times\ 0.04\\ \hline 0.0312\end{array}$$

Write a summary of the pattern. See if your pattern works for each example.

Multiplying decimals is very much like multiplying whole numbers. Study how multiplication is used to solve each of the problems in the introduction.

Example

a. 12 × 0.3 **b.** \$14.95 × 3 **c.** \$4.65 × 16.5

Solutions a. Start by estimating the answer. $\frac{1}{2}$ of 12 is about 6.

$$\begin{array}{r}12\\ \times\ 0.3\\ \hline \end{array}\qquad \begin{array}{r}12\\ \times\ 0.3\\ \hline 3.6\end{array}$$

How is the decimal point placed in the answer?

Think:
$$\begin{array}{r}12\\ \times\ 3\\ \hline 36\end{array}$$

Check: $\frac{3}{10}$ is less than $\frac{1}{2}$. 3.6 is less than 6. The answer is reasonable.

b. Estimate: \$15 × 3 = \$45

$$\begin{array}{r}\$14.95\\ \times\ 3\\ \hline \$44.85\end{array}$$

You could use a calculator. Is this answer reasonable?

c. Estimate: \$5 × 16 = \$80

$$\begin{array}{r}\$\ 4.65\\ \times\ 16.5\\ \hline \$76.725\end{array}$$

If you use paper and pencil, multiply as for whole numbers, then place the decimal point.

Now each of our original problems is solved—right? **WRONG!** Remember that when you are solving problems, you must reread the problem to be sure you have answered the question.

a. You were to find *how long* it would take to read the whole book. Your answer needs a label. The actual solution is 3.6 hours.

b. You were asked whether you will pay $55.25 for the 3 video tapes. $44.85 does not answer the question. The answer is "No!"

c. You were asked how much you earned. $76.725 is not the answer. When working with money, round your answer to the nearest hundredth. $76.73 is the answer.

Dividing decimals is also similar to working with whole numbers. Study these examples. Can you find the pattern that was used to place the decimal point in each answer?

$$2\overline{)2.50}^{\,1.25} \qquad 12\overline{)14.4}^{\,1.2} \qquad 3\overline{)2.1}^{\,0.7} \qquad 9\overline{)0.36}^{\,0.04}$$

Write a summary of the pattern. See if it works with these examples.

Remember that in working with fractions, you sometimes had to find an equivalent fraction. Decimals are fractions, too, and a fraction is one way to write a division.

To divide a decimal by a decimal, use equivalent fractions to express the divisor (the denominator) as a whole number.

$$\frac{6.97}{1.7} = \frac{6.97}{1.7} \times \frac{10}{10} = \frac{69.7}{17}$$

The divisor is now a whole number. Now divide. $17\overline{)69.7}^{\,4.1}$

EXERCISES

Which answer, **a** or **b**, is a reasonable estimate?

	a	b		a	b
1. 4.0 × 1.36	54.4	5.44	**2.** 1.8 × 3.8	6.84	0.684
3. 6.8 × 2.65	18.02	1.802	**4.** 0.8 × 0.365	0.0292	0.292
5. 3.18 ÷ 0.6	53	5.3	**6.** 2.92 ÷ 0.08	36.5	3.65
7. $0.007\overline{)0.0245}$	3.5	0.035	**8.** $0.3\overline{)0.1005}$	3.35	0.335

Find the cost of each purchase. (Remember, merchants round money amounts up.)

9. 8 bottles of orange juice at \$0.69 each

10. 6 rolls of tape at \$1.49 per roll

11. 18 m of wood trim at \$0.54/m

12. 6.8 kg of syrup at \$4.69/kg

13. 8 jars of pickles at \$1.49 each

Did you know some water is hard? When you take a bath, the ring around the bathtub is mostly caused by the calcium carbonate in the water.

14. One liter of hard water contains 0.25 g of calcium carbonate. How much is in a tub containing 350 L water?

15. Estimate how much calcium carbonate would be deposited in a year if you never scrubbed the bathtub. Suppose you take a bath 3 times a week.

Complete the following below by placing the decimal point in the product or quotient.

16. $9 \times 0.4 = 36$

17. $1.4 \div 2 = 7$

18. $2.6 \times 49.8 = 12948$

19. $7.74 \div 0.9 = 86$

20. $0.2 \times 0.3 = 6$

21. $0.138 \div 0.46 = 3$

22. $200 \times 1.23 = 24600$

23. $1.215 \div 13.5 = 9$

To get the best buy when you shop, you should compare prices. To compare prices you find the **unit price.** The unit price is the cost of one item. For the items below, find the unit price. Which is the better buy?

24. A package of 8 giant deluxe gift wrap bows for \$1.69 or a package of 2 giant deluxe bows for 59¢?

25. Ready toothpaste, 125 mL for \$1.29, or Tastee toothpaste, 200 mL for \$2.30?

26. Choose two different sizes of a food item or household supply. Find the unit prices. Which is the better buy?

SECTION 3–7

MULTIPLYING RATIONAL NUMBERS

WARM-UP ACTIVITY

5	10	3
4	6	1
9	2	7

In a magic square, all rows, columns, and diagonals have the same sum. One number in this magic square is not correct. Can you fix the magic square?

INTRODUCTION

In Section 3–2, you multiplied fractions when you found equivalent fractions.

$$\frac{1}{3} = \frac{1}{3} \times \frac{2}{2} = \frac{2}{6} \qquad \frac{3}{8} = \frac{3}{8} \times \frac{3}{3} = \frac{9}{24}$$

To multiply a whole number by a fraction, remember that a whole number is a rational number. It can be expressed as a fraction.

In a wooded area, $\frac{2}{3}$ of the trees are diseased. There are 60 trees. How many are diseased?

To find $\frac{2}{3}$ of 60 trees, find $\frac{2}{3} \times 60$.

$$\frac{2}{3} \times 60 = \frac{2}{3} \times \frac{60}{1} = \frac{120}{3} \text{ or } 40$$

Think:
$2 \times 60 = 120$
$3 \times 1 = 3$

There are 40 diseased trees.

SKILLS DEVELOPMENT

The same rules you learned for multiplying positive and negative integers apply to multiplying positive and negative rational numbers. Here are some examples.

Example 1

Find the product of $-\frac{1}{3}$ and $-\frac{4}{7}$.

Solution $\left(-\frac{1}{3}\right)\left(-\frac{4}{7}\right) = +\left(\frac{1}{3}\right)\left(\frac{4}{7}\right)$ → First, decide whether the product will be negative or positive.

$= +\frac{(1)(4)}{(3)(7)}$ → Multiply numerators. Multiply denominators.

$= +\frac{4}{21}$ or $\frac{4}{21}$ → Be sure the product is lowest terms.

Example 2

Find the product of $\frac{2}{3}$, $-\frac{9}{5}$, $\frac{5}{8}$.

Solution $\left(\frac{2}{3}\right)\left(-\frac{9}{5}\right)\left(\frac{5}{8}\right) = -\left(\frac{2}{3}\right)\left(\frac{9}{5}\right)\left(\frac{5}{8}\right)$

$= -\frac{(2)(9)(5)}{(3)(5)(8)}$

$= -\frac{90}{120} = -\frac{3}{4}$

Solution 2 You also can use your knowledge of equivalent fractions to find these products.

$-\left(\frac{2}{3}\right)\left(\frac{9}{5}\right)\left(\frac{5}{8}\right) = -\frac{\overset{1}{\cancel{(2)}}\,\overset{3}{\cancel{(9)}}\,\overset{1}{\cancel{(5)}}}{\underset{1}{\cancel{(3)}}\,\underset{1}{\cancel{(5)}}\,\underset{4}{\cancel{(8)}}}$ Discuss how this method works.

$= -\frac{3}{4}$

Example 3

Find the product of $\frac{2}{3}$ and $-1\frac{1}{5}$.

Solution $\left(\frac{2}{3}\right)\left(-1\frac{1}{5}\right) = -\left(\frac{2}{3}\right)\left(\frac{6}{5}\right)$ → Change the mixed number to an improper fraction.

$= -\frac{(2)\,\overset{2}{\cancel{(6)}}}{\underset{1}{\cancel{(3)}}\,(5)}$

$= -\frac{4}{5}$

EXERCISES

1. Calculate both of the following:

$-\frac{5}{8} \times \left(\frac{3}{4} \times \frac{4}{5}\right)$ $\qquad$ $\left(-\frac{5}{8} \times \frac{3}{4}\right) \times \frac{4}{5}$

2. Why are the answers found in Exercise 1 the same?

3. In Mickey's class of 32 students, a survey showed that $\frac{3}{4}$ of the class kept a pet. How many students had pets?

4. The survey also showed that there were 48 pets in all. If $\frac{1}{4}$ of all pets were dogs, how many dogs were there?

The membership fee for one year at a health club is \$150. The year begins in January and ends in December. You decide to join in the beginning of September.

5. What part of a year is September through December?

6. If your fee is based on the part of the year you use the facilities, how much is your fee?

An ad shows that a smoke sensor is for sale at $\frac{1}{4}$ to $\frac{1}{2}$ off. The regular cost is \$32.

7. What is the most you would pay?

8. What is the least you would pay?

Below are two ways to simplify an expression.

9. What property is used in Method 2?

Method 1

$$\begin{aligned} -\tfrac{4}{5}\left(-\tfrac{3}{4} + \tfrac{1}{3}\right) &= -\tfrac{4}{5}\left(-\tfrac{9}{12} + \tfrac{4}{12}\right) \\ &= -\tfrac{4}{5}\left(-\tfrac{5}{12}\right) \\ &= +\frac{\overset{1}{(\cancel{4})}\ \overset{1}{(\cancel{5})}}{\underset{1}{(\cancel{5})}\ \underset{3}{(\cancel{12})}} \\ &= \tfrac{1}{3} \end{aligned}$$

Method 2

$$\begin{aligned} -\tfrac{4}{5}\left(-\tfrac{3}{4} + \tfrac{1}{3}\right) &= \left(-\tfrac{4}{5}\right)\left(-\tfrac{3}{4}\right) + \left(-\tfrac{4}{5}\right)\left(\tfrac{1}{3}\right) \\ &= \tfrac{3}{5} + -\tfrac{4}{15} \\ &= \tfrac{9}{15} + -\tfrac{4}{15} \\ &= \tfrac{5}{15} \text{ or } \tfrac{1}{3} \end{aligned}$$

Simplify the following expressions. State whether you use Method 1 or Method 2 and why.

10. $\frac{3}{4}\left(\frac{3}{5} + -\frac{7}{5}\right)$

11. $-\frac{2}{3}\left(\frac{9}{10} - \frac{4}{5}\right)$

Evaluate. Remember the order of operations.

12. $\frac{1}{3} + \frac{2}{3} - \frac{1}{2}$

13. $\frac{3}{4} - \frac{1}{2} + \frac{1}{3}$

14. $\left(\frac{2}{3}\right)\left(-\frac{1}{4}\right)\left(-\frac{3}{4}\right)$

15. $\left(-\frac{2}{9}\right)\left(-\frac{3}{2}\right)\left(-\frac{6}{5}\right)$

Below is a copy of the stock market report at the close of today's trading.

Stock	Sales (hds)	High (yr.)	Low (yr.)	Close (today)	Net Change
Goetz	2283	$46\frac{3}{4}$	$45\frac{7}{8}$	$46\frac{1}{2}$	$-\frac{1}{2}$
Hoff	39	$17\frac{3}{8}$	17	17	$-\frac{3}{8}$

16. Express the high of Goetz in dollars and cents.

17. If Josefina bought 1000 shares of Hoff stock at yesterday's closing price, how has the value of her stock changed?

18. Josefina bought 2000 shares of Goetz at yesterday's closing price. At the end of today's trading, how much had the value of her stock changed?

19. Does the chart give you enough information to decide whether it would be better to invest in Goetz or Hoff? Explain your answer.

20. Explain how to find $\left(\frac{2}{3}\right)^2$.

Evaluate the following:

21. $\frac{1}{4} \times \frac{1}{4}$

22. $\frac{1}{5} \times \frac{1}{5}$

23. $\frac{1}{2} \times \frac{1}{2}$

24. $\frac{1}{3} \times \frac{1}{3}$

25. Look for a pattern in Exericses 21–24. What general rule can you state?

Use the distributive property to simplify the following multiplications:

26. $12\left(1\frac{1}{3}\right)$

27. $15\left(2\frac{1}{5}\right)$

28. $-\frac{1}{4}\left(8\frac{1}{2}\right)$

SECTION 3–8

DIVIDING RATIONAL NUMBERS

WARM-UP ACTIVITY

What happens when you multiply the numerator and denominator of a fraction by the denominator? Make up an example and try it. What did you find? Why?

Find a fraction that when you add its denominator to its numerator and denominator the result is double the original fraction.

INTRODUCTION

Mathematics is the study of patterns, as you saw in Chapter 1. Some patterns you already know can be used in new ways. Do you see a pattern in the examples below?

$$5 + (-5) = 0 \qquad \tfrac{5}{6} + \left(-\tfrac{5}{6}\right) = 0$$

The numbers 5 and -5, $\frac{5}{6}$ and $-\frac{5}{6}$, are called **opposites** because the sum of each pair is zero. Zero is the **additive identity** because when it is added to a number, the value of the number does not change, for example, $2 + 0 = 2$ and $\frac{1}{2} + 0 = \frac{1}{2}$.

SKILLS DEVELOPMENT

Adding -4 to 4 results in the additive identity. Does multiplication have a similar pattern? That is, is there a number that, when you multiply by it, you always get the multiplicative identity? What would the multiplicative identity be? We have to find that first.

When we add zero to a number, the number is unchanged, so zero is the additive identity. What number can we multiply any number by so that the value of the number is not changed? One, of course. The multiplicative identity is 1.

Now, what can we multiply a number by and get 1, the multiplicative identity?

Example 1

Use the rational numbers and multiply to fill in the □.

a. $5 \times \square = 1$ b. $\frac{5}{6} \times \square = 1$ c. $-2 \times \square = 1$

Solution

a. $5 \times \frac{1}{5} = 1$ b. $\frac{5}{6} \times \frac{6}{5} = 1$ c. $-2 \times -\frac{1}{2} = 1$

The pairs of numbers 5 and $\frac{1}{5}$, $\frac{5}{6}$ and $\frac{6}{5}$, and -2 and $-\frac{1}{2}$, are called **reciprocals** because the product of each pair is equal to 1.

Cher bought a block of cheese. She is deciding how to divide it. If she divides the block into halves, how many pieces are there?

If you *divide* by $\frac{1}{2}$, you get *twice* as many pieces. If you *divide* by $\frac{1}{4}$, you get four *times* as many, and if you *divide* by $\frac{1}{8}$, you get eight *times* as many. You know *times* means multiply. How are division and multiplication related?

Division	Multiplication
$1 \div \frac{1}{2} = 2$	$1 \times \frac{2}{1} = 2$
$1 \div \frac{1}{4} = 4$	$1 \times \frac{4}{1} = 4$
$1 \div \frac{1}{8} = 8$	$1 \times \frac{8}{1} = 8$

To divide by a fraction, you must multiply by the reciprocal of that fraction.

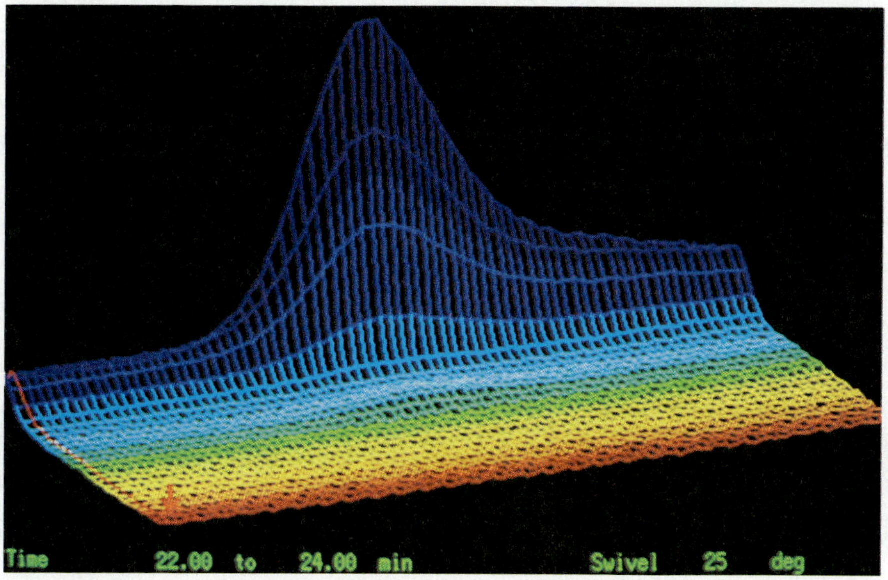

Example 2

Divide $12 \div \frac{1}{3}$.

Solution

Think what this would mean in terms of breaking up wholes, like Cher's block of cheese. Here we want to break up 12 wholes into thirds. We expected to get 3 times as many pieces, because we are dividing by 3. Three times as many pieces would be 36 pieces. Let's divide, using the pattern above, and check.

$12 \div \frac{1}{3} = 12 \times 3 = 36$. We do get 36 pieces.

Example 3

Divide $2\frac{1}{4} \div \frac{3}{4}$.

Solution $2\frac{1}{4} \div \frac{3}{4} = 2\frac{1}{4} \times \frac{4}{3}$

$$= \frac{9}{4} \times \frac{4}{3}$$
$$= \frac{36}{12}$$
$$= 3$$

(Remember, to multiply a mixed number, you must first write it as an improper fraction.)

Example 4

Divide $-2\frac{1}{2} \div -\frac{1}{2}$.

Solution $-2\frac{1}{2} \div -\frac{1}{2} = -2\frac{1}{2} \times -2$

$$= -\frac{5}{2} \times -2$$
$$= \frac{10}{2}$$
$$= 5$$

EXERCISES

Many calculators have a reciprocal key, [1/x]. Use your calculator to find these reciprocals. An example is shown first.

Example: $\frac{1}{4}$ [4] [(1/x)] [0.25]

1. $\frac{4}{5}$ **2.** $1\frac{2}{9}$ **3.** 0

4. Glen works in a shirt factory. It takes him $\frac{3}{4}$ h to make one shirt. How many shirts can he make in a $7\frac{1}{2}$-h shift?

5. Each time a bus covers its route, it uses $\frac{1}{12}$ tank of gas. How many times can it cover its route with $\frac{7}{8}$ tank of gas?

6. Tim can type a paragraph in $2\frac{1}{2}$ minutes. How many paragraphs can he type in $2\frac{1}{2}$ hours? Use mental math and logic to solve this problem.

When you multiply or divide fractions, you can simplify your work by using your knowledge of equivalent fractions.

$$\frac{\overset{1}{\cancel{16}}}{\underset{3}{\cancel{15}}} \times \frac{\overset{5}{\cancel{25}}}{\underset{2}{\cancel{32}}} \qquad \text{Divide 16 and 32 by 16. Divide 15 and 25 by 5.}$$

Use this method to simplify the following:

7. $\frac{15}{28} \div \frac{24}{49}$

8. $\frac{27}{55} \div \frac{45}{22}$

9. $\frac{26}{125} \div \frac{39}{20}$

10. $\frac{56}{63} \div \frac{70}{93}$

Simplify each of the following.

11. $\left(\frac{2}{3} - -\frac{1}{3}\right) \div \left(-\frac{3}{4} - -\frac{2}{3}\right)$

12. $\left(-\frac{1}{8} + -\frac{1}{4}\right)\left(\frac{1}{2} + -\frac{1}{6}\right)$

13. $\left(-\frac{2}{3} - -\frac{1}{2}\right) \div \left(-\frac{3}{2} - \frac{3}{2}\right)$

14. $\left(-\frac{4}{5} + \frac{1}{3}\right)\left(-\frac{1}{2} - \frac{1}{3}\right)$

Solve.

15. Lara has 6 blocks to walk to school. If she wants to get to school in $4\frac{1}{2}$ minutes, how fast must she walk?

16. A bag of chips fills $\frac{3}{4}$ of a bowl. How many bags are needed to fill 12 bowls?

17. The clerk at the cash register takes $2\frac{1}{4}$ min to serve a customer. How many customers can be served in 3 h?

18. *The White Coat* is a novel with 270 pages. Angela reads $\frac{3}{4}$ page per minute. How long will it take her to read the book?

19. To receive her swimming certificate, Sabrina must have 15 hours of instructions. Sabrina enrolled in a swimming certificate course. Each class is $\frac{3}{4}$ hour. How many classes are there?

20. Make up word problems similar to those above. Exchange with classmates and solve.

CHAPTER 3 SUMMARY

In this chapter you reviewed your knowledge of fractions and decimals. You combined your skills with these numbers and the skills with integers to develop skills working with rational numbers.

First, you explored both common and decimal fractions. You used equivalent fractions and decimals to make comparisons. You learned to express the same quantity as a common fraction and as a decimal. Then, in order for division of integers to be closed, you discovered rational numbers.

Second, you studied the model of addition and subtraction as it relates to rational numbers. Addition and subtraction of decimals was viewed as an extension of the operations with whole numbers. Addition and subtraction of fractions required finding a common denominator. Using the common denominator, equivalent fractions were found. Then the numerators were added or subtracted and placed over this common denominator. Decimals were examined first so you could use the calculator in your fraction work.

Third, you examined the model of multiplication and division of rational numbers. Again, multiplication of decimals was investigated as an extension of your work with whole numbers. Multiplication of fractions did **not** require finding a common denominator. You learned to recognize and find the reciprocal of a fraction. The product of a rational number and its reciprocal is always one. Dividing by a fraction was the same as multiplying by the reciprocal of that fraction. When multiplying and dividing rational numbers, you learned the sign of the answer was the same as for integers.

You were encouraged to use your calculator to simplify and check the problems in this chapter.

VOCABULARY

additive identity
common denominator
denominator
equivalent fractions
improper fraction
like fractions
lowest terms
mixed number
multiplicative identity
numerator
opposite
proper fraction
rational number
reciprocal
repeating decimal
terminating decimal

CHAPTER REVIEW

The word **fraction** has 8 letters and $\frac{3}{8}$ of them are vowels. What fraction of the letters in the following words are vowels?

1. denominator **2.** numerator

3. decimal

Write each fraction as a decimal.

4. $\frac{2}{5}$ **5.** $\frac{1}{8}$

6. $\frac{1}{3}$

Replace ○ by < or >.

7. $\frac{1}{3}$ ○ $\frac{1}{2}$ **8.** $\frac{3}{5}$ ○ $\frac{4}{7}$

9. $\frac{7}{4}$ ○ $\frac{7}{5}$ **10.** $\frac{-11}{3}$ ○ $\frac{5}{2}$

11. $\frac{-3}{5}$ ○ $\frac{4}{-7}$ **12.** $\frac{7}{4}$ ○ $\frac{-7}{5}$

Simplify.

13. $\frac{-4}{2}$ **14.** $\frac{20}{-4}$

15. $\frac{-8}{-9}$ **16.** $-\frac{3}{4} - -\frac{2}{3}$

17. $\frac{1}{3} - -\frac{1}{8}$ **18.** $-\frac{1}{3} - \frac{1}{4}$

19. $\left(-\frac{2}{3}\right)\left(-\frac{5}{4}\right)$ **20.** $-\frac{3}{4} \div \frac{1}{2}$

21. $1\frac{2}{3} \div -\frac{5}{6}$

Calculate.

22. $12.7 - 5.92$ **23.** $-6.9 + -0.89$

24. $8.3 - -9.85$ **25.** $(-3.4)(6.2)$

26. $7.2 \div 0.08$ **27.** $-3.5 \div -7$

Solve each problem.

28. John bought a bike for \$93.95. He sold it later for \$54.98. What was the net result of his transaction?

29. Three tapes costing \$6.98, \$7.99, and \$9.98 were bought for a collection.

a. How much in total was paid for the tapes?

b. What is the difference between the highest and lowest price?

30. The admission charge to see a performance by acrobats was as follows.

Adults	\$8.25	Students	\$5.75

Find the cost for the following.

a. 120 adults

b. 25 students

c. 240 adults and 130 students

31. Which is the better buy?

a. 5 bars of soap for 99 cents

b. 3 bars of soap for 58 cents

32. Crystal swam $5\frac{1}{3}$ lengths of the pool. Brenda swam $1\frac{1}{2}$ times as far as Crystal. How many lengths did Brenda swim?

33. Jackie bought a sweater marked \$34.99 at a $\frac{1}{3}$ off sale. How much did she pay for the sweater?

34. A film-processing machine takes $1\frac{1}{4}$ minutes to develop one picture. How long will it take to develop 30 pictures?

CHAPTER 3 TEST

Calculate.

1. $-\frac{1}{2} + \frac{1}{6}$
2. $\left(\frac{2}{3}\right)^2 - -\frac{4}{9}$
3. $-\frac{3}{10} - 1\frac{1}{5}$
4. $\left(\frac{2}{3}\right)\left(-1\frac{1}{2}\right)$
5. $\frac{4}{5} \div \frac{2}{3}$
6. $2\frac{5}{8} \div -3\frac{1}{2}$
7. $3.4 + 7.08$
8. $-0.65 - 1.36$
9. $7.8 - -5.3$
10. $(7.2)(-3.2)$
11. $5.49 \div 0.9$
12. $-22.1 \div -3.25$

Solve each problem.

13. A symphony is made up of a certain number of movements. The longest symphony takes about 99 minutes. The first movement is $\frac{1}{3}$ the total time. How many minutes is this?

14. The average "Top 40" video is $3\frac{1}{2}$ minutes long. How many "hits" can be shown during a one-hour, all-music video show?

15. Sacha decided to have a party and invite nine friends. She spends \$13.78 on snacks and her friends all donate \$1.00 for the party. How much does she pay?

16. An item priced \$45 is marked $\frac{1}{3}$ off. How much will you save?

17. Tickets at the school football game cost \$3.00. If there are 2000 people at the game, how much money was made on ticket receipts?

18. About $\frac{1}{5}$ of the population of the United States is of school age (5–17 years old). If there are 250,000,000 people in America, how many are of school age?

19. The Golden Gate Bridge is 1935 meters long. Wendy's step is 0.62 meters long. How many steps would she take in crossing the bridge?

20. Ken started school with a balance of \$156.78 in his checking account. Find his balance after making the following transactions.
    a. Withdrawal of \$45.00
    b. Deposit of \$20.00
    c. Withdrawal of \$67.50

CUMULATIVE REVIEW

Calculate.

1. $250 + 67 + 95$
2. $7825 - 976$
3. $(250)(675)$
4. $7500 \div 50$
5. $6.3 + 8.75$
6. $9.01 - 7.125$
7. $1.7 + 1.9 + 1.3$
8. $(6.5)(0.3)$
9. $0.75 \div 5$
10. $\frac{2}{3} + \frac{3}{4} + \frac{1}{2}$
11. $\frac{4}{5} - \frac{3}{10}$
12. $\left(\frac{3}{4}\right)\left(\frac{7}{8}\right)$
13. $\frac{1}{5} \div \frac{1}{4}$
14. $4\frac{1}{2} - 2\frac{3}{4}$
15. $3\frac{1}{3} \div \frac{7}{10}$
16. $-\frac{2}{3} - \frac{1}{2}$
17. $\left(7\frac{1}{3}\right)\left(-1\frac{1}{11}\right)$
18. $-1\frac{1}{3} \div -\frac{2}{3}$
19. $-6 + 7 + -5$
20. $-19 - -24$
21. $-4.8 \div 0.4$

Write the following numbers in decimal form.

22. two thousand, seven hundred
23. two thousand and seven hundredths

Graph the following sets of numbers on a number line.

24. Multiples of 4 between 10 and 30
25. Whole numbers less than 5
26. Prime numbers less than or equal to 11

Simplify using rules for order of operations.

27. $(8 - 2 \times -3) - (2^2 + 3 \times 5)$
28. $-\frac{3}{10} - \frac{2}{5} \times \left(-\frac{1}{2}\right)^2$

The stock market showed the following record for a certain commodity over 4 days: up 3 points, down 5 points, down 2 points, up 3 points.

29. Find the net gain or loss.
30. Find the average daily loss.
31. Construction on the pyramids of Egypt started about 2700 B.C. How old will they be in A.D. 2000?

Julio was only half-listening in his world history class. He heard 194 B.C., 276 B.C., and Eratosthenes. He reasoned these were the birth and death dates of Eratosthenes.

32. Which was his birth date?
33. How old was he when he died?
34. Deanne took part in a 10-question math challenge. For each correct answer she scored 10 points. However, for each incorrect answer she scored -15 points. Deanne's final score on the test was 0. How many of the 10 questions did she answer correctly?
35. Charlene went to a sale and bought a \$42.00 sweater marked $\frac{1}{3}$ off. How much did she pay for the sweater?
36. Eddie is cooking a special surprise. The recipe calls for $\frac{3}{8}$ teaspoon salt. If Paddy is doubling the recipe, how much salt should he use?

Today there is more and more information available to more and more people. As the complexity of the world increases, people need to be able to handle much more information in a quick and accurate way.

One way of handling such a flood of information is by using the latest in technology, such as calculators and computers. The impact of more advanced calculators and computers shows up each day with the ever-increasing display of factual information known as **data.**

Many important decisions are based on gathered data:

- When should a community be warned to take precautions because of the approach of a dangerous storm?
- Which television programs have the best ratings?
- What type of car would someone buy?

Ask Yourself

- *What information is important to me to decide what to do tonight?*
- *Where can I find information that would help my family decide which new appliance to buy?*
- *How can I find out if there is going to be a severe storm?*

CHAPTER 4

DATA ANALYSIS

SECTION 4–1

COLLECTING DATA

WARM-UP ACTIVITY

You have seen many slogans and phrases that advertisers write to help sell new products. Imagine that you have invented something new, such as a new breakfast food, a new frozen yogurt flavor, or a new computer game. Next write three different sales slogans. Then ask ten people which slogan they think is best. Will most of them agree with you?

INTRODUCTION

Pat Mullin daydreams about the perfect new car. It would have to accelerate quickly so that Pat could impress all the kids at school. How can Pat find information about acceleration? If Pat wants to know which car is most popular with students in the school, would Pat need different information?

SKILLS DEVELOPMENT

There are various methods of gathering data. Information can be obtained from:

- reference books
- observations
- surveys
- results of experiments
- measurements
- newspapers or magazines

Choosing the best method to gather data depends on the type of question. Think about Pat's first question: Which car has the best acceleration? Pat could find the information in a magazine about cars.

Example

Markada 960 Acceleration	Seconds	**Nobless 1400** Acceleration	Seconds
Zero to 30 mph	1.8	Zero to 30 mph	3.2
Zero to 40 mph	2.8	Zero to 40 mph	4.7
Zero to 50 mph	3.9	Zero to 50 mph	6.4
Zero to 60 mph	5.1	Zero to 60 mph	8.8
Zero to 70 mph	6.8	Zero to 70 mph	11.6
Zero to 80 mph	8.6	Zero to 80 mph	15.0

Solution From the information contained in the magazine, Pat can decide which car would fit the definition of "best" Pat made. What conclusion should Pat make? Pat wants acceleration. The Markada beats the Nobless.

Would this same car also be the favorite of the entire student body? There are two ways the question could be answered. One way is to ask everyone in school. When information is obtained from each member of the **population,** it is called **taking a census.** The second way is to **sample** the population—that is, ask only some of the population. But the sample should be representative of the whole school. Each student should have an equal chance of being chosen.

EXERCISES

Which method would you use to collect data to find each of the following? Give a reason for each method you choose.

1. the favorite rock group

2. the most popular television program

3. the favorite color

4. the number 1 song of the week

A chart of the "Top 10" of the week is based on sales of records at a local record store and on telephone requests to a local radio station.

A When I See your Smile	B Start the Music
C This Is for Children	D Forever Your Guy
E Sing Me a Song	F Just Like Junior
G Opposites Attract	H Downtown Train
I Don't Mind Me	J Hangin' On

5. Based on the following information from the sale of the records at Tylero's Top Tunes and Tapes, which record is number one?

B 251	C 256	A 223	I 227	G 257
F 243	E 257	H 265	J 278	D 225

6. Based on the following information from the sale of the records at Corky's Cassettes and Disc Center, which record is number one?

D 144	E 169	C 192	G 167	A 159
I 137	F 179	J 129	B 153	H 197

7. Based on the data taken at radio station WQUT, which record is the most requested?

B 658	D 225	H 858	J 834	G 776
A 439	E 790	I 258	C 732	F 508

8. Based on the data taken at radio station WMPI, which record is the most requested?

A 243	C 227	G 251	E 267	I 244
F 227	B 247	H 253	J 288	D 231

9. Predict which song is the "Top" for all the data.

10. Construct a chart to organize the information.

11. Based on the data in question 11, which song is the "Top"? Which is runner up?

A sample of 500 persons was asked, "What is your favorite breakfast juice?"

A Orange **B** Grapefruit **C** Tomato

The following data were obtained.

Juice	Replies
A Orange	183
B Grapefruit	48
C Tomato	123
D No Opinion	146

13. How many people had an opinion?

14. Based on the survey, which juice is liked the best?

15. Of the persons who replied, did more than half vote for the winner?

16. Suggest a method to find the answer for the members of your class. How do you think results would compare to those above?

Data were gathered using the methods described below. The following conclusions were made. Are any conclusions biased? Give reasons for your answers.

17. *Survey* The people living near the railroad tracks were asked if they wanted the tracks moved.

Conclusion Most people want the tracks moved.

18. *Survey* A telephone survey of teenagers was made in July.

Conclusion Most teenagers don't have summer jobs.

19. *Survey* People in the ticket line for "Threat of the Martians" were asked their favorite type of movie.

Conclusion Most people like science fiction movies best.

20. *Survey* Sales of jeans by a manufacturer over a one-year period were examined.

Conclusion Most people prefer to wear styled jeans.

SECTION 4–2

ORGANIZING DATA

WARM-UP ACTIVITY

NUMBER PUZZLER

Is 28,859 a prime number?

Use a calculator. Divide 28,859 by the primes, 2, 3, 5, 7, 11, and so on. How far do you need to go?

SKILLS DEVELOPMENT

Sometimes data are presented in a way that makes it difficult to pull out the information you need quickly. Consider the list of soccer scores shown below. How effective is the list if you want to know the total number of games the Pirates won?

Scores for Games Played			
Yanks 3	Pirates 1	Pandas 6	Pirates 1
Lions 2	Yanks 1	Generals 1	Yanks 0
Scorpions 3	Pandas 3	Yanks 5	Pandas 2
Pirates 2	Scorpions 2	Yanks 3	Scorpions 2
Pandas 3	Pirates 1	Lions 2	Generals 1
Scorpions 4	Generals 1	Pandas 3	Yanks 2
Lions 1	Generals 0	Scorpions 4	Pirates 3
Yanks 3	Lions 2	Generals 3	Pirates 1
Scorpions 3	Pandas 2	Scorpions 5	Lions 0
Generals 1	Pandas 1	Yanks 4	Generals 3
Pirates 4	Generals 0	Scorpions 3	Yanks 2
Pandas 6	Scorpions 5	Pirates 4	Lions 3
Generals 3	Pandas 2	Pandas 6	Generals 2
Pandas 4	Lions 4	Scorpions 6	Generals 5
Pirates 6	Yanks 2	Generals 5	Lions 4

The list can be used to make a table of game totals. One special kind of table is called a *matrix.* A matrix organizes numbers into rows and columns. A matrix that arranges the information is from the list of soccer scores.

	Games played	Games won	Games lost	Games tied	Goals for	Goals against
PIRATES	9	3	5	1	23	27
YANKS	10	5	5	0	25	25
GENERALS	12	4	7	1	25	34
LIONS	8	3	4	1	18	23
PANDAS	11	5	3	3	38	30
SCORPIONS	10	6	2	2	37	27

Now the question about the number of games the Pirates won is easily found: 3.

Example

a. Which team won the most games?

b. Which team lost the most games?

c. What was the average number of goals scored against each team?

Solutions a. Scorpions: 6 wins.

b. Generals: 7 losses.

c. Average number of goals scored against each team: (Use your calculator.)

$$\frac{27 + 25 + 34 + 23 + 30 + 27}{6} = \frac{166}{6} = 27.7$$

A cable television station asks its viewers to call the station to vote for one of three movies they prefer to watch. The votes are recorded in a *tally chart,* as shown below. If the tally chart also shows the frequencies, it is called a *frequency chart.*

Choice	Tally	Frequency
Movie 1	卌 卌 卌 //	17
Movie 2	卌 卌 卌 卌 卌 ///	28
Movie 3	卌 卌 卌 卌 /	21

EXERCISES

Copy and complete this table.

Type of Book Borrowed		
Type	**Tally**	**Frequency**
1. Science Fiction	卌 卌 ////	?
2. Mystery	?	22
3. Romance	?	7
4. Biography	////	?
5. Other	?	13

6. How many people liked mystery? romance? other?

7. How many people were surveyed in all?

8. Which type of book was the favorite?

Extend the frequency chart. Of the *Science Fiction* readers, 8 are ninth graders, 4 are tenth graders, and 2 are eleventh graders. For *Mystery,* 2 are in ninth grade, 7 are in tenth, and the rest in eleventh. *Romance* has 3 in ninth, 2 in tenth, and the remainder in eleventh. *Biography* has all 4 in eleventh grade. Finally, *Other* has 3 ninth graders, 6 tenth graders, and 4 eleventh graders. Use the additional information to refer to Questions 9–13.

9. Construct a matrix using the book types and grades as the titles for the rows and the columns. Does it matter which are rows and which are columns?

10. If the dimensions of a matrix are expressed as the number of rows by the number of columns, what are the dimensions of the matrix you have constructed in Exercise 9?

11. How many eleventh graders prefer mysteries?

12. How many people were surveyed? How many were in the ninth grade? tenth grade? eleventh grade?

13. Which type of book was the most popular with ninth graders?

The distances (kilometers) that various people jog each week are shown below.

11 18 10 25 6 15 21 6 16 19 8 20 20 25 15
21 9 17 13 20 11 20 11 22 12 14 17 19 17 22
14 23 4 25 8 16 32 18 15 33 23 14 9 23 30
30 16 18 5 12 27 5 30 7 18 19 24 28 8 22
8 17 10 26 24 11 35 17 22 10 11 15

14. Construct a frequency table for the data. Group data by 5: 0–5, 6–10, 11–15, 16–20, 21–25, 26–30, and 31–35. These groups are called **class intervals.**

15. In which two classes are most of the joggers?

16. Create a problem about the data. Write a solution to the problem.

The time taken (seconds) to run 100 meters was recorded for some students.

16.2 17.6 15.0 15.6 15.3 13.1 17.1 16.2 15.6 14.5 18.9
14.1 18.0 13.1 17.3 18.0 16.8 15.1 13.3 16.8 18.1 14.5
15.2 15.5 16.6 13.7 15.5 14.7 17.9 16.0 18.1 15.3 13.9
15.7 17.4 14.1 16.4 15.3 16.6 13.0 17.3 14.7 15.5
16.6 13.7 16.0 15.4 14.7 14.3 15.5 16.8 14.5 16.4
14.7 16.4 15.4 16.4 15.2 15.7 15.9 13.6 15.4 13.9

17. How many students participated in the test?

18. Construct a frequency table for the data. Use these class intervals: 13.0–13.9, 14.0–14.9, and so on to 18.0–18.9.

19. Estimate the average length of time taken to run 100 m. In which class interval does your estimate fall?

SECTION 4–3

STEM AND LEAF PLOTS

WARM-UP ACTIVITY

A list of random numbers has no pattern and each digit is as likely to appear as any other. One way to make such a list is to use a phone book.

Choose a page. Start anywhere and take the second to the last digit in each phone number.

Make a list of 50 digits. Does it look random? Which digit appears most? Which appears least?

Make another list from a different part of the phone book. How do your two lists compare?

INTRODUCTION

In our health-conscious world, it is important for us to know information about the foods we eat. In the matrix shown below, the calories, fat content, and carbohydrates are listed for some common foods. What kinds of questions can you answer from the information?

Item	Calories	Fat (grams)	Carbohydrates (grams)
Cottage Cheese	235	10	6
Whole Milk	88	8	291
Hard-Boiled Egg	80	6	1
Chili	340	16	31
Chicken Chow Mein	255	10	10
Banana	100	0	26
Apple	80	1	20
Bagel	165	2	158
Corn Flakes	95	0	21
Broccoli	45	1	8
Baked Potato	145	0	33

SKILLS DEVELOPMENT

One way to compare data is to set up a **stem and leaf plot.** In these numbers the ONES digit is called the **leaf** and the TENS digit is called the **stem.** Look closely at the stem and leaf plots shown.

Suppose you know that a group of foods have the following grams of fat: 6, 10, 21, 31, 1, 20, 26, 33, and 8. You can set up this display.

Tens	Ones
0	6 1 8
1	0
2	1 0 6
3	1 3

Note that the last row represents 31 and 33.

A telephone company is studying its long-distance charges. Part of the research contains data on the length of calls between two cities. The results of a sample of 30 calls are shown in minutes.

22	17	5	11	31	13	6	10	23	14	9	5	18	37	15
12	12	30	19	14	10	8	26	40	27	21	9	6	51	48

Here is the stem and leaf plot for all the data.

Tens	Ones											
0	5	6	9	5	8	9	6					
1	7	1	3	0	4	8	5	2	2	9	4	0
2	2	3	6	7	1							
3	1	7	0									
4	0	8										
5	1											

Using graph or grid paper helps keep the display neat, but is not essential.

Example

Use the data for the telephone research to answer each question.

a. How many telephone calls lasted exactly 10 minutes?

b. How many calls lasted 30 minutes or longer?

c. How many calls lasted less than 10 minutes?

d. Were most of the calls longer than 20 minutes?

Solution

a. 2 telephone calls lasted 10 minutes.

b. 6 calls lasted 30 minutes or longer.

c. 7 calls lasted less than 10 minutes.

d. No. Most calls were less than 20 minutes.

EXERCISES

The stem and leaf plot below shows the pulse rates, in beats per minute, of the students (at rest) in one class.

Tens	Ones
5	3 5 9 1
6	2 7 8 1 9 6 8 7 7 8 1 9
7	2 6 3 2 7 0 5 5 9
8	4 2 1 2 1 1 6
9	6 1 3

1. How many students had a pulse rate of 67 beats per minute?

2. How many students had a pulse rate of 81 beats per minute?

3. How many students had a pulse rate greater than or equal to 90 beats per minute?

4. How many students had a pulse rate lower than 70 beats per minute?

5. How many students were in the class?

A farm research station is developing a new variety of corn. The stem and leaf plot shows the height, in centimeters, of 50 test plants one month after being planted.

Tens	Ones
1	9
2	3 7 5 2
3	8 6 9 0 7 2
4	7 2 6 8 9 1 5
5	8 4 7 6 5 2 9 6 0 5
6	3 8 1 0 6 7 5 2 2 8
7	4 7 2 9 5 1 4 8
8	9 5 3 2

6. What was the height of the tallest plant?

7. What height was the shortest plant?

8. How many of the plants had a height less than 30 cm?

9. How many plants had a height of 60 cm or greater?

The number of super deluxe hamburgers sold each day at Divine Snacks was recorded over a period of 3 months. Use the information found in the Reference Section at the back of this text. Then construct a stem and leaf plot.

10. On how many days were fewer than 50 super deluxe hamburgers sold?

11. How many sales per day seem about average?

12. Why is it difficult to answer the above questions using the data in the given form?

13. List advantages of recording data as a stem and leaf plot to make conclusions.

14. Estimate the average number of sales each day based on your plot.

15. Use the data. Calculate the average number of sales each day. How close was your estimate?

16. Create and solve a problem based on your stem and leaf plot.

SECTION 4–4

DISPLAYING DATA

WARM-UP ACTIVITY

ALPHABET MATH:

A=1¢, B=2¢, C=3¢, D=4¢, E=5¢, . . . , Y=25¢, Z=26¢

What if each letter had a monetary value?

What is the value of your first name? (Use your calculator.) Which is worth more: your first name or your last name? How many words can you find that are worth exactly $1.00?

INTRODUCTION

Tally charts, frequency tables, matrices, stem and leaf plots, and bar graphs are used to display information. Graphs are also used to show data. We will examine several types of graphs.

SKILLS DEVELOPMENT

Consider the following matrix. The data represents a student council budget of $30,500 for the school year.

Activity	Amount Spent
Transportation	$4875
Musical	3275
School Clubs	8720
Yearbook	7650
Senior Prom	4225
Printing	1755

A **pictograph** uses symbols to represent data. It must have a title, a **key** or **legend,** and a labeled axis. A **vertical bar graph** uses bars to represent data. It must have a title, each axis labeled, and each axis identified. A **horizontal bar graph** uses bars drawn horizontally.

A **broken-line** graph uses points and the lines connecting them. Points on a broken-line graph can be thought of as ordered pairs. For example, the \$8720 cost in school clubs can be written as (School Clubs, \$8720).

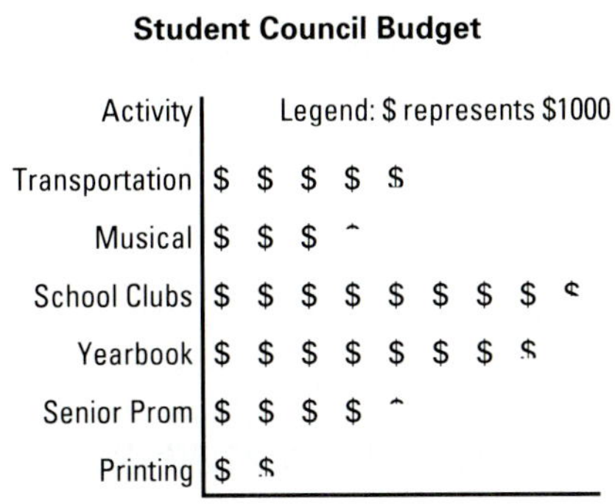

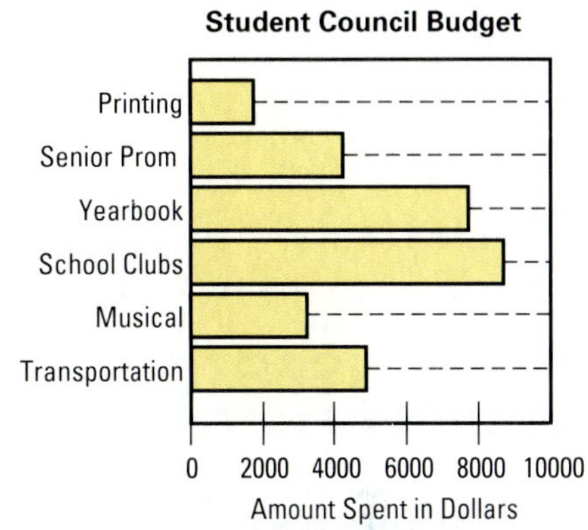

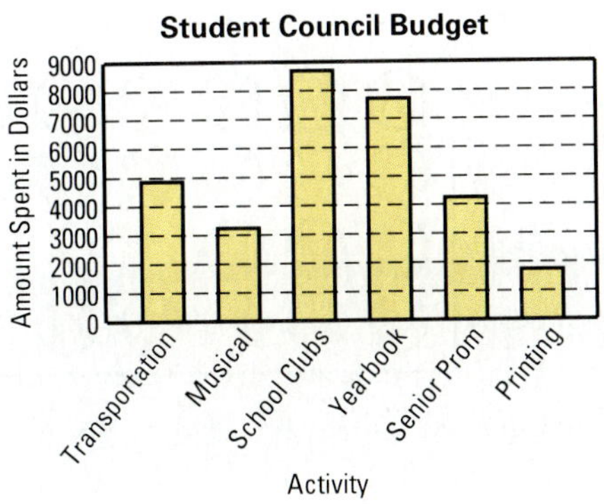

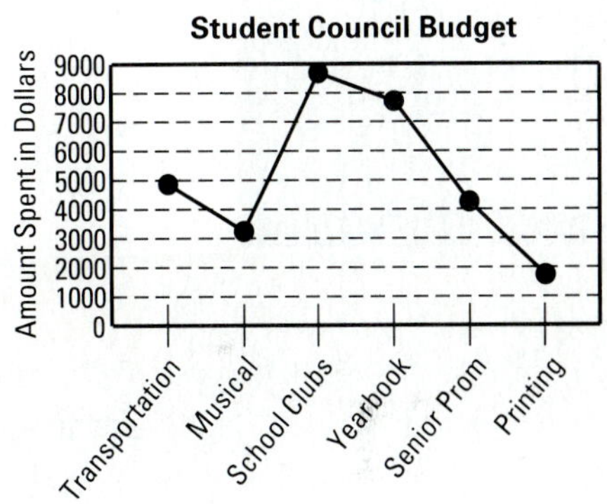

What kinds of observations can be made about the same information displayed in the different types of graphs?

For some kinds of information, a broken-line graph can be very powerful. Consider the graph giving the depth of the water level of a given reservoir. It shows the data on this chart. The data is collected on the first day of every month.

Month	Depth (meters)
Nov.	4
Dec.	3
Jan.	5
Feb.	6
Mar.	5.5
Apr.	7.5
May	8
June	9.5
July	6.5

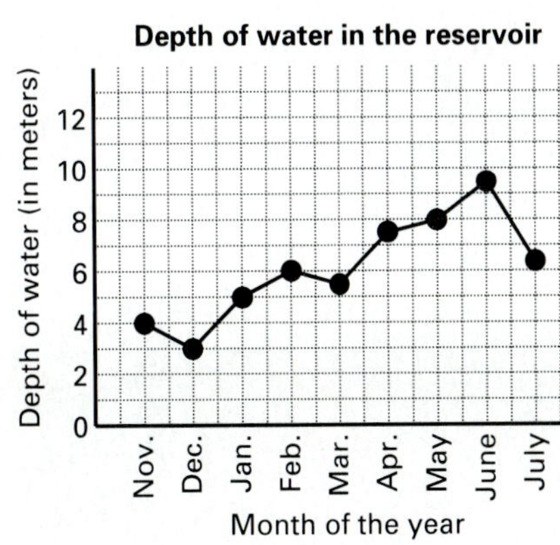

The graph on the last page clearly shows the changes in the water level over the period from November 1 to July 1. We can also estimate the water level on March 15. Obtaining information between the marked points is called **interpolating.**

If the graph were extended, then an approximation or estimate could be made to predict where the level might be in August. When you estimate a value that occurs beyond the last given value, then you are **extrapolating.**

EXERCISES

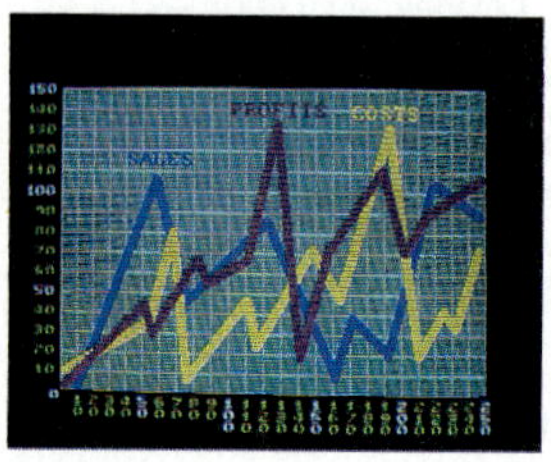

Data on the number of strikeouts by the following pitchers are displayed. Rena has 12 strikeouts.

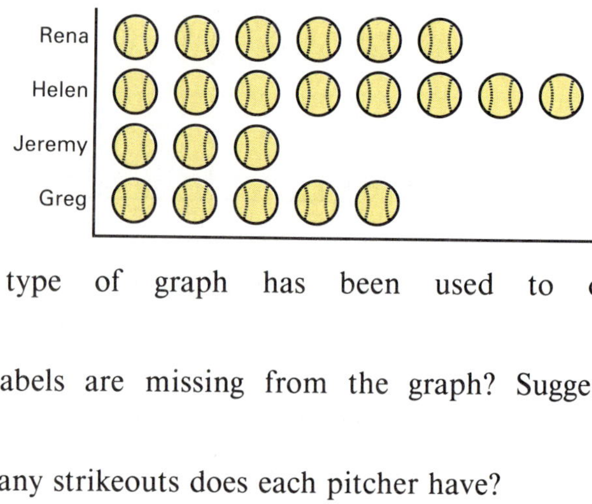

1. What type of graph has been used to display the data?

2. What labels are missing from the graph? Suggest labels to use.

3. How many strikeouts does each pitcher have?

4. What are some advantages and some disadvantages of using a pictograph to show these data?

Data on the sales of different-size pizzas are displayed.

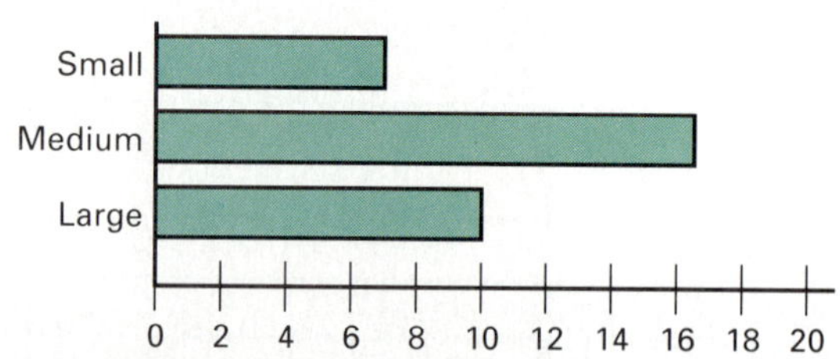

5. What type of graph has been used to display the data?

6. What features are missing from the graph?

7. How many large pizzas were sold?

8. Boxes need to be ordered for takeout orders. If 400 small boxes are ordered, about how many medium-size boxes should be ordered?

9. What are the advantages and disadvantages of using the bar graph?

A school is holding a month-long campaign to raise money for a swimming pool. The broken-line graph shows the total amount collected each day, for the first 10 days.

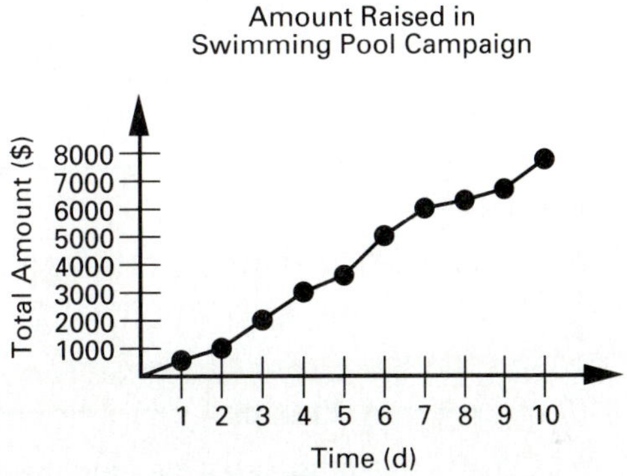

10. The goal for the end of the first week of the campaign was $5000. Did the school reach its goal? How much over or under the goal was it?

11. On which day was the greatest amount of money raised? About how much was collected during that day?

12. The goal for the end of the month is $25,000. Based on the amount raised so far, do you think the school will reach its goal in a month?

Choose 5 current popular songs. Conduct a survey of students to find out which of the songs most students prefer.

13. Organize the data in a frequency chart.

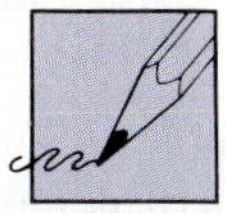

14. Construct a graph to display the data.

15. Make up some questions about your graph. Answer the questions.

SECTION 4–5

INTERPRETING DATA AND GRAPHS

WARM-UP ACTIVITY

PROBLEM-SOLVING: PUZZLER

A birthday cake for Mel was baked in the following shape. Eleven people came to his party. How can you cut the cake into congruent pieces so that each person gets the same amount of cake?

SKILLS DEVELOPMENT

One other kind of graph is a **circle graph.** It is useful when you want to show how a certain fixed quantity is divided into parts.

Example

This circle graph shows how park land is used in the city of Sunnyside.

1. How is most land used?
2. About what fraction of the land is used for soccer fields?
3. Is more space used for parking lots or for playgrounds?

Solution

1. It is wooded and undeveloped.
2. About one-fourth.
3. Parking takes about three times as much space as playgrounds.

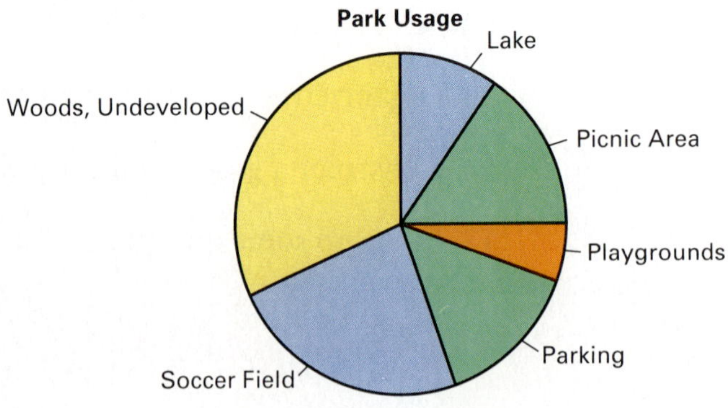

EXERCISES

In the Reference Section at the end of this text is a bar graph comparing lengths of rivers in the world. Use that graph to complete these exercises.

The bar graph shows that the Yangtze River is about twice as long as the Zambezi River. Compare the lengths of the following pairs of rivers.

1. Rio Grande and the St. Lawrence

2. Rio Grande and the Nile

3. Saskatchewan and the Nile

4. Saskatchewan and the Mississippi

5. Make at least two other comparisons of rivers.

Which river in each of these pairs is longer? About how much longer?

6. St. Lawrence and the Zambezi

7. St. Lawrence and the Saskatchewan

8. Mississippi and the Rio Grande

9. Yangtze and the Mississippi

Refer to the graph on river lengths in the Reference Section.

10. Find the average length of the rivers given in the bar graph.

11. Which rivers are shorter than the average?

12. Which rivers are longer than the average?

A survey was conducted to find out people's preference for different types of fruit. The following represents the results of the survey.

Fruit	Number who chose it
Apple	40
Banana	20
Orange	30
Peach	10

13. Construct a bar graph to show the information.

14. Conduct your own survey on a favorite food.

15. Make a bar graph to show the results.

Study the following pictograph.

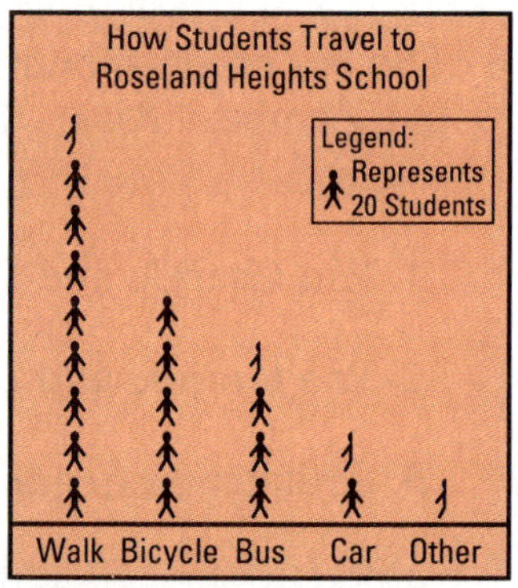

16. What do each of the symbols mean?

17. About how many students were asked?

18. Which method of travel is used by the most students? By the fewest students?

19. About how many students walk to school? How many bicycle?

20. About how many students go to Roseland Heights School?

21. What assumption did you make to answer Exercise 20?

22. Make up two other questions about the pictograph, and write answers to the questions.

The circle graph shows the sizes of pizzas ordered in one hour.

Pizzas Sold in 1 Hour

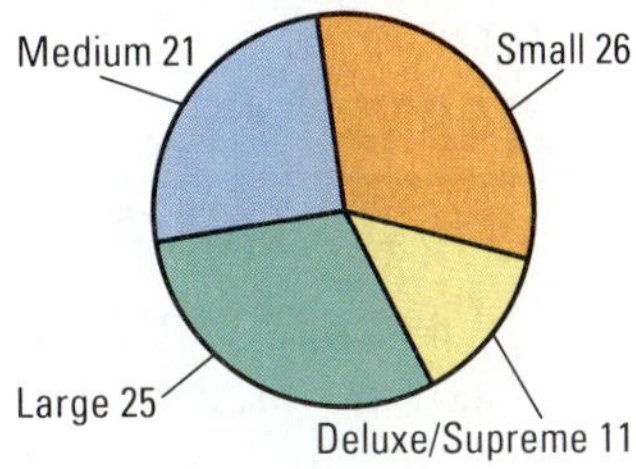

23. The average price of a medium pizza is $4.50. How much money was spent on medium pizzas?

24. The total number of pizzas sold in one working day is 830. How many large pizzas would you expect to sell?

25. What assumption did you make to obtain your answer in question 27?

The earth's population has been growing rapidly for several centuries.

Date	Population (estimated)
1650	550,000,000
1700	600,000,000
1750	725,000,000
1800	900,000,000
1850	1,200,000,000
1900	1,500,000,000
1950	4,340,000,000
1990	5,200,000,000

26. Draw a broken-line graph to show the data given for 1650 to 1950.

27. Extend your line graph to estimate the present population.

28. About how many times did the population increase between 1650 and 1850? between 1900 and 1990?

29. Estimate the world population by the year 2500.

SECTION 4–6

WHICH GRAPH IS MORE APPROPRIATE?

WARM-UP ACTIVITY

PUZZLE: DATA WORD SCRAMBLE!

Throughout this chapter, certain new words were introduced as part of the vocabulary of mathematics. Unscramble the letters to review these words.

1. i e c l r c p r h g a

2. t i r c h o p g p a

3. e e d n g l

4. r b a h r g a p

5. z r h i o a n o t l

6. u c q n e r y e f

7. c e t i l a v r

INTRODUCTION

Different types of graphs show information or data in different ways.

- pictograph
- stem and leaf plot
- bar graph
- broken-line graph
- circle graph
- other graphs

One kind of graph is often more suitable than another. To construct suitable graphs you will need to make many decisions.

SKILLS DEVELOPMENT

To select the best kind of graph, ask yourself:

- Which type of graph displays the information best?
- Is the information visual enough for a pictograph?
- What should the title be?
- How should the axes be labelled?
- What scale is best for the horizontal axes? for the vertical axes?
- What calculations are needed to construct the graph?

EXERCISES

Decide which kind of graph is best.

1. You want to tell about how many bones there are in the body.
2. You want to show what fraction of the bones are in the head and neck.
3. You want to calculate the fraction of the bones in the torso.
4. You want to calculate how many more bones are in the arms and hands than in the head and neck.
5. You want to tell quickly which parts of the body have about the same number of bones.

In a survey on doing sit-ups, 50 students participated. Each student completed as many sit-ups as possible. The data obtained follows.

Number of Sit-ups
51 43 76 26 63 49 12 32 59 42
48 45 34 27 12 91 56 57 13 24
66 37 27 45 53 47 65 54 15 18
56 55 83 35 75 18 39 72 10 23
28 52 36 67 46 58 38 48 68 51

6. What type of diagram or graph would you use to display the information? Why?
7. Construct a graph of your choice to show the data.

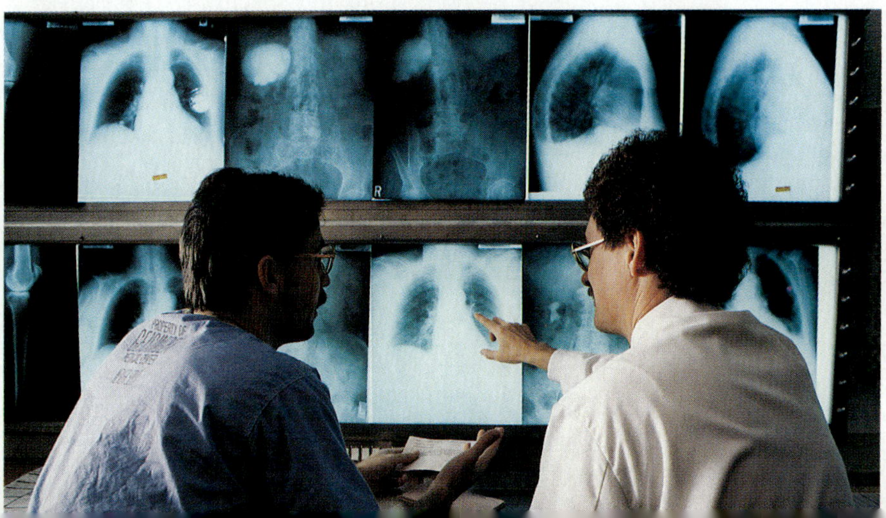

In an experiment, 100 g of a popular brand of corn flakes is analyzed. The following amounts of vitamins and minerals are shown in the chart.

8. Choose a suitable graph to show the information. Give reasons for your choice.

Vitamins and Minerals	
Niacinamide	21.2 mg
Riboflavin	3.5 mg
Thiamine	2.1 mg
Iron	14.1 mg

9. Use your graph to create three questions about the information.

10. Use the graph to answer the questions in Exercise 9.

The heights (cm) of dogs brought to a veterinarian were recorded.

44 51 94 21 61 84 86 81 74
56 72 55 25 46 63 32 74 78
65 39 66 11 33 23 91 57 69
28 68 76 49 54 52 66 47 58
48 38 77 43 86 15

11. Choose a suitable graph to show the information. Give reasons for your choice.

12. Use your graph to create three questions about the information.

13. Use the graph to answer the questions in Exercise 12.

The United States Constitution says that the president must be at least 35 years old.

14. Make a suitable Graph of the age at inauguration of the presidents since 1933.

Franklin D. Roosevelt	51
Harry S. Truman	60
Dwight D. Eisenhower	62
John F. Kennedy	43
Lyndon B. Johnson	55
Richard M. Nixon	56
Gerald Ford	61
Jimmy Carter	52
Ronald Reagan	69
George Bush	64

100 people were asked what kind of automobile they would probably buy in the next year. Their replies are shown in the table.

Sports car	5
Family sedan	20
Station wagon	12
Pick-up truck	17
Van	10
Other	4
None	33

15. Make a suitable graph of this data.

16. What are the different heights of the persons in your class?

17. Which of these colors do the members of your class like the best? blue, orange, yellow, green, red.

18. What are the top four TV shows watched by the class?

19. On a sunny day, the length of the shadow of a meterstick varies. How does it vary over one day?

20. How much does your temperature vary over a period of one month in your city? your state?

SECTION 4–7

DISTORTING DATA

WARM-UP ACTIVITY

PUZZLE: AMAZING CIRCLES!

For each maze, pass through the correct openings to arrive at the answer at the center by multiplying the numbers. Show all your work. Draw your route into the center.

SKILLS DEVELOPMENT

You may be aware of the expression "lying with statistics." Sometimes it is to a writer's advantage to adjust the picture or type of graph to allow a distorted view of the data to be presented. Consider the graph below.

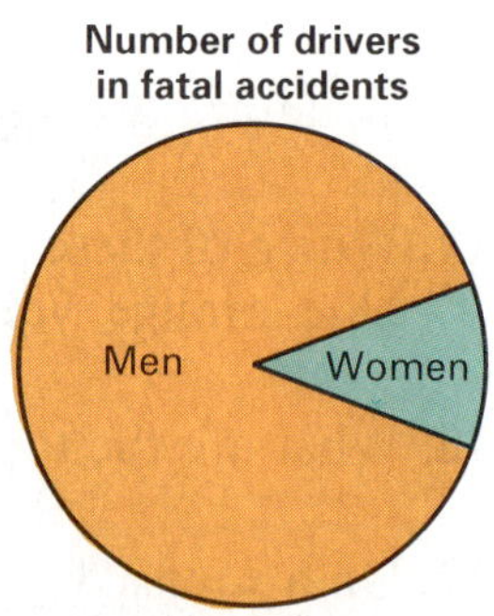

You might quickly conclude that men are worse drivers than women. To interpret the graph accurately and make valid conclusions, you often need more information. Consider: are there more men drivers than women drivers? Do men drivers travel more?

In this section we will examine situations that misuse data.

EXERCISES

A company asked two artists to draw a graph of the company's profits for a 6-month period.

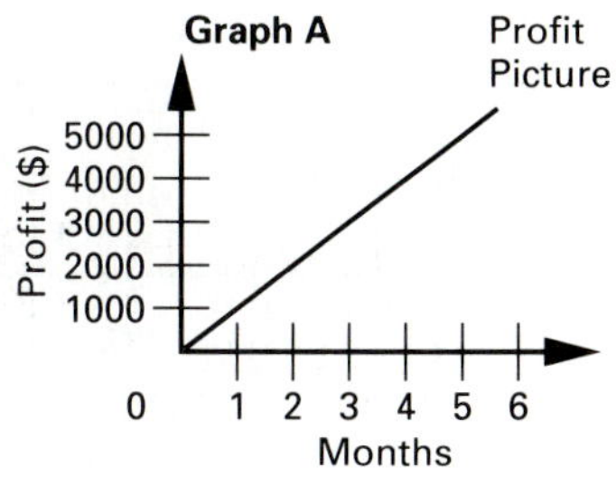

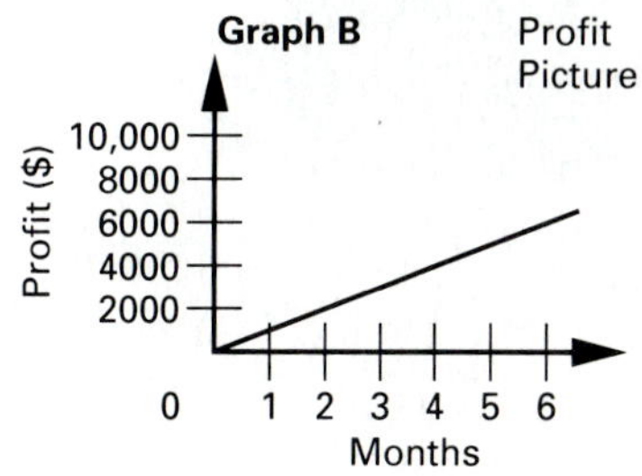

1. How are the graphs alike? How are they different?

2. If you were president of the company, which graph would you choose to publish? Explain why.

3. Predict the company's profit for the seventh month. What assumption do you make to obtain your answer?

The value of exports made by a company in two consecutive years is shown by the diagram.

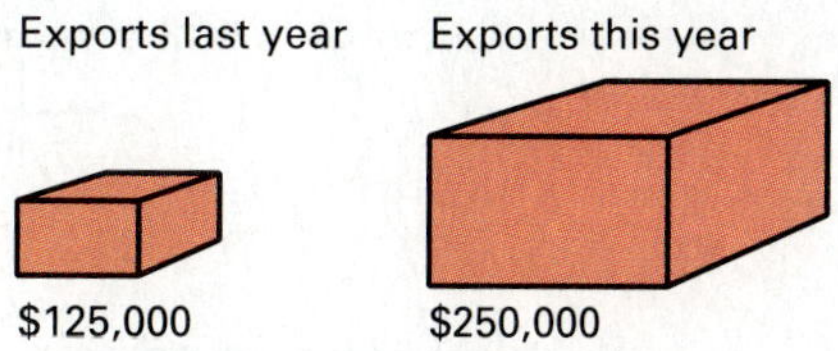

4. Which statement appears to be the best description of the information in the diagram?

   A Exports increased twice as much from last year.

   B Exports increased three times as much from last year.

   C Exports increased four times as much from last year.

5. Compare the volumes of the boxes. Each dimension of the small box has doubled. By how many times has the volume increased?

6. Use the dollar figures. By how much has the dollar value of the exports increased?

7. What false impression does the diagram give?

The expenses of a company during a 5-year period are shown by the bar graph.

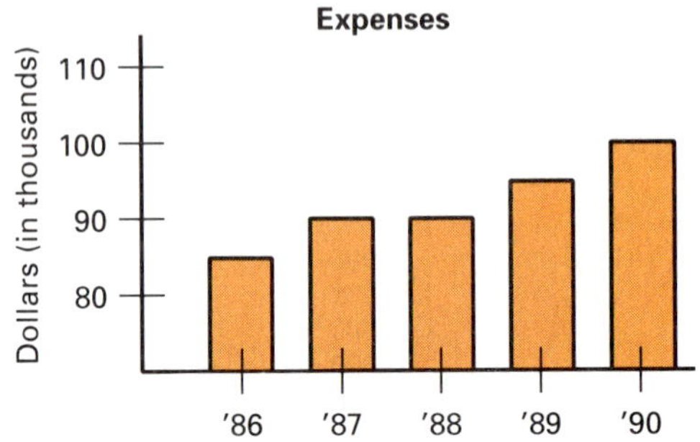

8. What false impression could the diagram give?

The increase in production by the apple juice company is shown as follows.

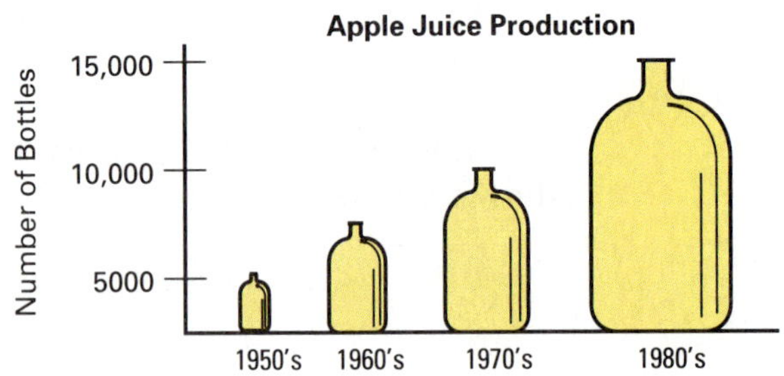

9. What false impression could the diagram give?

The sales of three apple juice manufacturers are shown in an advertisement.

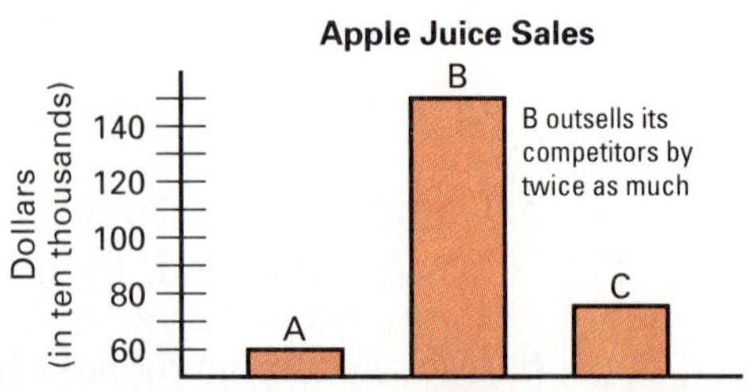

10. What false impression could the diagram give?

Use the following data.

The range (in miles) of various jetliners is shown.

Plane	Range (km)
B747	8000
L10–11	5900
DC8	8000
B727	4160
DC9	2500

11. Create a negative impression by drawing a graph to make the data appear distorted.

12. Then draw another graph to create a favorable impression.

13. Create a problem based on data from the graph, then solve the problem.

Use newspapers or magazines.

14. Find examples of graphs that have been used to create a false impression.

15. Find examples of data in the newspaper. Construct graphs to create a favorable impression, and also a false impression.

SECTION 4–8

PROBLEM SOLVING WITH DATA

WARM-UP ACTIVITY

PUZZLE: SOCCER SCORES

Three soccer teams played each other once. Afterward, a detailed table of the results was prepared. Unfortunately, some of the figures were missing.

Try to fill in the missing figures and determine the score for each of the three games.

Team	Games Played	Won	Lost	Tied	Total Goals For	Total Goals Against
A	2	0	1		3	4
B		0	1	1	3	4
C	2	2			2	0

INTRODUCTION

In previous sections you learned a number of skills which you can use to solve problems about data.

- Skills for collecting data
- Skills for organizing data
- Skills for displaying data
- Skills for interpreting data

In planning to solve problems with data, you will need to ask questions from a slightly different point of view, since the skills you are applying are different.

In your earlier work in this chapter you have already practiced problem solving with each skill. Now you will need to decide which skill to use to solve the problem.

SKILLS DEVELOPMENT

It is time to put all of the knowledge we have gained from this chapter to work for us in problem-solving situations. Often, to make predictions, look for a trend in the data.

Review the steps we can use to solve problems.

1. Ask yourself:

- What are you asked to find?
- What information is given?

In some problems you will need to obtain data. Use charts to organize your work. Record your data accurately.

2. Decide on a method.

- After you have collected your data, decide whether the problem can be better solved by drawing a graph, using a calculator, finding other data, and so on.

3. Do the work.

- Once you have planned which skills to use, record any assumptions you may make in finding an answer. Estimate the answer. Then check by calculating.

4. Check your work.

- Check to see whether your answer is reasonable.
- Does your answer make sense?

5. Make a final statement.

- You were asked to solve a problem about data. Thus, you need to make a final statement that shows you answered the question properly.

You can use your problem-solving skills with data to solve many types of problems. One of the most common types involves predictions.

- How does a TV network determine which of its TV programs are successful?
- How do hotels in vacation areas predict the temperature?
- How does a math teacher predict that a test is fair?
- How do politicians predict whether they can win an election?

Often, to make predictions, look for a trend.

EXERCISES

Sales figures from the Crystal Ski Shop are shown three ways.

Jan. 8, Ski Shop Sales	
Skis	$1600
Boots	$900
Poles	$400
Suits	$500
Accessories	$200

1. Do the sales figures appear to be exact? To what accuracy do you think the data are recorded?

2. What is the legend for the pictograph?

3. Using the same legend symbol, how would sales of $478 be shown?

4. What were the total sales from the sales of boots?

5. Which graph displays this information better?

6. From the circle graph, how can you tell that the value of the skis sold was less than $\frac{1}{2}$ of the total sales?

7. Draw a bar graph to display the data. What are the advantages and disadvantages of using the bar graph?

8. Which graph best displays the amount of sales for each item?

Sonny kept a record of the number and type of bicycles brought to school each day.

Day	Number of 10-speeds	Total number of bicycles
Mon	49	124
Tues	43	108
Wed	47	115
Thurs	52	132

9. Calculate the fraction of 10-speed bicycles brought to school each day. Round the numbers to obtain useful fractions.

10. Based on your results in Exercise 9, estimate the number of students out of 180 who have 10-speed bicycles.

11. If 125 bicycles were brought to school on Friday, predict how many would be 10-speeds.

At a Sports Shop the following running-shoe sales were recorded for six consecutive months.

Month	Total number of pairs of running shoes sold	Number of Orista brand sold
April	350	63
May	295	52
June	300	59
July	297	48
Aug	375	64
Sept	401	78

12. For each month shown do the Orista sales rise and fall in the same pattern as total shoes sold? What does this mean?

13. Out of a total sales of 450 pairs of running shoes in October, how many would you expect to be Oristas?

14. What assumption did you make in finding your answer in Exercise 13?

CHAPTER 4 SUMMARY

Think about all of the ideas you were introduced to in this chapter on data analysis. You certainly became aware of how much information you see or hear daily that needs to be organized and interpreted. You learned how to collect and organize data by taking surveys and recording the results on tally charts. You had opportunities to participate in some activities that gave you firsthand experience with gathering information.

You had many important decisions to make. After you decided how data was to be collected, there were several ways to report it: tables, stem and leaf plots, tallies, and matrices. After collecting and organizing, you had to decide which type of graph was best: pictographs, and bar, circle, and broken-line graphs. Sometimes it required going back to the questions you were trying to answer. For example, circle graphs were chosen when showing parts of the whole were important.

You noticed how easily data could be pictured to mislead the reader. You will probably take a closer look or listen more attentively next time you are exposed to a new graph or summary of data.

Finally, you worked to apply your knowledge to problem situations. It was interesting to be able to make predictions.

VOCABULARY

axis
broken-line graph
census
circle graph
class interval
column
data
extrapolating
frequency chart
horizontal bar graph
interpolating
key
leaves
legend
matrix
pictograph
population
prediction
row
sample
stems
tally chart
vertical bar graph

CHAPTER REVIEW

1. A survey on the number of viewers watching a TV miniseries was run. Describe the different methods that could be used and the advantages and disadvantages of each method.

2. The Laketown Bank has four branches in Holly Hills: Uptown (U), Downtown (D), Eastside (E), and Westside (W). Each location has three types of accounts—checking (c), savings (s), and market (m). The number of accounts at each location are: U has c = 36,790, s = 25,650, m = 380; D has c = 23,110, s = 17,410, m = 165; E has c = 11,250, s = 8792, m = 97; W has c = 32,113, s = 12,943, m = 258.
Organize the data in a matrix.

The following data show the students buying different sizes of sweaters.

Small (S) Medium (M) Large (L)
S M S L S M M S S L S M L S M
S S M S S L S M M M S S M S S

3. Construct a frequency table.

4. What fraction of the students bought medium sweaters?

The following graph shows the temperature over a 6-hour time period.

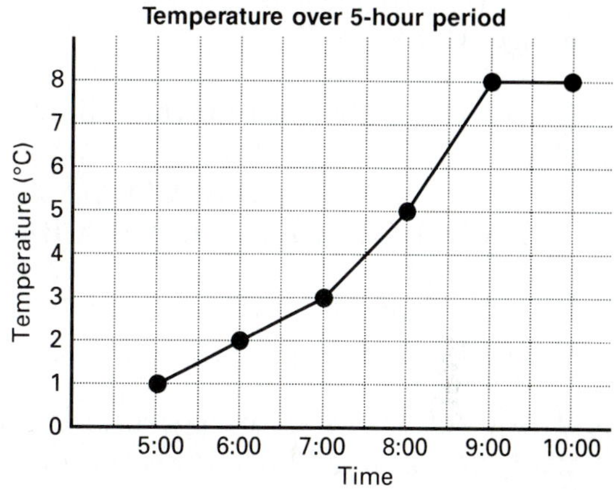

5. In what time interval did the temperature rise the most?

6. What was the approximate temperature at 7:30?

7. What would you predict would be the temperature at 11:00?

Average maximum temperatures for each month in Detroit, Michigan, are:

Jan −1°C	Apr 9°C	Jul 21°C	Oct 13°C
Feb 2°C	May 15°C	Aug 20°C	Nov 8°C
Mar 5°C	Jun 18°C	Sep 17°C	Dec 4° C

8. Display this data in a broken-line graph.

Over a 72-game schedule, Jill's team had the following record.

Wins 36 Losses 20 Ties 16

9. Display this data in a graph.

10. Explain why you chose the type of graph you used.

CHAPTER 4 TEST

In a survey, students were asked to name their favorite fruit.

Apples (A) Bananas (B)
Oranges (O) Pears (P)

1. Construct a tally chart for the following results.

   A B O A A O P A O A A O A A A
   O A B B A P A P A P B A B A B

2. Which fruit is preferred by the students?

3. If Michael wants to bring two kinds of fruit for a party, which kinds should he bring?

The following data are the number of minutes a sample of people use for daily exercise.

12	25	30	17	21	38	47	19	10	26
10	0	20	4	6	0	10	45	25	7
22	10	37	30	28	23	0	16	18	15
0	30	33	29	33	26	10	38	27	0
5	25	7	5	25	19	18	12	35	28

4. Construct a stem and leaf display for the data.

5. How many people exercise for 35 min or more each day?

A broiled hamburger is analyzed and the data are shown in the chart.

Substance	Mass
Water	44 g
Protein	22 g
Fat	22 g

6. Construct a circle graph for the data.

7. Based on the data, how much protein would be found in 1 kg of broiled hamburger?

The nutritive content of peanuts and pistachio nuts is shown in the double bar graph.

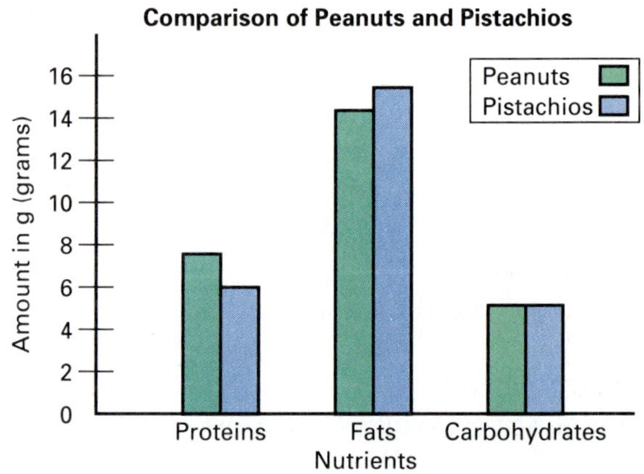

8. Which nutritive content is the same?

9. Which nut has more protein? About how much more?

In a tournament Kalana bowled 5 games with scores: 194, 215, 187, 152, and 220.

10. Construct a graph of your choice to display the data.

11. Give reasons for your choice.

12. Construct a graph that distorts the information.

SKILLS PREVIEW

Write in words.

1. $6 + 7$
2. $15 - 9$
3. 5×7
4. $8 + 8$
5. $16 \div 4$
6. $12 - 5$
7. 9×3
8. $36 \div 6$
9. $2 + 9$
10. $17 - 9$
11. 6×5
12. $45 \div 9$

Write a mathematical expression.

13. the difference between 9 and 3
14. two times seven
15. the product of 12 and nine
16. ten divided by five
17. three plus four
18. the quotient of six divided by 3

Simplify.

19. $6 + -4 + 5$
20. $15 - (3)(-6) + 4$
21. $\frac{8 + (4)(-6)}{3 + 9 - 4}$
22. $-1 + (4)^2 + -3$
23. $6 - (-5) - 7$
24. $\frac{-12 + (15 \div 3)}{10 - 3}$

Write each as a mathematical expression, then calculate.

25. Subtract 9 from the product of 6 and 3.
26. Add -18 to the quotient of 16 divided by 4.
27. Multiply the sum of 3, 2, and 6 by 12.
28. Find the difference of 17 and 8 multiplied by 4.
29. Divide the difference of -45 and 5 by 8.
30. Add 14 to the product of 8 and 8.

NASA
United States
NASA

Skills with algebra are used to solve many scientific problems. Advanced algebra is used to describe the paths travelled by space probes.

You can use algebra to solve problems that you might otherwise not be able to solve.

Arithmetic		Algebra
$3 + 3$		$x + x$
2^2	*How are each of these alike?*	x^2
$2(3) - 6$	*How are they different?*	$2x - 6$
$2(6) + 3(8)$		$2x + 3y$

In this chapter, you will develop skills with expressions in algebra. You will discover that the skills you learn in algebra are closely related to those you learned in arithmetic.

Ask Yourself

- *Where might you find a real life example of a variable expression?*
- *Why do you need a knowledge of algebra to solve problems when you already know arithmetic?*
- *Do you know the meaning of the word* ***variable?***

CHAPTER 5

WORKING WITH VARIABLES

SECTION 5–1

WHAT, LETTERS IN MATH?

WARM-UP ACTIVITY

Question: What did the mice say about the elephant?

To answer the question, find this sum on your calculator:

$$3097119 + 3088015$$

Turn your calculator upside down and read the words.

Digits on your calculator can form letters. Punch in each digit, then turn the calculator upside down to see what letter is formed.

INTRODUCTION

Toby is buying three compact disks. How much will he pay for them?

This question cannot be answered because we do not know how much each CD costs. When we have missing information, we use something to represent it. We use what is called a variable.

A **variable** is a letter that represents a number or numbers. It is customary to use lower case letters like x, y, or z as variables.

In prevous chapters you learned to change

7 times the sum of 1 and 5 to $7(1 + 5)$.

What if the phrase read "7 times the sum of some number and 5"?

In this case, we must change the phrase to a variable expression by replacing the "some number" with a variable. The phrase will become $7(x + 5)$.

Using variables, we can change a phrase to a variable expression.

6 times some number	$6x$	x is a variable
5 less than some number	$r - 5$	r is a variable

SKILLS DEVELOPMENT

When evaluating a variable expression you must be told what value to use for the variable. Then substitute that value into the expression and evaluate using the correct order of operations. A variable expression will change when the value of the variable changes.

To **substitute** means to put in place of.
To **evaluate** means to find the value.

Example 1

a. Evaluate $(2 + x)(3)$, where $x = -4$.

b. Evaluate $(2 + x)(3)$, where $x = 6$.

Solutions a. $(2 + x)(3)$ → Find the value.

$(2 + -4)(3) = (-2)(3)$ → Substitute -4 for x.

$= -6$

So, the value of $(2 + x)(3)$ is -6 when x is -4.

b. $(2 + x)(3)$ → Find the value.

$(2 + 6)(3) = (8)(3)$ → Substitute 6 for x.

$= 24$

So, the value of $(2 + x)(3)$ is 24 when x is 6.

You have learned several ways to write 7 times 8.

$$7 \times 8 \qquad 7 \cdot 8 \qquad 7(8)$$

What if the phrase about the cost of CDs read "the product of 3 and some number"?

When writing the product of x and y, we write xy. No operation sign is used. So, to write an expression for a phrase such as "the product of 3 and some number," we write, $3x$.

$3x$ means 3 times x.

So, Toby will pay 3 times the cost of a disk, or $3x$..

Example 2

Write a variable expression to represent each amount.

a. 2 less than a number

b. the product of 10 and some number

c. the sum of a number and 5

d. the quotient of x divided by 7

Solutions **a.** $x - 2$ **b.** $10y$

c. $z + 5$ **d.** $x \div 7$ or $\frac{x}{7}$

Variable expressions may include two or more variables.

Example 3

Evaluate $2m + 3n$, where $m = 4$, $n = 6$.

Solution $2m + 3n$ → Find the value.

$2(4) + 3(6) = 8 + 18$ → Substitute 4 for m, 6 for n.

$= 26$

So, the value of $2m + 3n$ is 26 when m is 4 and n is 6.

EXERCISES

Use $r = 3$. Match the expression with the value.

1. $r + 4$	**a.** 8
2. $4r$	**b.** 7
3. $4r - 4$	**c.** 2
4. $6r \div 9$	**d.** 12

Rewrite as a variable expression.

5. the sum of 12 and a number **6.** the product of x and 8

7. 14 minus y **8.** the quotient of 18 divided by y

Evaluate each expression if $s = 2$ and $t = 3$.

9. $3s - 2t$ **10.** $3t - 2s$ **11.** $4st$

12. $2t \div s$ **13.** $\frac{t + s}{t - s}$ **14.** $t + 2$

A team receives 2 points for a win and 1 point for a tie. The total number of points won by a team is given by the expression $2w + t$ where w is the number of games won and t is the number of games tied.

Calculate $2w + t$ for each of the following.

15. $w = 12, t = 5$ **16.** $w = 18, t = 8$ **17.** $w = 12, t = 0$

Refer to the expression in the previous question. Find the number of points that each team has in the Teen Football League.

TEEN FOOTBALL LEAGUE

	Team	Won	Lost	Tied	Total Points
18.	Concordes	12	3	1	
19.	B. C. Lions	7	8	1	
20.	Argonauts	12	2	2	
21.	Tiger Cats	11	3	1	
22.	Stampedes	8	6	2	

23. How many more wins would the Tiger Cats need to be in first place?

JUST FOR FUN

Do the Warm-Up Activity again. Make a list of the digits and the letters they form on the calculator. Then make up riddles so the answers can be found with the calculator.

SECTION 5–2

WORKING WITH VARIABLES

WARM-UP ACTIVITY

The length of this line is 5 centimeters. Look around the classroom and find at least 6 things that are about 5 cm long.

5 cm

Visualize two lengths of the same line:

5 cm 5 cm

Find at least six things in the classroom that are approximately that length. Use a metric ruler and check your guesses.

INTRODUCTION

In a university laboratory, students are working on new types of seeds to improve the quality of the plants grown from them. It is important that they keep each type of seed separate from others that they have made adjustments on. In mathematics we must do the same thing when combining variable expressions. It is important that we do not mix the different types of variables when adding.

SKILLS DEVELOPMENT

In the last section, $3a + 2b$ was called a variable expression. Here, it will be called an **algebraic expression**. It can be broken into two parts. Each part is called a **term**.

$$3a + 2b$$

$3a$ is called a term. $2b$ is called a term.

> $3a$
> 3 is the coefficient; a is the variable

Terms that have the same variables are called **like terms**. Terms with different variables are called **unlike terms**.

like terms	unlike terms
$3a$, $4a$, $5a$	$3a$, $4b$, $5c$
4, 5, 158	7 and $7x$

Polynomials is the general name for monomials, binomials, trinomials, etc. To **simplify polynomials** means to combine like terms. To add or subtract like terms, you add or subtract coefficients.

Polynomials	
Name	Examples
Monomial ↑ 1 term	$2x$; $3a$; -5
Binomial ↑ 2 terms	$2x + 3y$; $3a - 2b$
Trinomial ↑ 3 terms	$2x + 3y - 4k$; $-4a + 2b - c$

Example 1

Simplify the following.

a. $2c + 3c + 4$ **b.** $2x + y + 3x + 4y$

Solutions a. $2c + 3c + 4$ → Combine like terms.

$5c + 4$ → Add coefficients.

b. $2x + y + 3x + 4y$ → Combine like terms.

$2x + 3x + 1y + 4y$ → Rearrange like terms ($y = 1y$).

$5x + 5y$ → Add coefficients.

Notice, in each solution, the expressions have become unlike terms. You cannot simplify any further.

It is important to understand exponents of variables. **Exponents** of variables indicate a product of the variable. The exponent 2 in m^2 means the variable m occurs that many times as a factor in the product. That is, m is used 2 times as a factor in m^2.

Since m^1 means m, and
m^2 means $(m)(m)$, then the factors of m^2 are m and m.
Thus, m^2 and m are **unlike** terms.

Example 2

Write the prime factors of each expression.

a. $4a^3$ b. $2x(m + n)$

Solutions a. $4a^3$ → $2(2)\ (a)(a)(a)$

Prime factors of $4a^3$ are 2, 2, a, a, a.

b. $2x(m + n)$ → $2(x)\ (m + n)$

Prime factors of $2(x)(m + n)$ are 2, x, and $(m + n)$.

The degree of a monomial is the number of times a variable occurs as a factor of the monomial. For example,

$2x^3 = 2 \cdot x \cdot x \cdot x$ The variable is used 3 times. Its degree is 3.

$3abcd = 3 \cdot a \cdot b \cdot c \cdot d$ The monomial has 4 variables. Its degree is 4.

Example 3

Factor. Then tell the degree of the monomial.

a. $12ab$ b. xyz^2 c. $-5m^3$

Solutions

a. $12ab$ → $2(2)(3)(a)(b)$ The degree is 2.

b. xyz^2 → $(x)(y)(z)(z)$ The degree is 4.

c. $-5m^3$ → $(-1)(5)(m)(m)(m)$ The degree is 3.

EXERCISES

1. a. How are the terms $6x$ and $8x$ alike?
b. How are they different?
c. Why are they called like terms?

2. **a.** How are the terms $8x$ and $8y$ alike?
   b. How are they different?
   c. Why are they called unlike terms?

Identify each as M, monomial, or B, binomial.

3. $2x + 3y$ **4.** $4x$ **5.** $3b$

6. $a + 3b$ **7.** $4m + 3n$ **8.** y

Identify each as L, like terms, or U, unlike terms.

9. $a, 2a$ **10.** $3b, 3m$ **11.** $2x, 3x^2$ **12.** $5z, -3z$

Write the coefficient in each.

13. $11xy$ **14.** $7a^2$ **15.** mn **16.** $2ab^2c$

Simplify. (Recall: Simplify means to combine like terms.)

17. $a + b + a + b + a + b + a + b$

18. $a + 2a + b + b + b$

19. $a + b + a + c + b + a + c + b$

20. $m + 3m + n + 2n$

21. $k + m + 2k + 2m + k + 3m$

22. $x + y + x + y + z + y + z + x$

23. $2a + 3b + c + a + b + 2c$

24. $2a + 3b + 2a + 3b + 2a + b$

25. For each of the answers in Exercises 17–24, identify the expressions as monomials, binomials, or trinomials.

Tell the degree of each monomial.

26. $3a$ **27.** $4xyz$ **28.** $-8xy^2$ **29.** $6m^3$

Write the prime factors of each expression.

30. $3a$ **31.** $4a^2$ **32.** $2xy$

33. $-3x^2y$ **34.** $-8xy^2$ **35.** $4xyz$

SECTION 5–3

ADDING AND SUBTRACTING WITH VARIABLES

WARM-UP ACTIVITY

The Garza family raises horses and chickens. If the animals have a total of 20 heads and 64 feet, how many chickens do they have?

INTRODUCTION

When learning arithmetic, children begin by saying, "3 apples added to 2 apples is 5 apples." In algebra, that would be $3a + 2a = 5a$.

What if Leslie had 3 apples and 2 bananas and she combined them with 4 apples and 3 bananas that Ahmed had? They would have 7 apples and 5 bananas.

In algebra, that would be

$$(3a + 2b) + (4a + 3b) = 7a + 5b.$$

In this lesson, you will use earlier skills with integers to advance your knowledge of algebra.

SKILLS DEVELOPMENT

Adding polynomials is the same as combining like terms.

Example 1

Add the polynomials.

a. $(5x + 7y) + (3x + 2y)$ b. $(4a + 5) + (a + 1)$

Solutions a. $(5x + 7y) + (3x + 2y)$ → Collect like terms.
$= 5x + 3x + 7y + 2y$ → Add coefficients.
$= 8x + 9y$ → Sum

b. $(4a + 5) + (a + 1)$ → Collect like terms.
$= 4a + a + 5 + 1$ → Add coefficients.
$= 5a + 6$ → Sum.

To subtract polynomials, use your earlier skills with opposites. Recall: the **opposite** is also called the **additive inverse**.

Arithmetic	Algebra
The opposite of 3 is -3.	The opposite of $3a$ is $-3a$.
The opposite of -4 is 4.	The opposite of $-4x$ is $4x$.
The opposite of $6 - 2$ is $-6 + 2$.	The opposite of $2a - b$ is $-2a + b$.

Notice that the opposite of the polynomial, $2a - b$, is formed by replacing each term with its opposite, $-2a + b$.

Example 2

Write the opposite of each polynomial.

a. $-2a + 1$ **b.** $6x - 3y$ **c.** $-(-3x^2 + -2x)$

Solutions a. $2a - 1$

b. $-6x + 3y$

c. $3x^2 + 2x$

To subtract an integer, add its opposite. That is, you add the opposite of the subtrahend. For example,

To subtract positive 2, add negative 2.

$$6 - 2 = 6 + -2$$

Similarly, to subtract a polynomial, you add its opposite.

Example 3

Subtract the polynomials.

a. $3a - (-2a)$ b. $(2a + b) - (3a - 2b)$

Solutions a. $3a - (-2a)$ → Add the opposite of $-2a$.
$= 3a + 2a$ → *Add coefficients.*
$= 5a$ → Difference

b. $(2a + b) - (3a - 2b)$ → Add the oppposite of $3a - 2b$.
$= (2a + b) + (-3a + 2b)$ → Collect like terms.
$= 2a + -3a + b + 2b$ → Add coefficients.
$= -a + 3b$ → Difference

The **degree of a polynomial** is the degree of the term that has the greatest degree.

Example 4

Tell the degree of each polynomial.

a. $3x^2 + 4x + 5$ b. $7xy^2 - 11x^4$

Solutions a. $3x^2$ has a degree of 2.
$4x$ has a degree of 1.
So, the degree of
$3x^2 + 4x + 5$ is 2.

b. $7xy^2$ has a degree of 3.
$11x^4$ has a degree of 4.
So, the degree of
$7xy^2 - 11x^4$ is 4.

EXERCISES

Match each binomial with its opposite on the right.

1. $2x - 3y$ **a.** $-6 + 3n$

2. $-x^2 + y^2$ **b.** $2x + 3y$

3. $3n + 6$ **c.** $x^2 - y^2$

4. $6 - 3n$ **d.** $-2x + 3y$

5. $-2x - 3y$ **e.** $-3n - 6$

Add the polynomials.

6. $2m + (m - 3)$

7. $3x + (2x - y)$

8. $(3m + 2n) + (2m - 3n)$

9. $(3x - 4y) + (2x + 2y)$

Write the opposite of the polynomial.

10. $3x$

11. $-4x^2$

12. $a + b$

13. $6x^2$

14. $-(-e + f)$

15. $x^2 + 2x - 5$

16. $-(x^2 - 3x + 5)$

Subtract the polynomials.

17. $2a - 3b - (4a + b)$

18. $3a - 2b - (2a - b)$

19. $(6 - x) - (3 - x)$

20. $x - 2 - (x - 4)$

21. $6y - 12 - (y - 16)$

22. $9m - 3n - (-5m - 2n)$

Add or subtract.

23. $(2x - 3y) - (3x - 2y)$

24. $(4a - 2b) + (3a + b)$

25. $(2m - n) - (4m - n)$

26. $(3x^2 - 2x + 5) -$
$(2x^2 + x - 6)$

Write an example of each.

27. a binomial with degree of 2

28. a trinomial with degree of 3

Complete each of the following.

29. Find the sum of $a + 2b$, $6a - 3b$, and $-4a + 6b$.

30. Subtract the sum of $2a$ and $-3a + 4b$ from $-6a - 2b$.

31. How much less is $-3x + 2y$ than $4x - 2y$?

32. Use $m = 3$, $n = 2$.
Which trip is longer, A or B?
By how much?

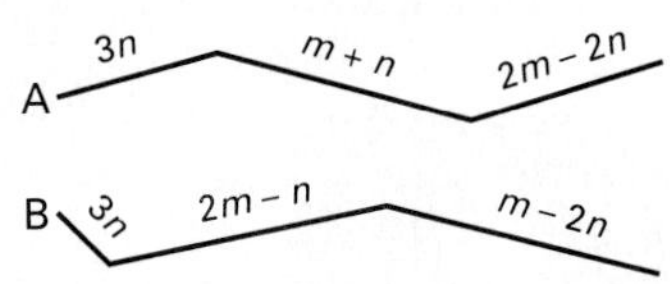

33. Use the photographs in this lesson to make up a subtraction problem. Write an algebraic expression like the one in the Introduction to describe your problem.

SECTION 5–4

MULTIPLYING WITH VARIABLES

WARM-UP ACTIVITY

Fill in the links in each chain with + or − to make true statements.

1. $a \;?\; a \;?\; 4b \;?\; 3b = 2a + b$
2. $3m \;?\; 3n \;?\; 2m \;?\; 2n = 5m - 5n$
3. $6b \;?\; 5b \;?\; 2b \;?\; 4b = 9b$
4. $4z \;?\; 3z \;?\; 3z \;?\; z = 3z$
5. $3a \;?\; a \;?\; 5a \;?\; 2a = 5a$
6. $3y \;?\; 2y \;?\; 3y^2 \;?\; 4y^2 = -y^2 + y$

INTRODUCTION

To plan the construction of a building, an engineer needs to use algebra.

To calculate the stresses and strains in the building, the engineer must simplify and evaluate many polynomials. But, before any calculations can be done, the engineer must be able to multiply polynomials.

In this lesson, you will multiply polynomials which will allow you to use arithmetic skills to advance your understanding of multiplying coefficients and variables with exponents.

SKILLS DEVELOPMENT

To find the product of two monomials, $(3m)(2n)$, use what you already know about finding area.

Remember:
mn means $(1m)(1n)$
$= 1mn$

Area = sum of the areas of the rectangles
$= mn + mn + mn + mn + mn + mn$
$= 6mn$
Area = product of the length and the width
$= (3m)(2n)$
$= (3 \cdot 2)(m \cdot n)$
$= 6mn$
So, this example suggests $(3m)(2n) = 6mn$.

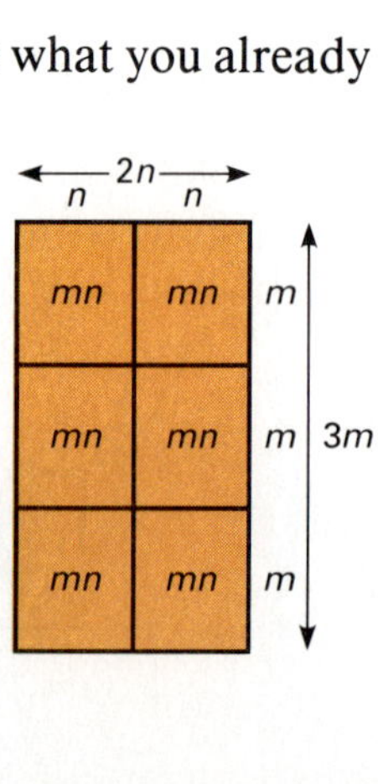

To find the product of two monomials, you also need to use your skills with exponents. For example, to simplify the following product, multiply the coefficients, and then multiply the variables.

$(2a^2)(3a^3)$

Think: $(2a^2)(3a^3) = (2)(a)(a)(3)(a)(a)(a)$
$= (2)(3)(a)(a)(a)(a)(a)$
$= 6a^5$

1. multiply coefficients

2. multiply the variables.

Exponents represent the number of times the factor occurs. $(a)(a)(a)(a)(a) = a^5$. So, $(2a^2)(3a^3) = 6a^5$.

These two models help to show rules for multiplying monomials:

Example 1

Find the product.

a. $(3a)(4ab)$ **b.** $(-3y^2)(4y^3)$

Solutions

a. $(3a)(4ab)$
$= (3)(4)(a)(a)(b)$
$= 12a^2b$ → So, $(3a)(4ab) = 12a^2b$.

b. $(-3y^2)(4y^3)$
$= (-3)(4)(y^2)(y^3)$ → Multiply.
$= -12y^5$ → So, $(-3y^2)(4y^3) = -12y^5$.

Using the distributive property, you can expand your knowledge from arithmetic to algebra.

Arithmetic	Algebra
$3(2 + 7) = 3(2) + 3(7)$	$3(x + 7) = 3(x) + 3(7)$
$= 6 + 21$	$= 3x + 21$

Example 2

Write the product without parentheses.

a. $3(x + 2y)$

b. $-2(x - 2)$

c. $x(5x + 1)$

Solutions

a. $3(x + 2y) = 3(x) + 3(2y)$ → Distribute 3. Simplify.
$= 3x + 6y$

b. $-2(x - 2) = -2(x + -2)$ → Rewrite $x - 2$ as addition;
$= -2(x) + -2(-2)$ Distribute -2. Simplify.
$= -2x + 4$

c. $x(5x + 1) = x(5x) + x(1)$ → Distribute x. Simplify.
$= 5x^2 + x$

EXERCISES

1. Simplified, the product, $(3ab)(-5a^2b) = -15a^3b^2$. How are the coefficients used to find the coefficient of $-15a^3b^2$?

2. How are the variables of the answer found?

Find the value of ? for the exponent in each of the following.

3. $(x^3)(x^3) = x^?$
4. $(a)(a^4) = a^?$
5. $(m^3)(m^2) = m^?$
6. $(x)(x^5) = x^?$
7. $(y^2)(y^3) = y^?$
8. $(y^2)(y) = y^?$
9. Use your skills with exponents to show that $(a^3)(a^2) = a^{3+2}$ is true for exponents in multiplication. Describe your conclusions.

Calculate each product of integers.

10. $(-2)4$ **11.** $5(-3)$ **12.** $(-6)(-2)$

13. $(-2) \times (-2) \times (-2)$ **14.** $(-2)^4$ **15.** $(-3)(-5)^2$

Find each product. Multiply the coefficients and the variables.

16. $(3x)(-2y)$ **17.** $(-5a)(4b)$ **18.** $(3ab)(-2a)$

19. $(2x^2)(3x^2)$ **20.** $(-4m^3)(-2m^3)$ **21.** $(-6y^3)(3a^2b^2)$

Find the value of each when $a = \frac{1}{2}$ and $b = \frac{1}{3}$.

22. $(2a)(3b)$ **23.** $(3ab)(a^2)$ **24.** $(6ab)(-2ab)$

Simplify each product. Write the expression without parentheses.

25. $a(3m + 2n)$ **26.** $-(y + 3)$ **27.** $-2(2y - 5)$

28. $-(2y - 4)$ **29.** $6a(3b - 2c)$ **30.** $-3k(2m + 2n)$

Write an expression for the area of each rectangle. [$A = lw$]

31.

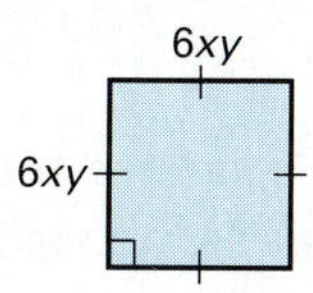

32.

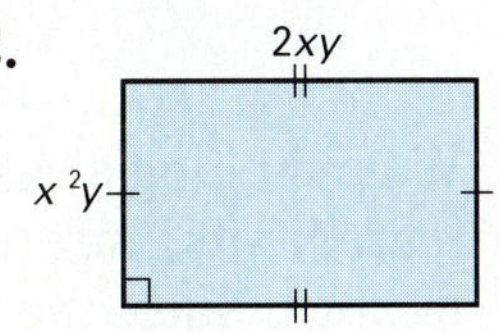

33.

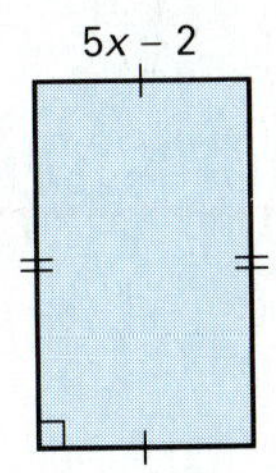

Engineers have planned to install a soundproof room. They will use this diagram.

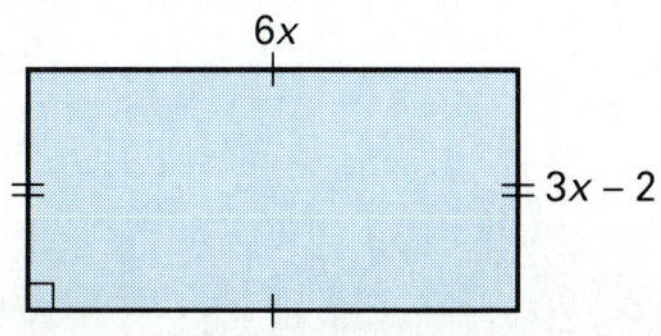

34. Write an expression for the area of the ceiling.

35. Find the area when $x = 6$. (Use a calculator.)

SECTION 5–5

SIMPLIFYING WITH VARIABLES

WARM-UP ACTIVITY

In this square there are really 5 squares hidden.

In this square there are really 14 squares hidden. Do you see them?

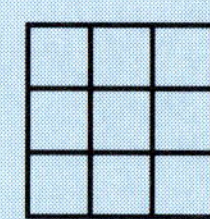

Look for a pattern or method to write the number of squares hidden in this figure.

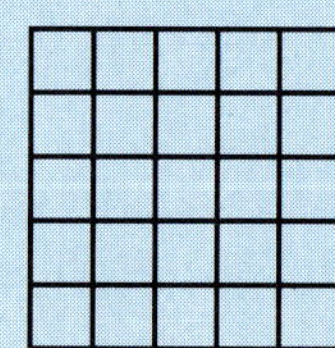

INTRODUCTION

When baking a cake, it is necessary to make sure you have the needed ingredients, mix the ingredients according to the recipe, pour the mixture into the properly sized pan, and bake it at the correct temperature for the appropriate time. If one of the steps is missing or done incorrectly, the cake could possibly turn into a disaster or at least one that no one wants to eat.

Simplifying polynomials is like baking a cake. The previous lessons in this chapter have given you the needed ingredients for simplifying.

In this lesson, we will mix the previously learned skills to prepare for the desired product.

SKILLS DEVELOPMENT

When simplifying polynomials be sure to follow the correct order of operations.

Example 1

Simplify.

a. $2(x + 3) + 4x$ **b.** $3(x - 2y) + 2(x + 4y)$

Solutions

a. $2(x + 3) + 4x$ → Distribute 2.
$= 2x + 6 + 4x$ → Combine like terms.
$= 6x + 6$
So, $2(x + 3) + 4x = 6x + 6$.

b. $3(x - 2y) + 2(x + 4y)$ → Distribute 3 and 2.
$= 3x - 6y + 2x + 8y$ → Combine like terms.
$= 5x + 2y$
So, $3(x - 2y) + 2(x + 4y) = 5x + 2y$.

When evaluating polynomial expressions, you may simplify first or substitute first.

Example 2

Find the value of $2x(x - y) - 3x(x - y)$ for $x = -1$, $y = -3$.

Solution 1

Simplify the expression, then substitute and evaluate.

$2x(x - y) - 3x(x - y)$ → Change $(x - y)$ to addition.
$= 2x^2 - 2xy - 3x^2 + 3xy$ → Distribute $2x$ and $3x$.
$= -x^2 + xy$ → Simplify.
$= -(-1)^2 + (-1)(-3)$ → Replace x with -1; $y = -3$.
$= -1 + 3$ → Combine like terms.
$= 2$
So, $2x(x - y) - 3x(x - y) = 2$ when $x = 1$, and $y = 3$.

Solution 2
Substitute, then simplify.

$2x(x - y) - 3x(x - y)$
$= 2(-1)(-1 - -3) - 3(-1)(-1 - -3)$ → Substitute.
$= (-2)(-1 + 3) - (-3)(-1 + 3)$ → Simplify.
$= -2(2) - (-3)(2)$
$= -4 - -6$
$= -4 + 6$
$= 2$

So, $2x(x - y) - 3x(x - y) = 2$ when $x = -1$, and $y = -3$.

EXERCISES

Tell what step would be done first to simplify the following. DO NOT SIMPLIFY.

1. $2x + 3(x + 2)$ **2.** $4y - 3(4 + 2y)$

3. $-6a - (2a - 5)$ **4.** $8(2a + 5) - 6a$

5. $3(x - y) - 2(2x - 3y)$ **6.** $-2x(x + 2) + 3$

Match the expression with its simplified form.

7. $3y + 2(y + 1)$ **a.** $-y + 3$

8. $4(2y - 3) + 4y$ **b.** $5y - 3$

9. $2y - 3(y - 1)$ **c.** $12y - 12$

10. $-(3y - 2) - 5y$ **d.** $5y + 2$

11. $2y + 3(y - 1)$ **e.** $-8y + 2$

Simplify each of the following.

12. $2m + 3 + 3(m - 4)$ **13.** $0 - 4m + 3(m + 6) + 5$

14. $0 + 6m - (8 - 3m) - 6$

15. $-8(9 - y) - y + 9$

16. $-8(9 - y) - (y + 9)$

17. $-(11y - 4) - 8y$

Evaluate each expression for the value given. Use a calculator or mental math.

$x = 5$
$n = -2$
$y = 3$
$s = -1$
$k = 2$

18. $-3x(x + 1)$

19. $n + 2n(n - 4)$

20. $2(y - 3) - (y - 3)$

21. $8(s - 2) + 3(s + 5)$

22. $-4(2 - k) - 2(k + 4)$

The profit made, in cents, on the sale of plants is given by the expression $0.05t(3t - 25)$ where t is the number of plants sold. Find the profit for each. Use a calculator.

23. $t = 10$

24. $t = 100$

25. $t = 50$

26. Find an expression for the area of each rectangle.

27. Predict which of the regions, A or B, has the greater area if $x = 5$.

28. Check your prediction in B by substituting $x = 5$.

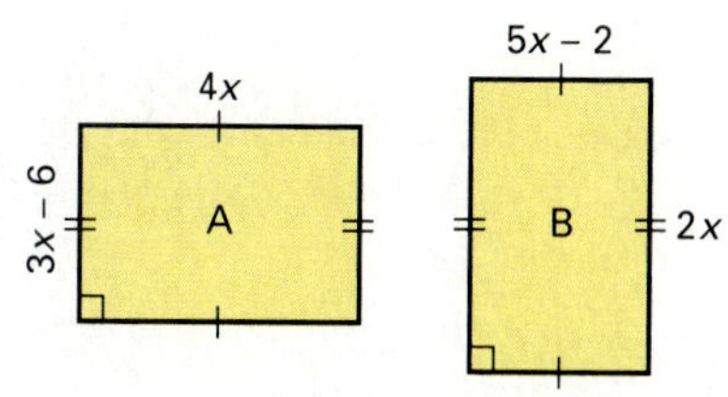

29. In the Warm-Up Activity, assume each □ = 1 cm^2. Find the area of the largest rectangle in each of the three squares.

SECTION 5–6

SIMPLIFYING BY DIVIDING

WARM-UP ACTIVITY

One hundred birds drink one hundred liters of water in one hundred days. How many liters of water will ten birds drink in ten days?

INTRODUCTION

The more algebra skills you learn, the more problems you will be able to solve. To master division of polynomials, you will use division skills with integers, rules for dividing with exponents, and multiplication for checking.

SKILLS DEVELOPMENT

In your work with numbers, you used your skill in factoring to simplify expressions. These skills extend to simplifying expressions in algebra. For example, to simplify when dividing in arithmetic or in algebra, use your skills with writing factors.

Arithmetic		Algebra
$\frac{75}{25} = \frac{(3)(5)(5)}{(5)(5)}$	Express the numerator and denominator as factors.	$\frac{50xyk}{25xk} = \frac{(2)(5)(5)(x)(y)(k)}{(5)(5)(x)(k)}$
$= \frac{(3)(\overset{1}{\cancel{5}})(\overset{1}{\cancel{5}})}{(\underset{1}{\cancel{5}})(\underset{1}{\cancel{5}})}$	Then divide by common factors.	$= \frac{(2)(\overset{1}{\cancel{5}})(\overset{1}{\cancel{5}})(\overset{1}{\cancel{x}})(y)(\overset{1}{\cancel{k}})}{(\underset{1}{\cancel{5}})(\underset{1}{\cancel{5}})(\underset{1}{\cancel{x}})(\underset{1}{\cancel{k}})}$
$= 3$		$= 2y$

To **divide monomials**, divide the coefficients, then the variables.

Example 1

Divide.

a. $\frac{6ab}{3a}$ b. $\frac{-18a^2b}{6ab}$

Solutions

a. $\frac{6ab}{3a} = 2b$

Think: $6 \div 3 = 2$
$ab \div a = b$

b. $\frac{-18a^2b}{6ab} = -3a$

Think:
$-18 \div 6 = -3$
$a^2b \div (ab) = a$

To find the result when you divide a polynomial by a monomial, use examples to which you know the answer. For example, use this known result.

$$3x(2a + 3b) = 6ax + 9bx \text{ means } \frac{6ax + 9bx}{3x} = 2a + 3b$$

Use the result to suggest a method of dividing a polynomial by a monomial. For example:

Think: $\frac{6ax + 9bx}{3x} = \frac{6ax}{3x} + \frac{9bx}{3x}$
$= 2a + 3b$ The result is the same as above.

Examples such as the previous one suggest a method of dividing a polynomial by a monomial. This method is used to divide binomials by monomials in the following example.

To **divide a polynomial by a monomial**, divide each term of the polynomial by the monomial.

Example 2

Simplify.

a. $\frac{4m + 8n}{4}$

b. $\frac{3kt - 6ks}{3k}$

Solutions

a. $\frac{4m + 8n}{4} = m + 2n$

Think: $\frac{4m}{4} = m \quad \frac{8n}{4} = 2n$

b. $\frac{3kt - 6ks}{3k} = t - 2s$

Think: $\frac{3kt}{3k} = t \quad \frac{-6ks}{3k} = -2s$

The method you have developed extends to dividing a trinomial by a monomial. For example:

$$\frac{6x^2 - 2x + 8}{2} = \frac{6x^2}{2} - \frac{2x}{2} + \frac{8}{2} \quad \text{Divide each term by 2.}$$
$$= 3x^2 - x + 4 \quad \text{Result is a trinomial.}$$

EXERCISES

Is the monomial in the correctly factored form? Write true or false.

1. $6xy = 2(3)(x)(x)(y)$

2. $25x^2y^2 = 5(5)(x)(x)(y)(y)$

3. $8xyz^2 = 2(2)(2)(x)(y)(z)(z)$

4. $-14a^2b^2 = 2(7)(a)(b)(b)$

Find the answer.

5. $\frac{18a^2b}{9ab} = ?$ Think: $\frac{18a^2b}{9ab} + \frac{(2)(3)(3)(a)(a)(b)}{(3)(3)(a)(b)} = ?$

6. $\frac{-25xy^2}{-5y} = ?$ Think: $\frac{-25xy^2}{-5y} = \frac{(-5)(5)(x)(y)(y)}{(-1)(-5)(y)} = ?$

7. $\frac{49m^2n}{-7m^2n} = ?$

8. $\frac{32a^3x^2}{8a^2x} = ?$

9. $\frac{15st^2}{3st} = ?$

Complete the missing steps for each division.

10. $\frac{-21a^2b}{7ab} = ?$ Think: $\frac{-21}{7} = ? \quad \frac{a^2b}{ab} = ?$

11. $\frac{-36x^3y^2}{-6xy^2} = ?$ Think: $\frac{-36}{-6} = ? \quad \frac{x^3y^2}{xy^2} = ?$

Simplify each of the following.

12. $\frac{9xy}{3x}$ **13.** $\frac{12x^2y}{4x^2}$ **14.** $\frac{4a^2}{a^2}$

15. $\frac{36mn}{-6m}$ **16.** $\frac{-32bc}{4b}$ **17.** $\frac{-16ab}{-2ab}$

18. $-24a^2b \div 6a$ **19.** $9x^2y^2 \div (-3xy)$

Simplify each of the following.

20. $\frac{4x + 12}{4} = ?$ Think: $\frac{4x}{4} = ?$ $\frac{12}{4} = ?$

21. $\frac{12m^3 - 6m^2}{-2m^2} = ?$ Think: $\frac{12m^3}{-2m^2} = ?$ $\frac{-6m^2}{-2m^2} = ?$

22. $\frac{3y + 18}{3}$ **23.** $\frac{4y - 16}{2}$ **24.** $\frac{-16y - 4}{-4}$

25. $(-48ax^2 + 24ay^2) \div (-12a)$ **26.** $(-75mn^2 - 50m^2n) \div (-25mn)$

Copy and complete the chart.

	Expression	Value of expression when $x = 1$	$x = -1$	$x = 10$
27.	$\frac{6x + 12}{3}$	?	?	?
28.	$\frac{15x^2 - 10x}{5x}$	?	?	?

Simplify each of the following.

29. $\frac{-4x^3 + x^2 + 3x}{x}$ **30.** $\frac{8k^3 - k^2 + 2k}{k}$ **31.** $\frac{9x^3 + 6x^2 - 15x}{3x}$

Write an algebraic expression for each.

32. Last year, a homeowner paid $1285 for gas and a surcharge of $97. What was the average monthly bill?

33. A runner from Germany completed the 400-meter run in 51.08 seconds. How many meters per second was this?

SECTION 5–7

SIMPLIFYING USING COMMON FACTORS

WARM-UP ACTIVITY

Using sixes and any of the four operation signs for addition, subtraction, multiplication, and division, write problems that have an answer of 1 through 10.

Example: $(6 + 6 + 6) \div (6 + 6 + 6) = 1$

INTRODUCTION

In a hockey game, players must have a variety of skills that are offensive and defensive. During a game, a player may skate towards the net, and immediately have to switch directions

In mathematics, too, there are often opposite directions of play. In this lesson, reverse operations are studied.

SKILLS DEVELOPMENT

The application that reverses or undoes multiplication is called **factoring.**

Multiplication: Start with $41 \cdot 5$, end up with 205.
Factoring: Start with 205, end up with $41 \cdot 5$.

The same holds true for polynomials. On the next page, the distributive property of multiplication is used to end up with the product, $3x + 12$.

Factoring reverses this operation. You begin with the terms of the product, $3x + 12$, and end up with the factors that were multiplied to obtain it, $3(x + 4)$.

Multiplication: Start with $3(x + 4)$, end up with $3x + 12$.
Factoring: Start with $3x + 12$, end up with $3(x + 4)$.

Let's review some terminology.

$$\underset{\text{factors}}{3\,(x\ +\ 4)} = \underset{\text{terms}}{3x + 12}$$

3 is a **common monomial factor** of $3x + 12$.

To factor using common monomial factors, you must find the Greatest Common Factor of the terms.

Example 1

Find the Greatest Common Factor of the terms.

a. $12x$, 18 **b.** $4a$, $12ab$ **c.** $-6x^3y$, $24x^2y^2$

Solutions a. Factors of $12x$: $2(2)(3)(x)$
Factors of 18: $2(3)(3)$
The Greatest Common Factor is $2(3)$ or 6.

b. Factors of $4a$: $2(2)(a)$
Factors of $12ab$: $2(2)(3)(a)(b)$
The Greatest Common Factor is $2(2)(a)$ or $4a$.

c. Factors of $-6x^3y$: $-1(2)(3)(x)(x)(x)(y)$
Factors of $24x^2y^2$: $2(2)(2)(3)(x)(x)(y)(y)$
The Greatest Common Factor is $2(3)(x)(x)(y)$ or $6x^2y$.

To factor a polynomial, begin by looking for a factor common to all the terms. If you can identify the Greatest Common Factor, you will be able to factor the polynomial in one step. Otherwise, it may take you more than one step. Keep factoring out common factors until the terms no longer have a common factor.

To factor the polynomial $3x^2 + 6x - 12$, first notice that 3 is the greatest common factor. Now, take out the common factor, and use the distributive property to write the remaining factors of each term.

$$3x^2 + 6x - 12 = 3(x^2 + 2x - 4)$$

Multiply to check: $3(x^2 + 2x - 4) = 3x^2 + 6x - 12$

Example 2

Factor and check by multiplication.

a. $2m + 4$ **b.** $6x^2y - 15x$ **c.** $25x^3 + 50x^2 + 5x$

Solutions

a. Factor: $2m + 4 = 2(m + 2)$ 2 is the common factor.

Check: $2(m + 2) = 2m + 4$

b. Factor: $6x^2y - 15x = 3x(2xy - 5)$ $3x$ is the common monomial factor.

Check: $3x(2xy - 5) = 6x^2y - 15x$

c. Factor: $25x^3 + 50x^2 + 5x$
$= 5x(5x^2 + 10x + 1)$ $5x$ is the common monomial factor.

Check: $5x(5x^2 + 10x + 1) = 25x^3 + 50x^2 + 5x$

Factoring can reduce the number of calculations needed to evaluate an expression. For example, to calculate $6a + 6b + 6c$ when $a = 3$, $b = -2$, $c = -4$, compare the following.

A

$$\begin{aligned} &6a + 6b + 6c \\ &= 6(3) + 6(-2) + 6(-4) \\ &= 18 + -12 + -24 \\ &= 6 + -24 \\ &= -18 \end{aligned}$$

B

$$\begin{aligned} &6a + 6b + 6c \\ &= 6(a + b + c) \\ &= 6(3 - 2 - 4) \\ &= 6(-3) \\ &= -18 \end{aligned}$$

How many calculations did you need for A and for B?

EXERCISES

Express each of the following in the simplest factored form. The first one is done for you.

1. $6x^2y$ $(2)(3)(x)(x)(y)$

2. $6abc$ **3.** $8xy^2$ **4.** $20x^2y$

5. $4a^2b$ **6.** $-16a^2b$ **7.** $-9ax^2b$

Find the missing factor.

8. $16ab = (?)(4a)$ **9.** $12mn = (?)(6n)$ **10.** $25y^3 = (?)(-5y^2)$

Match the terms with their Greatest Common Factor.

11. $ax^2, 2bx^2, 3cx^2$	**a.** x
12. $3x^3, 3x^2y, 3x^2z$	**b.** $2xy$
13. $x^3, -8x^2, + 16x$	**c.** $3x^2$
14. $6x^2y, -4xy^2, 10x^2y^2$	**d.** xy
15. $x^3y, -2x^2y, -xy^2$	**e.** x^2

Factor each of the following.

16. $6 - 12x$ **17.** $5x^2 - 3x$ **18.** $2x^2 - 6x$

19. $4ab - b^2$ **20.** $a^2 - 2a$ **21.** $4y^2 - 16$

22. $28a^2 - 14ab$ **23.** $36mn - 25m^2n^2$ **24.** $6x^2 - 12x + 15$

25. If $x = 2$ and $y = -3$, evaluate
  a. $3xy^2 - 3x^2y$ **b.** $3xy(y - x)$
  c. Why are your answers the same in a and b?
  d. Which answer required fewer calculations?

JUST FOR FUN

- Use four 5s to write 100.
- Use four 9s to write 100.

SECTION 5–8

SOLVING PROBLEMS USING FORMULAS

WARM-UP ACTIVITY

How many birthdays will you have celebrated after you are one million minutes old? Use a calculator.

INTRODUCTION

When bowling in a league, it is sometimes customary to figure handicaps for new bowlers. One formula for doing this is

$$H = 0.8(200 - a)$$

where H = handicap and a = average score.

Problems in this lesson use this and other formulas as the topics for replacement of variables.

SKILLS DEVELOPMENT

Some problems use formulas. But not all problems can be solved by formulas. When a formula is used in a problem, you must evaluate the formula.

Use the following steps as a basic strategy for solving problems with formulas.

Steps for evaluating formulas.

1. Find the values for each variable by carefully reading the problem.
2. Substitute the known values for the variables.
3. Solve using the correct order of operations.
4. Check for the reasonableness of your answer.

Example 1

The formula for the perimeter of a rectangle is $P = 2l + 2w$, where P is the perimeter, l is the length of the rectangle, and w is the width of the rectangle. Find the perimeter of a rectangle 4 inches by 9 inches.

Solution

$P = 2l + 2w$	or $P = 2(l + w)$	Formula
$P = 2(4) + 2(9)$	$P = 2(4 + 9)$	Substitute values.
$P = 8 + 18$	$P = 2(13)$	$l = 4, w = 9$
$P = 26$ in	$P = 26$ in	

So, the perimeter is 26 inches when $l = 4$ and $w = 9$.

Check: Does 26 inches sound reasonable?

Example 2

The Ohans have decided to buy their son a car. They have borrowed \$4395 for three years. The annual interest is 12%. What amount will they owe in three years? Use the formula:

$A = p + prt$ where A = total amount due

p = the amount borrowed (principal)

r = interest rate

t = time borrowed

Solution

$A = p + prt$	Formula
$A = 4395 + 4395(0.12)(3)$	Substitute values 12% = 0.12.
$A = 4395 + 1582.20$	Multiply first.
$A = \$5977.20$	Add.

So, the Ohans will owe $5977.20 in 3 years.

Check: Does $5977.20 sound reasonable?

EXERCISES

An arena makes $2.09 on each red ticket and $1.56 on each blue ticket. The total profit, P, in dollars is given by $P = 2.09r + 1.56b$, where r = the number of red tickets, and b = the number of blue tickets.

1. Calculate, to the nearest dollar, the profit for 200 red tickets and 160 blue tickets.
2. Estimate which of the following attendances gives more profit, A or B.
   A: 435 reds and 180 blues B: 265 reds and 250 blues
3. Calculate the profit for A and B. How close was your estimate?

The Print Right Copy Company uses the formula $C = S + 0.05n$ for figuring the cost in dollars for duplicating materials. C = total costs, S = the set-up charge, n = the number of copies.

Copy and complete the chart. Use a set-up charge of $12.50.

	Number of copies n	Total Cost	Average cost per copy (Round to nearest cent.)
4.	10	$13.00	13 ÷ 10 = $1.30
5.	100	$17.50	17.5 ÷ 100 = $0.18
6.	1000	$62.50	62.5 ÷ 1000 = $0.06

7. What do you notice about the average cost per copy?

Solve each problem.

8. To find average speed, use the formula $r = \frac{d}{t}$ where r = the speed (rate), d = distance, and t = time. On a camping trip, Ryan's family drove 90 miles in 1.5 hours. What was the average speed of the car during the trip?

9. The Langleys have a rectangular swimming pool that measures 12 feet by 22 feet. The Hogues have a swimming pool that measures 15 feet by 15 feet. Both pools are six feet deep. Which pool holds more water? How much more? Use the formula $V = lwh$, where V = the volume, l = length, w = width, and h = height.

10. Mrs. Wortan borrows \$3500 from the First Term Bank. The annual interest rate is 9%. She repays the loan six months later. What is the interest at this time? Use the formula $A = p + prt$.

11. Alissa plays city league softball. So far this season, she has had 51 hits in 267 times at bat. She walked 14 times. What is her batting average? (Round to the nearest thousandths.) Use the formula $B = h \div (t - w)$, where B = batting average, h = hits, t = times at bat, and w = walks.

12. Laura works for one of the local paper printers and makes 25 cents for each paper she delivers, plus a weekly salary of \$25. This is written as the formula $I = S + 0.25n$, where I = Income, S = weekly salary, and n = the number of paper delivered. How much will Laura make if she delivers 225 papers?

13. Find out about bowling. What is meant by "handicaps" for bowlers?

14. Is the concept of handicaps used in other sports?

15. Use the formula at the beginning of this lesson to make up problems about two new bowlers. Use your calculator to figure handicaps for the two bowlers.

SECTION 5–9

WRITING WORD PHRASES IN ALGEBRA

WARM-UP ACTIVITY

Paula is going to redecorate her room. She is going to paint the walls, replace the carpet, and get a new bedspread. Her choices have been narrowed down to:

Colors		Bedspread Names
Paint	Carpet	
white snowflake creme frost	tan mauve teal	Springtime Rosebud Sunbeam Wheatfield

How many different combinations can Paula select?

INTRODUCTION

When visiting a foreign country, it is best if you can read the language. However, there are times when travelers must depend on their ability to translate signs into words. If the wrong translation is made for any given sign, the result could be embarrassing or possibly even life threatening.

SKILLS DEVELOPMENT

In algebra, it is necessary to be able to translate words and numbers into algebraic expressions. This enables you to understand what is to be done with the numbers.

The charts on the top of the next page review some of the key words used to represent the arithmetic operations.

Words that suggest addition +	Words that suggest subtraction −
addend ? sum plus total ?	difference decrease ? exceeds ?

Words that suggest multiplication ×	Words that suggest division ÷
times product ? area ?	share ? each quotient ? average

Example 1

Write an expression for each.

a. a number increased by 3.

b. 4 less than 5 times a number.

Solutions Let n represent the number.

a. the number increased by 3 → Increased by means add.

$$\underbrace{\text{number}}_{n}\ \underbrace{\text{increased by}}_{+}\ \underbrace{3}_{3}$$

So, the expression is $n + 3$.

b. 4 less than 5 times n → Times means X. Write $5 \cdot n$ as $5n$.

$5 \cdot n - 4$ → Less than means subtract.

So, the expression is $5n - 4$.

There are some expressions that can be simplified after the symbols are inserted.

Example 2

Multiply the sum of $-3x$ and $2x$ by $-2x$.

Solution $-2x(-3x + 2x) \rightarrow$ Add first.

$-2x\,(-x) = 2x^2$

So, the sum of $-3x$ and $2x$ multiplied by $-2x$ is $2x^2$.

Example 3

Find the average of $3x$, $-6x$, and $9x$.

Solution $\frac{3x + -6x + 9x}{3}$ → Average means first add, then divide by the number of terms.

$\frac{6x}{3} = 2x$

So, the average of $3x$, $-6x$, and $9x$ is $2x$.

We can translate a mathematical phrase in a word problem to an algebraic expression so that the numbers become more meaningful.

The small town of Halifax had a number of different road signs. As the town grew, *the number doubled and then increased by 5.*

Example 4

a. Write an expression for a number doubled and then increased by 5.

b. Evaluate $2n + 5$ when n is 12.

Solution Let n represent the number.

a. n doubled then increased by 5 → A number doubled is $n \cdot 2$ or $2n$.

$2n \quad + \quad 5$ → Increased by means add.

So, the expression becomes $2n + 5$

b. $2n + 5$ → Substitute 12 for n.

$2(12) + 5$ → Multiply, then

$24 + 5$ → add.

So, $2n + 5$ is $24 + 5$ or 29 when n is 12.

EXERCISES

Match each with an expression on the right.

1. n doubled	**a.** $n - 4$
2. n increased by five	**b.** $n + 2$
3. n decreased by four	**c.** $\frac{1}{4}n$
4. two more than n	**d.** $2n$
5. three less than n	**e.** $n + 5$
6. one fourth of n	**f.** $n - 3$

Write a word expression for each.

7. $n + 4$ **8.** $2t + 1$ **9.** $3t$ **10.** $\frac{2}{3}d$

Write an algebraic expression. Simplify.

11. Add $3x + 2$ to $4x - 6$.

12. Decrease $3x - 4y$ by $-2x - 6y$.

13. Subtract $3x^2 - 2y^2$ from $2x^2 - 6y^2$.

14. Subtract $6a - 3b$ from the sum of $2a - 6b$ and $-3a + 2b$.

15. Find the product of $-3x$ and $2xy$.

16. Find the product of $2x$, $3x$, and $-4x$.

17. Divide $6a^2b$ by $-3ab$.

18. Divide $6x^2 - 9xy$ by $-3x$.

19. Divide the product of $12x$ and $-2x^2$ by $-6x$.

20. Find the average of $12a^2$, $6ab$, and $-27b^2$.

21. Multiply the sum of $3x^2 - 2x$ and $2x^2 - 4x$ by $3x$.

Paul packs oranges in boxes by the dozen. In three hours, he packed a different number of boxes each hour. He represented this as 3x, −5x, and 8x for each of the three hours.

22. Write an expression for the average number of oranges Paul packed.

23. Evaluate the expression when x = 12.

CHAPTER 5 SUMMARY

The word *algebra* conjures up notions of letters and symbols. It is the language of numbers, operations, and variables. This chapter has introduced the idea of using letters (variables) in mathematics. You substituted values for them, simplified by combining like terms, added, subtracted, multiplied, and divided. Using common monomial factoring, you learned another method of simplifying. At the end of the chapter you worked with variables in formulas that are used in many job situations. The final lesson prepared you to set up equations, changing words to algebra.

You have to get used to the idea that variables are like numbers in that each letter represents a different number, but cannot be combined to make one specific number. This is why we use the phrase *like terms*. Remember, only like terms can be added or subtracted. Now that you are familiar with algebraic thinking, you can solve many problems using more than one method. Have fun using different methods. Challenge yourself!

VOCABULARY

additive inverse
algebraic expression
binomials
coefficient
degree of monomial
degree of polynomial
evaluate
exponents
factoring
like terms
monomials
opposite
polynomials
simplify
substitute
term
trinomials
unlike terms
variable
variable expression

CHAPTER 5 REVIEW

Evaluate each expression. Use $t = 3$ and $n = 7$.

1. $t + 8$
2. $n - 4$
3. $t + n$
4. $n - t$
5. $3n - 5$
6. $4n - 5t$

Name the coefficient and tell the degree in each.

7. $3k$
8. $-4xyz$
9. $-rw^2$

Name two terms that are alike.

10. $3d, 3k, -2d$
11. $-4n, 4n^2, 2n$
12. $3mn^2, -4mn, -3mn$

Identify each as M (monomial) or B (binomial).

13. $3h + 4$
14. $5xy$
15. 2

Simplify. (Add or subtract.)

16. $(2x - 3r) + (2r - x)$
17. $3m + 2n + m + 2n$
18. $6 - (3c - 5)$

Find each product.

19. $(4a)(2a)$
20. $6ac(-2a)$
21. $-y^2(-2y^3)$
22. $2(x + 3)$
23. $-4(m - 2)$
24. $-3c(c + 2)$

Simplify.

25. $2g + 2(4 - 3g)$
26. $2c - 3(c + 4)$
27. $3x - 6(2x - 3)$
28. $\frac{15y^2}{y}$
29. $\frac{4n + 12}{4}$
30. $\frac{-8h^2k^2}{-4hk}$
31. $\frac{4ab - 8ac}{-a}$
32. $\frac{15x}{3}$
33. $\frac{-12x^2y}{3xy}$

Find the missing factor.

34. $4x + 8 = ?(x + 2)$

35. The Spiffy Car Rental Company uses the formula $C = 10d + 0.2m$ to find the total cost for renting one of their cars. C = total cost, d = days rented, and m = miles driven. Find the total cost for renting a car for 4 days and driving 300 miles.

Write an algebraic expression for each.

36. 3 more than x
37. subtract the product of $3k$ and $-4n$ from $3kn$
38. add the sum of $2k$ and $5k$ to $-13k$

CHAPTER 5 TEST

Find the value when $w = 15$ and $x = 4$.

1. $w - x$

2. $3w + 14$

3. $x^2 - 2w$

Identify each as M (monomial), B (binomial), or T (trinomial).

4. $7x + 2y$

5. $-21abc$

6. $-m + 2n - 5$

Give an example of each.

7. three like terms **8.** two unlike terms

Simplify. (Add or subtract.)

9. $4d + 3 - 2d - 1$

10. $3g - 2(5g - 4)$

11. $2w^2 - 3w + w^2 + 2w$

12. $(2x + 3y) + (-3x + -2y)$

Find the product.

13. $3d(-2d)$

14. $-2a^3(-4a)$

15. $-4(2x - 3)$

Simplify.

16. $2(x - 2y) - 4(3y - 2x)$

17. $2a(a - 3b) - 3a(a + 2b)$

18. $\frac{24x}{-3}$

19. $\frac{36x^3}{-4x}$

20. $\frac{-12m + 8mn}{-4mn}$

21. What is the greatest common factor in $-8k^3$, $4k^2$, and $16k$?

Find the missing factor.

22. $4r - 6t = \square(2r - 3t)$

23. $3rd^2 - 6rd = 3rd(\square - \square)$

The formula for finding the volume, V, of a rectangle is $V = lwh$, where l = length, w = width, and h = height.

24. Find the volume of a box that is 4 inches wide, 5 inches tall, and 8 inches long.

Write an algebraic expression for each.

25. The sum of x and 12.

26. n is doubled and then subtracted from 20.

27. Increase the product of z and 5 by $9z$.

28. The product of $4x$ and $5h$ squared.

SKILLS PREVIEW

To be successful in the next two chapters, you should review the skills presented here. If you cannot complete any of the exercises, review them from previous chapters or ask your teacher.

Add or subtract.

1. $173 + 248$ **2.** $538 - 67$

3. $54.3 - 8.2$ **4.** $22.6 + 17.9$

5. $-7 + 8$ **6.** $-1 - -3$

7. $\frac{5}{8} - \frac{3}{5}$ **8.** $4\frac{2}{3} - 3\frac{1}{2}$

9. $2\frac{1}{4} + 1\frac{1}{7}$

Multiply or divide.

10. 46×3 **11.** $51 \div 3$

12. $16.3 \div 2$ **13.** 2.6×0.04

14. $(-3)(-8)$ **15.** $-18 \div 2$

16. $\frac{5}{8} \div \frac{1}{4}$ **17.** $1\frac{1}{2} \div \frac{2}{3}$

18. $2\frac{1}{3} \times 3\frac{1}{2}$

Use $<$, $>$, or $=$ to fill each ○.

19. $\frac{1}{2}$ ○ $\frac{4}{8}$ **20.** $\frac{-2}{3}$ ○ $\frac{1}{3}$

21. $\frac{-5}{6}$ ○ $\frac{-2}{3}$ **22.** -8 ○ -1

23. 4.3 ○ 2.1 **24.** -1.1 ○ -1.5

Graph each integer on a number line.

25. 3

26. -2

27. 0

Write an algebraic phrase.

28. 4 less than y

29. Twice the sum of 2 and 7

Evaluate each expression. Use $n = 5$ and $t = 2$.

30. $6t - n$

31. $3(n - t)$

32. $n^2 + t$

Solve this problem.

33. Jose won the bowling tournament. His bowling scores were 125, 147, and 136. What was his average score for the three games?

Mathematics and equations can play many roles. Think about the equations needed to build bridges. First, you need to know the distance the bridge must span. Suppose you know that a river is 500 feet wide at the point where the bridge is to be built. An engineer cannot just plan to build a 500-foot bridge. There must be allowances made for the river edges. The engineer must plan how far to extend the bridge so it is stable enough to hold the bridge footings.

Suppose the bridge contains various triangular cross sections for stability. You can use an equation to find how long particular beams of the bridge should be.

Solving equations is one of the most important skills you will learn in algebra. Many problems in everyday life can be solved by solving equations.

Ask Yourself

- *What equation am I solving when I decide if I have enough money for everything I need to buy?*
- *When consumers apply for an auto loan, how do banks process the application and figure out the amount of interest to be paid?*
- *How do I calculate how much a credit card really costs?*

CHAPTER 6

SOLVING EQUATIONS AND INEQUALITIES

SECTION 6–1

WRITING ALGEBRAIC EQUATIONS

WARM-UP ACTIVITY

A double-pan balance has a 10-gram mass on one side. On the other it has a 6-gram mass and a 4-gram mass.

1. Is the scale in balance? How can you tell?
2. Could you replace the 10-gram mass with two other masses without losing balance? Give an example. Could you do this with three masses? Give an example.
3. What will happen if you add a 2-gram mass to one side? To both sides? What if you add a 2-gram mass to one side, and a pair of 1-gram masses to the other side?

INTRODUCTION

In your language arts classes, you learned to write complete sentences. You learned to connect your phrases or expressions with verbs.

In the last chapter you learned to change word phrases into algebraic expressions. In this chapter you will learn to write complete algebraic statements. You will connect your expressions with equal signs. Equal signs act like verbs in algebra.

SKILLS DEVELOPMENT

An **equation** is an algebraic statement that says two expressions have the same value. We say the expressions are equal. In algebra, you can use an equal sign (=) to mean "the total is," "the result is," "the sum is," or "is equal to."

Example 1

Translate the following word statement into an algebraic equation.

When a number is increased by 10, the result is 26.

Solution Let x represent the number. You know that "a number increased by 10" is written $x + 10$. Now you use an equal sign to say that that expression is equal to 26.

$$x + 10 = 26$$

When you change word statements to algebraic equations, translate the expressions first. Then connect the expressions with equal signs.

Example 2

Translate the word statement into an algebraic equation.

If Betsy Sue's age is doubled and added to 50, the sum is 78.

Solution Let x represent Betsy Sue's age. Then her age doubled is $2x$.

$$2x + 50 = 78$$

Example 3

Translate the word statement into an algebraic equation.

Simone worked h hours and Carey worked 3 hours longer than Simone. The total hours they worked together is 28.

Solution Simone worked h hours
Carey worked $h + 3$ hours

$$h + (h + 3) = 28$$

EXERCISES

Match each word sentence with the correct equation.

1. 3 more than a number divided by 3 is 7.	**a.** $3n = 7$
2. A number increased by 3 equals 7.	**b.** $13 - n = 7$
3. When a number is subtracted from 13 the result is 7.	**c.** $\frac{n}{3} = 7$
4. A number multiplied by 3 equals 7.	**d.** $n - 3 = 7$
5. 13 is 7 less than some number.	**e.** $n + 3 = 7$
6. A number decreased by 3 is equal to 7.	**f.** $7n = 3$
7. A number divided by 3 is equal to 7.	**g.** $\frac{n}{3} + 3 = 7$
8. A number times 7 is equal to 3.	**h.** $13 = n - 7$

Translate each of the following into an algebraic equation. Let x represent the number.

9. A number increased by 4 is 12.

10. 6 times a number is equal to 36.

11. 4 more than some number is 16.

12. One third of a number which is then decreased by 5 is 11.

13. 16 is 4 less than some number.

14. If some number is added to 12, the result is 48.

15. 25 is 5 more than some number.

Select a variable and write an algebraic equation for the following.

16. Six more than the number of seats is 98.

17. 12 less than twice the distance is 25.

18. Five more than 3 times the cost is $150.

19. Yesterday's average temperature minus 3 is 83.

20. \$500 more than triple the investment is \$1700.

21. 5 more than the sum of 6 and a number is 18.

Choose variables and tell what they represent. Then write the equations.

22. Helen is h years old. Lesley is 2 years older than Helen. The sum of their ages is 30.

23. The number of student tickets sold was n. The number of parent tickets sold was three more than twice the number of student tickets. The total number of tickets sold was 213.

24. One book costs d dollars. Another book costs twice as much. It costs \$27 to buy the two books.

Now translate back. Can you write word sentences for these equations? For example: $5n - n = 60$ could be Nell's age subtracted from five times her age is 60.

25. $2n + 5 = 17$

26. $x - 8 = 21$

27. $3s - 1 = 2$

28. $\frac{1}{2}(n - 1) = 12$

SECTION 6–2

PROPERTIES OF EQUALITY

WARM-UP ACTIVITY

Write a two- or three-digit whole number.	561
Find the sum of the squares of its digits.	$5^2 + 6^2 + 1^2 = 62$
Repeat until the sum is equal to 58 or 1. Follow the example on the right.	$6^2 + 2^2 = 40$
	$4^2 + 0^2 = 16$
	$1^2 + 6^2 = 37$
	$3^2 + 7^2 = 58$
Try three other numbers.	

INTRODUCTION

To keep a pan scale balanced, while adding more weight, each pan must gain the same amount of weight. If weight is added to only one pan the scale will become unbalanced.

Equations are like pan balances. The equation is balanced only when both sides have the same value. To **solve** an equation means to find the value of the variable that makes the statement true. To find this value, you perform operations on the equation, always keeping the equation in balance.

SKILLS DEVELOPMENT

The **addition property of equality** states that if you add the same number to both sides of an equation, the equation will still balance.

$$\begin{aligned} &\text{If} & x &= y, \\ &\text{then} & x + z &= y + z. \end{aligned}$$

Check out the addition property for yourself using arithmetic.

$$\begin{aligned} 8 &= 8 \\ 8 + 5 &= 8 + 5 \\ 13 &= 13 \end{aligned}$$

Try some other numbers to convince yourself of the addition property of equality.

Now use algebra.

Suppose	$x = 8.$
Then	$x + 5 = 8 + 5,$
or	$x + 5 = 13.$

Because we know $x = 8$, the addition property of equality says we can add 5 to each side. The sum on the left, $x + 5$, must equal the sum on the right, $8 + 5$ or 13.

You know that subtracting a number is adding the opposite of that number. The addition property of equality holds for adding negative numbers too.

$$\begin{aligned} x &= 8 \\ x + -5 &= 8 + -5 \\ x - 5 &= 8 - 5 \\ x - 5 &= 3 \end{aligned}$$

The **multiplication property of equality** states that you can multiply both sides of an equation by the same number and the products on each side are equal.

$$\begin{aligned} &\text{If} \quad & x &= y, \\ &\text{then} \quad & xz &= yz. \end{aligned}$$

Study the two examples below.

$$\begin{aligned} 8 &= 8 \qquad & x &= 8 \\ 8(5) &= 8(5) \qquad & 5x &= 8(5) \\ 40 &= 40 \qquad & 5x &= 40 \end{aligned}$$

There is no need for a division property of equality because dividing by a number is the same as multiplying by its reciprocal. Study the examples below.

$$\begin{aligned} 8 &= 8 \qquad & x &= 9 \\ 8 \div 2 &= 8\left(\tfrac{1}{2}\right) \qquad & x \div 3 &= 9 \div 3 \\ 4 &= 4 \qquad & x\left(\tfrac{1}{3}\right) &= 9\left(\tfrac{1}{3}\right) \\ & & \tfrac{x}{3} &= 3 \end{aligned}$$

Examples

Using the properties of equality, tell what number was used to build the new equation.

a. $x = 8$
$x + \square = 14$

b. $x = 15$
$x - \square = 11$

c. $x = -3$
$x(\square) = -12$

d. $x = 20$
$x(\square) = 4$

Solution Check these values of $\square$ in the equations.

a. $\square = 6$ **b.** $\square = 4$ **c.** $\square = 4$ **d.** $\square = \frac{1}{5}$

EXERCISES

For each equation, add 7 to build an equivalent equation.

1. $x = 3$ **2.** $y = 5$ **3.** $m = 6$

For each equation, subtract 5 to build an equivalent equation.

4. $b = 6$ **5.** $s = 5$ **6.** $r = 4$

For each equation, multiply by 4 to build an equivalent equation.

7. $n = 8$ **8.** $t = 21$ **9.** $z = 3$

The equations in each pair are equivalent. How was the second equation built from the first?

10. $y = 3, y + 9 = 12$

11. $t = 3, t - 5 = -2$

12. $x = -1, x + 6 = 5$

13. $m = 18, 2m = 36$

14. $r = -5, \frac{1}{5}r = -1$

15. $p = 24, \frac{1}{3}p = 8$

Has this equation been built from $p = 3$?

16. $p + 6 = 9$

17. $6p = 12$

18. $p - 8 = -5$

19. $-1p = 2$

20. $p + 9 = 10$

21. $p - 7 = -4$

22. $p + 9 = 12$

23. $5p = 15$

Let $k = 9$. Is this equation equivalent?

24. $k + 3 = 10$

25. $8 + k = 17$

26. $k - 9 = 0$

27. $6.3 + k = 15.3$

28. $11 + k = 21$

29. $\frac{1k}{3} = 3$

30. $3k = -27$

31. $2k = 18$

32. Write the two properties of equality that you have learned. Give an algebraic example for each one.

SECTION 6–3

UNDO WITH ADDITION

WARM-UP ACTIVITY

You have landed at the airport in the city of your best friend LaDonna, who moved last year. The only directions that you have to get to her house are in a letter she wrote. The letter says "From my house I travel three blocks south and then 11 blocks west to get on Highway 100. Five miles south on Highway 100 is the exit to Airport Road. The airport is 7 miles east from this intersection."

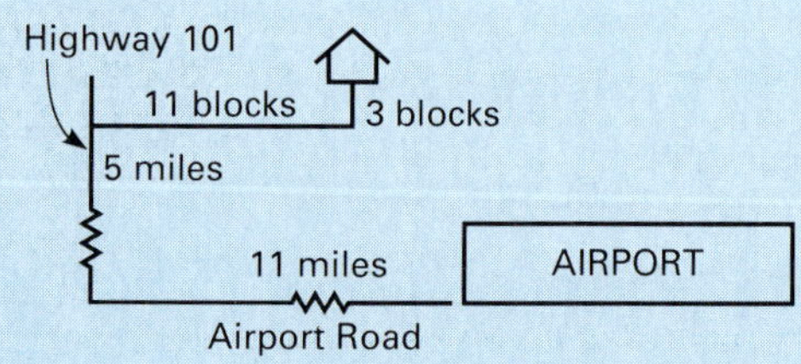

LaDonna has told you how to get from her house to the airport. How can you get to LaDonna's house from the airport?

INTRODUCTION

To get to LaDonna's house, you have to work backwards from the information she sent. To solve the equation you must work backwards. You undo the operations until you find the value of the variable that make the equation true. The number you find is called the **solution.**

To work backwards, use the properties of equality and inverse operations to find the solution.

SKILLS DEVELOPMENT

To solve an equation, ask yourself what operation was performed on the variable. Then ask how you can undo the operation. After you have solved the equation, use substitution to check your solution.

Example 1

Solve $x - 10 = 23$.

Solution

$$x - 10 = 23$$
$$x - 10 + 10 = 23 + 10$$
$$x + 0 = 33$$
$$x = 33$$

Check:

$$33 - 10 \stackrel{?}{=} 23$$
$$23 = 23$$

Example 2

Solve $x + 12 = 6$.

Solution

$$x + 12 = 6$$
$$x + 12 - 12 = 6 - 12$$
$$x + 0 = -6$$
$$x = -6$$

Check:

$$-6 + 12 \stackrel{?}{=} 6$$
$$6 = 6$$

Example 3 is a word problem. You have to write an equation, solve it, and check it. Since this is a word problem, after you solve your equation you must go back and reread the problem. Make sure that your solution is reasonable and that you answered the *question* to the problem asked. For example, if the solution was -12 or $\frac{1}{2}$ and you were asked for the number of chairs, your solution would not be reasonable. Also, if the question is, "Do you have enough money?" you would answer yes or no. You would not give a number solution.

Example 3

The custodian said that seven more than the number of chairs in the lunchroom is equal to 297. How many chairs are in the lunchroom?

Solution Let c = the number of chairs.

$$c + 7 = 297$$

$$c + 7 - 7 = 297 - 7$$

$$c + 0 = 290$$

$$c = 290$$

There are 290 chairs in the lunchroom. (Because this is a word problem, you must give your solution in words. You weren't asked to "solve for c.")

Check:

$$c + 7 = 297$$

$$290 + 7 \stackrel{?}{=} 297$$

$$297 = 297$$

Reread the problem. Is your answer reasonable?

EXERCISES

What is the inverse of each step?

1. add 8

2. subtract 6

3. subtract 1.5

4. add −3

5. subtract −10

6. add −4.6

Tell what number must be added or subtracted to solve the equation. Do not solve.

7. $x + 5 = 9$

8. $x - 0.3 = 0.7$

9. $5 + x = -4$

10. $-5 + x = 4$

11. $2 = m - 10$

12. $0.2 = 0.10 + m$

The final two steps are missing from these problems. Finish them.

13. $y + 6 = 15$
$y + 6 - 6 = 15 - 6$

14. $k - 8 = 23$
$k - 8 + 8 = 23 + 8$

The solutions for each equation are shown. Check each solution. If the solution is not correct, find the correct solution.

15. $y + 8 = 16 \quad y = 8$

16. $k - 3 = 9 \quad k = 12$

17. $m - 2 = 19 \quad m = 17$

18. $k + 8 = 16 \quad k = 24$

Solve. Check your solutions.

19. $y - 6 = 4$

20. $8 + k = 12$

21. $15 + z = 23$

22. $11 = s + 3$

23. $s - 3 = 5$

24. $5 + t = 18$

25. $3 = 6 + r$

26. $x - 12 = -6$

27. $7.3 = m + 2.5$

28. $9.6 + p = 12.5$

Solve.

29. $x - 7 = 9$

30. $x + 11 = 13$

31. $x + 14 = -2$

32. $x - 12 = 11$

Write an equation and solve it. Remember to check your solution in the equation and check your answer by rereading the problem.

33. If Joe would have gotten 3 more problems correct, he would have had a total of 20 correct problems. How many problems did Joe answer correctly?

34. Ten less than the number of cars in the caravan to the game was 25. How many cars were in the caravan?

35. If a student had 2 more letters in his last name, his name would have as many letters as the word "equation." How many letters are in the student's last name? What might the name be?

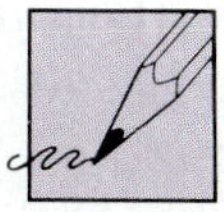

36. Remember an actual incident when you used mathematics to work backward from what you knew to what you wanted to find out. Write a word problem about it. Exchange problems with a classmate and solve each other's word problem.

SECTION 6–4

UNDO WITH MULTIPLICATION

WARM-UP ACTIVITY

I have a dollar's worth of change in my pocket. There are 35 coins in all. There are as many nickels as dimes. What coins are in my pocket?

INTRODUCTION

The Woody Creek Bicycle Club spent Saturday afternoon riding along their usual trail. The trail was 15 miles long. They spent 5 hours riding the entire length. They had been repeating this ride for three months to determine their average speed.

SKILLS DEVELOPMENT

LouAnne remembered the formula for speed that she learned in math class.

$$\text{Distance traveled} = (\text{rate of speed})\ (\text{time traveled})$$
$$d = rt$$

The Woody Creek bikers knew $d = 15$ miles and $t = 5$ hours. They could easily solve $15 \text{ mi} = r(5h)$ just by looking at the equation. They knew their rate was 3 mi/h or 3 mph.

The bikers solved their problem easily because the numbers were small and 5 is a factor of 15. But the numbers are not always nice and easy. In this lesson you will learn to solve equations by undoing multiplication.

Study the equations below to see how they were built.

$$\begin{aligned} x &= 8 \\ 2x &= 2(8) \\ 2x &= 16 \end{aligned} \qquad \begin{aligned} y &= -2 \\ 3(y) &= 3(-2) \\ 3y &= -6 \end{aligned}$$

You can solve the equations by using the multiplication property of equality and the reciprocal.

$$\begin{aligned} 2x &= 16 \\ \left(\tfrac{1}{2}\right)2x &= \left(\tfrac{1}{2}\right)16 \\ x &= 8 \end{aligned} \qquad \begin{aligned} 3y &= -6 \\ \left(\tfrac{1}{3}\right)3y &= \tfrac{1}{3}(-6) \\ y &= -2 \end{aligned}$$

Example 1

Solve and check $6x = 42$.

Solution

$$\begin{aligned} 6x &= 42 \\ \tfrac{1}{6}(6x) &= \tfrac{1}{6}(42) \\ x &= 7 \end{aligned}$$

Check:

$$\begin{aligned} 6x &= 42 \\ 6(7) &\stackrel{?}{=} 42 \\ 42 &= 42 \end{aligned}$$

Example 2

Solve and check $\frac{1}{3}x = 9$.

Solution

$$\frac{1}{3}x = 9$$
$$3\left(\frac{1}{3}\right)x = 3(9)$$
$$(1)x = 27$$
$$x = 27$$

Check:

$$\frac{1}{3}x = 9$$
$$\frac{1}{3}(27) \stackrel{?}{=} 9$$
$$9 = 9$$

Example 3

If the Allisons have traveled 440 miles at an average of 55 mph, how long have they been traveling?

Solution

$$d = rt$$
$$440 = (55)t$$
$$\frac{440}{55} = t$$
$$8 = t$$

They have been traveling for 8 hours. Notice that dividing by 55 is the same as multiplying by the reciprocal, $\frac{1}{55}$.

Check:

$$440 \stackrel{?}{=} (8)(55)$$
$$440 = 440$$

EXERCISES

For each equation decide by what number you need to multiply. Do not solve.

1. $4x = 12$ **2.** $\frac{y}{3} = 4$ **3.** $\frac{1}{3} = 2y$

4. $\frac{m}{8} = 3$ **5.** $6p = 12$ **6.** $6 = \frac{m}{4}$

7. $-3k = 18$ **8.** $\frac{x}{3} = -5$ **9.** $\frac{1}{3}p = \frac{1}{5}$

Solutions are shown for each equation. Check each solution. If the solution is not correct, find the correct solution.

10. $3y = 12 \quad y = 4$

11. $\frac{1}{2}m = 18 \quad m = 9$

12. $5t = 25 \quad t = 125$

13. $\frac{2}{3}x = 12 \quad x = 8$

Solve and check.

14. $\frac{a}{3} = -6$

15. $4a = 24$

16. $\frac{k}{3} = 3.2$

17. $3k = 9 + 12$

18. $8y = -64$

19. $\frac{2}{3}x = 4$

20. $k \div 6 = -18$

21. $4k = 6.4$

22. $\frac{4}{5}y = 8$

23. $\frac{5}{4}m = 20$

24. A car travels at 60 mph. Calculate how long it will take the car to travel 330 miles.

25. A horse trots 46 miles in 5 hours. What is its average speed?

26. How long does it take a train to travel 1020 km at an average speed of 120 km/h?

27. At the 1936 Olympics Jesse Owens ran 100 meters in 10.3 seconds. Calculate his speed in meters per second, to the nearest tenth.

28. George drives his all-terrain vehicle for 37.5 miles at 15 mph. How long did he ride?

29. In $\frac{3}{4}$ of an hour a bus travels 39 miles. At what average speed was the bus traveling.

SECTION 6–5

GRAPHING THE SOLUTION SET

WARM-UP ACTIVITY

Mr. Martin had the following exercise on the overhead projector for his class to do when they came in Monday morning. What was his message to the class?

Draw this number line and fill in the letters given in the instructions. The first one is done for you.

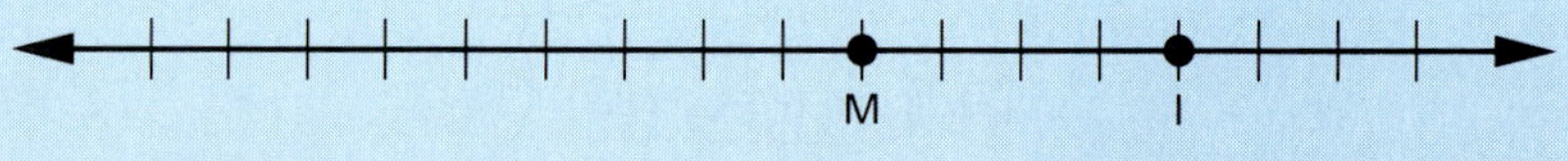

M to I is +4	O_2 to O_3 is +4
I to O_1 is −8	O_3 to G_1 is −6
O_1 to N_1 is +7	G_1 to D is +3
N_1 to N_2 is +2	D to G_2 is +8
N_2 to O_2 is −8	G_2 to R is −4

INTRODUCTION

In this section you will solve equations and graph the solutions on a number line.

SKILLS DEVELOPMENT

To graph a single value on a number line, place a solid dot at the location that corresponds to the value. Label at least two numbers on each side of the value you locate.

Example 1

Solve and graph $x + 11 = 15$.

Solution

$$x + 11 = 15$$
$$x + 11 - 11 = 15 - 11$$
$$x = 4$$

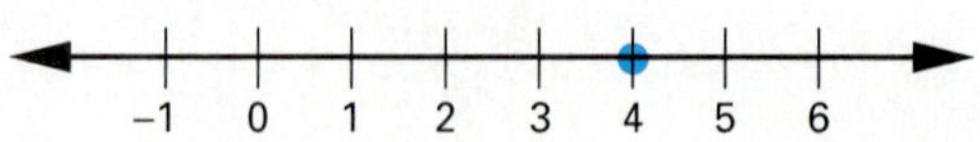

Example 2

Solve and graph $3\mathbf{n} = 15$.

Solution

$$3n = 15$$
$$\tfrac{1}{3}(3n) = \tfrac{1}{3}(15)$$
$$n = 5$$

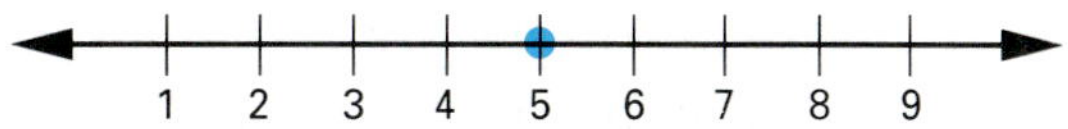

Example 3

Solve and graph $\frac{y}{2} = \frac{1}{3}$.

Solution

$$\frac{y}{2} = \frac{1}{3}$$
$$2\left(\frac{y}{2}\right) = 2\left(\frac{1}{3}\right)$$
$$y = \frac{2}{3}$$

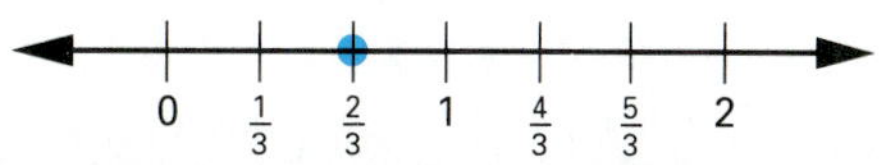

Example 4

Solve and graph $\frac{4}{5}x = 3$.

Solution

$$\frac{4}{5}x = 3$$
$$\left(\frac{5}{4}\right)\frac{4}{5}x = 3\left(\frac{5}{4}\right)$$
$$x = \frac{15}{4} \text{ or } 3\frac{3}{4}$$

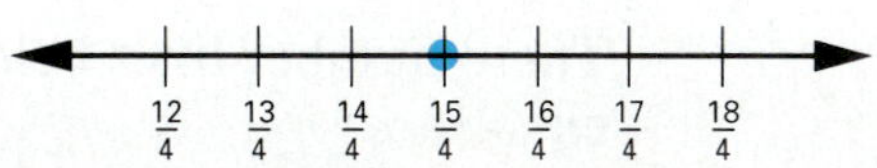

EXERCISES

1. Solve the equations and show all the solutions on one number line. Label the points with the letters in front of the equations.

L $\frac{4}{5}x = \frac{4}{5}$ S $\frac{1}{2}x = -1$

I $4x = 0$ E $x + 7 = 9$

M $x - 2 = -3$

Match the equation to its graphed solution.

2. $1 + x = 6$ **a.** 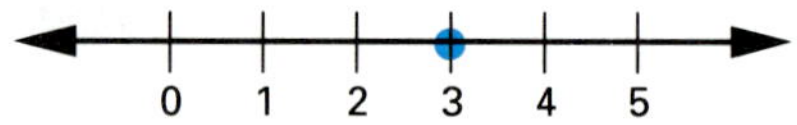

3. $x + -2 = 5$ **b.** 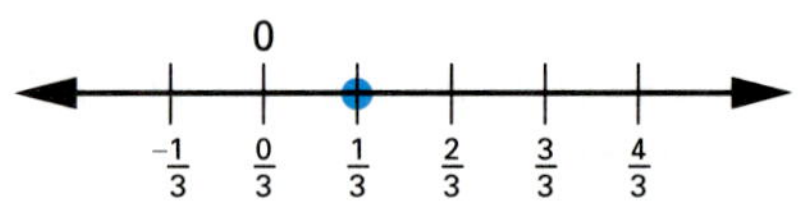

4. $x - 3 = 0$ **c.** 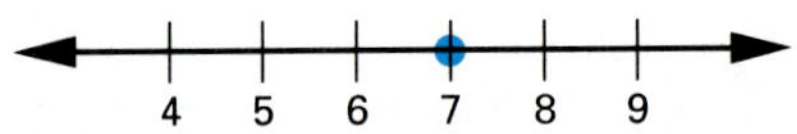

5. $\frac{1}{3}x = -1$ **d.** 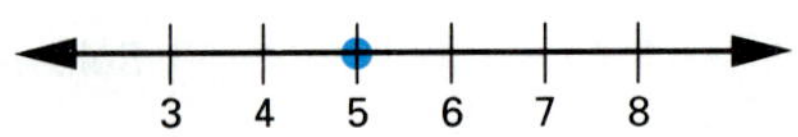

6. $6x = 2$ **e.** 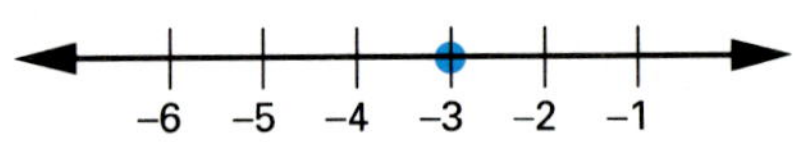

These number lines below show the graphs of the solutions. Find the equations.

7. 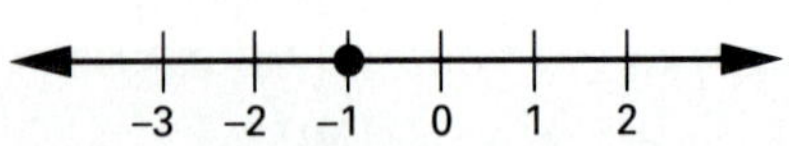

A $x + 5 = 3$ B $x - 4 = -3$ C $2x = -2$

8. 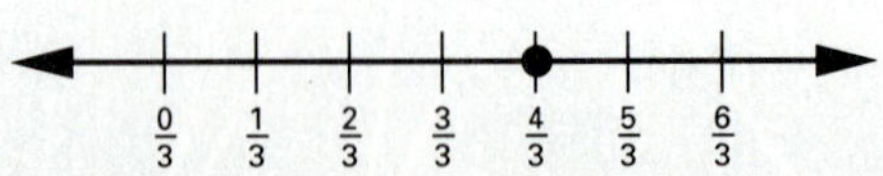

A $2x = 3$ B $\frac{x}{3} = \frac{4}{9}$ C $x = \frac{8}{3}$

Solve and graph.

9. $n + 5 = 8$

10. $3 + t = 14$

11. $7 + x = 7$

12. $\frac{2}{5} x = 1$

13. $\frac{1}{4} x = 2$

14. $x - 1 = -5$

15. $4x = -2$

16. $3x = -6$

Tell which letter gives the approximate location of the solution for each equation. Use a calculator if you wish.

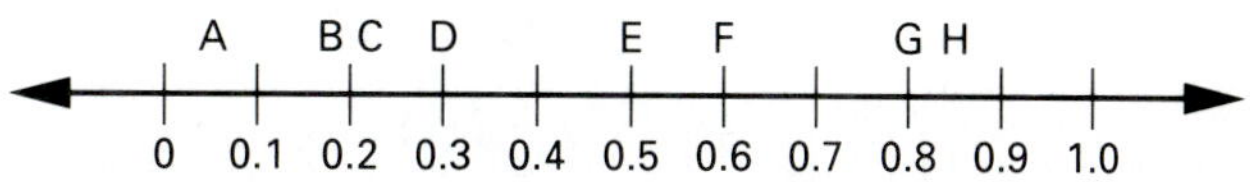

17. $x + 0.25 = 0.75$

18. $x - 0.01 = 0.3$

19. $x - 1.2 = -0.4$

20. $x + 0.15 = 1$

21. $x - 0.1 = -0.05$

22. $x + 0.2 = 0.38$

23. $x + 0.03 = 0.25$

24. $x - 0.11 = 0.5$

25. According to the log of the *Mary Ellen Carter* (the year before she went down in a legendary storm), her bearings were 40 degrees north latitude and 76 degrees west longitude. A later entry gave her bearings as 28 degrees north latitude and 60 west longitude. How many degrees (state north or south) latitude had the *Mary Ellen Carter* sailed? How many degrees (state east or west) longitude?

SECTION 6–6

SOLVING TWO-STEP EQUATIONS

WARM-UP ACTIVITY

Evaluate each expression. Write the value in the corresponding box.

If your solutions are correct, the sum of the numbers in one arm of the X should total the sum of the numbers in the other arm.

1. $4y - 3y$, if $y = 2$

2. $6 + 3a + 1$, if $a = 3$

3. $(9)(4y) + 2$, if $y = 1$

4. $\frac{3b \div 1}{2}$ if $b = 8$

5. $3 + \frac{3}{4}c$, if $c = 4$

6. $\frac{4m}{m}$, if $m = 6$

7. $2 \times d - 2$, if $d = 8$

8. $\frac{10 \times 10 + 20}{4y}$, if $y = 3$

9. $15 \div 3 + (12 - f)$, if $f = 9$

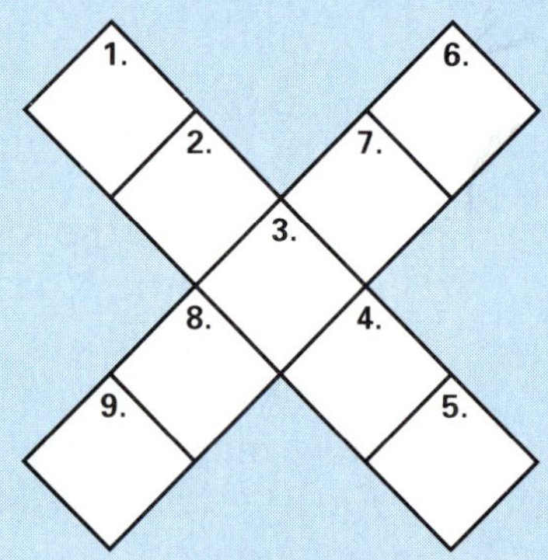

INTRODUCTION

Did you ever play "Mother May I"? Suppose "Mother" told you to stand up, take 5 steps forward, 3 steps to the right, and 4 steps backward. What steps would you need to take to get back to your starting place?

SKILLS DEVELOPMENT

Sometimes several operations have been performed to build an equation.

$$\begin{aligned} x &= 8 \\ (2)x &= 8(2) \\ 2x &= 16 \\ 2x + 1 &= 16 + 1 \\ 2x + 1 &= 17 \end{aligned} \qquad \begin{aligned} x &= 3 \\ \tfrac{1}{3}(x) &= \tfrac{1}{3}(3) \\ \tfrac{1}{3}x &= 1 \\ \tfrac{1}{3}x - 4 &= 1 - 4 \\ \tfrac{1}{3}x - 4 &= -3 \end{aligned}$$

To solve these equations, perform the inverse of the steps used to build them. You must think of the steps in reverse order too.

Example 1

Solve $2x + 1 = 17$.

Solution $2x + 1 = 17$ → has been added, so subtract 1.

$$2x + 1 - 1 = 17 - 1$$
$$2x = 16$$ → 2 has been multiplied, so multiply by $\frac{1}{2}$.
$$\tfrac{1}{2}(2x) = 16\left(\tfrac{1}{2}\right)$$
$$x = 8$$

Example 2

Solve $\frac{1}{3}x - 4 = -3$.

Solution $\frac{1}{3}x - 4 = -3$ → has been subtracted, so add 4.

$$\tfrac{1}{3}x - 4 + 4 = -3 + 4$$
$$\tfrac{1}{3}x = 1$$ → $\frac{1}{3}$ has been multiplied, so multiply by 3.
$$x = 3$$

Don't forget to check your work.

EXERCISES

Tell what operation is performed in each step.

1. $3x + 4 = 16$
$3x + 4 - 4 = 16 - 4$
$3x = 12$

2. $2a - 1 = 5$
$2a - 1 + 1 = 5 + 1$
$2a = 6$

3. Starting with $x = 2$, Jean built the equation $5x + 1 = 11$. What steps did she use?

Identify the first step in solving each equation.

4. $2f + 3 = 15$

5. $5e - 8 = 22$

6. $6 + d = 12$

7. $12 - 3g = 6$

Match each equation in the first column with one from the second column that is obtained from the first by using an inverse operation.

8. $3x + 1 = 8$

9. $3x - 4 = 7$

10. $3x - 2 = 7$

11. $3x + 5 = 10$

a. $3x = 11$

b. $3x = 5$

c. $3x = 9$

d. $3x = 7$

The equations in each of the following pairs are related. Tell how the second equation is obtained from the first.

12. $4x - 2 = 18 \quad 4x = 20$

13. $19 = 8y + 3 \quad 16 = 8y$

14. $\frac{1}{2}x + 8 = 16 \quad \frac{1}{2}x = 8$

15. $-18 = 3x - 12 \quad -6 = 3x$

Solve each equation. Remember to check.

16. $24 = 3x - 3$

17. $\frac{4}{5}x - 3 = 5$

18. $\frac{2x}{3} + \frac{1}{3} = 2$

19. $2y - 13 = 25$

20. $5 + 2a = 25$

21. $\frac{y}{3} + 2 = 8$

22. $3a - 2 = -17$

23. $x + 4.5 = 7.6$

Solve each equation. Graph the solution.

24. $7 = 1 + \frac{2}{3}x$

25. $0.3x = 1.2$

26. $4 - 3m = 10$

27. $0.2x + 4.5 = 4.9$

The perimeter of a rectangle is given by the formula $P = 2l + 2w$, where l and w are the dimensions of the length and the width.

28. Find P if $l = 6$, $w = 3$.

29. Find l if $w = 8$, $P = 24$.

30. Find w if $l = 16$, $P = 48$.

31. Find l if $w = 7$, $P = 84$.

32. The cost C in dollars, to produce one kind of poster is $C = 15 + 8n$, where n is the number of posters.

a. Find C if $n = 36$.

b. Find n if $C = 79$.

33. A number n, when doubled and increased by 3, is equal to 27. What is the number?

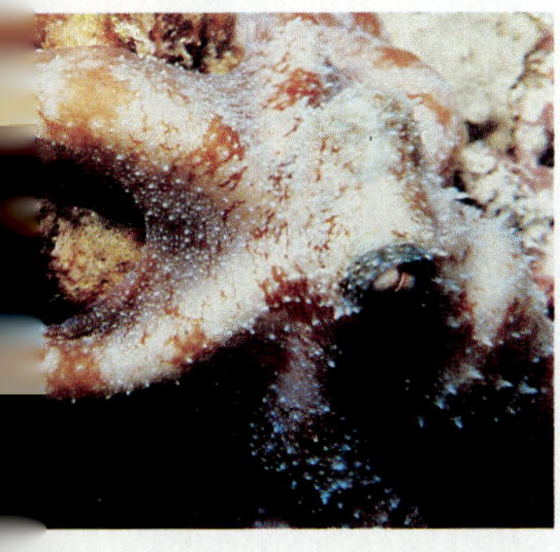

34. A number p, when doubled and subtracted from 96, is equal to 60. What is the number?

35. The octopus has eight arms. To find the number of hearts it has, multiply the solution of this equation by -1. $6 = 2h + 12$

36. How many sides does a snowflake have? Solve the equation to find out. $4s + 4 = 28$

37. Find the length of the first-move bonus word in the game of Scrabble by solving this equation. $2x + 4 = 8$

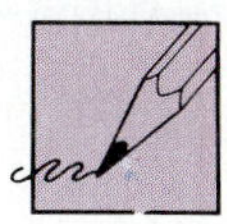

38. Many everyday activities must be performed in reverse order when we want to undo them. For example, when you get dressed, you put on your socks, then your shoes. To undo this, first you remove your shoe, then your sock. Give another example and explain.

SECTION 6–7

VARIABLES ON BOTH SIDES

WARM-UP ACTIVITY

Follow the instructions.

Choose a number.

Add 15 to it.

Double your answer.

Add your original number.

Divide by 3.

Subtract the number of pennies in a dime. What result did you get?

Can you explain using algebra why your answer was the same number you started with?

Build another number trick using algebra. Exchange with another student and see if the tricks work.

INTRODUCTION

After Margaret closes the concession stand at the ballpark, she counts the money received during the game. She has found that it works better to group the coins and currency. That is, she counts the number of quarters, the number of dimes, the number of \$1 bills, etc., and then multiplies to find the total amount.

SKILLS DEVELOPMENT

You can use the same approach as Margaret when solving equations. The best approach is to group the variables first.

Example 1

Solve $x + 3x - 1 = 11$.

Solution

$$\begin{aligned} x + 3x - 1 &= 11 \\ 4x - 1 &= 11 && \rightarrow \text{Combine the } x\text{-terms.} \\ 4x - 1 + 1 &= 11 + 1 && \rightarrow \text{Perform the inverse.} \\ 4x &= 12 && \rightarrow \text{Simplify.} \\ \tfrac{1}{4}(4x) &= 12\left(\tfrac{1}{4}\right) && \rightarrow \text{Perform the inverse.} \\ x &= 3 && \rightarrow \text{Simplify.} \end{aligned}$$

Sometimes the variable appears on both sides of the equation. Use the properties of equality and inverse operations to group the variables onto one side.

Example 2

Solve $4x + 3 = x - 6$.

Solution

$$\begin{aligned} 4x + 3 &= x - 6 \\ 4x + 3 - x &= x - 6 - x && \rightarrow \text{Think: the inverse of } x \text{ is } -x. \\ 4x - x + 3 &= x - x - 6 && \rightarrow \text{Use the commutative property.} \\ 3x + 3 &= -6 && \rightarrow \text{Simplify.} \\ 3x + 3 - 3 &= -6 - 3 && \rightarrow \text{Perform the inverse.} \\ 3x &= -9 && \rightarrow \text{Simplify.} \\ \tfrac{1}{3}(3x) &= -9\left(\tfrac{1}{3}\right) && \rightarrow \text{Perform the inverse.} \\ x &= -3 && \rightarrow \text{Simplify.} \end{aligned}$$

EXERCISES

Match each expression with its simplified form.

1. $8x + -3x$	**a.** $6x$
2. $3x - x$	**b.** $2x$
3. $9x + -3x$	**c.** $7x$
4. $6x - 2x$	**d.** $5x$
5. $8x + 3x - 4x$	**e.** $10x$
6. $9x - 2x + 3x$	**f.** $4x$

Tell what operation you would perform to group the variables in each equation.

7. $9x = 5x + 45$

8. $11k + 1 = -3k - 13$

Solve and check.

9. $6 - 2m = m$

10. $20 + 7y = 12y$

11. $5k + 8 = 3k + 26$

12. $4k + 11 = 6k + 17$

13. $6y + 3 = 3y - 3$

14. $8y = 16.5 + 3y$

To answer the following questions, solve the equation given.

15. Today more than 100 nations compete in the Olympics. How many nations competed in the first Olympics in 1896? Solve $6n - 3 = 5n + 6$.

16. You may have seen a performance of the Ringling Brothers Circus. How many actual Ringling brothers were there? Solve $16 - 3m = 11 - 2m$.

17. How many Great Lakes are there? Solve $3x - 8 = 2x - 3$.

18. In the famous book by Daniel Defoe, Robinson Crusoe was shipwrecked on an island. How many years did he spend there? Solve $8 - 2y = -20 - y$.

19. The pentathlon is a group of events in the Olympics. How many events are in the pentathlon? Solve $5y + 3 = 3y + 13$.

20. To find the percentage of the earth's water that is drinkable, Peter solved the equation below and multiplied the solution by -1. $4w - 7 = 6w - 5$

21. Only part of an iceberg shows above the water. Solve the equation to find the fraction. $3c - 6 = -5 - 6c$

Solve and graph on a number line.

22. $8x - 3x = 15$

23. $7y = 2y + 10$

24. $18 + 3k = k + 30$

25. $30 - 3n = 3n$

26. The steps in the Warm-Up Activity are written below in algebraic form. Can you design a set of steps that starts with any number and always ends with 7? Design your steps using algebra, then rewrite them in word sentences like the steps in the Warm-Up Activity. Then give your steps to a classmate to try.

n
$n + 15$
$2(n + 15)$
$2(n + 15) + n$
$[2(n + 15) + n] \div 3$
$[2(n + 15) + n] \div 3 - 10$

SECTION 6–8

UNDOING PARENTHESES

WARM-UP ACTIVITY

Solve the equations.

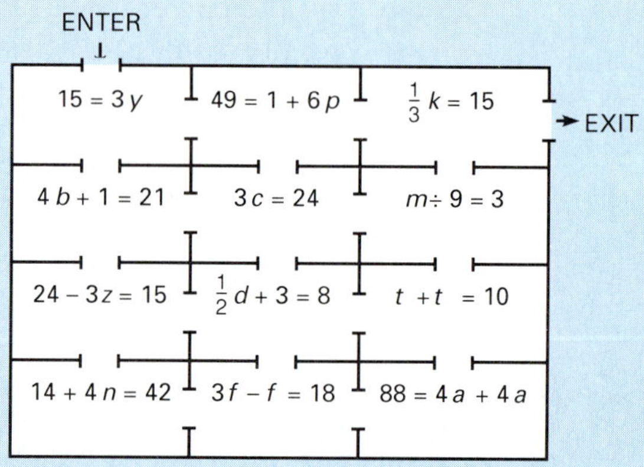

Find a path through the maze so that the solution for the equation in each square on the path is an odd number.

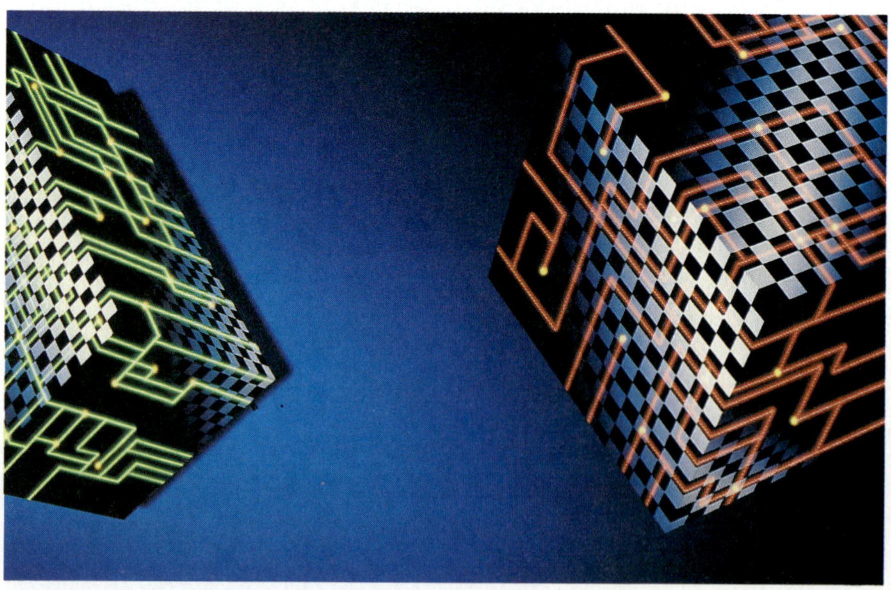

INTRODUCTION

The equation that relates the temperature in Celsius degrees to the temperature in Fahrenheit degrees is

$$C = \tfrac{5}{9}(F - 32)$$

You can use this equation when you know the Fahrenheit degrees and want to find the Celsius degrees, but what if you know the Celsius degrees and want to find the Fahrenheit degrees? You would have to solve an equation that has parentheses. In this section you will learn to solve equations that have parentheses.

SKILLS DEVELOPMENT

To solve an equation that has parentheses, use the distributive property.

Example 1

Solve $4x + 3(20 - x) = 80$.

Solution

$$\begin{aligned}
4x + 3(20 - x) &= 80 \\
4x + 60 - 3x &= 80 && \rightarrow \text{Use the distributive property.} \\
x + 60 &= 80 && \rightarrow \text{Simplify.} \\
x + 60 - 60 &= 80 - 60 && \rightarrow \text{Perform the inverse.} \\
x &= 20 && \rightarrow \text{Simplify.}
\end{aligned}$$

Check:

$$\begin{aligned}
4(20) + 3(20 - 20) &\stackrel{?}{=} 80 && \rightarrow \text{Substitute into both } x\text{-terms.} \\
80 + 3(0) &\stackrel{?}{=} 80 \\
80 &= 80 && \rightarrow \text{The sides are equal, so } x = 20 \text{ is correct.}
\end{aligned}$$

As you solve more equations, you will be able to perform some steps mentally.

Example 2

Solve $y + 6(y - 3) = 5(y + 2)$.

Solution

$$\begin{aligned}
y + 6(y - 3) &= 5(y + 2) \\
y + 6y - 18 &= 5y + 10 && \rightarrow \text{Use the distributive property on both sides.} \\
7y - 18 &= 5y + 10 \\
2y &= 28 \\
y &= 14
\end{aligned}$$

Check:

$$\begin{aligned}
14 + 6(14 - 3) &\stackrel{?}{=} 5(14 + 2) \\
14 + 6(11) &\stackrel{?}{=} 5(16) \\
14 + 66 &\stackrel{?}{=} 80 \\
80 &= 80
\end{aligned}$$

When the equation contains fractions, it is easier to solve if you clear fractions by multiplying both sides by the least common denominator.

Example 3

Solve $\frac{3}{4}t - 2 = \frac{1}{2}(t + 2)$.

Solution $4\left(\frac{3}{4}t - 2\right) = 4[\frac{1}{2}(t + 2)]$

Think: multiply both sides of the equation by 4, the LCD.

$$3t - 8 = 2(t + 2)$$
$$3t - 8 = 2t + 4$$
$$3t - 2t = 4 + 8$$
$$t = 12$$

Think: apply the distributive property.

Check: $\frac{3}{4}(12) - 2 \stackrel{?}{=} \frac{1}{2}(12 + 2)$
$$9 - 2 \stackrel{?}{=} \frac{1}{2}(14)$$
$$7 = 7$$

EXERCISES

Simplify each of the following.

1. $2(y - 3)$ **2.** $4(x - 5) - 7$

3. $-2(x - 6)$ **4.** $-2(x - 5) - 6$

5. $3(3 - 2y)$ **6.** $5x - 2(x + 5)$

7. $\frac{1}{2}x - \frac{1}{3}x$ **8.** $\frac{4y}{5} - \frac{2y}{3}$

Find the lowest common denominator and multiply both sides of the equation by it to eliminate the fractions. Do not solve the equations.

9. $3y + \frac{3}{4} = \frac{9}{4}$ **10.** $\frac{2}{3}x + 7 = 4$

11. $\frac{1}{2}x - \frac{1}{3}x = 2$ **12.** $\frac{y}{3} + \frac{y}{4} = -1$

For each equation, give the first step to solve it.

13. $5(x + 2) = 15$ **14.** $3(2n - 1) = 9$

15. $10 = 2(y + 3)$ **16.** $2(2m + 3) = -10$

Solve and check.

17. $3(2y - 4) = 9$

18. $3(m - 8) = 0$

19. $3(x + 2) = 2x$

20. $2y = 3(y - 1)$

21. $8(n - 1) = 4(n + 4)$

22. $2(x + 3) = 3(x - 3)$

23. $4(3 - m) = 5(2m + 1)$

24. $12(2m - 3) = 2(m + 4)$

25. $2y + \frac{1}{2} = \frac{2}{3}$

26. $\frac{7}{6}x - 2 = \frac{1}{3}$

27. $\frac{3}{5}n - 4 = 14$

28. $3 - \frac{m}{2} = 5 - \frac{m}{3}$

29. $\frac{3m}{2} = 15$

30. $\frac{1}{2}z - 3 = z + 3$

31. $\frac{1}{3}(2x - 1) = 1$

32. $\frac{1}{6}(6 + 2y) = 2$

33. $\frac{1}{2}(x - 5) = \frac{x}{4}$

34. $x - 1 = \frac{x}{2}$

35. $\frac{1}{2}(y + 2) = \frac{y}{3}$

36. $\frac{3}{5}(2x + 15) = 3$

Use the formula in this section to find the Fahrenheit temperature.

37. The Celsius temperature is 35°.

38. The Celsius temperature is 50°.

Math Tip

If you solve the equation $6x + 2(2x - 1) = 10x - 2$, you get:

$$10x = 10x.$$

This equation is true for every value of x. Whenever an equation is true for every value of the variable, it is called an **identity.**

If you solve $6x - 2(2x + 5) = 2x + 2$, you get:

$$2x - 10 = 2x + 2$$
$$-10 = 2.$$

This statement is never true. No value of x satisfies this equation. We say the solution set is the **empty set,** shown by $\{\ \}$.

SECTION 6–9

GRAPHING INEQUALITIES

WARM-UP ACTIVITY

The following diagrams represent the first four rectangular numbers. Why are they called rectangular?

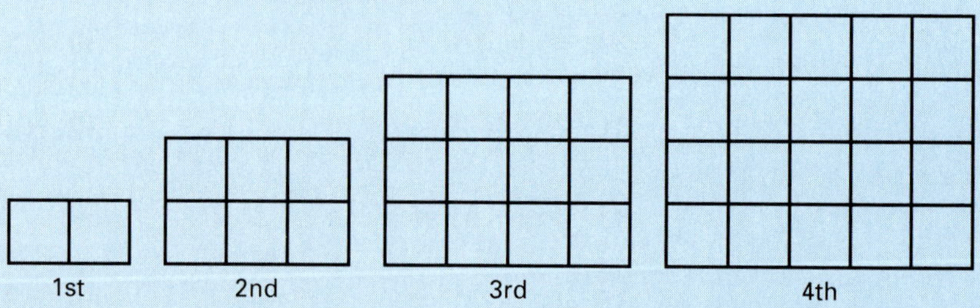

What are the fifth and sixth rectangular numbers?

Find a general rule to help you tell the value of a particular rectangular number.

INTRODUCTION

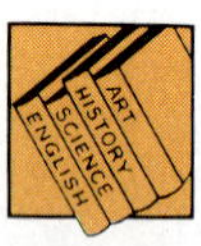

During droughts (serious dry spells), scientists may try to cause rain by spraying (seeding) clouds from an airplane with tiny crystals of dry ice. For clouds to form, the temperature of the clouds must be greater than $-15°C$ and less than $-10°C$.

SKILLS DEVELOPMENT

The integers that represent these temperatures are -14, -13, -12, -11, as shown on the number line below.

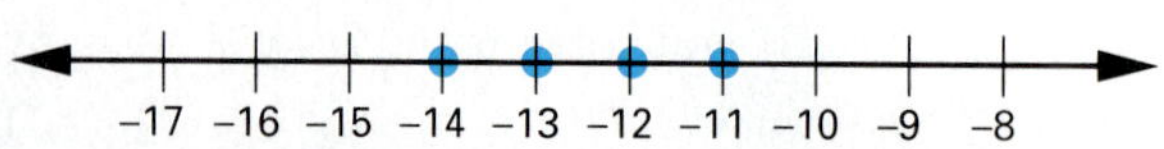

Since a temperature measurement is not usually a whole number of degrees, a number line for temperature must show all real values, numbers greater than -15, and numbers less than -10 too. This number line is shown below. The shaded part shows the correct temperatures for seeding.

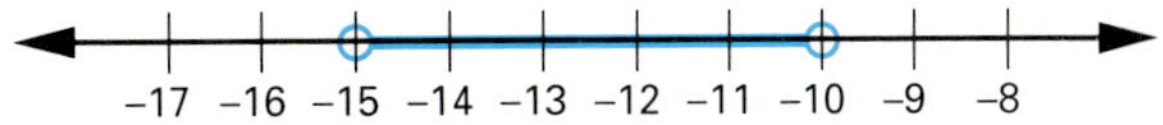

Compare the two number lines. How are they alike? How are they different?

Notice that when the solution is chosen from the set of real numbers, a solid line is used on the graph. The open ends on the shading tell us that the values at the ends are not part of the solution. That is, -15 is not in the solution set, but -14.999 is.

What if the solution set includes all numbers less than zero? It is not possible to graph all the numbers in the solution set. Place a solid arrow at the left end of the number line. This shows the shading continues in that direction indefinitely and all the numbers in that direction are included in the solution set.

Example 1

Graph $x < 3$.

Solution

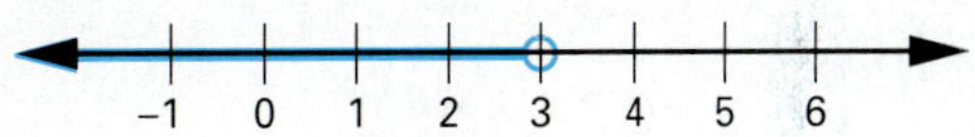

Since 3 is not in the solution set, an open circle is placed at 3. The numbers to the left of 3 are less than 3; therefore, shade to the left. The arrow shows that the solution set continues to the left.

The symbol $\leq$ means "is less than *or* equal to." The symbol $\geq$ means "is greater than *or* equal to."

To graph the solution set of an **inequality** with $\geq$ or $\leq$, the circle is solid to show that the number is included in the solution.

Example 2

Graph $x \geq 5$.

Solution

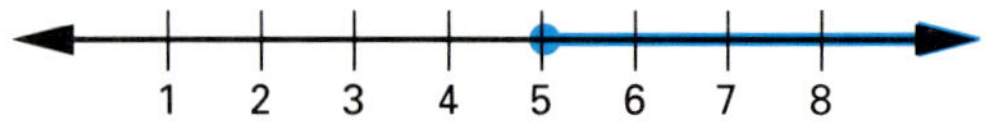

Since 5 is in the solution set, the circle at 5 is solid. All the numbers to the right of 5 are in the solution set so the shading continues to the right.

EXERCISES

Match the inequality with its graph.

1. $x < 3$

2. $x \leq 3$

3. $x > 3$

4. $x \geq 3$

a. 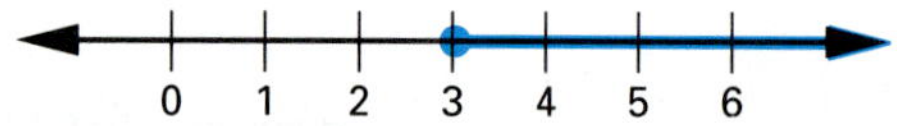

b. 

c. 

d. 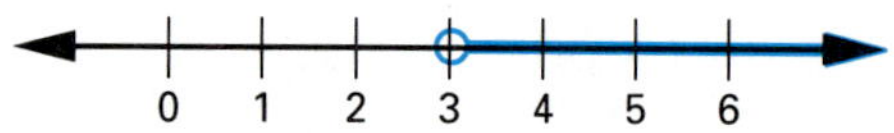

5. When does the graph of an inequality on a number line have an open circle?

6. What does a solid circle in a graph on the number line mean?

Give the inequality that each number line represents.

7. 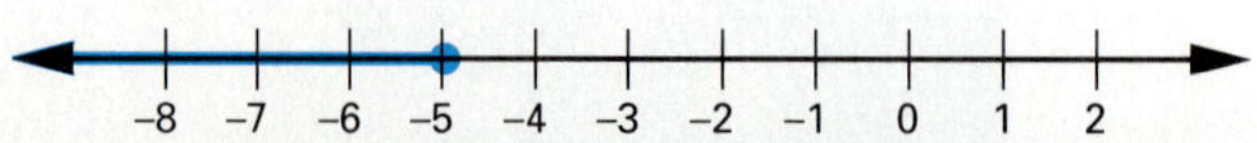

8. 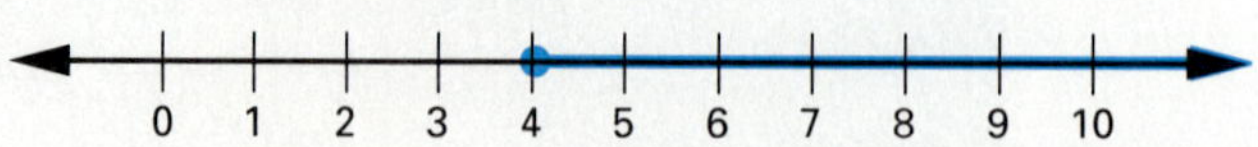

9. 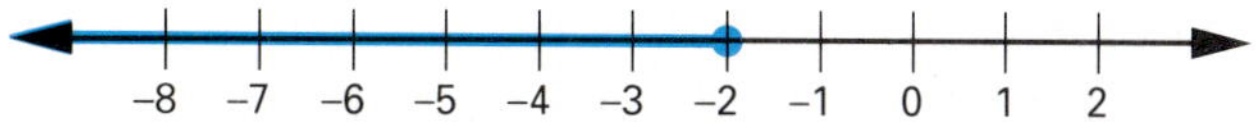

10. 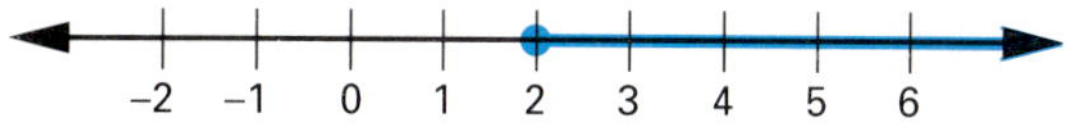

11. 

Graph the inequality on a number line.

12. $y > -1$ **13.** $q < 0$ **14.** $x \geq -2$

15. $b < -4$ **16.** $e \leq 4$ **17.** $t < 2$

18. $m \geq -5$ **19.** $n > 0$ **20.** $b < -2$

21. All of the people at the concert were over age 14. Draw a number line to represent the ages of the people at the concert.

22. On January 1, the bank balance of each student in the ninth grade at Greenback High School was $210 or less. Show their bank balance on a number line.

23. For cross-country skis, blue ski wax should be used when the temperature is between 0°C and −10°C. Use a number line to represent the temperature range when blue wax should be used.

24. Engine oil manufacturers recommend that 10W20 oil be used when the temperature is between −10°C and −30°C. Use a number line to represent this temperature range.

25. To store meat for an extended period of time in a freezer, the temperature should be kept in the range −5°C to −10°C. Show these temperatures on a number line.

26. Give four word sentences that can be written as algebraic sentences to show that you understand the symbols ≤, <, ≥, and >. Give the algebraic statements too.

SECTION 6–10

SOLVING INEQUALITIES

WARM-UP ACTIVITY

Find a path through the squares so that each integer on the path is more than the next integer. Each move along the path is made horizontally or vertically, not diagonally.

Start. (above second column, top row)

+5	+4	+15	−8	+3
−3	0	+3	−25	−60
−13	+1	−50	−58	−75
−20	−35	−40	−25	−85
−8	+6	−38	−100	−90

Finish. (below fourth column, bottom row)

INTRODUCTION

The following numbers are in order from smallest to largest.

$$-4 \quad -1 \quad 3 \quad 6 \quad 10$$

Notice what happens in each case.

Add 1 to each.	Add −1 to each.	Multiply each by −1.
−3 0 4 7 11	−5 −2 2 5 9	4 1 −3 −6 −10

When adding, the order stays the same. When multiplying by a negative, the order is reversed.

What do you think happens if you multiply by a positive number? Divide by a negative number? Check these out yourself.

SKILLS DEVELOPMENT

The Horse and Rodeo Club budgeted \$450 for food for their parent appreciation banquet. The meals cost \$8.25 each. How many persons can attend without going over the budget?

Since they do not have to spend all of the \$450, this problem would not be set up as an equation but as an inequality.

Let x = number of persons

$$8.25x \le 450$$

To solve inequalities you will need to use properties, just as you used the properties of equality to solve equations.

Because the order is reversed when multiplying the numbers by a negative number, you must reverse the inequality sign when you multiply or divide both sides of an inequality by a negative number. For example,

$$-3 < 5 \text{ but } -3(-2) > 5(-2) \text{ because } 6 > -10.$$

Except for multiplying or dividing by a negative number, you use the properties to solve inequalities the same way that you have solved equations.

Example 1

Solve $x - 5 > 1$.

Solution

$$x - 5 > 1$$
$$x - 5 + 5 > 1 + 5$$
$$x > 6$$

Example 2

Solve $-4x < 10$.

Solution

$$-4x < 10$$
$$-\tfrac{1}{4}(-4x) < -\tfrac{1}{4}(10)$$
$$x > -\tfrac{10}{4}$$
$$x > -\tfrac{5}{2}$$

Checking inequalities has two steps. First, substitute the number from the solution. It should make the two sides equal. Second, substitute another number from the solution set to check the direction of the inequality sign.

Check for Example 1:

Step 1 Does $6 - 5 = 1$? Yes, so 6 is correct.

Step 2 Select another number from the solution set of $x > 6$. We select 10.

Is $10 - 5 > 1$? Yes, so $>$ is the correct direction. Both steps are correct, so $x > -\frac{5}{2}$ is correct.

Example 3

Set up an inequality and solve the problem about the Horse and Rodeo Club mentioned previously.

Solution Let x be the number of persons attending the banquet.

$$8.25x \leq 450$$
$$x \leq \frac{450}{8.25}$$
$$x \leq 54.54$$

Since x must be an integer, we answer the question by saying that if at most 54 persons attend the banquet, the club will not exceed its food budget.

Example 4

Solve and graph $-2x + 3 \leq 9$.

Solution $-2x + 3 \leq 9$

$-2x + 3 - 3 \leq 9 - 3$ Subtract 3 from both sides.

$-2x \leq 6$

$(-\frac{1}{2})(-2x) \geq 6(-\frac{1}{2})$ Multiply by $-\frac{1}{2}$. The inequality *reverses*.

$x \geq -3$

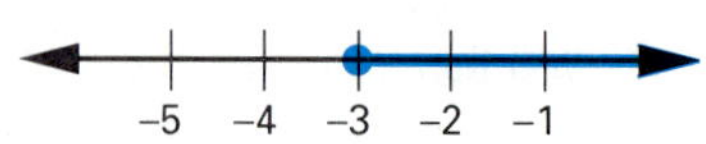

Check: Does $-2(-3) + 3 = 9$?
$6 + 3 = 9$ Yes, so -3 is correct.

Select another number from the solution set. We select 1.

Is $-2(1) + 3 \leq 9$?
$-2 + 3 \leq 9$ Yes, so $\geq$ is correct.

EXERCISES

Give the first step in solving and tell if the inequality sign will reverse.

1. $d + 5 < 11$ **2.** $-2n > 5$ **3.** $\frac{1}{2}n > -5$

4. $15 + m < 10$ **5.** $-14 > 3p - 8$ **6.** $20 \leq 7t + 6$

Match each inequality with the correct solution.

7. $\frac{1}{3}x + 5 < 7$ **a.** $x < -4$

8. $-2x - 3 < 5$ **b.** $x < 6$

9. $4x < -16$ **c.** $x > -4$

10. $3(x + 4) - 5(x - 1) < 5$ **d.** $x > 6$

Solve each inequality.

11. $n + 3 < 12$ **12.** $-m + 1 > 5$

13. $t - 3 \leq 8$ **14.** $t + -2t \leq 18$

15. $3m - 1 < 11$ **16.** $-6a - 3 < 16 - 2a$

Solve and graph each inequality.

17. $18 + 3x < x + 30$ **18.** $5m - 9 > 9 + 4m + 2m$

19. $-1 - 4y \leq 6 + 3(5 - 2y)$ **20.** $2(x - 3) + 3 > 15 - x$

Write an algebraic statement and solve.

21. k is doubled and is less than 6.

22. m is increased by 4 and is less than or equal to 8.

23. 5 more than the product of x and -3 is greater than 11.

24. The range of a Frisbee is never more than 10 seconds times the speed of release. If a Frisbee is thrown horizontally at 6 meters/second, what is its maximum range?

CHAPTER 6 SUMMARY

This chapter has been preparing you to become an expert problem solver. You first learned to translate words into numbers and symbols. Then you learned to solve equations by using properties of equality. You also learned how to locate solutions of equations on number lines. You also learned to solve inequalities and to graph those solutions.

Mastering the ability to solve equations and inequalities will allow you to expand your knowledge in algebra. You will use that knowledge in future chapters to solve many problems.

VOCABULARY

addition property of equality
commutative property
distributive property
empty
equation
identity
inequality
inverse
multiplication property of equality
parentheses
solution
solve

CHAPTER REVIEW

Translate each into an algebraic equation.

1. A number multiplied by 3 is 12.

2. When a number is decreased by 2 the result is 4.

3. A number divided by 5 and decreased by 3 is equal to 11.

Tell what the variable represents and write the equation for this problem.

4. The average of one student is 5 points higher than that of another. The sum of their average is 177.

Which property has been used? Use A for Addition Property and M for Multiplication Property.

5. $n = 16$, $\frac{1}{2}n = 8$

6. $k = 12$, $k + 5 = 17$

7. $x = -3$, $x - 7 = -10$

8. $y = 4$, $-5y = -20$

Solve using the properties of equality.

9. $k - 3 = 10$

10. $n + 8 = 13$

11. $n - 17 = 2$

12. $13 + t = 25$

13. $x + 7 = 25$

14. $t - 3 = 19$

Write an equation and solve it.

15. The number of days until the next vacation added to 15 is 28. How many days are there until the next vacation?

Solve using the multiplication property of equality.

16. $\frac{n}{2} = -8$ **17.** $\frac{2x}{3} = 14$

18. $7z = 21$ **19.** $-2.5y = 20$

20. $\frac{c}{1.2} = -4$ **21.** $-4t = -48$

Solve using the formula $d = rt$.

22. The football field is 2 miles from the school. How fast must the team run to reach the field in $\frac{3}{4}$ hour?

Solve and graph on a number line.

23. $x + 5 = 12$ **24.** $y - 7 = -4$

25. $3n = 12$ **26.** $\frac{k}{2} = -3$

27. $2x + 5 = 27$ **28.** $6 - 3t = 21$

29. $3x - 3.4 = 24.5$ **30.** $\frac{3}{2}x - 6 = 1$

31. Do you know how many horses are on a polo team? Solve the problem to find the answer. Solve $6x + 8 = 56$.

Solve.

32. $4c + 3c - 5c = 16$ **33.** $8n - 3 + 4n = 21$

34. $4x + 11 = 6x - 5$ **35.** $4x - 20 = 5x + 5$

36. $5y + 4 = 3(y - 4)$ **37.** $6(x - 3) = 2(3 - x)$

38. $2(n + 5) = 13$ **39.** $\frac{2}{5}r + \frac{3}{10} = \frac{r}{2}$

40. $\frac{x}{3} - \frac{1}{2} = \frac{5x}{6}$ **41.** $\frac{k}{2} - \frac{3}{10} = \frac{4}{5}$

Graph the inequalities on a number line.

42. $x > 2$ **43.** $y \leq -5$

44. $n \geq 0$

45. Use a number line to represent the numbers between -12 and -6.

Solve and graph each inequality.

46. $c + 4 > -2$ **47.** $2n - 7 > 3$

48. $-6x \geq 3x - 27$ **49.** $\frac{x}{-3} \leq 11$

50. For a children's play, Albert wants to dress the seven dwarfs alike. Their hats require $\frac{3}{4}$ yard of felt. Albert has a 6-yard bolt of green felt that he would like to use. Can he make all the hats green?

CHAPTER 6 TEST

Translate each into an algebraic equation.

1. When a number is decreased by 4 the result is 12.

2. The sum of twice a number and 7 is 43.

3. The price of one record is \$1.35 more than the price of another. The total price of the two records is \$21.75.

Solve.

4. $x + 4 = 28$
5. $x - \frac{2}{3} = 5$
6. $4x = 32$
7. $\frac{1}{2}x = 14$
8. $\frac{3}{5}x = 9$
9. $5x - 4 = 16$
10. $\frac{2}{5}x + 3 = 43$
11. $0.2x + 1 = 4$
12. $2.5x - 9.1 = 6.9$
13. $x + 3x = 56$
14. $2x - 12 = x + 3$
15. $3(x + 2) = -3$
16. $2(x + 8) - 6 = -4$
17. $\frac{1}{2}(x - 3) = 7$

Use the formula $d = rt$.

18. The distance from BJ's house to her grandmother's house is 224 miles. How long will BJ have to travel if she drives an average of 56 mph?

Graph on a number line.

19. $x > 3$
20. $x \le 0$
21. numbers between -1 and 4
22. numbers between 0 and -3

Solve the inequalities.

23. $x - 1 > 9$
24. $5x \le -10$
25. $-3x + 2 < 11$
26. $3(x + 1) \le 3$
27. Depending on conditions, an automobile may get as many as 15 kilometers for each liter of gas. How far might the car go on 55 liters of gas? Write an inequality and solve.

CUMULATIVE REVIEW CHAPTERS 4–6

A class of students took a math quiz and received the following scores.

7	5	6	8	10	10	7	10	8	9
8	7	10	6	8	7	9	7	8	5
9	7	8	8	7	9	3	8	8	6
6	8	9	9	10	7	7	8	9	9
7	7	9	7	8	7	9	10	10	7

1. Construct a frequency chart for these scores.

2. How many students took the quiz?

3. How many students received a score of 8?

4. Which was the most frequent score?

Find the value when $a = 3$ and $b = -4$.

5. $a + b$

6. $3a + 2b$

7. $2a - 6$

8. $a^2 + 3b$

Simplify.

9. $3x + 8x + 12 - 6$

10. $(3st + 2s) - (s + 5st)$

11. $d(2d + 6) + 3d(d - 4)$

12. $\frac{9b + 27}{-9}$

13. $\frac{5xy - 20x}{5x}$

Translate each into an algebraic equation. Then solve the equation.

14. The sum of twice a number and 9 is 41.

15. A number divided by 3 and decreased by 1 is equal to 8.

16. $5x \leq -35$

17. $x + 7 \geq -11$

18. $2(x - 4) > 14$

19. $-4x + 3 < 15$

Think of the problems that the ancient Egyptians had to solve to build the pyramids. Not only did they have to plan how to build the pyramids, they had to solve the problem of transporting the heavy stones over many miles.

The problems you have to solve in everyday life may not be as complicated as those the Egyptians solved, but they are just as important. Many times common sense is needed to solve problems.

You have already used many methods to solve problems. In this chapter you will learn new methods and practice with ones you already know. You will learn to decide whether you should use a calculator, computer, paper and pencil, or mental math. You will learn to make a diagram, chart, or graph.

To be successful in our society, you need to develop problem-solving skills. In this chapter, you will study the answers to these questions and others.

Ask Yourself

- *What are relevant facts?*
- *What strategies help solve certain types of problems?*
- *What is a model for solving problems?*
- *What does it mean to generalize?*

CHAPTER 7
PROBLEM SOLVING

SECTION 7–1

LET'S BEGIN

WARM-UP ACTIVITY

For each pair, tell which segment is longer.

Measure to check your answer. Were you correct?

INTRODUCTION

The most important application of mathematics is in the solution of problems. To solve problems you can use the same steps over and over again, although the problems may be different. Begin by reading the problem carefully. Then ask yourself these two questions: What am I to find? What do I know?

SKILLS DEVELOPMENT

At 12:00 noon Bev went into the Tops Record Stop and bought 2 compact disks. The CDs were on sale for \$5.99 each. She bought 6 CDs last year at \$4.99 each and 3 at \$4.50 each the year before. How much did she pay for the CDs she bought this year?

Example 1

Answer these questions for the problem about Bev buying CDs:

a. What are you to find?

b. What do you know?

Solutions

a. The cost of the CDs Bev bought when she went into the store this year.

b. Bev went into the store at 12:00 noon. The name of the store is Tops Record Stop. Bev bought two CDs. The CDs are on sale for \$5.99 each. Last year Bev bought six CDs at \$4.99 each. Two years ago she bought three CDs at \$4.50 each.

Example 2

Edgar jogged for 1.5 hours. How far did he jog?

Answer these questions about this problem:

a. What are you to find?

b. What do you know?

c. What known facts will help you find a solution?

d. What additional facts, if any, are needed to solve this problem?

Solutions

a. How far Edgar jogged.

b. Edgar jogged for 1.5 hours.

c. It would be appropriate to solve this problem using the distance formula (distance = rate × time). The time, 1.5 hours, is a relevant known fact.

d. The **relevant missing fact** is Edgar's rate of jogging. Until this information is supplied, the problem cannot be solved.

Quality problem-solving consists of five steps—Begin, Strategy, Method, Solution, and Evaluate—as shown in the diagram below. In this section we considered the first step.

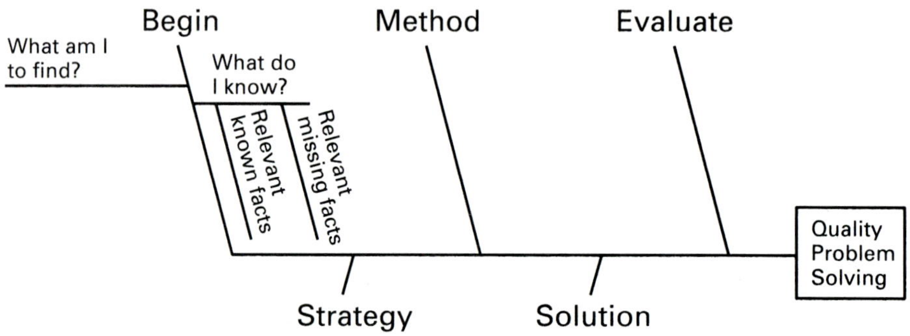

EXERCISES

For each problem ask yourself the questions shown in the chart above. Then try to solve the problem. If the problem cannot be solved, tell why.

1. Virja has $500 in the bank. Yannak and Virja are to share the $1000 profit from their summer lawn-care service. If Yannak is to receive $100 less than Virja, what is each person's share?

2. Last summer Sylvia earned 3 times as much as Susan. Together they earned $96. How much did each girl earn?

3. Each week Maureen works 12 hours. If she is paid $5.25 per hour, how much does she make in a year?

4. Marc bought 12 ounces of raisins and 1 quart of buttermilk to bake muffins for the school bake sale. The muffins sold for $0.50 each. How much did Marc spend on ingredients for the muffins?

5. The plane traveled 1065 meters on the runway before taking off. The runway is 2500 meters long and 320 meters wide. Today the plane has 35 passengers. How much extra runway did the pilot have?

A chart often contains more information than you need. You need to be able to choose only that information that is relevant to the problem. Use the information in this chart of the attendance at the Gardens during a 12-month period.

Jan. 4653	May 2196	Sept. 2865
Feb. 3196	June 1896	Oct. 4696
Mar. 6938	July 1283	Nov. 5698
Apr. 5216	Aug. 1492	Dec. 7893

6. The summer months are June, July, and August. Which month had the greatest summer attendance? The least winter attendance?

7. Last year, the circus was at the Gardens during the months of October, November, and December, and 3695 people saw it. How many people were in attendance at other events at the Gardens during the time of the circus?

8. List the information in the chart that was not used to solve the problems in Exercises 6 and 7.

9. Write a problem based on the information you listed in Exercise 8. Then solve it.

These problems cannot be solved because they do not contain all the needed information. Make up reasonable information for what is missing. Then solve the problems.

10. The odometer read 46504.7 when the Jacksons began their trip. How far did the family travel during the trip?

11. Michael's car can travel 20.9 miles on 1 gallon of gas. How far can it travel on a tank of gas?

12. After the basketball game, the coach took the players to lunch and left a \$4.50 tip. What was the total amount the coach paid?

13. A stereo is advertised for \$295. If Janet pays one-third cash and the rest in monthly payments, how much is each monthly payment?

SECTION 7–2

CHOOSE YOUR METHOD—EVALUATE YOUR WORK

WARM-UP ACTIVITY

Cathy found cubes fascinating. She spent many hours building big cubes with smaller cubes. For her birthday, her grandmother gave her 500 new cubes. She already has 370 of the same size. What are the dimensions of the biggest cube that Cathy can build? How many small cubes would she use? How many small cubes will she have left? What is the biggest cube she can build with the remaining cubes?

SKILLS DEVELOPMENT

Besides the methods of mental math and paper-and-pencil, technology has given us new methods to help perform calculations. The calculator is an especially useful tool for handling calculations with large numbers.

Example 1

Find the cost of 24 liters of fruit juice if 13 liters are $1.19 each and the rest are $0.89 each.

Solution

To begin, you want to find the total cost of the fruit juice. You know 13 liters cost $1.19 each and 11 liters cost $0.89 each. A calculator would be a good choice for the calculation.

$$(13 \times \$1.19) + (11 \times \$0.89) = \$15.47 + \$9.79 = \$25.26$$

The cost of the juice is $25.26.

Once you have found a solution, evaluate it. First consider the answer to your beginning question, What am I to find? Does your solution match what you were supposed to find? Then determine whether your solution is reasonable. Estimate the solution. Then compare your actual solution to your estimate.

Example 2

Is the solution to the problem in Example 1 reasonable?

Solution

Round the cost of each liter of juice to $1. Then estimate the total cost of the juice.

$$24 \times \$1 = \$24$$

$25.26 is close to $24, so the solution is reasonable.

The computer is useful for doing several repetitive calculations quickly. A computer operates following a set of instructions. Sometimes these instructions are on a disk you put into the computer. These disks are referred to as **software.** Sometimes these instructions are programs you can type into the computer yourself. The following program, AVERAGE, will help you use the computer for problem solving. This program finds the average of *n* numbers.

```
10 REM FINDING AVERAGES
15 SUM = 0
20 INPUT "HOW MANY NUMBERS DO YOU HAVE?"; N
25 FOR X = 1 TO N
30 INPUT "ENTER YOUR NUMBERS. PRESS THE ENTER
KEY AFTER EACH NUMBER."; T
35 SUM = SUM + T
40 NEXT X
45 PRINT "AVERAGE IS"; SUM/N
50 END
```

Example 3

This chart shows the points scored by five basketball players in the first six games of the season.

	Game					
Player	#1	#2	#3	#4	#5	#6
Mercedes	12	10	8	16	2	4
Sonia	6	8	2	10	4	9
Raquel	11	6	5	9	6	7
Isabel	5	3	7	10	12	8
Jen	10	17	12	15	18	11

Find the average points scored by each player. Then evaluate your work.

Solution

The computer would be a good choice of method for this problem. Using the AVERAGE program, type RUN, then enter 6 followed by the points scored by Mercedes. Repeat this process for each player.

Player	Average points scored per game
Mercedes	8.7
Sonia	6.5
Raquel	7.3
Isabel	7.5
Jen	13.8

You were to find the average points scored by each player. You did this. For each player, the average points calculated is greater than that player's lowest score and less than that player's highest score. Therefore the solution is reasonable.

The problem-solving steps featured in this section have been added to the problem-solving diagram introduced in Section 7–1.

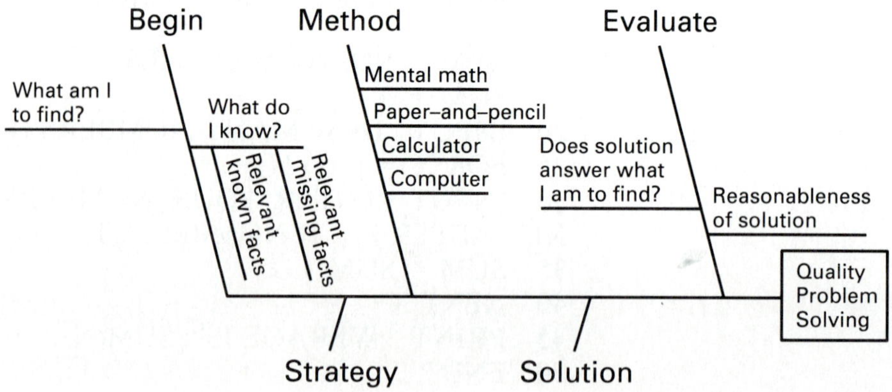

EXERCISES

Choose the method appropriate for each of the following problems. There may be more than one possible answer.

A Mental **B** Pencil-and-Paper
C Calculator **D** Computer

1. Which numbers are divisible by 3?

2. Calculate the value of $3m + 2n$ where m is 36 and n is 24.

3. Is 3781 a prime number?

4. Have each student in your class count up the total number of letters in their name. What is the average number of letters in a name?

5. Have you ever heard of the "midnight sun"? In lands that lie north of the Arctic Circle, there is constant daylight. If there are 1752 hours of constant daylight, how many consecutive days is this?

Use the pictograph for the next two exercises.

Bones in Parts of the Human Body

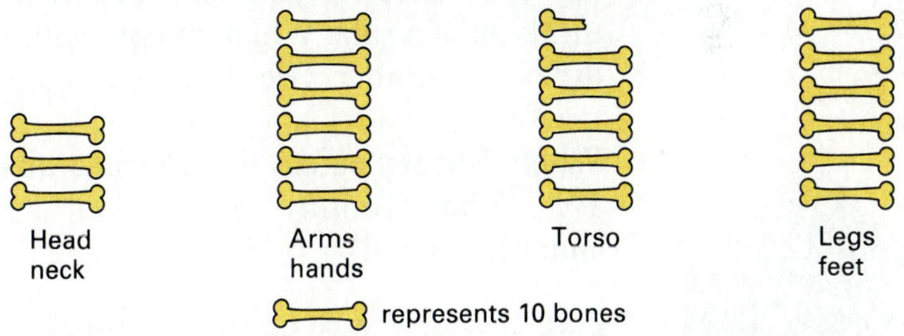

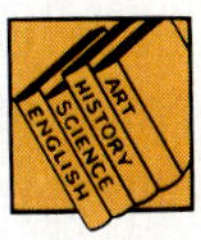

6. How many bones are there in the legs and feet?

7. What fraction of the bones are in the head and neck?

8. Louis loves to read about Egypt. He heard that a pyramid is for sale at a price of 10 million dollars, and he got an idea to raise the money. He plans to ask people for \$1.00 donations. If every person Louis asks gives him \$1.00 and he asks someone every minute, how long will it take Louis to raise the money?

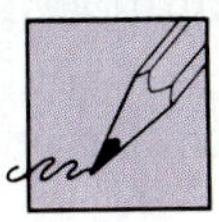

9. Write four word problems each of which can best be solved using a different method: mental, pencil-and-paper, calculator, or computer.

SECTION 7–3

MAKE A DIAGRAM, GRAPH, OR CHART

WARM-UP ACTIVITY

An equilateral triangle is a triangle with all three of its sides equal. Use six straight lines to draw a figure containing eight equilateral triangles.

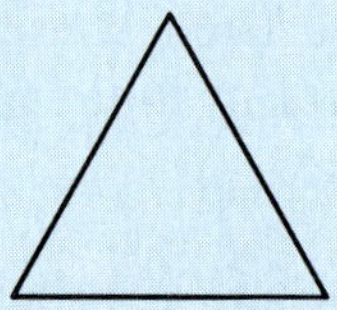

INTRODUCTION

Clarissa enjoys tutoring Tommy, a fourth grader, in mathematics. One day she made up a special problem. Clarissa said, "Tommy, I am thinking of a rule. When you put a number into my rule, it comes out a different number. See if you can guess my rule."

When Tommy said, "7," Clarissa answered, "The rule changes it to 14." When Tommy asked about 2, Clarissa answered, "The rule changes it to 9."

One strategy for solving problems is to make a chart. Clarissa suggested to Tommy that it might help him figure out the rule if he made a chart. It would help solve the problem if he found the outcomes of consecutive numbers and ranked them. Tommy guessed several more numbers and made the following chart.

Tommy's numbers	2	3	4	5	6	7
Outcome of Clarissa's rule	9	10	11	12	13	14

"I see," Tommy said. "You are adding 7 to my number."

"Yes," Clarissa answered. "If we call your number x, my rule is $x + 7$."

SKILLS DEVELOPMENT

Inherent in problem solving is the idea that a pattern exists and you are trying to find it. To help you find this pattern, you can use the strategy of making a diagram, a graph, or a chart.

Example 1

Jonas has 3 different colored flags. If he displays all 3 flags in a row, how many different ways can he arrange them?

A7.3.2

Solution

Some problems are easier to analyze using a **tree diagram.**

Begin by thinking how many flags Jonas could put first. (Any one of the three.)

Now Jonas has chosen one flag. How many choices does he have for the second flag?

Jonas has now chosen two flags. How many choices are left for the third flag?

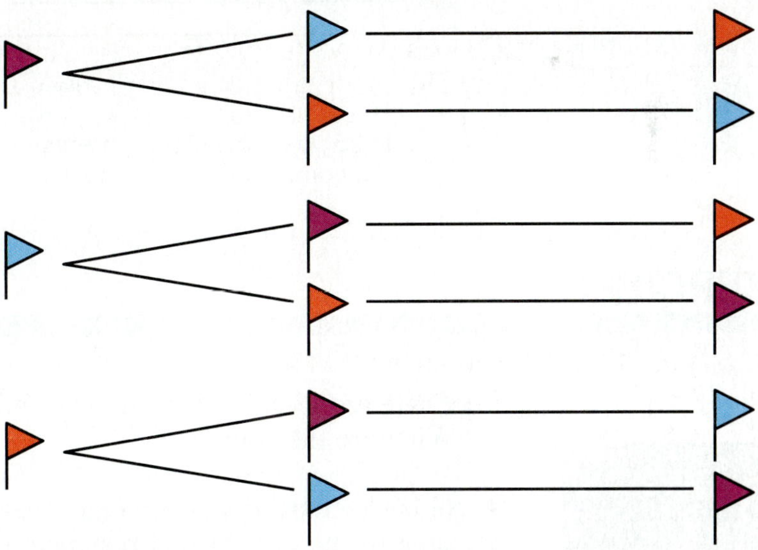

There are 6 different ways to display the flags.

Example 2

Max was studying the difference between simple interest and compound interest. The book he was reading showed the total savings on an investment of $1000 at 10% simple interest and 10% interest compounded quarterly over the same period of years.

TOTAL SAVINGS ($)

Years	With simple interest	With compound interest
5	1500	1639
10	2000	2685
15	2500	4400
20	3000	7210

Max wanted to know how much money he would have in 25 years under each plan. How can he predict the amount?

Solution Max made a broken-line graph to display the data. Then he extended each graph to make a prediction for 25 years.

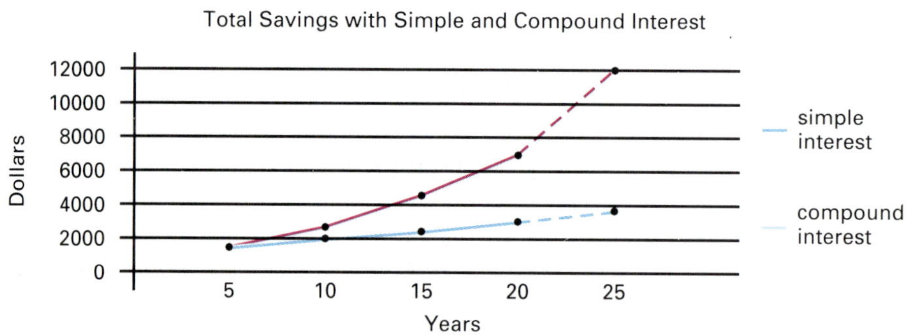

In 25 years the simple interest would produce $3500, while the compound interest plan would produce about $12,000.

EXERCISES

1. There are 13 coins in a jar. The value of the coins is a half dollar. What are the coins?

A golden rectangle is a rectangle that the Greeks considered most pleasing to the eye. In this rectangle the length is about 1.62 times longer than the width.

2. Draw a golden rectangle with a width of 5 cm.

3. Find the perimeter of a golden rectangle with a width of x.

4. Ida and Ali live in the same apartment building. Ida leaves the apartment building and walks 5 blocks north. Ali also leaves the building, but he walks 3 blocks east and 2 blocks south. What is the shortest distance in blocks between them?

5. Tommy thought of a rule for Clarissa to guess. The chart shows Clarissa's numbers and the outcomes using Tommy's rule. What is Tommy's rule?

Clarissa's numbers	1	2	3	4	5
Outcomes of Tommy's rule	1	3	5	7	9

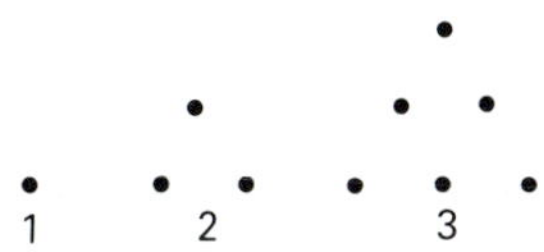

A triangular number is the number of dots which form a triangle. The first 3 triangular numbers are shown at left.

6. Find the first ten triangular numbers.

7. Find the sum of the first ten triangular numbers.

8. Point A is located at -3 on the number line. Point B is 5 units from Point A. What are the possible locations of B?

Some figures are traceable. They can be traced without backtracking or drawing over a segment twice. Tell whether each figure is traceable.

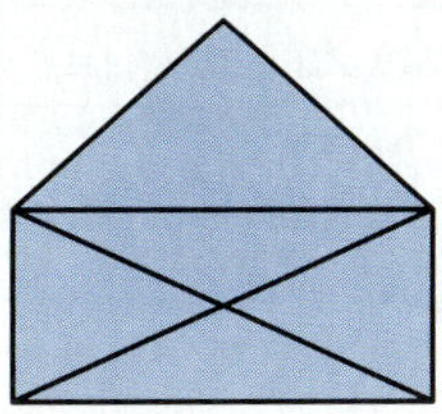

9.

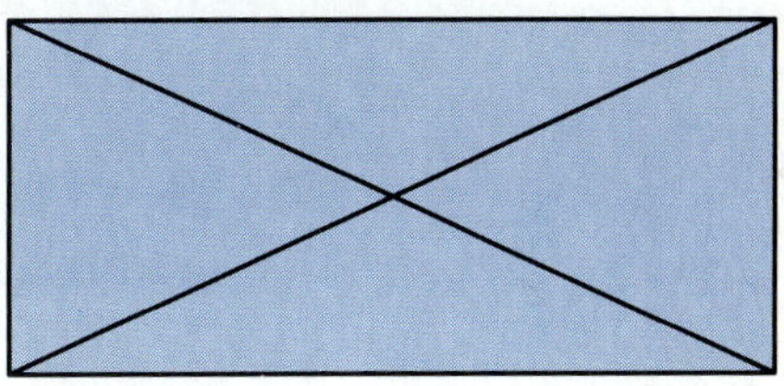

10.

11. An octagon is an eight-sided figure as shown at the right. What is the smallest number of triangles into which the area of an octagon can be divided?

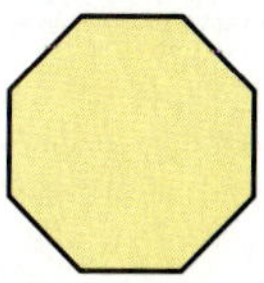

12. Jay has 10 cans with a diameter of 7.5 cm and a height of 12 cm. He plans to pack them in 2 rows of 5 each. What are the dimensions of the smallest box he can use?

13. Sondra has 3 shirts, each of which she can wear with 2 pairs of jeans. How many different outfits can Sondra make with these shirts and jeans?

14. Work with a partner. Think of a rule. As your partner names numbers, tell the outcome of applying your rule to each number. Continue until your partner can figure out your rule. Then trade roles.

SECTION 7–4

MAKE IT SIMPLER

WARM-UP ACTIVITY

Arnold, Jim, and Joe are married to Barbara, Sheila, and Chris. Sheila's sister is married to Joe. Arnold has never met Chris. Chris is an only child. Who is married to whom?

Use the following chart and eliminate possible matches to solve this problem.

	Arnold	Jim	Joe
Barbara			
Sheila			
Chris			

INTRODUCTION

How many squares are in this 6×6 square?

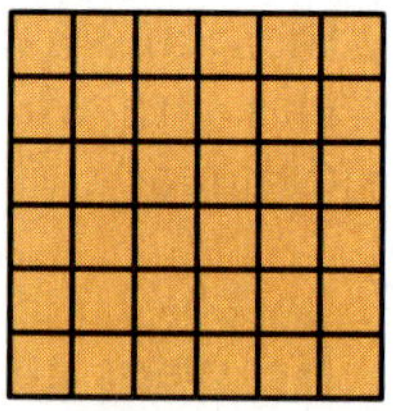

The obvious answer is 36. But that is wrong. There are actually 91 squares. Can you figure out how to get that many? In this lesson you will learn how to make a difficult problem easy!

SKILLS DEVELOPMENT

Sometimes by making the problem simpler it is possible to see a pattern. Extending this pattern helps you find the answer. Organizing your work in a chart is often helpful.

Example 1

How many squares are in a 6 × 6 square?

Solution Begin with a 1 × 1 square. It has just 1 square in it. Then investigate a 2 × 2 square. It has 5 squares in all: one large 2 × 2 square and four smaller 1 × 1 squares.

Now try a 3 × 3 square. It has one large 3 × 3 square and nine smaller 1 × 1 squares. It also has four 2 × 2 squares as shown in the drawing. So, a 3 × 3 square contains 14 squares in all.

For this problem, look for the pattern and extend it to the 6 × 6 square.

1 × 1 square:	1
2 × 2 square:	1 + 4
3 × 3 square:	1 + 4 + 9
4 × 4 square:	1 + 4 + 9 + 16
5 × 5 square:	1 + 4 + 9 + 16 + 25
6 × 6 square:	1 + 4 + 9 + 16 + 25 + 36

There are 91 squares in the 6 × 6 square.

Sometimes it is easier to think about a statement if you use a specific example.

Example 2

Is the sum of two odd numbers an odd number?

Solution To solve this problem, choose two odd numbers such as 3 and 5. The sum of 3 + 5 is 8. This number is even, so the answer is No.

An example that does not work will verify that a statement is false. Such an example is called a **counterexample.** However, finding many examples that do work does not assure that the statement is true.

Using specific examples to test the truth of a statement can be dangerous in two ways:

- if you do not choose your examples carefully; and
- if you jump to a conclusion before trying enough different possibilities.

Example 3

Is this statement true or false? $a^2 \geq a$

Solution To avoid overlooking counterexamples, do the following:

a. Try a negative number: $(-3)^2 \overset{?}{\geq} 3$, $9 \geq 3$ is true.

b. Try a positive number: $(4)^2 \overset{?}{\geq} 4$, $16 \geq 4$ is true.

c. Try zero: $0^2 \overset{?}{\geq} 0$, $0 \geq 0$ is true.

d. Try a fraction between 0 and 1: $\left(\frac{1}{2}\right)^2 \overset{?}{\geq} \frac{1}{2}$, $\frac{1}{4} \geq \frac{1}{2}$ is false.

Therefore, $a^2 \geq a$ is a false statement.

EXERCISES

Is each of the following statements true or false? Justify your answers by checking specific examples.

1. $1/a + 1/b = 1/a + b$
$a \neq 0, b \neq 0$

2. $(1/a)(1/b) = 1/ab$
$a \neq 0, b \neq 0$

3. $a(b + c) = ab + ac$

4. $a(bc) = (ab)(ac)$

5. What is the sum of the first 100 odd numbers $(1 + 3 + 5 + 7 + \ldots + 99)$?

6. How many squares are there in the left figure?

7. How many triangles are there in the right figure?

Is each of the following statements always true, sometimes true, or never true?

8. The product of two odd numbers is odd.

9. The sum of two prime numbers is prime.

10. The sum of an unknown number of odd numbers is odd.

Rectangle *ABCD* has been subdivided into two rectangles. The areas of the rectangles are given.

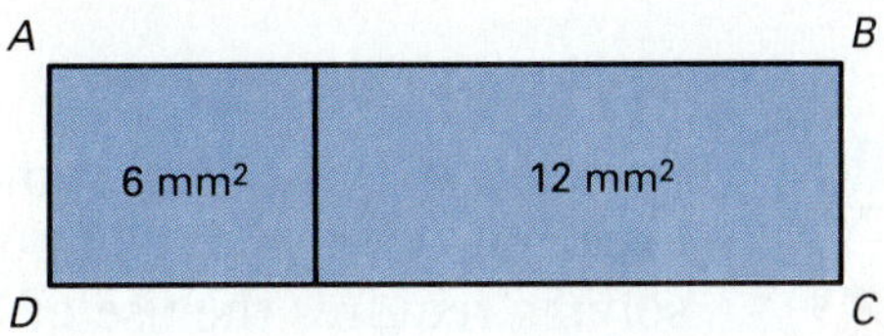

11. Find possible dimensions of rectangle *ABCD.*

12. Find a possible perimeter of rectangle *ABCD.*

13. Find the area of rectangle *ABCD.*

14. Margie, Rose Mary, and Shirley are dating Alex, Frank, and John, but Shirley does not like John. Rose Mary is dating John's brother. Alex is dating Rose Mary's sister. Who is dating whom? (Assume that people who date like each other!)

15. Write a problem like the one in Exercise 14. Be sure to include enough information, but not too much. Exchange problems with another student and try to solve each other's problems. If you find that your problem is too easy or that you left out some information, fix it up.

SECTION 7–5

PREDICT AND TEST

WARM-UP ACTIVITY

Arrange each of the digits 1 through 7 in the circles so that the sum along any line is 10.

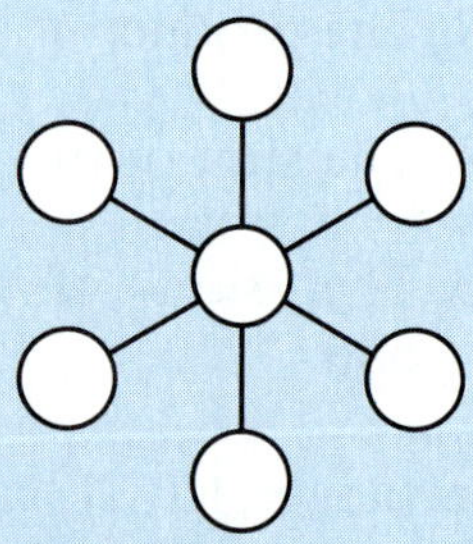

INTRODUCTION

In science class, you sometimes use laboratory equipment in an experiment to check a possible problem solution. Some problem solving requires the use of logical reasoning. In this section, you will learn to use logical reasoning to predict and test possible solutions to problems. Solving problems that require logical reasoning can be a lot of fun. In fact, there are puzzle magazines available that contain these types of problems.

SKILLS DEVELOPMENT

Omni found this puzzle in a magazine. To complete this addition problem, Omni is allowed to use each of the ten digits exactly once.

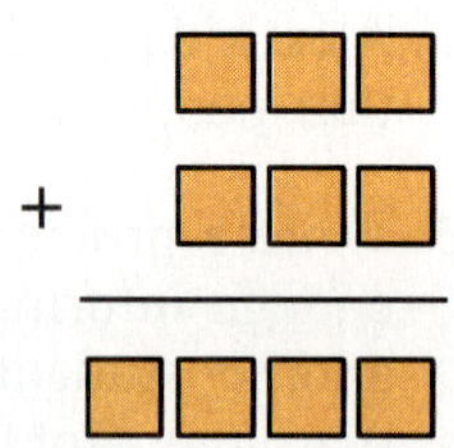

Omni began the problem. She guessed digits for the ones and tens places in both and found the answer.

$$\begin{array}{r} 32 \\ +\ 54 \\ \hline 86 \end{array}$$

"Good guess," said Omni. "No repeating numbers." Then she predicted the hundreds digits.

$$\begin{array}{r} 132 \\ +\ 854 \\ \hline 986 \end{array}$$

"Whoops!" Omni has problems. She has used 8 two times. The only digits she has not used are 0 and 7. But she cannot begin a number with 0.

Omni also notices that to get a 4-digit answer she will need to have a sum larger than 9 in the hundreds place. She needs to consider how the sums of digits are related to each other to efficiently solve the problem.

By thinking about what she had guessed already and trying some more number combinations, Omni solved the puzzle.

$$\begin{array}{r} 652 \\ +\ 437 \\ \hline 1089 \end{array}$$

The strategy of **predict and test** means guessing a solution and then checking to see if it works. Sometimes, you will find the correct solution on the first or second guess. More often, as you predict and test you will begin to see a pattern or a relationship that helps you solve the problem.

Example 1

Find 3 consecutive integers such that the sum of their squares totals 365.

Solution Use the predict and test strategy. Record the results of each test in a chart.

Numbers	Squares	Sum
3, 4, 5	9, 16, 25	50—way too small
13, 14, 15	169, 196, 225	590—too large
10, 11, 12	100, 121, 144	365—correct

You have found 3 consecutive integers such that the sum of their squares equals 365. They are 10, 11, and 12. But, **always consider the possibility of more than one correct answer.** Since the 3 digits are squared, the original numbers could be negative and still result in a positive sum.

Therefore, 3 consecutive integers such that the sum of their squares equals 365, are either 10, 11, and 12 or −12, −11, and −10.

In your earlier work you learned a way to solve linear equations. Another way to solve equations is to use the strategy of predict and test.

Example 2

The cost, C, in dollars, of producing a newsletter is given by $C = 25 + 2n$, where \$25 is the initial cost in dollars and n is the number of newsletters produced.

How many newsletters can be produced at a cost of \$41?

Solution When you substitute 41 for C, you get the equation $41 = 25 + 2n$.

To use predict and test, systematically try different values. This is called **systematic trial.**

First estimate a value.	$35 > 41$, so estimate a greater value.
Try $n = 5$.	Try $n = 8$.
$41 = 25 + 2(5)$	$41 = 25 + 2(8)$
$41 = 25 + 10$	$41 = 25 + 16$
$41 = 35$ False	$41 = 41$ True

So, 8 newsletters can be produced at a cost of \$41.

EXERCISES

In each of the following, the letter **A** stands for a digit. Find all the possible values of **A.**

1. A1AA2 is divisible by 36.

2. AA3AA is divisible by 9.

3. A234A is not divisible by 5.

4. A73AA is not divisible by 9.

Copy and complete the chart.

	n	$6n + 2$	$5n + 6$
5.	2	?	?
6.	3	?	?
7.	4	?	?

8. Use your answers for Exercises 5–7. What is the solution of the equation $6n + 2 = 5n + 6$?

9. Use systematic trial to solve the equation $5n - 3 = 3n + 7$.

10. The commission, C, in dollars, for selling magazines is given by the equation $C = 10 + 3n$, where n is the number of magazines sold. Write and solve an equation to find how many magazines a person sold who earned a commission of $46.

11. Estimate the cube root of 10. Round your answer to the nearest hundredth. (Hint: The cube root of 10 is the number which when used as a factor three times is equal to 10.)

12. Find two consecutive whole numbers the difference of whose squares is 31.

13. Each different letter in this addition problem represents a different digit. Find the value of each letter and rewrite the problem using numbers.

```
  THIS
  SURE
+   IS
------
  FINE
```

14. Work with a small group of students to write a 1-page *Puzzle Newsletter.* Be sure to include problems like those in the Warm-Up Activity and Exercises 1–4, 12, and 13.

SECTION 7–6

MAKE A MODEL

WARM-UP ACTIVITY

Ann, Bob, and Cheryl live in the same neighborhood. Ann goes to the middle school. Bob goes to the high school. Cheryl goes to the elementary school. When they travel to and from school, they take routes that never cross and that do not go behind Bob's house or the schools. Trace the houses and find their routes.

Which School Route?

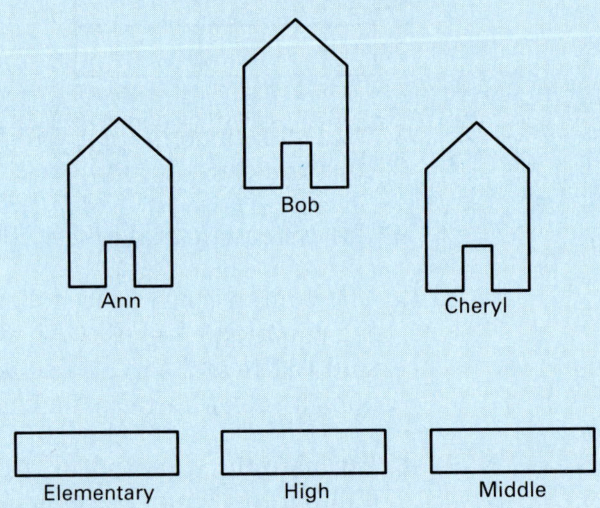

INTRODUCTION

a. Sam employs three salespersons. Sales have increased, and he plans to give each salesperson a $50 bonus. How much money must he budget to give them the bonus?

b. Nikko works for a landscape designer. She is planning a walkway that runs parallel to a building. If the walkway is 3 feet wide and 50 feet long, what is its area?

SKILLS DEVELOPMENT

One of the main uses of mathematics is to solve problems using mathematical models. You have already learned to use many of these models: addition, subtraction, multiplication, division, and exponentiation.

Look at the Problems A and B in the Introduction. How are the problems the same? Is there one way that can be used to solve both problems? Yes, both problems can be solved using the multiplication model.

Example

Solve Problems **a** and **b**.

Solutions a. money budgeted = bonus per person × number of people

money budgeted = (\$50)(3) = \$150

b. The area of a rectangle can be found by using the formula *area* = *length* × *width*, or $A = lw$

$A = (50)(3) = 150$ square feet

The computer club wants to visit a manufacturing plant to see computer technology at work. However, they have no money in their treasury. The members decide to have a fund-raiser rather than pay dues. They have information on fund-raisers. One brochure says the club would make half the amount of money they have in sales but would have to pay shipping charges of \$3.50 for each case of items they order. It also says that each case contains 60 items and would sell for a total of \$300. A method for finding an equation to model the computer club's profit is given on the next page.

One way to find a modeling sentence is to use systematic trials and make a chart. Continue trying examples until you see a pattern. Then write the mathematical equation that describes the pattern.

Number of Cases	Amount Made on Cases	Total Freight Bill	Club Profit
1	\$150	\$3.50	\$146.50
2	2(\$150) = \$300	2(\$3.50) = \$7	\$293.00
3	3(\$150) = \$450	3(\$3.50) = \$10.50	\$439.50

Think about what you are doing each time. Let t = number of cases.

- You are multiplying the number of cases times $\frac{\$300}{2}$, or \$150. This can be written $150t$.
- Next you are multiplying the number of cases times \$3.50. This can be written $3.50t$.
- Then you are subtracting these two quantities. This can be written $150t - 3.50t$, or $146.50t$.

The modeling sentence is profit $= 146.50t$.

Letting p = profit, the equation becomes $p = 146.50t$.

EXERCISES

Match Exercises 1–4 to one of the mathematical models given below.
I. $a = b + c$ II. $a = b - c$ III. $a = bc$ IV. $a = \frac{b}{c}$

1. Georgia jogged 6 blocks on Monday and 5 blocks on Tuesday. How far did she jog in all on those two days?

2. Jack went up the hill at 2 meters per second. If the path up the hill is 10 meters, how long did the trip take?

3. Lucinda deposited \$400 in her checking account on February 1. Three days later she withdrew \$150. How much money did she have left in her account?

4. Kyle wants to buy 10 baseball cards. If the cards cost 5 cents each, how much money does he need?

5. Using the mathematical models you matched to Exercises 1–4, identify *a, b,* and *c* for each and then solve the problems.

Petra's father collects antique cars. He often drives one of his cars in a parade. His average speed is 5 miles per hour, and the parades usually last $1\frac{1}{2}$ hours.

6. What mathematical model can be used to find the length of the parade routes?

7. How long are the parade routes?

At 20 degrees Celsius, sound travels at a speed of 344 meters per second. During a thunderstorm, sound from a clap of thunder is heard a few seconds after the flash of lightning is seen.

8. Write a modeling sentence describing how far away a storm is.

9. If Jeremy hears the thunder 2 seconds after he sees the lightning, how far away is the thunderstorm?

10. Carmelo is listening to the weather station. The announcer says that a thunderstorm is 2 kilometers away. If Carmelo sees lightning, how long will it be before he will hear the thunder?

Amelia owns a dress shop. The dresses in the shop are priced to sell at twice their cost. During the inventory sale, the dresses are marked $\frac{1}{4}$ off, which means $\frac{1}{4}$ off the selling price.

11. What is the usual selling price of a dress that costs $80?

12. What is the sale price of the dress in Exercise 11?

13. How much profit is made on the dress in Exercise 11 when it is on sale?

14. Write an equation that could be used to find the sale price of a dress if the cost is known.

15. What was the cost of a dress sale priced at $60?

16. What is the largest amount Amelia could mark off on a sale and not lose money?

17. Using the mathematical model $a = b + c$, write two different problems.

SECTION 7–7

ACT IT OUT

WARM-UP ACTIVITY

Raul was touring the United States on the train. As he passed the many farms in Nebraska, he remembered a classic mathematics problem.

"Farmer Jones looked at the cows and ducks in the farmyard. He counted 85 heads and 290 legs. How many cows and ducks were there?"

Raul can solve this problem. See if you can also.

INTRODUCTION

Chikaro said, "With four parallel cuts I can divide a piece of paper into five separate pieces."

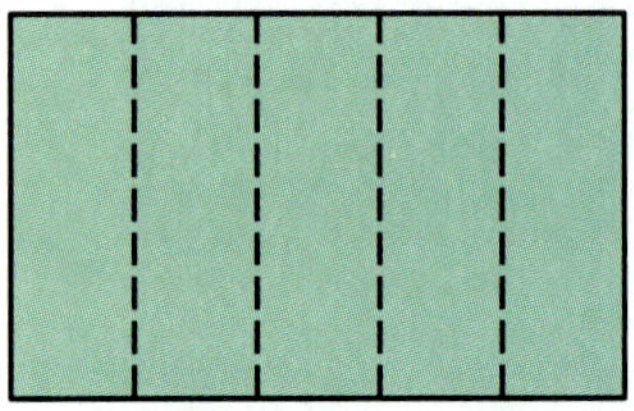

Oki thought about the problem. She found scissors and paper and tried different ways to cut the paper. She said, "With four cuts I can divide a piece of paper into nine pieces."

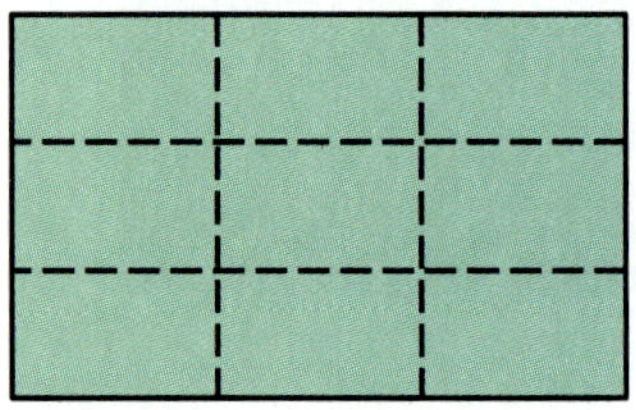

To find the greatest number of pieces the paper can be divided into with four cuts, you would need to experiment. When doing experiments, it is important to record your findings. You might record the number of pieces you got on each try.

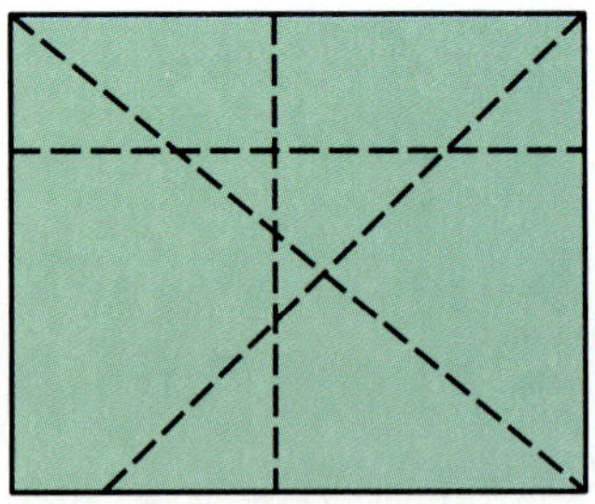

The diagram shows that the greatest number of pieces is 11.

SKILLS DEVELOPMENT

Sometimes a problem is easier to understand and solve if you act it out or experiment using equipment.

Acting out a problem to find a solution usually means you start at the beginning of the problem and experiment until you find outcomes which satisfy the conditions of the problem. Sometimes, though, the given information tells you the outcome. A strategy for solving these types of problems is working backwards.

Example 1

Luisa works at a child-care center. Each week she receives a rating of 1 to 10 from her supervisor, with 10 being the highest rating. If she averages 8 or above for a 6-week period, she receives a $0.25 per-hour pay increase. So far Luisa has received ratings of 6, 9, 8, 7, and 10. What rating must she receive for the sixth week to get the pay raise?

Solution Luisa decides to work backwards to find the rating she needs for the sixth week.

She thinks, To average 8 for 6 weeks I would need a total number of 48 points, because $\frac{\text{48 points}}{\text{6 weeks}}$ is an average of 8 points. I have a total to date of 6 + 9 + 8 + 7 + 10, or 40 points. I need at least 48. Since 48 − 40 is 8, if I get a rating of 8 or above in the sixth week, I will get the raise.

Example 2

Wade was reading the puzzle page in Thursday's newspaper. He saw a picture of a die and was asked to find the number on the bottom.

Solution Wade got a die and positioned it the same as the die in the picture. When he looked at the bottom of the die he found 6 dots.

The number on the bottom of the die in the picture is six.

EXERCISES

1. This is the beginning of a design composed of a circle in the center and rings formed around the center using circles. How many circles will be in the sixth ring out from the center?

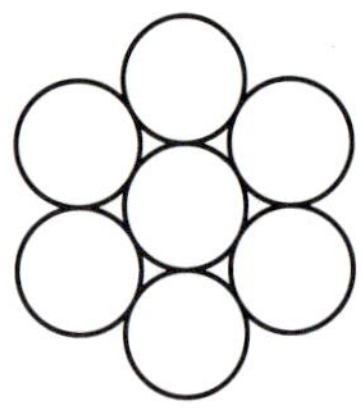

To solve this problem, you will need to use disks. As you make the design using the disks, fill in the chart below and look for a pattern.

	Center	Ring					
		1	2	3	4	5	6
Number of Disks	1	6	?	?	?	?	?

At the right is an arrangement of toothpicks. Make a model of this arrangement to help you solve Exercises 2 and 3.

2. Take away 5 toothpicks and leave exactly 3 squares.

3. Take away 4 toothpicks and leave exactly 4 squares.

4. Arrange 10 dots to make 5 rows with 4 in each row.

5. Steve, Molly, Frank, and Darcy went to an amusement park. When they entered the gate, a photographer asked them to stand in a row to have their picture taken. How many different arrangements of the friends are possible?

6. Twelve people are at a party. Every person shakes hands with every other person one time. How many handshakes take place?

7. The answers that Sharon, Zelda, and Barbara gave on a true/false quiz are shown in the chart. If one student had all the right answers and no student missed all the questions, what were the girls' scores?

Question	Sharon	Zelda	Barbara
1	F	T	T
2	T	T	F
3	F	F	T

8. There are 9 basketball teams in the city youth league. Make up a schedule of games so that each team will play every other team on 9 consecutive Saturdays.

9. How many squares of different sizes can be drawn on a 5-by-5 grid?

10. Using 5 square tiles, how many different shapes can be made? For this problem, tiles must be edge to edge and not corner to corner nor matching partway.

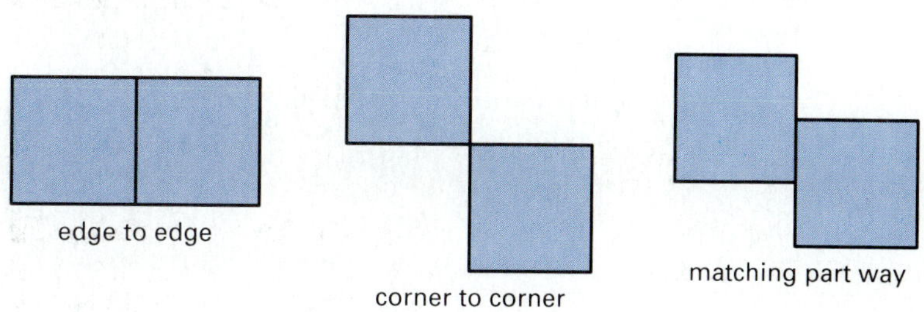

SECTION 7–8

GENERALIZE AND CREATE

WARM-UP ACTIVITY

Follow each instruction in order. Use the numbers in the box.

1. Cross out 2 numbers that total 98.
2. Cross out 2 numbers whose product is 500.
3. Cross out 2 numbers whose difference is 4.
4. Cross out 2 numbers for which the sum is a prime number.
5. The remaining number should be divisible by 8.
6. List the prime factors of the remaining number.

35		**25**
	28	
15		**14**
	96	
63		**32**
	20	

INTRODUCTION

Molly had several pieces of kite string and a pair of scissors. She made an *M* with a piece of string and cut through it once. She ended up with 5 pieces of string.

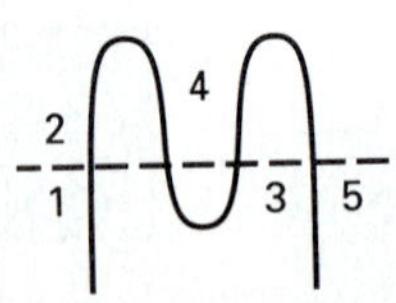

Molly wondered how many pieces she would end up with if she made 10 cuts.

SKILLS DEVELOPMENT

As you have been solving the problems in this chapter, have you noticed that finding a pattern was often the secret to solving the problem? Many times you had to extend the pattern to find the answer. Perhaps as you made a chart you recognized a pattern and described it in words.

Consider Molly's problem. As she made a chart, she found the following pattern.

Cuts	Pieces of String
1	5 +4
2	9 +4
3	13 +4
4	17 +4
5	21

Each number in the second column is 4 more than the number before it.

Molly predicted the tenth cut would result in 41 pieces of string.

To **generalize** a problem means to see a pattern and express it mathematically. To look at a pattern and describe what is happening as you move from one term to the next term is describing the pattern **recursively.**

Example 1

How can Molly's pattern be described as a recursive formula?

Solution

First, think about it using words. Then look at what parts repeat and consider how to rewrite them mathematically. The phrase "4 more than" is the same each time. In mathematics that would be written "+ 4".

- The first term is 5. $a_1 = 5$
- The second term is 4 more than the first term. $a_2 = a_1 + 4$
- The third term is 4 more than the second term. $a_3 = a_2 + 4$
- The fourth term is 4 more than the third term. $a_4 = a_3 + 4$
- In general $a_n = a_{n-1} + 4$

The mathematical way to express a recursive formula has two parts: an initial value and a sentence that generates the pattern.

For this example, the initial value is written as $a_1 = 5$. The sentence that generates the pattern is

$$a_n = a_{n-1} + 4$$

where n is a member of the set $\{2, 3, 4, \ldots\}$.

Molly saw that the number of pieces of string was 1 more than 4 times the number of cuts.

Example 2

How can the pattern Molly saw be generalized?

Solution In general, $a_n = 4n + 1$ where n is a member of the set $\{1, 2, 3, \ldots\}$.

Describing the pattern of how the number of each term is related to the term is called using a **rule.** A rule is an efficient way to show relationships, but it is not always possible to find one.

As you work towards finding the solution to a problem, you may see a pattern. Always try to generalize the pattern using either a recursion formula or a rule.

EXERCISES

Below are three buildings made with cubes. Use these drawings to help you answer Exercises 1–3.

Building 1 Building 2 Building 3

1. Complete the chart.

Building Number	Number of Cubes Used
1	4
2	
3	

2. Generalize the pattern using a recursion formula.

3. How many cubes are needed to build the sixth building?

Rita had numerous rhombuses from which she built larger rhombuses. As she built these, she noticed a pattern.

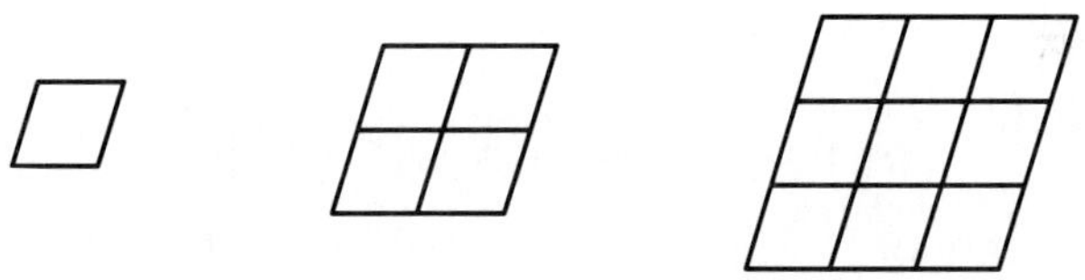

4. Make a chart showing the pattern and generalize it by writing a rule.

5. How many rhombuses did Rita need to make a 4-by-4 rhombus?

Aurillio used these tile arrangements in art class to make a three-dimensional collage.

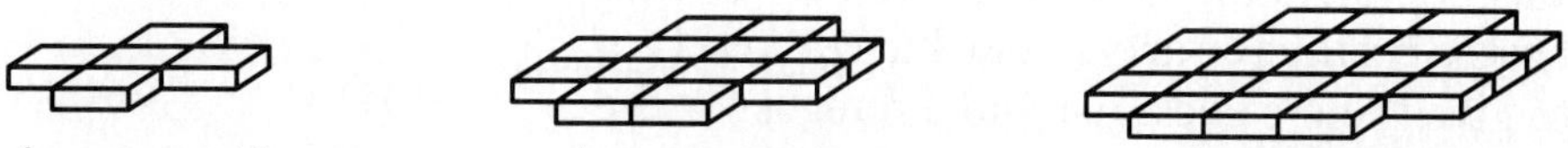

Arrangement 1 Arrangement 2 Arrangement 3

6. Make a chart showing the number of tiles he used in each arrangement.

7. Look for a pattern in the chart you made for Exercise 6. Generalize this pattern using a rule.

CHAPTER 7 SUMMARY

In this chapter you developed a five-step method of problem solving. When you began, you learned to ask yourself what you are to find and what facts are relevant. Once you have selected a strategy, you learned to choose an appropriate method. The final steps are using the strategy and method to find a solution and then evaluating that solution to determine whether it is what you were asked to find and if it is reasonable. Following these steps will help you to work smart and become a quality problem solver.

VOCABULARY

counterexample
generalize
mathematical model
predict and test
recursively
relevant facts
rule
software
tree diagram

CHAPTER REVIEW

Dominick will be 16 on his next birthday, July 31. His parents have agreed to let him have a birthday party. They have asked him to give them a guest list and a budget for the party. He plans to invite 11 friends. How much will the party cost?

Answer the following questions about this problem.

1. What are you to find?

2. What do you know?

3. What are the relevant known facts?

4. What additional facts are needed?

5. Bill wants to completely fence a field of 100 square feet. If the fenceposts are placed 2 feet apart, what is the fewest number of posts he will need?

Tell whether each statement is always true, never true, or sometimes true.

6. The sum of two numbers is divisible by their least common factor.

7. The quotient of two whole numbers is a rational number.

8. I am the remainder you get when you choose a prime number greater than 3, square it, add 17, and then divide by 12. What number am I?

9. Geraldo saw part of a license plate of a driver leaving the scene of an accident. He remembered LFO 4 . He also remembered the last two symbols were numbers and that the last number was either a 3 or 5. How many different license plates could he have seen?

A group of students are comparing fuel mileage in three different models of cars. Car 1 traveled 100 miles on 5 gallons of gas. Car 2 traveled 120 miles on 5.6 gallons of gas. Car 3 traveled 90 miles on 4.6 gallons of gas.

10. Find the gas mileage for each car. Round your answers to the nearest tenth.

11. Which car got the best gas mileage?

12. A lamp is priced to sell at twice its cost to the store of $50. Later it is put on sale for 30% off the selling price. What is the sale price of the lamp?

13. Margie placed one hexagon on the table and made a ring of hexagons around it. How many hexagons will she need to complete the fifth ring?

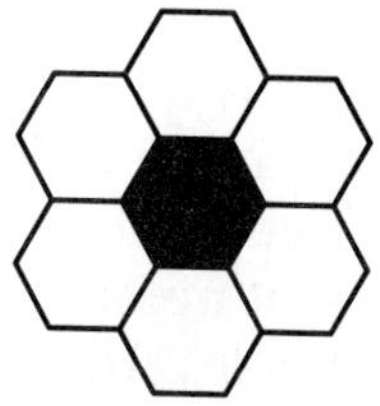

14. Using a geoboard, how many different shapes can you make having a perimeter of 12? Draw each of your shapes on grid paper. Shapes that are flipped or turned do not count as different shapes.

Consider the following figures. Look for a pattern in their perimeters. Remember, perimeter is the distance around a figure.

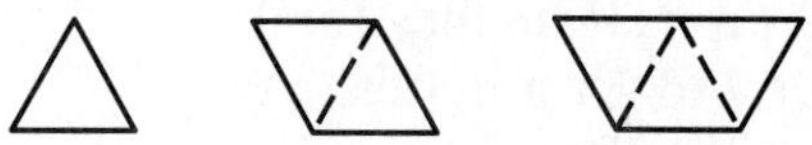

Figure 1 Figure 2 Figure 3

15. Generalize the pattern using a recursion formula.

16. Generalize the pattern using a rule.

17. What would the perimeter of a figure composed of 10 triangles be?

CHAPTER 7 TEST

Match each problem in Column A with the mathematical model in Column B.

Column A	Column B
1. Randi left her house at 9 A.M. She arrived at her grandparents' at noon. If she lives 150 miles from her grandparents, how fast did she travel?	**I.** $d = rt$ **II.** $t = \frac{d}{r}$ **III.** $r = \frac{d}{t}$

2. A racer averages 180 mph during a 500-mile race. Assuming that average speed includes yellow flags and pit stops, how long will he take to complete the race?

3. Jim flies his airplane about 150 mph. If he flies for 1 hour and 15 minutes, how far will he travel?

4. Which of the following is a solution for x and y in the equation $x - 3y = 8$?

a. $x = 3$ and $y = 0$
b. $x = 2$ and $y = 2$
c. $x = -1$ and $y = -3$
d. $x = -2$ and $y = 2$

5. One hundred fifteen people attended the end-of-baseball-season party. The committee provided 2 hot dogs and 2 orange drinks per person. The committee of 15 people gave $5 to each person to help pay for the party. How many hot dogs were needed to feed all the people?

6. How many cuts will it take to cut a 4 m log into 10 shorter logs with the same diameter as the original log?

7. Copy and complete the chart.

n	$2n^2$	$3n + 5$
3		
2.5		
0		
−1		
−1.5		

8. Use the chart you completed in Question 7. What are the solutions of the equation $2n^2 = 3n + 5$?

Which method would be best to perform each calculation?

9. $\frac{4400}{11}$

10. $36.50 + $5.52 − $4.75

11. $1\frac{2}{3} - 5\frac{1}{3}$

12. 10% of 900

13. Generalize the pattern shown in this chart using a recursion formula.

Number of Miles	Cost of Taxi
1	$5.00
2	$5.25
3	$5.60
4	$5.75
5	$6.00

SKILLS PREVIEW

This rectangle has 8 equal parts in all. Some parts are black, some are white, and some are striped.

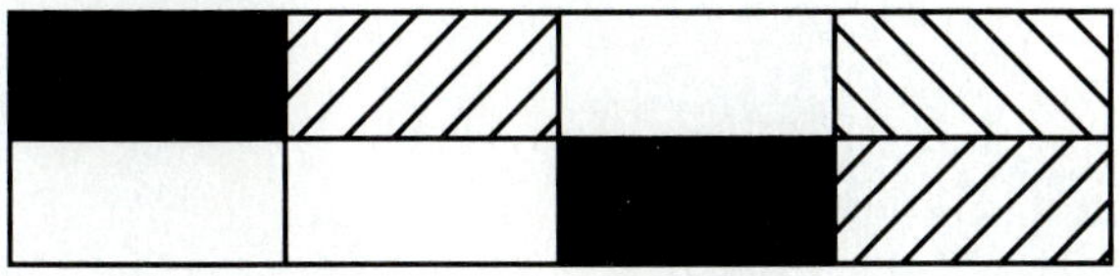

1. Write a fraction to show what part of the rectangle is black.
2. Write a fraction to show what part of the rectangle is white.
3. Write a fraction to show what part of the rectangle is neither black nor white.
4. Write a ratio to compare the number of black parts to the number of white parts.
5. Write a ratio to compare the number of black parts to the entire rectangle.
6. Write a ratio to compare the number of white parts to the entire rectangle.

One week Margaret earned \$50 at a parttime job. She spent \$20 on clothes, \$10 on a movie and dinner, and saved the rest.

7. What fraction of her earnings did she spend on clothes?
8. What fraction did she save?
9. Write a ratio to compare her savings to the amount she spent.
10. Write a ratio to compare her savings to her total earnings.

Use this number line for the problems below. Write both a fraction and a ratio for each comparison.

11. Compare $\overline{AB}$ to $\overline{BC}$.
12. Compare $\overline{BC}$ to $\overline{AB}$.
13. Compare $\overline{AB}$ to $\overline{AC}$.
14. Compare $\overline{BC}$ to $\overline{AC}$.
15. Compare $\overline{AC}$ to $\overline{BC}$.
16. Compare $\overline{AC}$ to $\overline{AB}$.
17. Which is equivalent to $\frac{2}{3}$, $\frac{4}{3}$ or $\frac{4}{6}$?
18. Which is equivalent to $\frac{3}{2}$, $\frac{2}{3}$ or $1\frac{1}{2}$?
19. Which is equivalent to $\frac{3}{4}$, 0.75 or 0.7?
20. Which is equivalent to $\frac{2}{5}$, 0.2 or 0.4?

Write five fractions equivalent to each of the following fractions.

21. $\frac{1}{2}$
22. $\frac{2}{3}$
23. $-\frac{3}{5}$
24. $-\frac{1}{4}$
25. $\frac{7}{10}$
26. $\frac{1}{100}$

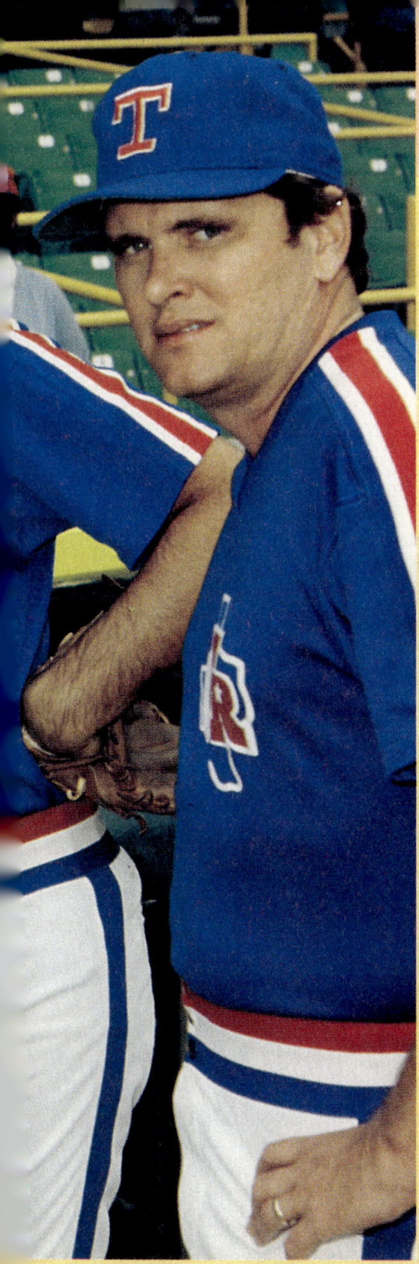

You make comparisons everyday. You can run faster than your brother. Your father is taller than you. Your aunt earns more than your uncle.

Sometimes you can make direct comparisons. Stand back-to-back with a friend. Who is taller? But, who is the tallest student in your school? That question would take a long time to answer if you compared students one-to-one. Mathematics allows us to compare such quantities. You can measure every student's height and make a list that shows the tallest student at the top of the list.

You can look at tall buildings and decide which is tallest. Does looking always give an accurate view? What happens when one building is farther than another building?

In this chapter, you will learn to use ratios, proportions, and percents to accurately compare quantities.

Ask Yourself

- *How can quantities be compared?*
- *How are proportions used to solve problems?*
- *How are ratios, proportions, and percents related?*
- *How do percents help to solve problems?*

CHAPTER 8

RATIOS, PROPORTIONS, AND PERCENTS

8–1: Ratios
8–2: Equivalent Ratios and Proportions
8–3: Ratio and Proportion: Consumer Applications
8–4: Using Proportions to Solve Problems
8–5: Percent
8–6: Finding Percents
8–7: Finding the Percent of a Number
8–8: Finding the Whole Number
8–9: Deciding Which Skill to Use

SECTION 8–1

RATIOS

WARM-UP ACTIVITY

A. Sam is taller than Calvin. Calvin is shorter than Orville. Orville is taller than Len. Len is taller than Sam.
Who is taller, Sam or Orville?
Who is shorter, Len or Calvin?

B. Rita is shorter than Frieda, as is Irma. Erica is shorter than both Rita and Irma. Rita is taller than Irma.
Who is the tallest?
Who is the shortest?

INTRODUCTION

You often need to compare one quantity with another. For example, the season record for a school basketball team is shown. You can compare the number of games won to the number of games played by using a ratio. A **ratio** is a comparison of quantities expressed in the same units.

	Cedar Heights
Games won	31
Games played	46

In working with ratios, you must know the **order** in which the terms are to be written. For example, 31:46 means

number of games won → compared to → number of games played
31:46

The above ratio 31:46 is read "31 is compared to 46." The symbol : means is compared to. The numbers 31 and 46 are called the **terms** of the ratio.

SKILLS DEVELOPMENT

Example

Of the 15 bicycles in the bicycle rack, 11 are ten-speed.

a. Write a ratio comparing the number of ten-speeds to the number of bicycles.

b. Interpret the meaning of the ratio 15:11 for the bicycles.

Solutions

a. number of ten-speeds → 11:15 ← number of bicycles
(compared to)

b. 15:11
number of bicycles → 15 compared to 11 ← number of ten-speeds

The order in which the terms of a ratio are written is very important. If you change the order of the terms, you change the meaning of the ratio.

Often in a recipe, you may need to compare more than two ingredients. For example, in a recipe 6 cups of flour, 3 cups of milk, and 1 cup of butter are used. To compare the amount of flour to milk to butter, you could write the three-term ratio:

6:3:1
(flour, milk, butter)

A ratio compares quantities expressed in the same units. To compare measures, they must be expressed in the same units.

EXERCISES

The diagram shows triangles and squares.

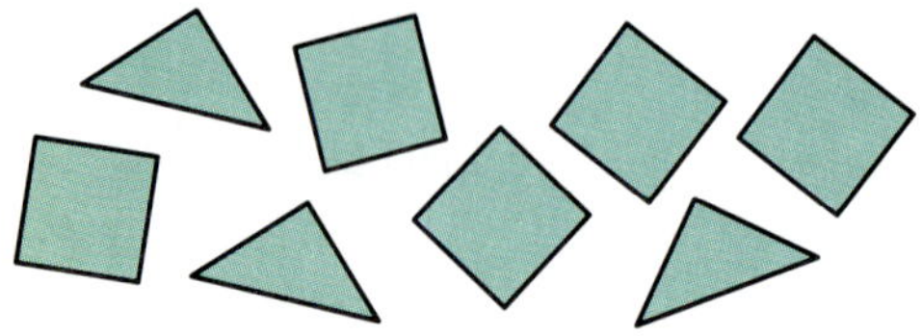

1. Write a ratio comparing the number of triangles to the number of squares.

2. Write a ratio comparing the number of squares to the number of triangles.

3. Write a ratio comparing the total number of figures to the number of triangles.

Write a ratio for each of the following:

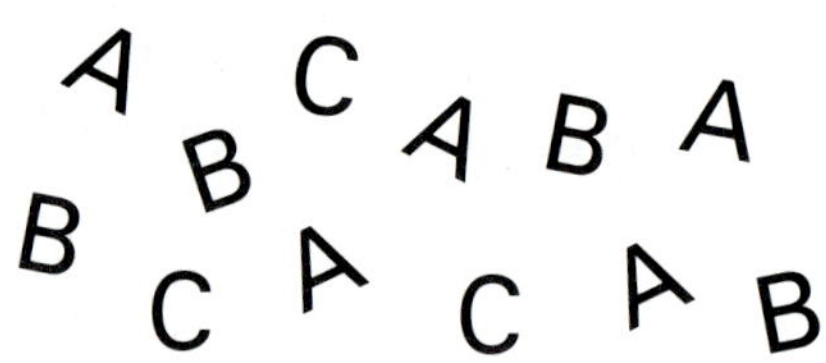

4. the number of As to the number of Bs.

5. the number of Bs to the number of Cs.

6. the number of Cs to the number of Bs.

7. What other ratios can you write for the letters A, B, C?

Interpret these ratios for the letters A, B, and C above.

8. 4:3 **9.** 4:5 **10.** 3:5

11. One aquarium has 40 guppies and another has 67. Write a ratio to compare the number of guppies in the first aquarium to that in the second.

12. In a video game, Tanya scored 11,525 points and Mark scored 9083. Write a ratio to compare Tanya's score to Mark's score.

The chart shows the number of games played and goals scored by various hockey players.

	Martin	Blanden	Jackson
Games played	45	39	42
Goals scored	23	22	25

Write a ratio to compare the following:

13. goals scored to games played for each player.

14. games played to goals scored for each player.

In a survey of Birchmont Park the number of trees found were:

Type of tree	Number of trees
maple	9
elm	4
beech	1
walnut	2
oak	5
chestnut	6
pine	8

Write a ratio to compare the numbers of the following types of trees:

15. oak to chestnut.

16. maple to elm.

17. beech to walnut.

18. elm to pine.

Use the chart in the previous question.

19. Find the total number of trees.

20. Write a ratio to compare the number of elm trees to the total number of trees.

21. Write a ratio to compare the number of walnut trees to the total number of trees.

SECTION 8–2

EQUIVALENT RATIOS AND PROPORTIONS

WARM-UP ACTIVITY

Su-Lyn found she had spent the following amounts of time during one week.

Sleeping	50 hours
Eating	14 hours
At school	32 hours
Free time and entertainment	48 hours
Homework	14 hours
Work	10 hours

Start today. Keep track of how you spend your time for one week.

INTRODUCTION

The Owens are planning to enlarge the size of their garden. Last year they used a mixture of two bags of peat moss for every three bags of special plant soil. This year they will need to use nine bags of special plant soil. If they increase the amount of peat moss in the same way, how many bags will they need?

One way to solve the problem for the Owens is to work with **proportions**. This section will show you how.

SKILLS DEVELOPMENT

The ratios 1:3 and 2:6 are called **equivalent ratios**. An equation that shows two ratios are equivalent is called a **proportion**.

To obtain equivalent ratios, you multiply both terms by the same number.

Ratio form	Fractional form
1:3 = 2:6 1 × 2 3 ×2	$\frac{1}{3} = \frac{1 \times 2}{3 \times 2} = \frac{2}{6}$

Ratios equivalent to 1:3 are shown.

1:3, 2:6, 3:9, 4:12, . . . $\frac{1}{3}, \frac{2}{6}, \frac{3}{9}, \frac{4}{12}, \ldots$

The ratio 1:3 is said to be in the lowest terms since the only common factor of each term is 1.

2:6 is not in lowest terms. $\frac{2}{6}$ is not in lowest terms.

To solve problems involving proportion, you need to find the missing terms of ratios, as shown in the next example.

Example 1

In the swim class, for every 2 students who chose backstroke, 3 chose crawl. 96 students chose the crawl. How many chose backstroke?

Solution Use n to represent the number of students choosing the backstroke.

$2:3 = n:96$ Write as a proportion.

$\frac{2}{3} = \frac{n}{96}$ Express as fractions.

$96(\frac{2}{3}) = 96(\frac{n}{96})$ Multiply by the LCD.

$64 = n$

64 students chose the backstroke.

Another way to solve proportion problems is called **cross multiplication**. Example 2 shows this method.

Example 2

Carlton Discs sells 2 cassette tapes for every 7 CDs sold. On Wednesday the store sold 49 CDs. How many tapes were also sold that day?

Solution Let x = the number of tapes sold.

$2:7 = x:49$ Write as a proportion.

$\frac{2}{7} = \frac{x}{49}$ Rewrite as fractions.

$2(49) = 7x$ *Cross multiply.*

$98 = 7x$

$14 = x$

The store sold 14 cassettes on Wednesday.

EXERCISES

Solve the following problems about equivalent ratios.

1. Which of the following are equivalent ratios?
6:8 9:15 12:16 15:20

2. Which ratios are equivalent to 3:5?
9:15 12:20 20:25

3. Which ratio is not equivalent to 4:3?
8:6 12:8 20:15

Write an equivalent ratio for each.

4. 2:5 **5.** 6:5 **6.** 3:8

7. $\frac{1}{2}$ **8.** $\frac{3}{2}$ **9.** $\frac{2}{7}$

Write each ratio in lowest terms.

10. 4:8 **11.** $\frac{4}{6}$

12. $\frac{6}{10}$ **13.** 10:5

14. $\frac{25}{100}$ **15.** 100:10

16. 85:20 **17.** 8:12

Write a ratio in lowest terms for each statement.

18. From a survey, Jean found that 20 out of 25 students walk to school. Write a ratio of students who walk to school compared to the total number of students.

19. In a mixture of garden soil there are 4 pails of peat moss to 6 pails of earth.

20. In Harry's class 3 out of 9 students have brown eyes.

21. There were 12 maples and 16 evergreens in the park.

Find the missing term by solving the proportion.

22. $3:6 = k:12$ To solve this proportion, you can also use the fractional form. $\frac{3}{6} = \frac{k}{12}$

23. $9:12 = 3:k$

24. $\frac{2}{3} = \frac{n}{15}$

25. $\frac{s}{12} = \frac{3}{4}$

26. $8:k = 4:3$

Solve these problems by using proportions.

27. In a recent poll the ratio of boys to girls interviewed was 15:5. If 102 boys were interviewed, how many girls would you expect to be interviewed?

28. Free-range farm eggs are shipped in boxes containing 180 eggs. Anita opened a carton of one dozen of the eggs and found that 2 were brown eggs. How many brown eggs would you expect to find in a full box?

29. At the corn roast, 3 cobs were cooked for every 7 persons. If 147 cobs were cooked, how many people were at the roast?

30. A survey showed that 4 persons out of every 9 persons wear seat belts. Out of 180 persons, how many persons would you expect to wear seat belts?

31. At an air show, there were 2 Piper Cubs for every 3 Hawks. There were 36 Hawks. How many Piper Cubs were there?

32. For a quarterback, the ratio of number of passes completed to number of passes attempted is 2:5. In a game, 20 passes were attempted. How many would you expect were completed?

33. Use proportions to solve. Find how many bags of peat moss the Owens need, as set up in the Introduction to this section.

SECTION 8–3

RATIO AND PROPORTION: CONSUMER APPLICATIONS

WARM-UP ACTIVITY

What should you keep after you have given it to someone else?

- Write each ratio in lowest terms.
- Fill the blanks with the matching letter.

___ ___ ___ ___ ___ ___ ___ ___

$\frac{7}{5}$ $\frac{10}{1}$ 5:7 1:2 3:1 21:8 $\frac{4}{1}$ $\frac{3}{4}$

1. 3:6 = ? (O)

2. $\frac{16}{4}$ = ? (S)

3. $\frac{35}{25}$ = ? (A)

4. 15:21 = ? (R)

5. $\frac{18}{24}$ = ? (E)

6. 9:3 = ? (M)

7. 42:16 = ? (I)

8. $\frac{100}{10}$ = ? (P)

INTRODUCTION

Barry and his little sister went to the grocery store to purchase 14 apples for lunches. Using the skills he had learned about ratios, Barry was able to decide which apples were the best buy.

SKILLS DEVELOPMENT

To compare prices in order to find the best buy, calculate the unit price. The **unit price** is the price for one item or unit. Barry had to think, "If 5 apples cost $1.15, and it is written as a ratio in the form of 5:1.15, I can write a proportion to find the cost for 1 apple."

Barry set up this proportion: $\frac{5}{1.15} = \frac{1}{x}$.

Since this type of ratio does not compare the same kind of units, it is sometimes referred to as **rate**.

rate	in ratio form	in fraction form
6 bottles for $2	6:2 or 3:1	$\frac{6}{2}$ or 3
200 m in 25 s	200:25 or 8:1	$\frac{200}{25}$ or 8
$399 for 3 weeks	399:3 or 133:1	$\frac{399}{3}$ or 133

Follow these steps to find the unit price.

1. Write the rate as a ratio in fraction form.

2. Set up a proportion using the fraction from Step 1 equal to 1 divided by x.

3. Solve the proportion.

4. Write the result as money.

Example 1

Help Barry find which apples are the better buy. He can buy apples at 5 for $1.15 or 8 for $1.76.

Solution

$\frac{5}{1.15}$	Write ratios.	$\frac{8}{1.76}$
$\frac{5}{1.15} = \frac{1}{x}$	Set up proportions to find the cost of one apple.	$\frac{8}{1.76} = \frac{1}{x}$
$5x = 1.15$	Cross multiply.	$8x = 1.76$
$x = 0.23$	Solve.	$x = 0.22$

The unit price at 5 for $1.15 is $0.23. The unit price at 8 for $1.76 is $0.22. You know $0.22 < $0.23, so 8 for $1.76 is the better buy.

Sometimes the unit price must be carried out three or four decimal places in order to find the better buy.

Example 2

Find the better buy: a package of 8 wieners that costs $1.69 or a package of 12 wieners that costs $2.55.

Solution package of 8: $\frac{8}{1.69} = \frac{1}{x}$

$$8x = 1.69$$

$$x = 0.21125$$

package of 12: $\frac{12}{2.55} = \frac{1}{x}$

$$12x = 2.55$$

$$x = 0.2125$$

Unit prices: package of 8 is 21.13 cents
package of 12 is 21.25 cents

The package of 8 wieners is the better buy.

EXERCISES

Find the unit price of each item. Round each answer to 0.01 of a cent.

1. A package of 52 almonds costs 84 cents.
2. 4 mini pizzas cost $2.89.
3. 24 cans of juice cost $6.89.
4. 200 paper clips cost 96 cents.
5. A case of 12 cans of oil costs $20.28.
6. 12 nectarines cost $2.29.
7. 450 napkins cost $2.66.

A 30 m size tape dispenser costs 98 cents.

8. Find the cost of 1 m of tape to the nearest 0.1 of a cent.
9. Find the cost of 1 cm of tape.

Henry wanted to buy some special nylon string for his kite. Find the cost per meter for each store.

10. Store A: 50 m/$1.80

11. Store B: 100 m/$3.40

12. Store C: 60 m/$2.70

13. Store D: 75 m/$3.00

14. Which store has the best buy on kite string?

Four stores advertised specials on fancy dinner rolls Friday afternoon. Find the cost per roll.

15. A: 1 dozen for $3.36

16. B: 6 for $1.44

17. C: 3 dozen for $9.36

18. D: 2 dozen for $6.00

19. Which store has the best buy on fancy dinner rolls?

Use these advertisements for the exercises below.

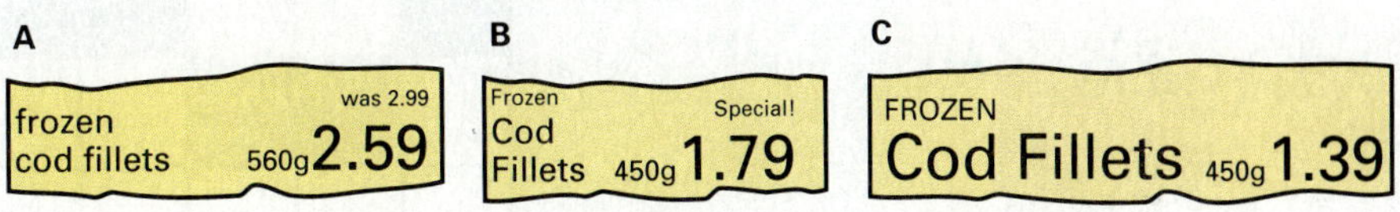

20. Find the unit cost for each ad.

21. Which ad shows the best buy?

The local video stores are trying to attract new business. Find the cost per movie rental for each plan. Round answer to the nearest cent.

22. Plan A: Rent 5 movies in any week for only $6.00

23. Plan B: Rent 8 movies in any week for $7.00

24. Plan C: Rent 10 movies in any month for $9.00.

25. Plan D: Rent 25 movies in any month for $21.50.

26. Which plan is the best buy?

27. Your family can only watch movies on weekends. Which plan would you choose? Why?

SECTION 8–4

USING PROPORTIONS TO SOLVE PROBLEMS

WARM-UP ACTIVITY

Use your number sense to complete this number puzzle. Copy the grid and complete. The clues give the **sum** of the digits which fill the squares. No zeros are used. The digits 1–9 cannot repeat in any group.

Across
1. 16
3. 7
6. 39
8. 16
10. 17
12. 21
16. 6
17. 11

Down
1. 17
2. 22
3. 5
4. 7
5. 20
7. 16
9. 16
10. 14
11. 23
13. 3
14. 5
15. 3

INTRODUCTION

Asham can type 96 words on a 2-minute timed test. To apply for a job at an attorney's office, he must take a typing examination that will last 5 minutes. Assuming that Asham will type at the same rate, a proportion can be set up comparing words to minutes in order to find if he will reach the 200-word minimum in 5 minutes.

SKILLS DEVELOPMENT

When using proportions to solve problems, there are two things you must remember.

#1 — You often make assumptions.
#2 — The two sides of the proportion must compare in the same order.

Using the problem about Asham, we assume that he can type at the same rate for 5 minutes. When setting up the proportion, compare words to minutes both times.

Let's see if Asham will be able to have an interview. Since the number of words is not known in the second ratio, we will use the variable x to represent it.

- Write a proportion. $\frac{96}{2} = \frac{x}{5}$ $\frac{\text{words}}{\text{minutes}} = \frac{\text{words}}{\text{minutes}}$
- Cross multiply. $2x = 480$
- Solve. $x = 240$

Assuming he does type at the same rate, Asham will be able to have an interview.

Proportions can also be used to find the missing length when given two similar triangles. Similar triangles have the same shape but are not the same size. When solving for the missing length, write a proportion that compares corresponding sides.

EXERCISES

Select a comparison, in words, that can be used to set up a proportion.

1. Eighteen inches of snow fell in six hours.

2. The goalie allowed 2 goals out of 11 shots.

3. In 45 minutes 413 pounds of potatoes were peeled.

4. A bobsled travels 128 kilometers in one hour.

Solve each of the following problems. List any assumptions you made to obtain the answer.

5. Nathan types at a rate of 45 words per minute. How long will it take him to type 900 words?

6. Steven scored 7 goals in 14 games. How many goals would you expect him to score in 36 games?

7. Nyghi walked 10 blocks in 25 minutes. How many blocks could she walk in 2 hours?

8. Susan scored 28 points in 4 games of basketball. How many points would you expect her to score in 20 games?

Solve.

9. One week Jon worked 18 hours washing cars and earned \$58.50. What is his rate of pay per hour?

10. If Jon worked 65 hours in July, how much did he earn?

Use these ads to answer the following questions.

Store A	Store B
3 Badminton birds for \$1.99	5 Badminton birds for \$2.77

11. Calculate the unit price at Store A.

12. Calculate the unit price at Store B.

13. Which store has the better rate?

Find the length of the missing side. $\triangle ABC$ and $\triangle DEF$ are similar.

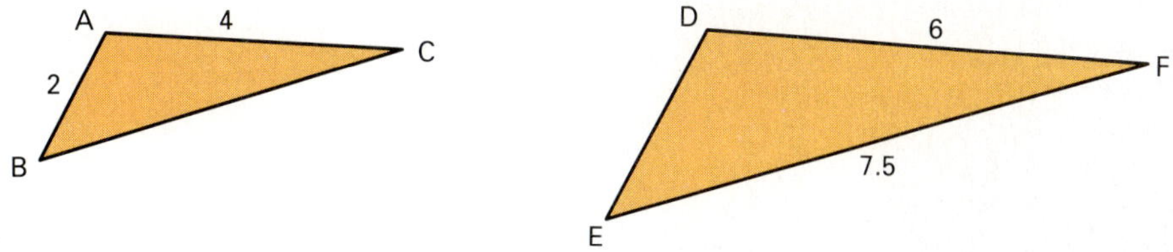

14. Write the corresponding sides as ratios.

15. Write a suitable proportion to find the length of BC. What is the length of BC?

16. Find the length of DE.

A computer program can solve proportions that are in the form $\frac{A}{B} = \frac{x}{D}$, where A, B, and D are known.

```
10  PRINT "INPUT KNOWN QUANTITIES, A, B, AND D
20  INPUT A, B, D
30  PRINT "X = ", A * D/B
40  END
```

Use this program for the following exercises.

17. Explain how this program solves proportions.

18. Can this program be used to solve $\frac{14}{23} = \frac{20}{x}$? Explain your answer.

19. If $A = 18$, $B = 3$, and $D = 6$, what would the computer print as the output?

20. A student can complete 36 problems in one hour. How many problems can that student complete in 45 minutes?

21. Gina got an answer of 1620 problems for the last exercise. She knew that answer was not reasonable, but does not know what she did wrong. Tell what Gina did wrong.

SECTION 8–5

PERCENT

WARM-UP ACTIVITY

Find the pattern and then fill the blanks.

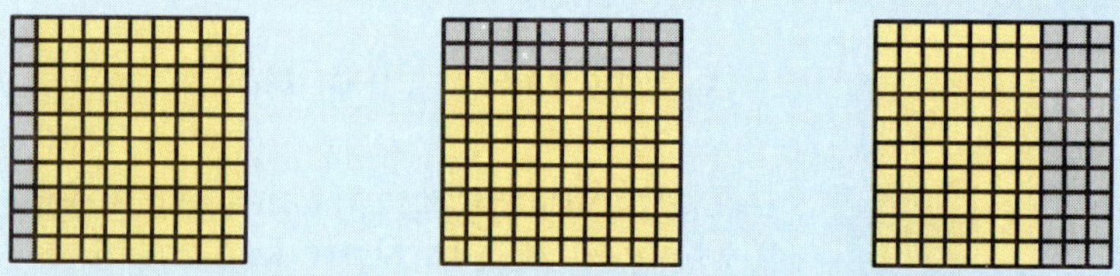

____ out of 100 are shaded.	____ out of 100 are shaded.	____ out of 100 are shaded.
____ are shaded.	____ are shaded.	____ are shaded.

INTRODUCTION

In all aspects of our lives we encounter ratios written as percents. From previous work you know that ratios can be written in fraction or decimal forms. Fractions, decimals, and percents are all closely related and often are rewritten in one of the other forms as needed to solve problems.

SKILLS DEVELOPMENT

To record the information in the ad, you could write a ratio.

> 45 of 100 ARE GONE!
> Hurry get the best
> buy on bicycles today!

45:100

number sold total number

The ratio of bicycles sold to the total number of bicycles can also be written in fractional form as $\frac{45}{100}$. When the second term is 100, the symbol % is often used.

$\frac{45}{100}$ can be written as 45% and is read **45 percent.**

Since percent means per hundred, $\frac{45}{100}$ could be interpreted as 45 out of 100, or 45:100.

Example 1

Write 23 of 50 as a percent.

Solution 23 of 50 can be written 23:50.

23:50 can be written $\frac{23}{50}$.

Find an equivalent ratio that makes the second term 100 by multiplying by 2.

$$\frac{23 \times 2}{50 \times 2} = \frac{46}{100}$$

$$\frac{46}{100} = 46\%$$

Thus, 23 of 50 represents 46%.

Example 2

Write $\frac{4}{5}$ as a percent.

Solution $\frac{4}{5} = \frac{x}{100}$ Find the missing term.

$400 = 5x;\ x = 80$ Cross multiply and solve.

$\frac{4}{5} = \frac{80}{100}$

Thus, in percent form,

$$\frac{4}{5} = \frac{80}{100} = 80\%$$

The pattern in the following examples shows a way to change a decimal to a percent.

Example 3

Write each as a percent.

a. 0.75 **b.** 0.8 **c.** 0.435

Solutions a. $0.75 = \frac{75}{100} = 75\%$

b. $0.8 = \frac{8}{10} = \frac{80}{100} = 80\%$ Think $\frac{8 \times 10}{10 \times 10} = \frac{80}{100}$

c. $0.435 = \frac{435}{1000} = \frac{435 \div 10}{1000 \div 10} = \frac{43.5}{100} = 43.5\%$

To change a decimal to a percent, multiply the decimal by 100, or move the decimal point 2 places to the right.

To write a fraction as a decimal, divide the numerator by the denominator. If your calculator has fraction keys, they may help you. Then change the decimal to a percent.

Example 4

Write 60% as a fraction in lowest terms.

Solution $60\% = \frac{60}{100}$ Percent means per hundred.

$= \frac{60 \div 20}{100 \div 20} = \frac{3}{5}$

Often you will see percents that involve fractions or decimals. To write a percent as a decimal numeral, remember that percent means per hundred, as shown.

$37\frac{1}{2}\% = 37.5\%$ $\frac{1}{2} = 0.5$, so $37\frac{1}{2} = 37.5$

$= \frac{37.5}{100}$ or 0.375

EXERCISES

Each diagram is made from 100 squares. Write a fraction, a decimal, and a percent to show what part has been shaded.

1. 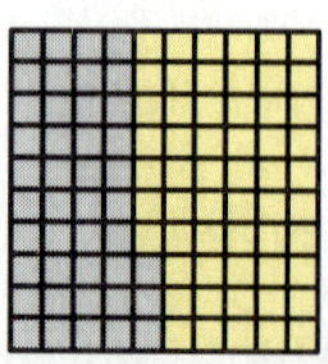

2. 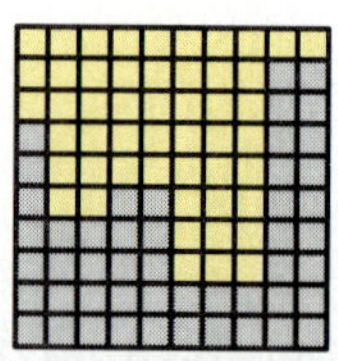

3. 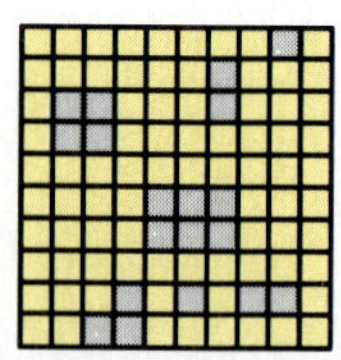

4. For each design above, what percent is not shaded?

Express each of the following as a percent.

5. $\frac{25}{100}$

6. $\frac{27}{100}$

7. $\frac{10}{100}$

8. $\frac{39.5}{100}$

9. $\frac{800}{100}$

10. $\frac{85}{100}$

Write each decimal as a percent.

11. 0.73

12. 0.48

13. 0.09

14. 0.3

15. 0.125

16. 0.625

Write each fraction as a percent. Round answers to one decimal place as needed.

17. $\frac{7}{10}$

18. $\frac{11}{20}$

19. $\frac{1}{10}$

20. $\frac{1}{50}$

21. $\frac{2}{5}$

22. $\frac{3}{8}$

23. $\frac{2}{3}$

24. $\frac{1}{9}$

25. $\frac{9}{11}$

In each statement a ratio occurs. Express the ratio as a percent.

26. In Allan's class 6 out of 10 students have blue eyes.

27. 4 out of 5 students take a bus to school.

28. Out of every 20 households, 15 have a dog.

Out of 500 students interviewed, 375 said they have too much homework.

29. Write a percent for the information.

30. Write a similar problem and write a percent.

Skiing is a popular sport in the United States. At a ski lodge 40 skiers were interviewed, with the following results. Copy and complete the table.

		Number	Percent
31.	Cross-country racing?	8	
32.	Downhill racing?	4	
33.	Cross-country pleasure?	12	
34.	Downhill pleasure?	16	

SECTION 8–6

FINDING PERCENTS

WARM-UP ACTIVITY

Copy and label blanks as shown below. Perform each calculation. Use the chart. Write the letter for each answer above the correct blank. What can be right but never wrong?

____ ____   ____ ____ ____ ____ ____
4.   **1.**   **7.**   **3.**   **6.**   **2.**   **5.**

Remember to use the correct order of operation.

1. $\frac{3}{4} \times \frac{5}{6} + \frac{1}{8}$ **2.** $1\frac{3}{4} + \frac{1}{2} \times 4$ **3.** $2\frac{5}{8} \div \frac{7}{8} - \frac{2}{3}$

4. $\frac{3}{4} \times \frac{5}{8} \div \frac{3}{16}$ **5.** $2 + 3 \times \frac{5}{8}$ **6.** $(2 + 3) \times \frac{5}{8}$

7. $1\frac{2}{3} \div \frac{5}{6} + 1\frac{1}{3}$

$\frac{3}{4}$	$2\frac{1}{3}$	$2\frac{1}{2}$	$3\frac{1}{8}$	$3\frac{1}{3}$	$3\frac{3}{4}$	$3\frac{7}{8}$
N	N	A	G	A	L	E

INTRODUCTION

Student	Ed	Oni	Ali	Mona	Inez
Number correct	72	35	16	18	81
Number of problems	100	50	25	25	100

Which student scored higher on the different versions of a quiz?

It is difficult to compare the results because the total number of questions is different.

If all results are expressed as percents, they can be compared easily because they would all be parts of 100. By changing comparisons into percents, you make the results more meaningful.

SKILLS DEVELOPMENT

Example 1

Rank the scores of the students, as given in the Introduction to this section.

Solution Ed and Inez each completed 100 problems. Write their scores as a fraction or decimal, then change to a percent.

Ed: $\frac{72}{100}$ or $0.72 = 72\%$ Inez: $\frac{81}{100}$ or $0.81 = 81\%$

Use proportions to find the percent correct.

Oni: $\frac{35}{50} = \frac{x}{100}$ Ali: $\frac{16}{25} = \frac{x}{100}$ Mona: $\frac{18}{25} = \frac{x}{100}$

$50x = 3500$ $25x = 1600$ $25x = 1800$

$x = 70$ $x = 64$ $x = 72$

Oni scored 70%. Ali scored 64%. Mona scored 72%.

Now rank the scores in order. Inez 81%; Ed and Mona 72%; Oni 70%; Ali 64%.

In sports, the performances of players are often compared, using percent. A coach can then use these percents to make important decisions.

Example 2

In a basketball game a technical foul has been called by the referee. Which player should shoot the free throw?

Player	Free throws scored	Free throws attempted
Stein	41	77
Marshall	39	74
Steadman	45	83

Solution Express each player's record as a percent. You can use your calculator to change each ratio to a decimal. Round your answers to the nearest tenth of one percent. The ratio is "number of free throws scored" to "number of free throws attempted."

Stein $\frac{41}{77} = 0.532467532$ or 53.2%

Marshall $\frac{39}{74} = 0.527027027$ or 52.7%

Steadman $\frac{45}{83} = 0.542168674$ or 54.2%

Compare the percents, and see that Steadman has the highest percent. He should shoot the free throw.

EXERCISES

Write a percent for each of the following. Which is greater?

1. $\frac{7}{25}, \frac{7}{10}$ **2.** $\frac{17}{50}, \frac{3}{20}$ **3.** $\frac{4}{5}, \frac{15}{25}$

4. $\frac{29}{50}, \frac{26}{45}$ **5.** $\frac{52}{75}, \frac{66}{95}$ **6.** $\frac{12}{15}, \frac{19}{25}$

7. 28 trees out of 156 or 36 trees out of 210

8. 36 packages out of 125 or 70 packages out of 245

Write a percent for each of the following.

9. 20 cm out of 125 cm **10.** 25 mm of 5 cm

11. 25 cm out of 2.5 m **12.** 30 s out of 4 min

13. \$60 out of \$250 **14.** 2 m of 8 m

15. A macaroni and cheese dish had 250 grams of macaroni and 45 grams of cheese. What percent of the dish was cheese?

16. Terry picked 8 baskets of tomatoes. His sister threw away 1 of the baskets due to bruising. What percent were good?

17. A mini-keyboard has 8 black keys and 12 white keys. What percent of the keys are black? white? Which percent is greater?

18. Who saves the greater percent of her earnings? (Estimate first, then calculate.)
- Sara earns $20 a week part-time and saves $12
- Miriam earns $23 a week part-time and saves $14.

19. On a test of 30 questions, Harriet correctly answered 20 of them. In her next test she answered 15 out of 20 correctly. On which test did she do better?

20. The chart shows the results for the question asked in a survey. On which day did the greater percent say yes?

"Do you enjoy winter sports more than summer sports?"

Day	Number of persons	Yes
Sat.	112	86
Sun.	89	64

21. The Garfields spend $450 a month on housing.
- **a.** If the income is $2550 each month, what percent of money is spent on housing?
- **b.** What percent is not spent on housing?
- **c.** What do you notice about the total of **a** & **b**?

22. In a poll, 18 out of 45 people enjoyed the new pizza restaurant in town. In another poll taken, 26 out of 59 people liked the restaurant. Which poll has a greater percent of people who liked the restaurant? Give some reasons why the difference may have occurred.

23. Two lakes have different amounts of fish. For every 145 fish in Rice Lake, 76 are bass. For every 200 fish in Go Home Bay, 98 are bass. Use your skills with percent to help you decide which lake has the best bass fishing.

SECTION 8–7

FINDING THE PERCENT OF A NUMBER

WARM-UP ACTIVITY

Write the missing term from each proportion in the proper square. If you answer each question correctly, then you will have a magic square.

1. $\frac{2}{3} = \frac{x}{6}$ **2.** $\frac{36}{18} = \frac{x}{1}$ **3.** $\frac{27}{15} = \frac{x}{5}$

4. $80{:}8 = x{:}1$ **5.** $1{:}5 = x{:}25$ **6.** $0{:}1 = x{:}2$

7. $\frac{15}{30} = \frac{x}{2}$ **8.** $\frac{7}{63} = \frac{x}{72}$ **9.** $\frac{3}{2} = \frac{x}{4}$

1.	2.	3.
4.	5.	6.
7.	8.	9.

The magic number is ______.

INTRODUCTION

In the previous section you found the percent of two numbers by writing a ratio that compared the part to the whole. Sometimes the percent is given, but the part or the whole is missing. Using what you know about proportions and percent, you will learn how to find the missing value. In this section you will find the missing part.

SKILLS DEVELOPMENT

The proportion we use to find the missing part is:

$$\frac{\text{percent}}{100} = \frac{\text{part}}{\text{whole}}.$$

Percent is the number that is written in front of the symbol %. **100** is always 100. **Part** is the number that usually follows the words "is". **Whole** is the number that usually follows the word "of". **Remember**: The percent of the whole is the part.

Example 1

a. Find 1% of 250. **b.** Find 25% of 136.

Solutions a. **percent** — 1

100 — 100 $\frac{1}{100} = \frac{x}{250}$

part — unknown $100x = (1)250$

whole — 250 $x = 2.5$

b. **percent** — 25

100 — 100 $\frac{25}{100} = \frac{x}{136}$

part — unknown $100x = 25(136)$

whole — 136 $x = 34$

Some calculators have the % symbol on them. If you have a calculator, try this:

To find 25% of 136, press these keys.

[1] [3] [6] [X] [2] [5] [%]

Did you get an answer of 34? Try using your calculator for other examples.

When the percent is a decimal or fraction, use the same procedure.

Example 2

a. Find 2.3% of 46. **b.** Find $1\frac{1}{4}$% of 68.

Solutions a. **percent** — 2.3 $\frac{2.3}{100} = \frac{x}{46}$

part — x $100x = 2.3(46)$

whole — 46 $100x = 105.8$

$x = 1.058$

b. **percent** — $1\frac{1}{4}$ $1\frac{1}{4} = \frac{x}{68}$

part — x $100x = 1\frac{1}{4}(68)$

whole — 68 $100x = 85$

$x = 0.85$

Example 3

The original price of a bicycle is $149.95. During a sale it is reduced 10%. How much will the discount be during the sale?

Solution **percent** — 10 $\frac{10}{100} = \frac{x}{149.95}$

part(discount) — x $100x = 1499.50$

whole(original) — 149.95 $x = 14.995$

The discount will be $15.

EXERCISES

Match each expression with its proportion.

1. 12% of 80 is what? **a.** $\frac{x}{100} = \frac{12}{80}$
2. 12% of what is 80? **b.** $\frac{12}{100} = \frac{80}{x}$
3. What percent of 80 is 12? **c.** $\frac{x}{100} = \frac{80}{12}$
4. What percent of 12 is 80? **d.** $\frac{12}{100} = \frac{x}{80}$

Calculate 1% of each of the following.

5. 100 m **6.** 1000 km **7.** 3000 cars

8. 30 cm **9.** 150 g **10.** 35 in

Calculate 10% of each of the following.

11. 120 oz **12.** 600 m **13.** 45 km

14. 130¢ **15.** 160 people **16.** 480 beads

Calculate 50% of each of the following.

17. 210 **18.** 38 **19.** 105.2

Calculate each of the following.

20. 100% of 120 **21.** 100% of 360

For each of the following estimate the value, calculate the value, and round to a reasonable value.

22. 45% of 16 people **23.** 78% of 24 stamps

24. 39% of 18 tires **25.** 11% of 32 computers

26. 17% of $77 **27.** 2.5% of 34¢

28. 12.5% of 19 words **29.** 15% of 23 songs

To find the sale price of an item, first find the amount of discount. Then subtract the discount from the original price.

Find 25% off $120.
Discount is 25% of $120, or $30.
Sale price is $120–$30, or $90.

Each article has a 15% discount. (The price is reduced by 15%.) Copy and complete the chart.

	Original price	Discount	Selling price
30.	$120	?	?
31.	$342	?	?
32.	$96	?	?

SECTION 8–8

FINDING THE WHOLE NUMBER

WARM-UP ACTIVITY

How does a shark keep its teeth sharp?

- Find the answer to the question by answering each percent question that follows. Example: The first 50% of the word ITEM is IT.
- Write the letters in order.
- Put in spaces so the answer makes sense.

1. first 50% of ITEM

2. 100% of SHE

3. first or last 25% of DEED

4. first $33\frac{1}{3}$% of STREET

5. 100% of HEM.

6. first 70% of CONTINUING

7. last 75% of FALL

8. last 25% of TIDY

9. last 60% of GRAND

10. 100% of GROWS

11. last 60% of RENEW

12. first 50% of RAIN

13. first 60% of ZORRO

14. last 80% of HEDGE

15. 100% of DONE

16. first 75% of SING

17. last 60% of LATHE

18. last 50% of STIR

19. first 75% of PLAY

20. last 50% of FACE

___ ___   ___ ___ ___ ___ ___   ___ ___ ___ ___

___ ___ ___ ___ ___ ___ ___ ___ ___ ___ ___   ___ ___ ___

___ ___ ___ ___ ___   ___ ___ ___   ___ ___ ___ ___ ___

___ ___ ___ ___ ___   ___ ___ ___ ___   ___ ___

___ ___ ___ ___ ___   ___ ___ ___ ___ ___

INTRODUCTION

A stereo has been reduced \$259.00, which is a 30% reduction. What was the original price?

Merchants use this type of advertising to entice customers to come into the store and ask the price, giving the salesperson an opportunity to sell the item. People who know how to work with percents can figure the original price themselves.

SKILLS DEVELOPMENT

In this section the unknown will be the whole, or the original amount. Recall from Section 8–7 that the number written in place of **part** in the proportion usually follows the word "is" and is the discount.

Example 1

68% of what number is 17?

Solution **percent** — 68 $\frac{68}{100} = \frac{17}{x}$

part — 17 $68x = 17(100)$

whole — x $68x = 1700$

$x = 25$

Example 2

Find the original price of the clock radio if it is marked 20% off, so you save \$7.25.

Solution **percent** — 20 $\frac{20}{100} = \frac{7.25}{x}$

part — 7.25 $20x = 100(7.25)$

whole — x $x = 36.25$

The original price was \$36.25.

EXERCISES

In each problem, tell what the 3 represents: the percent, the part, or the whole.

1. 50% of 6 is 3.

2. 3% of 60 is 1.8.

3. 90% of 3 is 2.7.

4. 2% of 150 is 3.

Write a proportion for each problem. DO NOT SOLVE.

5. 1% of a number is 12?

6. 50% of what is 80%?

7. 6% of the people at the beach have umbrellas. 21 people have umbrellas. How many people are at the beach?

8. Jeff saved $98 on a tennis racquet at a 25% off sale. How much did the racquet originally cost?

Find each number.

9. 1% of a number is 16.

10. 10% of a number is 6.3.

11. 50% of a number is 140.

12. 100% of a number is 250.

Find *n* in the following.

13. 10% of *n* is 3.

14. 40% of *n* is 10.

15. 50% of *n* is 16.

16. 6% of *n* is 24.

Find the number in each of the following.

17. 8% of the number is 32.

18. 25% of the number is 5.

19. 13% of the number is 65.

20. 8% of the number is 64.

Find each of the following to one decimal place.

21. 40 is 75% of what number?

22. 75 is 18% of what number?

23. 28 is 35% of what number?

24. 77 is 44% of what number?

For each of the following, express your answers to one decimal place or the nearest cent.

25. About 15% of the tree seedlings planted grew. If 375 seedlings grew, how many were planted?

26. Sasha hit 20% of the pitches thrown to her. If she had 8 hits, how many pitches were thrown?

27. 30% of the students had no cavities in the toothpaste test. There were 135 students that had no cavities. How many were tested?

28. You can save \$13.50 if you buy the running shoes now. They are 30% off. What is the regular price of the shoes?

29. A rubber ball bounces to 70% of the height from which it was dropped. A ball bounced up to 9 ft after it was dropped. From what height was the ball dropped?

30. The amount of an iceberg showing above the water is 10% of the total mass. The amount calculated above the water is 285 tons. What mass of the iceberg is below the water?

31. During a pollution survey, 12.2% of the garbage was glass. The amount of glass was 100 lbs. How much garbage was collected in all?

32. On a pizza order you save 10% if you pick it up yourself. Andrea saved \$1.35 on her order. How much did the pizza cost originally?

33. Did you know that TIP means To Improve Performance? A customer often leaves a tip of 15% if the service is good. How much was the bill if \$1.65 was left as the tip?

34. A bicycle store is selling bicycles for a 45% reduction. The savings on one bicycle is \$86. How much did the bicycle cost before the sale?

35. At the Olympic Games 85% of the stadium was full. There were 35,700 people. What is the seating capacity of the stadium?

SECTION 8–9

DECIDING WHICH SKILL TO USE

WARM-UP ACTIVITY

1. Find $5^1, 5^2, 5^3, 5^4, 5^5, 5^6, 5^7, 5^8, 5^9$. What pattern is formed by the ones digits in the answers? Predict what the ones digit in 5^{100} will be.

2. Calculate $2^1, 2^2, 2^3, 2^4, 2^5, 2^6, 2^7, 2^8, 2^9$. Note the pattern and tell what the ones digit will be in 2^{100}.

3. Repeat the previous question for $7^1, 7^2, 7^3, 7^4, 7^5, 7^6, 7^7, 7^8, 7^9$, and 7^{100}.

INTRODUCTION

Good problem-solvers use the available information to narrow the possibilities and identify a solution method. All percent problems can be solved using the proportion from this statement: the percent of the whole is the part.

$$\frac{\text{percent}}{100} = \frac{\text{part}}{\text{whole}}$$

SKILLS DEVELOPMENT

When solving problems involving percent, you need to decide which number from the proportion is unknown. Compare the three statements below to determine the unknown.

a. Find what percent 6 is of 24.
b. Find 40% of 35.
c. 20% of what number is 15?

Step 1 Rewrite each statement in the form you have learned.
Step 2 Set up the proportion.
Step 3 Solve the proportion.
Step 4 Check that the answer is reasonable.

Example 1

Find what percent 6 is of 24.

Solution x% of 24 is 6. The percent is unknown.

$$\frac{x}{100} = \frac{6}{24}$$

$$24x = 600$$

$$x = 25 \qquad 25\% \text{ of } 24 \text{ is } 6.$$

Example 2

Find 40% of 35.

Solution 40% of 35 is x. The part is unknown.

$$\frac{40}{100} = \frac{x}{35}$$

$$100x = 1400$$

$$x = 14 \qquad 40\% \text{ of } 35 \text{ is } 14.$$

Example 3

20% of what number is 15?

Solution 20% of x is 15. The whole is unknown.

$$\frac{20}{100} = \frac{15}{x}$$

$$20x = 1500$$

$$x = 75 \qquad 20\% \text{ of } 75 \text{ is } 15.$$

Sometimes finding percents is only one step in solving a problem, as in the following example.

In most states, you must pay sales tax on certain goods.

Example 4

A bicycle is on sale for $149.50. Find the total cost of the bicycle that has 4% sales tax.

Solution

Selling price of bicycle	$149.50
Sales tax 4% (4% of 149.50)	+ 5.98
Total cost	$155.48

You pay $155.48 for the bicycle.

EXERCISES

For each problem tell whether the unknown is the percent, the part, or the whole.

1. 60% of 240 is what?

2. 33 is what percent of 165?

3. What is 43% of 275 points?

4. 50% of how many fish is 350 fish?

5. $\frac{25}{100} = \frac{x}{33}$

6. $\frac{x}{100} = \frac{15}{45}$

Write a problem to go with each proportion.

7. $\frac{x}{100} = \frac{60}{80}$

8. $\frac{40}{100} = \frac{x}{30}$

9. $\frac{33}{100} = \frac{60}{x}$

Solve each problem. Estimate your answer between steps 1 and 2.

10. 50% of what number is 37?

11. 30 is what percent of 120?

12. What is 12.5% of 32?

13. 60% of 80 is what number?

14. 40 is what percent of 75?

15. 80% of what number is 48?

16. 75 is what percent of 200?

17. What percent of 600 is 150?

For each of the following problems, round percents to the nearest whole number percent, and money to the nearest cent.

18. Linda correctly answered 80% of the 20 questions in a quiz.

a. How many did she answer correctly?

b. How many did she miss?

19. The recent Community Fund Drive set its goal at \$45,000. If \$35,000 was collected, what percent of its goal was reached?

20. Because of a big windstorm, 12% of a farmer's 625 fruit trees were ruined. How many were ruined?

21. A baseball glove costs Jeff \$29.50. Calculate his total cost if the sales tax is 8%.

Three students at Oakwood School did an environment survey. One question they asked was: Would you be willing to buy milk and other drinks in returnable containers? They found that 83% were willing to do so. Suppose 249 persons said yes.

22. How many persons were interviewed?

23. How many said no?

The skin is about 16% of a person's total body mass.

24. If Caroline's mass is 90 lbs, calculate the mass of her skin.

25. Gina's mass is 108 lbs. Calculate the mass of her skin.

26. If the mass of Noele's skin is 17 lbs, calculate her mass.

Read this article.

> What Part of the Human Body Is Water?
>
> A recent study was made to determine the amount of water in the human body. Experiments show that the average person, by mass, is 65% water. During the study, the results also showed that this figure varied from person to person. A lean person may have as much as . . .

27. George worked out how much water was in his body. His calculations showed he has 104 lbs of water. What is George's mass?

28. Susan found she has 66 lbs of water in her body. What is her mass?

CHAPTER 8 SUMMARY

In this chapter you learned how to express comparisons as ratios. Ratios can be set up using a colon or as a fraction. To solve problems, you learned to use equivalent ratios to set up proportions. By solving the proportion, you solved the problem. Ratios and proportions are very useful in solving consumer problems involving rate, unit prices, and similar figures.

You also learned how to solve problems involving percents by recognizing whether you were to find the part, the whole, or the percent. Percents are used in many ways in real life. You have learned to solve many such problems.

VOCABULARY

equivalent ratios
part
percent
proportion
rate
ratio
terms
unit price
whole

CHAPTER REVIEW

Use the table to write the ratios.

Name	Number Sold
Kyle	25
Kayla	33
Kern	28
Total	86

1. Kyle's sales to Kern's sales
2. Kayla's sales to the total sales
3. Kyle's sales to Kayla's sales to Kern's sales
4. What is the ratio of Kern's sales to total sales, written as a fraction in lowest terms?

Tell whether the ratios are equivalent.

5. $\frac{2}{3} = \frac{12}{15}$
6. $\frac{5}{10} = \frac{1}{2}$
7. $\frac{8}{9} = \frac{89}{99}$

Solve each proportion.

8. $\frac{3}{4} = \frac{x}{12}$

9. $\frac{4}{5} = \frac{x}{25}$

10. $\frac{8}{9} = \frac{x}{126}$

11. $\frac{1}{x} = \frac{10}{30}$

12. $\frac{x}{22} = \frac{12}{33}$

13. $\frac{7}{8} = \frac{63}{x}$

Find the unit price for each product. Then decide which is the better buy.

14. 2 bottles of juice for $3.98 or 3 bottles of juice for $5.70

15. 3 dozen rolls for $2.68 or 4 dozen rolls for $3.60

Use proportions to solve each problem.

16. Sahara made 16 of 24 free throws. If she shoots at that rate, how many free throws will she make if she shoots 42 free throws?

17. A train traveled 180 miles in 4 hours. At that rate, how long will it take for the train to go 540 miles?

Write each of the following as a percent.

18. $\frac{1}{2}$

19. $\frac{3}{8}$

20. 0.75

21. 0.054

22. In one class there are 18 girls and 12 boys. What percent are boys?

Write each as a percent. Tell which is greater.

23. $15 out of $20 or 35¢ out of 50¢

24. 6 wins out of 18 games or 9 wins out of 24 games

Solve each of the following.

25. 18% of 250 is what?

26. What percent of 20 is 5?

27. 100% of 567 is what?

28. 18 is 50% of what number?

29. 3 is what percent of 6?

30. What percent of 45 is 9?

31. You left a 15% tip at a restaurant. If you left a $3 tip, what was the total bill?

32. Juan bought a radio for $20. The sales tax was $1.20. What was the rate of the sales tax?

33. The goal of the fund-raising committee is $40,000. They have collected 73% of their goal. How much have they collected?

34. A car normally sells for $9,565. It is on sale for 15.5% off. To the nearest whole dollar, what is the sale price?

CHAPTER 8 TEST

Write a ratio for each situation.

1. Angelica ran 8 blocks. Andy ran 7 blocks. What is the ratio of the distances Andy and Angelica ran?

2. Bernie read 150 pages while Barbara read 200 pages. Write a ratio to compare Bernie's number of pages read to Barbara's. Express the ratio as a fraction in lowest terms.

Tell whether the ratios are equivalent.

3. $\frac{8}{11} = \frac{11}{8}$

4. $\frac{2}{5} = \frac{32}{75}$

5. $\frac{3}{10} = \frac{60}{200}$

Solve each of the following.

6. $\frac{7}{10} = \frac{x}{120}$

7. $\frac{x}{8} = \frac{3}{4}$

8. $\frac{5}{x} = \frac{35}{42}$

Write each of the following as a percent.

9. 0.95

10. $\frac{3}{5}$

11. $\frac{3}{16}$

Solve each of the following.

12. 22% of $180 is what?

13. 15 is what percent of 75?

14. What number is 40% of 80?

15. 25% of what number is 80?

16. You can buy 18 tickets for $10 or 30 tickets for $21. Find the unit price for each. Tell which is the better buy.

17. A plane traveled 650 miles in 2 hours. At that rate, how far will the plane travel in 5 hours?

18. There were 25 adults and 25 students on a bus. What percent of the people on the bus are students?

19. You want to buy a coat that costs $45. The sales tax is 7%. What will the total cost be?

20. The school's fund-raising committee wants to buy computers for the math laboratory. They need to raise $5270. They have raised 23%. To the nearest whole dollar, how much have they raised?

SKILLS PREVIEW

Find each product.

1. 4×4
2. $2 \times 2 \times 2$
3. $3 \times 3 \times 3 \times 3$
4. $1 \times 1 \times 1 \times 1 \times 1$

Add.

5. $46 + 32 + 9 + 16$
6. $13 + 13 + 8 + 8$
7. $35 + 67 + 29$
8. $129 + 110 + 98$
9. $54 + 31 + 16 + 22 + 42$
10. $75 + 75 + 34 + 34 + 15$
11. $2.69 + 11.09 + 0.64$
12. $30.32 + 16.9 + 21.18$
13. $49.64 + 39.08 + 52.4$
14. $109.5 + 75.64 + 80.3$

Divide.

15. $296 \div 4$
16. $549 \div 6$
17. $355 \div 5$
18. $627 \div 3$

Multiply.

19. 32×21
20. 42×134
21. $9 \times 12 \times 22$
22. $25 \times 19 \times 35$
23. 4.2×3.9
24. 9.23×0.6
25. $1.5 \times 0.4 \times 0.72$
26. $25.1 \times 3.8 \times 1.1$

Calculate.

27. $2(3.8) + 2(4.5)$
28. $2(12.03) + 2(24.8)$
29. $2(31.6) + 2(14.53) + 2(14.65)$
30. $2(46.81) + 2(26.18) + 2(33.33)$
31. $(3 + 2x)(3x) + 2(3x)$
32. $4(3x + 1) + x(3x + 2)$

Everyone is concerned with measurement. In modern-day technology, the length of time it takes for some events to happen is measured in milli and micro seconds.

Measurement plays a vital part in all your activities. Many of the sports you enjoy depend on the measurement of distance. You use units of measurement when you buy clothing and food. Your parents need to know the dimensions of a room before they redecorate it. Measurement is also needed to build aircraft, boats, bridges, and buildings.

You may measure distances in inches, feet, and miles or in centimeters, meters, and kilometers. In this chapter you will work with exponents and the metric and customary measurement systems. You will learn skills needed for working with measurement, and you will apply these skills to measuring various figures.

Ask Yourself

- *What is scientific notation?*
- *What makes the metric system so easy to use?*
- *What happens when you square a binomial?*
- *What are volume, capacity, and mass?*
- *What are cubic units?*

CHAPTER 9

EXPONENTS AND MEASUREMENT

SECTION 9–1

LAWS OF EXPONENTS

WARM-UP ACTIVITY

Place parentheses around groups of numbers to make a true statement.

a. $8 \div 4 - 3 \times 6 + 2 = 64$
b. $4 - 3^2 + 2 \times 5 - 1 = 3$

INTRODUCTION

Scientists, engineers, and other people may design objects in two, three, and even four dimensions. A **tesseract** is a square shown in four dimensions.

The figure shown here is a tesseract. You can indicate a tesseract by using exponents. You would write it as x^4.

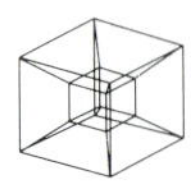

SKILLS DEVELOPMENT

You have already studied many of the laws of exponents. In this lesson you will review these laws. Study the pattern below.

$3^3 \times 3^5 = 3^{3+5} = 3^8 \qquad 3^2 \times 3^6 = 3^{2+6} = 3^8$

$3 \times 3^7 = 3^{1+7} = 3^8$ (3 means 3^1)

You can extend the pattern to a more general result. When you multiply powers with like bases, you add the exponents.

$$(a^m)(a^n) = a^{m+n}$$

Remember to use the correct vocabulary.

power a^3 ← 3 is the **exponent**
← a is the **base**

a^3 is a power and represents repeated multiplication.

$a^3 = a \times a \times a$ (3 factors) — a^3 is said to be in exponential form.

In earlier chapters you studied exponents. Try to find the patterns in the examples below.

Example 1

$$(3y^3)^2 = (3y^3)(3y^3) = 9y^6$$

Use your calculator to simplify.

a. $(5x)^3$ b. $(2x)^4$ c. $(10y^3)^3$

d. Write a general law represented by the equations.

Solutions

a. $(5x)(5x)(5x)$ $125x^3$ b. $(2x)(2x)(2x)(2x)$ $16x^4$ c. $(10y^3)(10y^3)(10y^3)$ $1000y^9$

d. $(xy)^n = x^n y^n$

Example 2

$$\left(\frac{3}{4}\right)^2 = \frac{(3)(3)}{(4)(4)} = \frac{9}{16}$$

Use your calculator to simplify.

a. $\left(\frac{7}{10}\right)^3$ b. $\left(\frac{8}{9}\right)^3$ c. $\left(\frac{4}{5}\right)^3$

d. Write a general law represented by the equations.

Solutions a. $\frac{343}{1000}$ b. $\frac{512}{729}$ c. $\frac{64}{125}$

d. $\left(\frac{x}{y}\right)^n = \frac{x^n}{y^n}$

Example 3

$$\frac{a^7}{a^4} = \frac{a \times a \times a \times a \times a \times a \times a}{a \times a \times a \times a}$$

$$= a^3$$

$$\frac{2^3}{2^2} = \frac{2 \times 2 \times 2}{2 \times 2}$$

$$= 2$$

a. Write a general law represented by these equations.

b. Evaluate and simplify the following expression using the laws of exponents.

$\frac{(xy)^4}{x^2}$ for $x = 2$ and $y = 3$

Solutions a. $\frac{x^m}{x^n} = x^{m-n}$

b. $\frac{(xy)^4}{x^2} = \frac{(2 \times 3)^4}{2^2}$

$= \frac{6^4}{2^2}$ Use your calculator.

$= 324$

EXERCISES

Remember: If you need to check your understanding of any of the laws of exponents, use a numerical example.

Express your answers in exponential form and simplify.

1. $(10^3)(10^2)$ **2.** $(10^4)^3$ **3.** $\frac{4^5}{4^3}$

4. $(x^3)(x^2)$ **5.** $(m^4)^3$ **6.** $\frac{n^5}{n^3}$

7. $(2y)^3$ **8.** $(3x^2)^2$ **9.** $\left(\frac{m}{n}\right)^4$

Express the following laws of exponents in your own words.

10. $(a^m)(a^n) = a^{m+n}$ **11.** $\frac{a^m}{a^n} = a^{m-n}$ **12.** $(a^m)^n = a^{mn}$

13. Give two numerical examples for each law in Exercises 10–12.

Copy and complete the following pattern.

14. $(-2)(-2) = 4$ $\qquad (-2)^2 = 4$
$(-2)(-2)(-2) = -8$ $\qquad (-2)^3 = -8$
$(-2)(-2)(-2)(-2) = ?$ $\qquad (-2)^4 = ?$
$(-2)(-2)(-2)(-2)(-2) = ?$ $\qquad (-2)^5 = ?$
$(-2)(-2)(-2)(-2)(-2)\ (-2) = ?$ $\qquad (-2)^6 = ?$

Use the pattern in Exercise 14. Tell whether each of the following will be negative or positive.

15. $(-1)^{36}$ **16.** $(-2)^7$ **17.** $(-2)^4$

18. $(-1)^{100}$ **19.** $(-1)^{89}$ **20.** $(-5)^{88}$

Evaluate the following if $x = 3$ and $y = 2$.

21. $-2xy^2$ **22.** $-(2xy)^2$ **23.** $(-2xy)^2$

24. Why are your answers in Exercises 21–23 not all the same?

Replace ○ with >, <, or =.

25. 2^3 ○ 3^2 **26.** 3^2 ○ $2 + 5$ **27.** 4^2 ○ 2^4

Identify the law of exponents you need to use to solve each equation below. Then solve each equation.

28. $2^x = (2^3)(2^5)$ **29.** $8^3 \div 8^2 = 8^k$

30. $5^y = (5^2)^3$ **31.** $\frac{3^6}{3^x} = 3^2$

32. The area of a circle (A) is represented by the equation $A = \pi r^2$, where $\pi = 3.14$ and $r =$ the radius. Find the area of a circle that has a radius of 9 in.

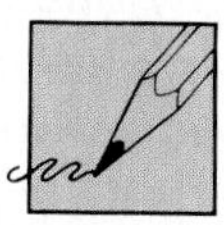

33. The tesseract at the beginning of this section is only one of many kinds of multidimensional figures that can be studied on a computer. Use references to find out about other kinds of computer-generated figures and the role these figures play in today's technology.

SECTION 9–2

SCIENTIFIC NOTATION

WARM-UP ACTIVITY

Use your calculator. Evaluate the powers in each circle. Then draw a trail from start to finish so the line passes through a total of 102.

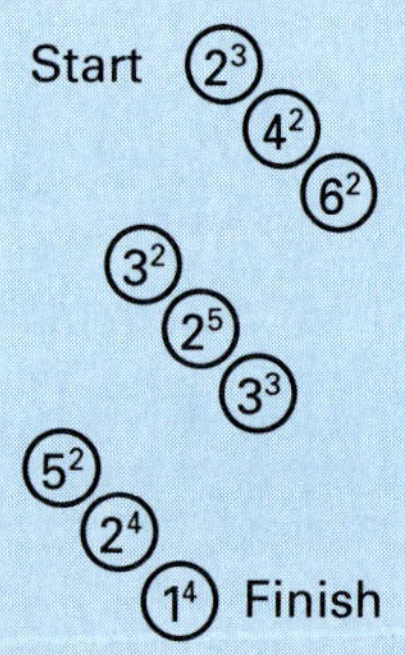

INTRODUCTION

Corianne noticed that all of the exponents she had been working with were positive integers. She wondered what would happen if exponents were negative integers. Look at the following pattern and make a conjecture as to how to continue it.

$10^5 = 100{,}000$
$10^4 = 10{,}000$
$10^3 = 1000$
$10^2 = 100$
$10^1 = 10$

SKILLS DEVELOPMENT

Corianne noticed the exponents were decreasing by one. She noticed each number had been divided by 10. Use your calculator to experiment with numbers to extend Corianne's list. What do you need to do to get to 10^6 from 10^5? From 100,000 to 1,000,000? What do you need to do to get to 10^0 from 10^1?

Corianne generalized this pattern to extend the laws of exponents.

$x^0 = 1$ where $x \neq 0$ $\quad x^{-m} = \frac{1}{x^m}$ where $x \neq 0$

Example 1

Al used a calculator to solve the following problems.

35,000,000 × 2000 = E700 0.0000025 ÷ 50 = 0

Do his answers seem reasonable? Explain.

Solution No. The product of 35,000,000 and 2000 should be greater than 700. Also, only $0 \div x$, $x \neq 0$, will give an answer of 0.

Calculators and computers are limited in that they are only able to indicate numbers that are so large or so small. Many hand-held calculators can show only 8 digits. If you try to multiply numbers whose product is greater than 99,999,999 your calculator will indicate an error in some way. Your calculator may indicate this with an "E" or another symbol in front of the product. How does your calculator indicate such errors?

To find answers to problems like those above, you can use **scientific notation.**

To write 150,000,000 in scientific notation, write:

$$\begin{aligned} & 150{,}000{,}000 \\ &= 1.5 \times 100{,}000{,}000 \\ &= 1.5 \times 10^8 \leftarrow \text{Powers of 10 are used.} \end{aligned}$$

↑

This number is always greater than or equal to 1 and less than 10.

To write 0.000025 in scientific notation, write:

$$\begin{aligned} & 0.000025 \\ &= 2.5 \times 0.00001 \\ &= 2.5 \times 10^{-5} \end{aligned}$$

Remember, this number is greater than or equal to 1 and less than 10.

When working with scientific facts, often very large or very small numbers are used.

Example 2

Write each number in scientific notation.

a. The average distance from Earth to the sun is about 150,000,000 km.

b. The greatest distance from the planet Pluto to the sun is about 7,320,000,000 km.

c. The average diameter of the virus that causes polio is 0.000025 mm.

d. The average length of bacteria is 0.0015 mm.

Solutions

a. $150{,}000{,}000$
$= 1.5 \times 100{,}000{,}000$
$= 1.5 \times 10^8$

b. $7{,}320{,}000{,}000$
$= 7.32 \times 1{,}000{,}000{,}000$
$= 7.32 \times 10^9$

c. 0.000025
$= 2.5 \times 0.00001$
$= 2.5 \times 10^{-5}$

d. 0.0015
$= 1.5 \times 0.001$
$= 1.5 \times 10^{-3}$

Example 3

Use scientific notation to solve each problem. Express each answer in scientific notation.

a. $35{,}000{,}000 \times 2000$ **b.** $0.0000025 \div 50$

Solutions a. $(3.5 \times 10^7)(2.0 \times 10^3)$ Use scientific notation.
$(3.5 \times 2.0)(10^7 \times 10^3)$ Use associative property.
7×10^{10} Multiply.

b. $(2.5 \times 10^{-6}) \div (5.0 \times 10^1)$
$(2.5 \div 5.0) \times (10^{-6} \div 10^1)$
0.5×10^{-7} ← This is not in scientific notation.
5.0×10^{-8}

EXERCISES

Complete each pattern.

1. $3^3 =$
$3^2 =$
$3^1 =$
$3^0 =$
$3^{-1} =$
$3^{-2} =$

2. $5^3 =$
$5^2 =$
$5^1 =$
$5^0 =$
$5^{-1} =$
$5^{-2} =$

Study this pattern. Find the missing values.

3.	$\frac{1}{10}$		10^{-1}
4.	$\frac{1}{100}$	0.01	
5.	$\frac{1}{1000}$	0.001	
6.	$\frac{1}{10,000}$	0.0001	
7.	$\frac{1}{100,000}$		10^{-5}
8.	$\frac{1}{1,000,000}$		10^{-6}
9.	$\frac{1}{10,000,000}$		10^{-7}

Find the exponent in each of the following.

10. $4630 = 4.63 \times 10^{?}$

11. $0.00345 = 3.45 \times 10^{?}$

Find the missing values of *n*.

12. $46,900 = n \times 10^4$

13. $0.0038 = n \times 10^{-3}$

A computer uses the following form to display 1250 in scientific notation.

1.25 + E03

The number is greater than or equal to 1 and less than 10. — means 10^3

Write each number in scientific notation and as a decimal.

14. 6.83 + E02 **15.** 9.25 − E03 **16.** 2.02 − E05 **17.** 4.51 + E06

Each number is to be converted to scientific notation on a computer. What will be displayed?

18. 28,600 **19.** 0.000321 **20.** 0.00345 **21.** 69,600,000

For each of the following questions, express the numbers in scientific notation. Then solve the problem. Express your answers in scientific notation.

22. The thickness of a page in a book is 0.00008 m. How thick is a 1500-page book?

23. A microsecond means 0.000001 second. How many microseconds are there in an hour?

SECTION 9–3

MEASURING LENGTH

WARM-UP ACTIVITY

Work each part correctly and you will have a magic square. What is the sum of each row, column, and diagonal?

$6 + 2 \times 5$	0.2×10	$81 \div 3^3$	$169 \div 13$
$5^2 - 20$	One more than $10^2 \div 10$	$2a^2 - 8$ when $a = 3$	2^3
$\frac{27}{3}$	The sum of the prime factors of 10	$b^2 + 2$ if $b = 2$	The common denominator for thirds and fourths
$4^2 - 2^4 + 4$	$18.5 - 4.5$	The average of 18 and 12	$\frac{5^2 - 4^2}{3^2}$

INTRODUCTION

In the last lesson you learned about how very large distances can be shown with scientific notation. Now you will learn more about distance and length. Measurement plays an important part in all your activities. Measurement involves comparing. A **standard unit** of measurement is used so that no matter who measures an object the results are the same. This unit is compared to the object being measured to see how many of these units there are. For example, to measure the length of the segment AB, choose a unit of measure for length, and count how many of these units there are. What unit of measurement would you use to find the length?

SKILLS DEVELOPMENT

Measurements of length are often called **linear measurements**. In the United States, sometimes lengths are measured with customary units of measure and other times they are measured using metric units.

Before it was necessary to measure with a high degree of accuracy, people measured lengths using parts of their body. The length of one's thumb was about one inch. Since different people had different measurements, their answers varied. Why do you think standard units were developed?

Most reference books contain a "Table of Measures." Use the Reference Section in the back of this book to complete this table.

1 foot = _____ inches
_____ feet = 1 yard
_____ yards = 1 mile

The inch is divided into the fractional parts of halves, fourths, eighths, sixteenths, and so on.

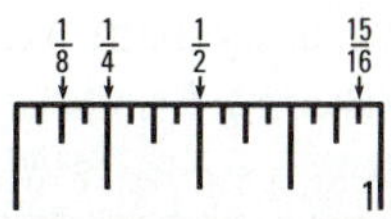

Using inches, the segment AB is 2 inches in length.

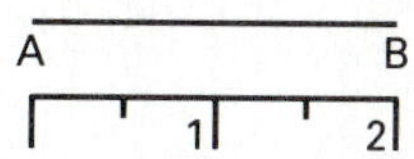

Example 1

How many inches are in a yard?

Solution One way to change units is called **dimensional analysis**. You begin with what you know and make unit changes until you have what you want.

$$1 \text{ yard} \times \frac{? \text{ feet}}{? \text{ yard}} \times \frac{? \text{ inches}}{? \text{ feet}}$$

Equivalent values replace the question marks. Like units can be reduced.

$$1 \cancel{\text{yard}} \times \frac{3 \cancel{\text{feet}}^{1}}{1 \cancel{\text{yard}}_{1}} \times \frac{12 \text{ inches}^{1}}{1 \cancel{\text{foot}}_{1}} = 36 \text{ inches}$$

These expressions are equal to one and do not change the value of the original measurement.

There are 36 inches in 1 yard.

About 1790, France designed a system of measurement based on powers of ten. This system is known as the metric system. In the metric system, the base unit for measuring length is the meter. Other units are formed by attaching a prefix to this base unit.

place value	1000	100	10	1	.1	.01	.001
power of 10	10^3	10^2	10^1	10^0	10^{-1}	10^{-2}	10^{-3}
prefix	kilo	hecto	deka		deci	centi	milli
unit of length	kilometer	hectometer	dekameter	meter	decimeter	centimeter	millimeter
symbol	km	hm	dkm	m	dm	cm	mm
memory aid	0.6 of a mile			little longer than a yard		 width of a paper clip	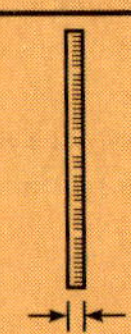 thickness of a dime

Using metric measurement, segment AB is about 5 centimeters in length.

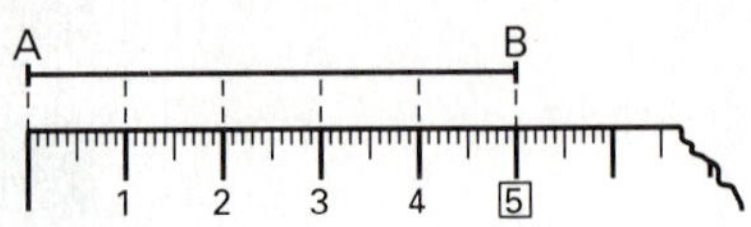

Because the relationships in the metric system are based on tens, a unit of measurement may be easily changed to another. This is done by multiplying or dividing by some power of 10, namely, 10, 100, 1000. Notice that the symbols do not use periods.

1 kilometer (km) = 1000 meters (m)	1 meter = 10 decimeters (dm)
1 hectometer (hm) = 100 meters	1 meter = 100 centimeters (cm)
1 dekameter (dkm) = 10 meters	1 meter = 1000 millimeters (mm)

Example 2

How many meters are in 5.5 km?

Solution $5.5 \text{ km} \times \frac{? \text{ m}}{? \text{ km}} = 5.5 \overset{1}{\cancel{\text{km}}} \times \frac{1000 \text{ m}}{1 \underset{1}{\cancel{\text{km}}}} = 5500 \text{ m}$

There are 5500 m in 5.5 km.

EXERCISES

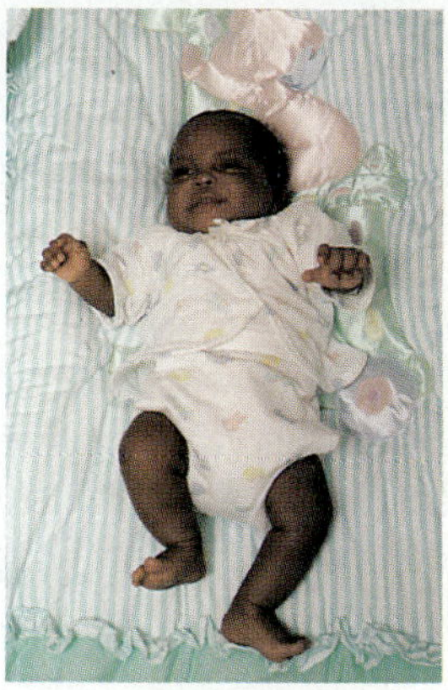

1. A newborn is about 20 inches long. Express this length in feet.

2. Measure your height and express it in feet and inches.

Convert.

3. 440 yards to feet

4. 36 inches to feet

5. The smallest bird is the bee hummingbird, which is 57 mm long. How many centimeters is that?

6. The world's highest waterfall is Angel Falls in Venezuela, with a height of 980 m. How many kilometers is that?

Which customary unit—inches, feet, or miles—would you use to measure the length of each item?

7. pencil

8. pin

9. fence

10. highway

11. pole

12. knife

13. Look at the items in Exercises 7–12. Which metric unit would you use to measure the length of each of these items?

SECTION 9–4

MEASURING PERIMETER

WARM-UP ACTIVITY

April and Bart each make a trail through the woods.

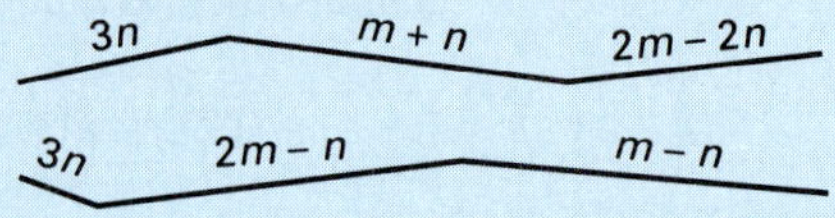

Use $m = 3$ m and $n = 2$ m. Which trip is longer? By how much?

INTRODUCTION

Kimitaka is helping his father panel a recreation room. His father asks him to find the perimeter of the room so they can decide how many panels to buy.

SKILLS DEVELOPMENT

One of the important applications of measurement is finding the distance around a certain area or region. This distance is called the **perimeter**. Kimitaka can find the perimeter of the room by measuring the length of each side and adding the measurements.

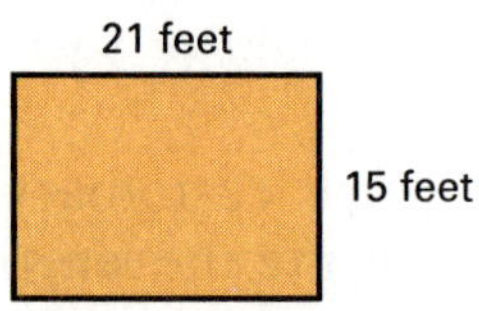

$$\begin{aligned}\text{perimeter} &= 2(15) + 2(21) \\ &= 30 + 42 \\ &= 72\end{aligned}$$

The perimeter of the recreation room is 72 feet.

Example 1

Find the cost of fencing the rectangular parking lot shown if the installed fencing costs \$9.20 per meter.

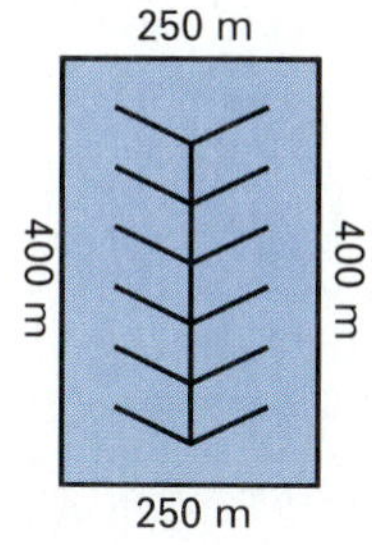

Solution Step 1 Find the perimeter.

$P = 400 + 250 + 400 + 250$

$P = 1300$

The perimeter is 1300 m.

Step 2 Find the cost of the fence.

1 m costs \$9.20.

1300 m × cost = 1300 × \$9.20 or \$11,960.

The cost of the installed fencing is \$11,960.

Example 2

Find the perimeter of the following play area.

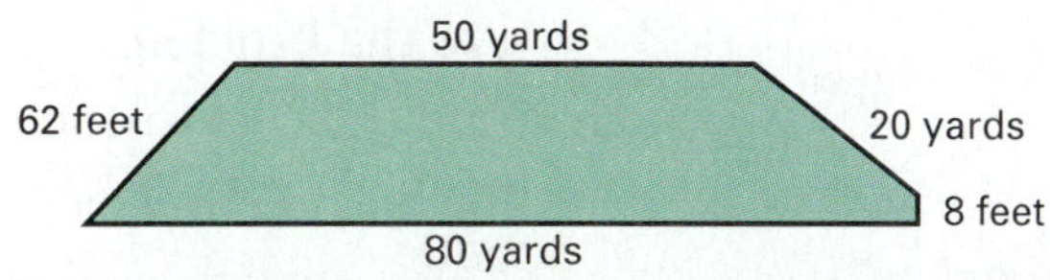

Solution Step 1 Change all measurements to the same units. Generally, you would choose the unit most likely to be commonly used. In this case, you would change yards to feet.

$50 \cancel{yd} \times \frac{3 \text{ ft}}{1 \cancel{yd}} = 150 \text{ ft}$ $\quad 20 \cancel{yd} \times \frac{3 \text{ ft}}{1 \cancel{yd}} = 60 \text{ ft}$

$80 \cancel{yd} \times \frac{3 \text{ ft}}{1 \cancel{yd}} = 240 \text{ ft}$

Step 2 To find the perimeter, add all the lengths.

$P = 62 + 150 + 60 + 8 + 240$

$P = 520$

The perimeter of the play area is 520 feet.

EXERCISES

1. You want to frame the poster. Find the perimeter.

2. a. Kirsten keeps her horse in a rectangular field 88 yards long and 100 feet wide. Find the perimeter. Draw a diagram to help you picture the problem.
   b. Find the cost to enclose the field if fencing costs \$2.59 per running foot.

Find the perimeter of the following figures.

3. 4. 5.

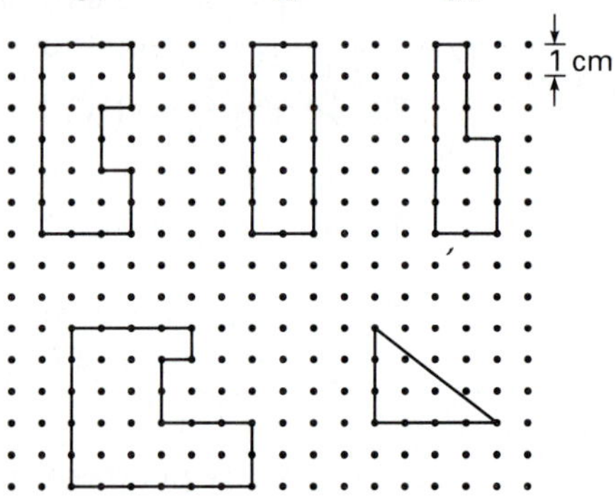

6. 7.

Find the perimeter of each figure.

8.

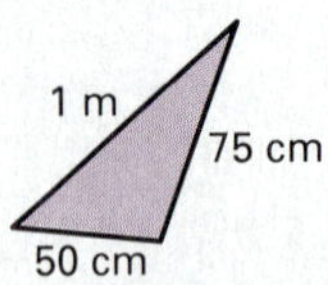

9.

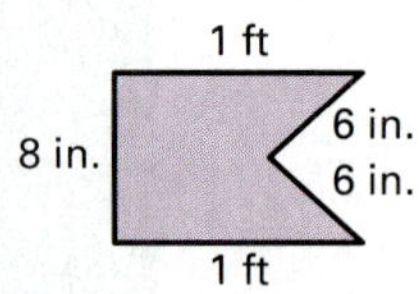

10.

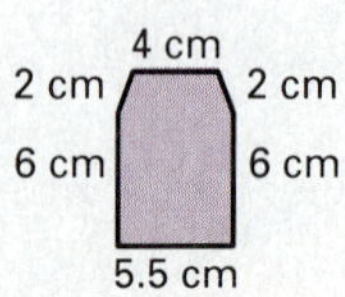

11.

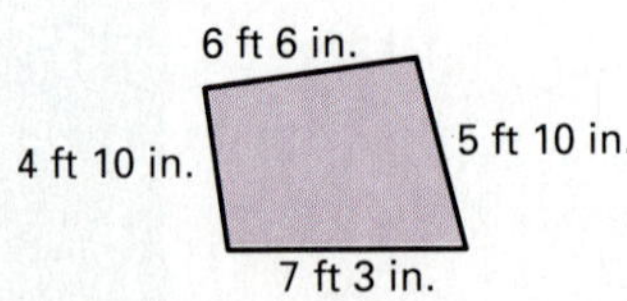

12. Which has the greatest perimeter, **a** or **b**? Estimate first, then measure.

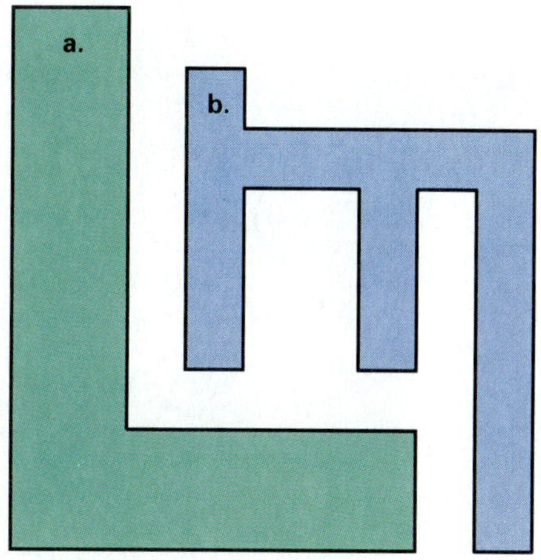

Each calculation finds the perimeter of a figure. Match the calculation and the figure.

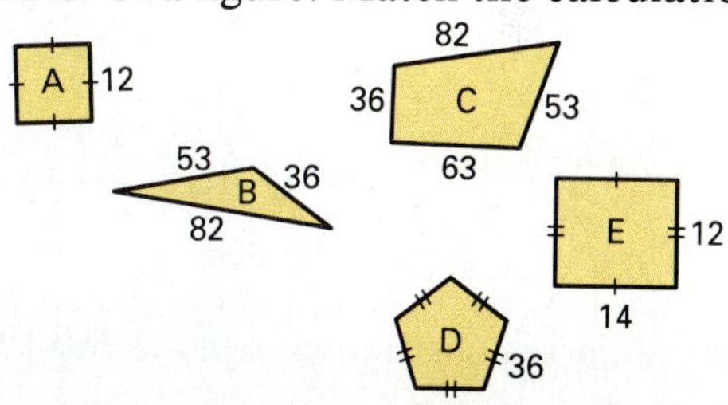

13. 53 + 36 + 82

14. 36 + 82 + 53 + 63

15. 4 × 12

16. 2 × (14 + 12)

17. 5 × 36

AN INVESTIGATION

For this investigation you will need several different sizes of cans or circles cut from paper. You will also need string, a ruler, and a calculator. Use the string to help you find the **circumference**, distance around, of each object. Estimate where the diameter of the circle would be and measure it. Complete the following chart.

Circular Object	Measure of Circumference C	Measure of Diameter d	Calculation to Hundredths $\frac{C}{d}$

- What do you notice about your answers for the calculation $\frac{C}{d}$?
- Finding an exact value for the ratio $\frac{C}{d}$ has intrigued people for centuries. This ratio is represented by the Greek letter π (pi) and is pronounced the same as the word "pie." To the nearest hundredth, what value did you find for π?
- Write an equation for finding the circumference of a circle if you know the diameter.
- Write word problems that require finding the circumference of circles.

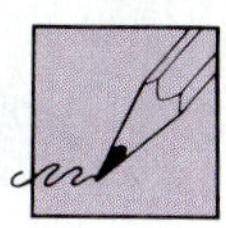

SECTION 9–5

MEASURING AREA

WARM-UP ACTIVITY

a. Draw an 8 x 8 square as shown in Figure A.

b. Cut along the lines drawn and form a rectangle as in Figure B.

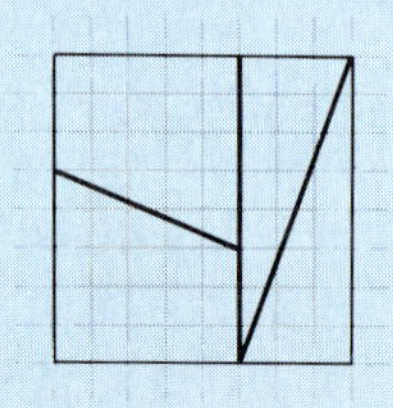

Figure A

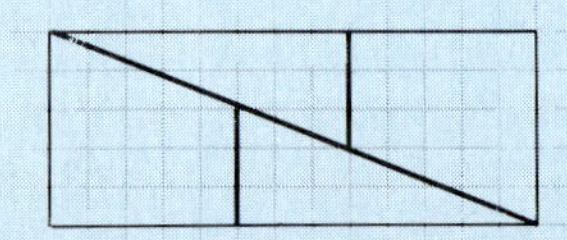

Figure B

Area of Fig. A = ?

Area of Fig. B = ?

Explain your findings.

INTRODUCTION

Ernie played baseball, but he was in a batter's slump. Some of his friends suggested that he crouch more and decrease his strike zone. A pitcher must throw the ball into the area that lies as shown in the diagram. To calculate the area of the strike zone, you need to know the width and length of the rectangular-shaped strike zone.

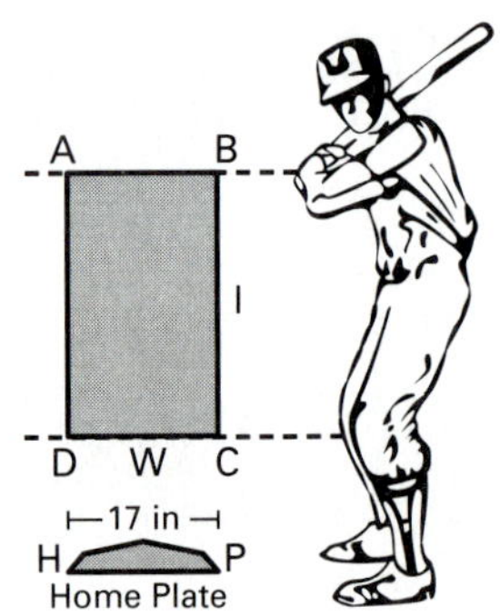

SKILLS DEVELOPMENT

There are many times you need to use skills involving area. To find the area of a region, a unit of measure is chosen and the surface is "tiled" or covered. The number of units needed to cover the surface is its **area**.

To find the area of a right triangle find one-half the area of the rectangle made up of two of the triangles.

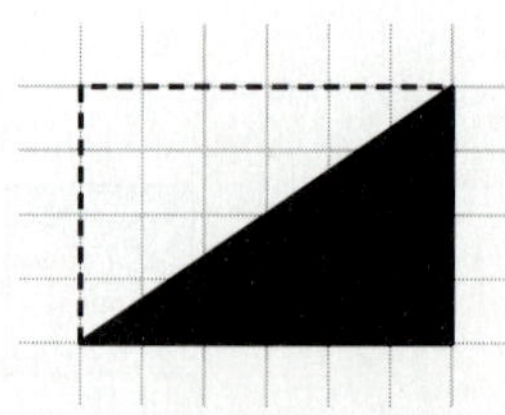

Each unit of area shown on the grid below represents 1 cm^2.

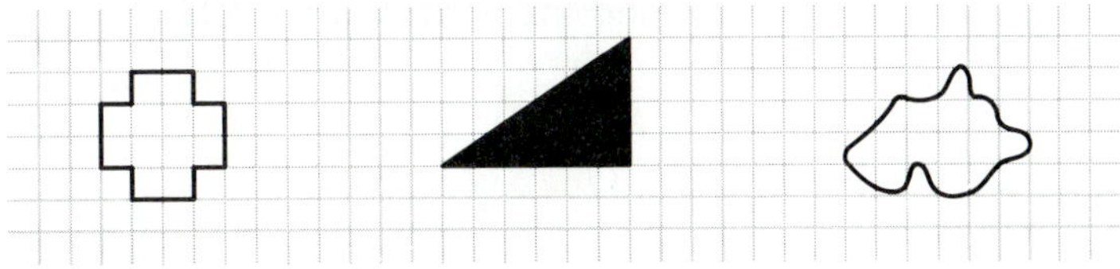

The area of the region is 12 cm^2.

The area of the region is 12 cm^2.

The area of the region is about 12 cm^2.

To find the area of a rectangular region, you could look for a pattern.

Count the number of square units. The area is 18 m^2.

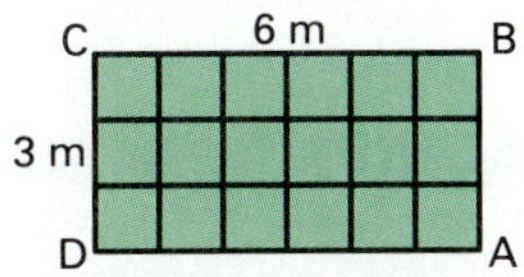

Look for a pattern to find the area:

$$\underset{\substack{\text{length} \\ \text{(m)}}}{6} \times \underset{\substack{\text{width} \\ \text{(m)}}}{3} = \underset{\substack{\text{area} \\ (\text{m}^2)}}{18}$$

The area is 18 m^2.

You could calculate the areas of other rectangles by using the formula, $A = l \times w$.

Example 1

Find the area of each figure.

a.

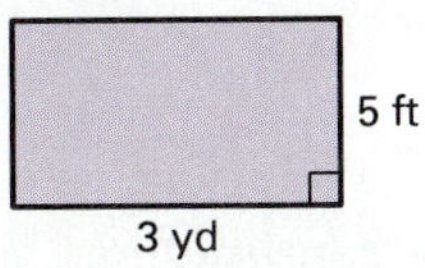

b.

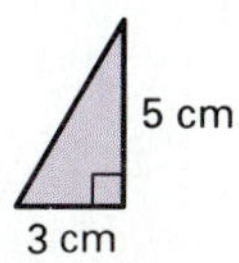

Solutions a. $A = l \times w$

Change yards to feet.

$A = 9 \times 5$

$A = 45$

The area is 45 ft^2.

b. $A = \frac{1}{2} \times l \times w$

$A = \frac{1}{2} \times 3 \times 5$

$A = 7.5$

The area is 7.5 cm^2.

Example 2

Find the area of the circle.

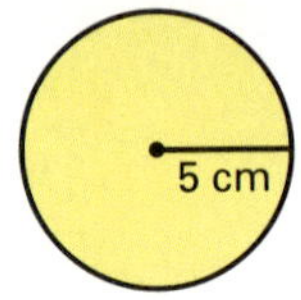

Solution By using the Reference Section of this text, you know the formula for the area of a circle is $A = \pi r^2$. Use 3.14 for π.

$A = (3.14)(5)^2$
$= 78.5$

The area of the circle is 78.5 cm^2.

EXERCISES

I am the word SQUARE. Draw me in four different ways. Match the description with the drawings.

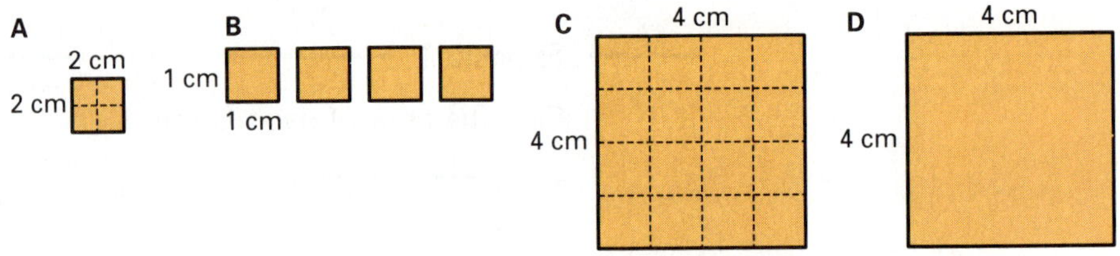

1. four square centimeters
2. four centimeters squared
3. four centimeter squares
4. four-centimeter square

Find the area of each region.

represents $1m^2$

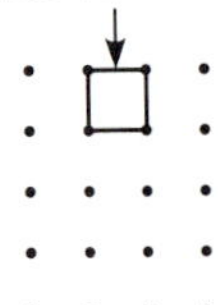

5. 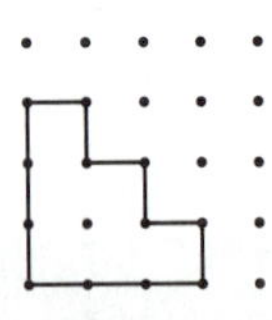

6. 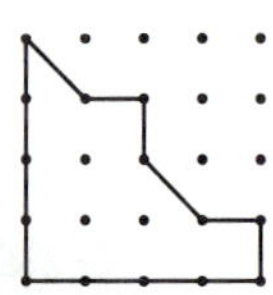

7. 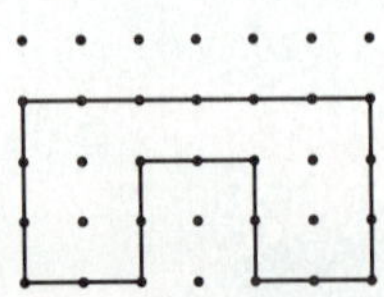

8. 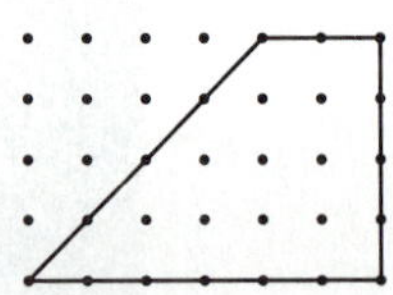

For each region, estimate its area. Then find the area.

9.

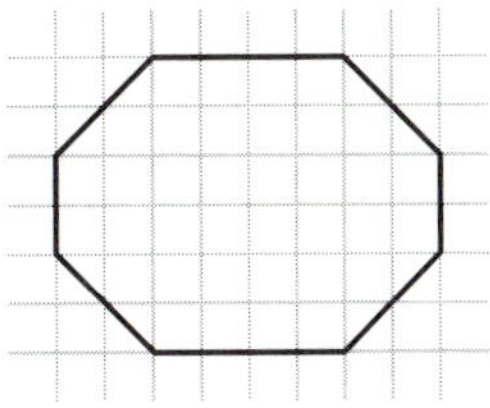

10.

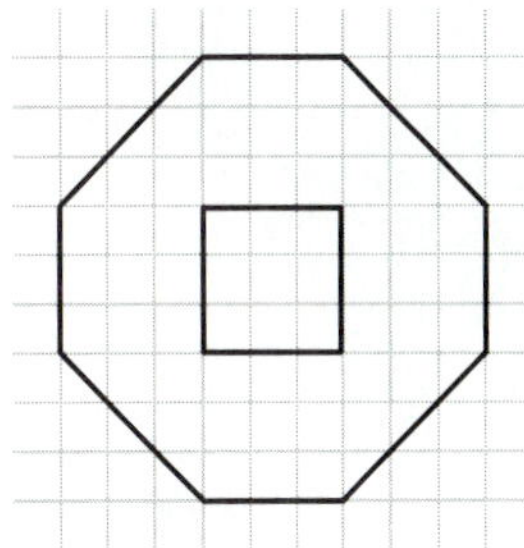

Using a geoboard or grid paper, construct rectangles of the following area.

11. 1 square unit

12. 2 square units

13. 4 square units

14. 5 square units

Find the area of each circle.

15.

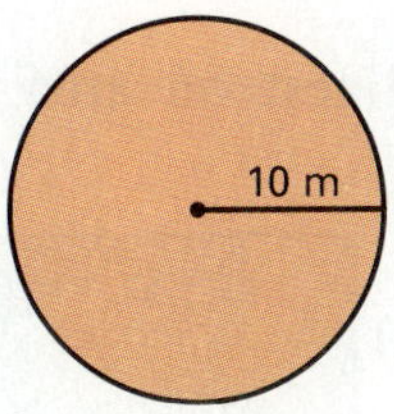

16.

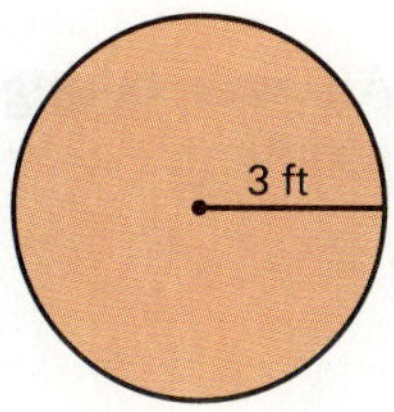

17.

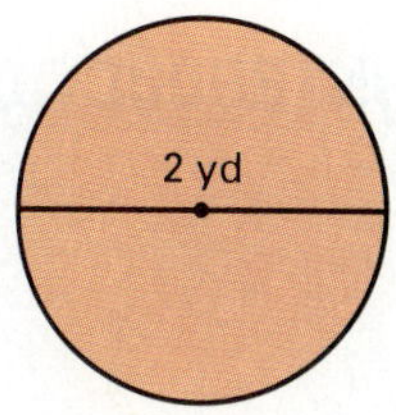

Estimate each area. Then find each area to the nearest tenth of a meter.

18. Field A

19. Field B

20. Field C, including pond

21. Pond in Field C if it has a diameter of 21 meters.

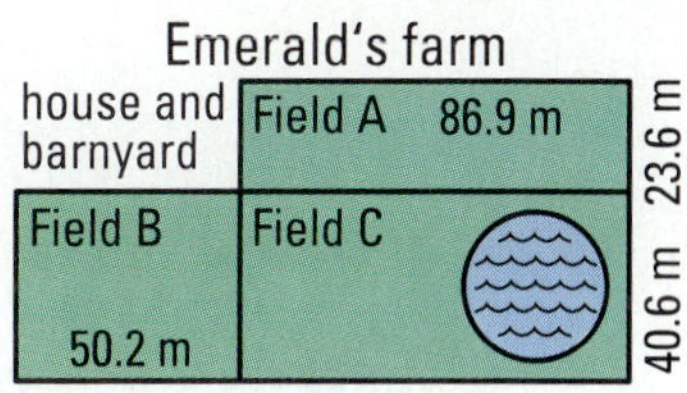

Complete the following statements.

22. 1 square foot = ? square inches

23. 1 square yard = ? square feet

24. 1 square meter = ? square centimeters

SECTION 9–6

POLYNOMIALS AND AREA

WARM-UP ACTIVITY

Develop a rule to find these products mentally. The first three products are given.

$15^2 = 225$	$45^2 = ?$	$75^2 = ?$
$25^2 = 625$	$55^2 = ?$	$85^2 = ?$
$35^2 = 1225$	$65^2 = ?$	$95^2 = ?$

INTRODUCTION

Ali was taking a skill test on area. He was doing the following problem.

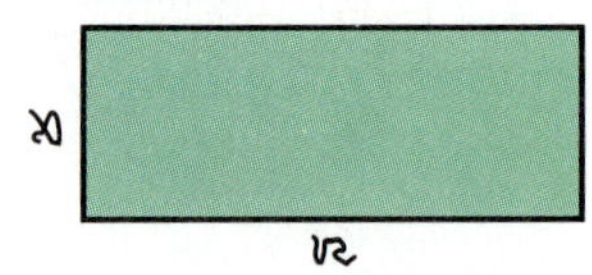

However, Ali could not read the numbers for the length and width. When he asked the test instructor what the numbers were, she replied, "The length is 2 more than the width." How can Ali express the area of this rectangle?

SKILLS DEVELOPMENT

The area of the rectangle, in square units, is given by the product. Think of a diagram.

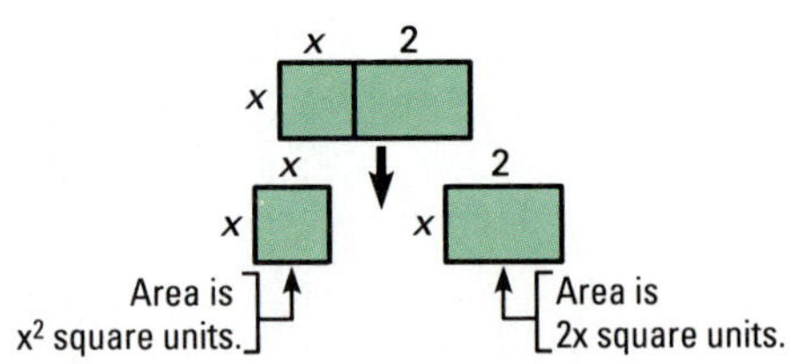

$$x(x + 2)$$

represents width   represents length

To find the area of the rectangle, Ali could find the sum of the parts.

Area $= x(x + 2) = x^2 + 2x$

This equation illustrates the distributive property. You can extend the use of the distributive property to learn to multiply binomials.

______$(x + 2) = ($______$)x + ($______$)2$

If you use $x + 3$ for ______ then you obtain

$$\begin{aligned}(x + 3)(x + 2) &= (x + 3)x + (x + 3)2\\ &= x^2 + 3x + 2x + 6\\ &= x^2 + 5x + 6\end{aligned}$$

Thus $(x + 3)\ (x + 2) = x^2 + 5x + 6$

factors terms

Think of a diagram.

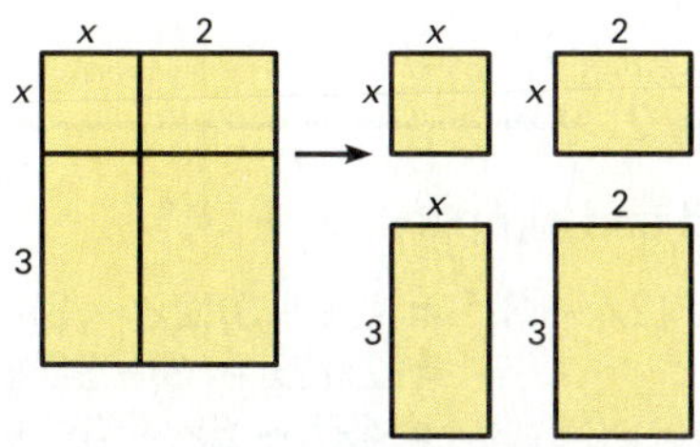

Area is given by $(x + 3)(x + 2)$.

Area is given by
$x^2 + (2x + 3x) + 6$
$= x^2 + 5x + 6$

Example 1

Expand and simplify.

a. $(x + 6)(x + 4)$ **b.** $(2y - 3)(3y + 2)$

Solutions a. $(x + 6)(x + 4)$

$= \underline{(x + 6)}x + \underline{(x + 6)}4$

$= x^2 + 6x + 4x + 24$

$= x^2 + 10x + 24$

b. $(2y - 3)(3y + 2)$

$= \underline{(2y - 3)}3y + \underline{(2y - 3)}2$

$= 6y^2 - 9y + 4y - 6$

$= 6y^2 - 5y - 6$

When the factors of a product are the same, you are able to use exponents.

$(x + 2)^2$ means $(x + 2)(x + 2)$

To find the square of a binomial, you can again use the distributive property.

$(x + 2)^2 = (x + 2)(x + 2)$

$= x(x + 2) + 2(x + 2)$
$= x^2 + 2x + 2x + 4$
$= x^2 + 4x + 4$

The square $(x + 2)^2$ can be illustrated by this model.

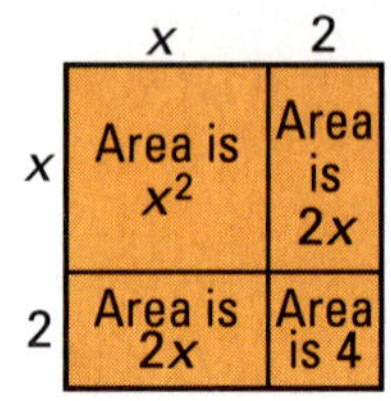

Area of square =
$x^2 + 2x + 2x + 4$
So $(x + 2)^2 = x^2 + 4x + 4$

Example 2

Expand and simplify $(2x - 3)^2$.

Solution $(2x - 3)^2 = (2x - 3)(2x - 3)$
$= 2x(2x - 3) - 3(2x - 3)$
$= 4x^2 - 6x - 6x + 9$
$= 4x^2 - 12x + 9$

EXERCISES

Simplify each of the following.

1. $-3(-4)$ **2.** $-8(4y)$ **3.** $3m(2m)$ **4.** $4y(-5y)$

Find each of the products.

5. $y(y - 2)$ **6.** $-2(2 - x)(-2)$ **7.** $2m(3m - 1)$

8. $-6(-4 - 3k)$ **9.** $-n(3n - 4)$ **10.** $n^2(2n^2 - 1)$

11. To find the product $(x + 4)(x + 5)$, Carol wrote $(x + 4)(x + 5) = x(x + 5) + 4(x + 5)$. Complete her solution to find the product.

12. To find the same product, Andy wrote $(x + 4)(x + 5) = (x + 4)(x) + (x + 4)(5)$. Complete his solution to find the product.

13. What do you notice about the answers in Exercises 11 and 12? Why?

14. The product $(x + 4)(2x + 1)$ is shown by the diagram below. Find the missing values. Then use the diagram to find the product $(x + 4)(2x + 1)$.

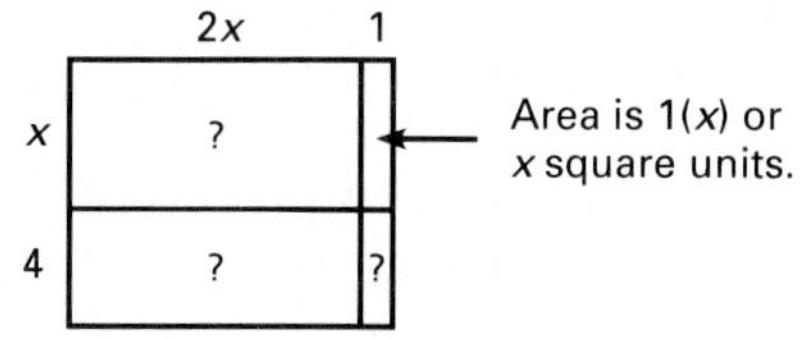

The distributive property is used to expand the product of two binomials. Complete each of the following and simplify your answer.

15. $(x + 4)(x + 5) = x(?) + 4(?)$

16. $(y - 3)(y - 6) = y(?) - 3(y - 6) = y^2 - 6y - 3y + 18 = y^2 - 9y + 18$

17. $(y + 6)(y - 3) = y(y - 3) + ?(?)$

18. Copy and complete the chart.

Binomials	Calculation	Answer
	$x(x + 8) + 9(x + 8)$	
	$m(m - 2) + 4(m - 2)$	
	$x(x - 3) + 6(x - 3)$	
	$3y(2y + 1) - 4(2y + 1)$	

19. Jennifer traveled for $(x + 8)$ hours at a speed of $(x - 3)$ km/h. Find and simplify an expression to show the distance traveled.

20. **a.** Find the area of a rectangle whose sides, in meters, measure $3x - 2$ by $4x - 5$.

b. Find the perimeter of the above rectangle.

c. What operation do you use to find the area in **a**?

d. What operation do you use to find the perimeter in **b**?

Expand and simplify.

21. $(x - 2y)(x + 2y)$

22. $(3m + 2y)(3m - 2y)$

23. $(5a - 2b)(5a + 2b)$

24. $(7p - 2q)(7p + 2q)$

25. What do you notice about your answers to Exercises 21–24?

26. Evaluate $2(2x + y)^2 - 2(x + y)^2$, when $x = -1$ and $y = 3$.

SECTION 9–7

MEASURING VOLUME

WARM-UP ACTIVITY

Which container holds the most rice?

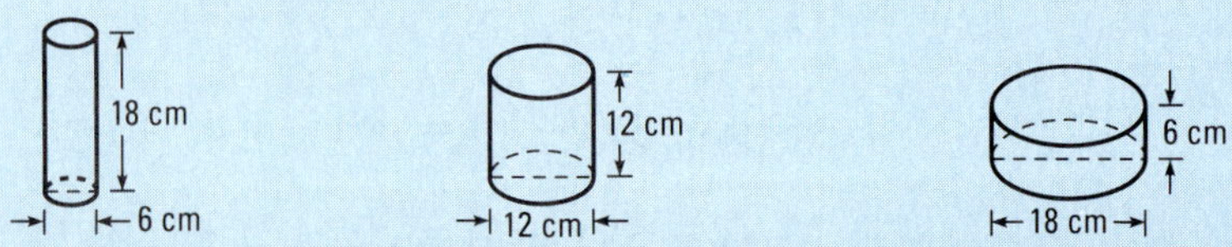

To solve this problem: 1) make a model of each cylinder, 2) fill each model with rice, and 3) compare the amount of rice needed for each cylinder.

INTRODUCTION

Containers and packages occur in many different sizes and shapes. The shapes shown have something in common. They are all prisms.

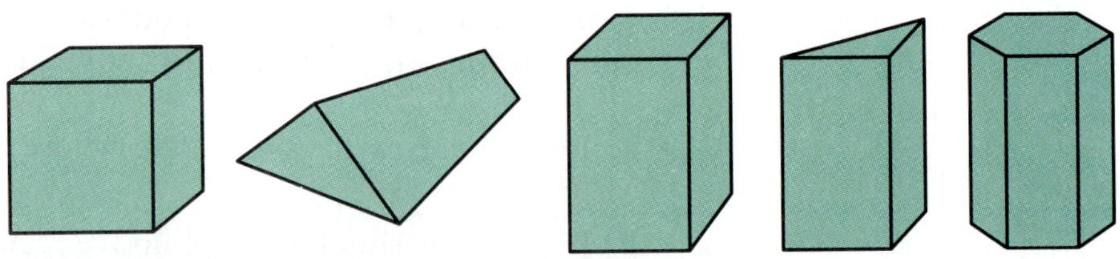

How are the prisms alike? How are they different?

SKILLS DEVELOPMENT

The **volume** of the solid is the amount of space it occupies. Volume is measured in cubic units.

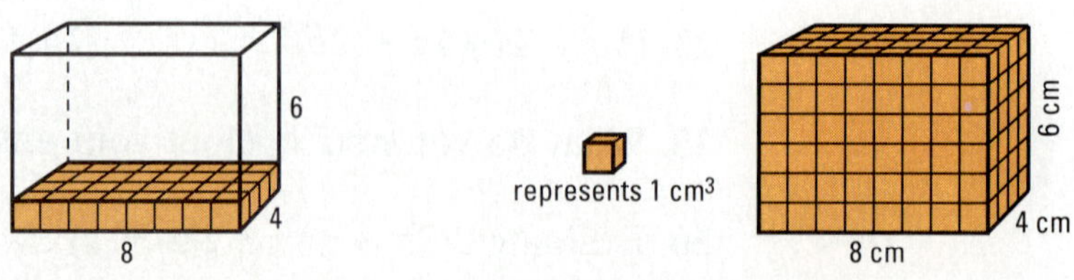

A rectangular solid is one kind of prism. To find the volume of a rectangular solid, you can count the number of cubic units in one layer. Then count the number of layers.

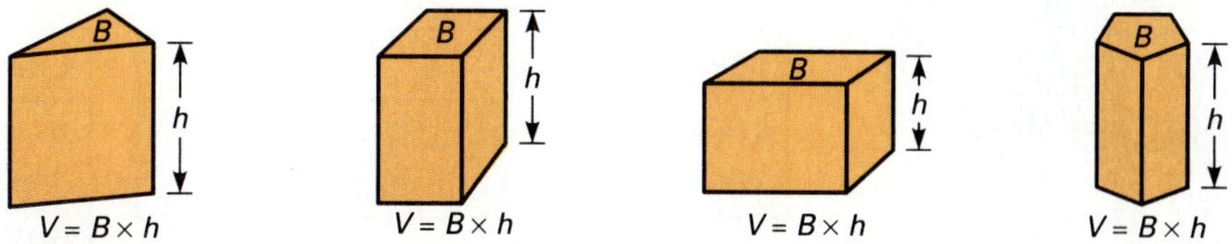

There are 8 × 4 or 32 cubic units in one layer. There are six layers. So the volume is 6 × 32 cm^3 or 192 cm^3.

To count the number of cubic units each time is time-consuming. You can also use a pattern or formula to find the volume.

Volume = length × width × height
$V = l \times w \times h$

You can use the following formula for any prism.

Volume = area of base × height
$V = B \times h$

Example 1

Calculate the volume of each prism. Express your answer to one decimal place. (Note: Dimensions must be in the same unit.)

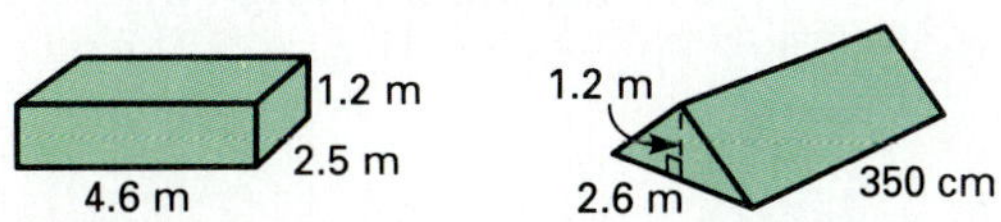

Note:
350 cm = 3.5 m

Solutions

a. $V = l \times w \times h$
$V = 4.6 \times 2.5 \times 1.2$
$V = 13.8$
The volume is 13.8 m^3.

b. $V = B \times h$
$B = 0.5 \times 2.6 \times 1.2 = 1.56$
$V = 1.56 \times 3.5$
$V = 5.46$
The volume is 5.46 m^3.

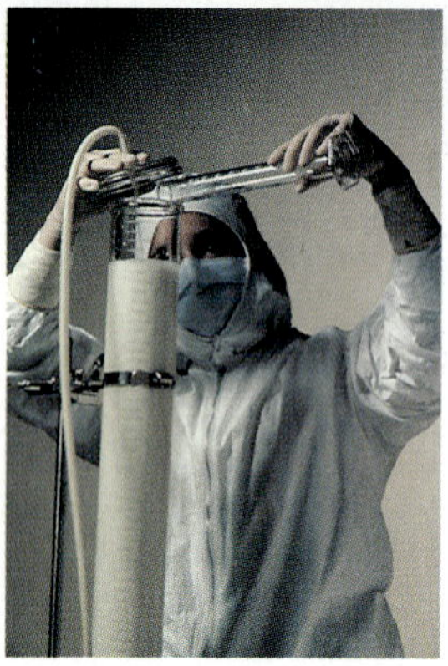

The **capacity** of a container is the amount of substance it is able to hold. Common units of capacity in both metric and customary units are shown in the table below.

Metric Capacity
1000 milliliters (mL) = 1 liter (L)
1000 liters = 1 kiloliter (kL)

Customary Capacity
2 cups (c) = 1 pint (pt)
2 pints = 1 quart (qt)
4 quarts = 1 gallon (gal)

Sometimes rather than measuring the amount of space a substance occupies, we are interested in its mass. Common units are shown below.

Metric Mass
1000 milligrams (mg) = 1 gram (g)
1000 grams = 1 kilogram (kg)

Customary Mass
16 ounces (oz) = 1 pound (lb)
2000 pounds = 1 ton

Example 2

Jeremy weighs 59,100 grams. What would be his mass in kilograms?

Solution

$$59{,}100 \text{ g} \times \frac{1 \text{ kg}}{1000 \text{ g}} = 59.1 \text{ kg}$$

Jeremy would weigh 59.1 kg.

EXERCISES

Find the volume of each solid. Each cube represents 1 cm^3.

1. 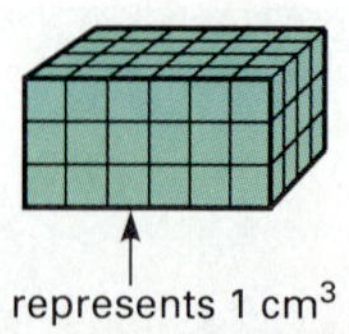

2. 

3. 

4. Farmer Owen has a storage bin that measures 4.6 m by 3.2 m by 4.9 m. He had it built on a square slab of concrete that was 5.0 m on each side. Calculate the capacity of the bin to the nearest cubic meter.

Which is greater?

5. 114 mg, 2 g

6. 3 c, 1 pt

7. 12 kg, 128 g

8. 20 cu ft, 1 cu yd

9. 362 mL, 1 kL

10. 6 qt, 2 gal

Calculate the volume of each prism.

11.

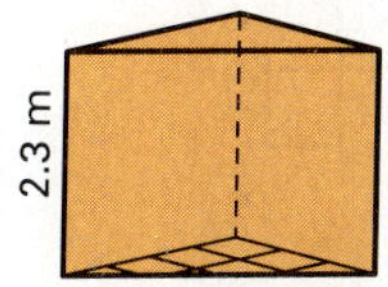

12.

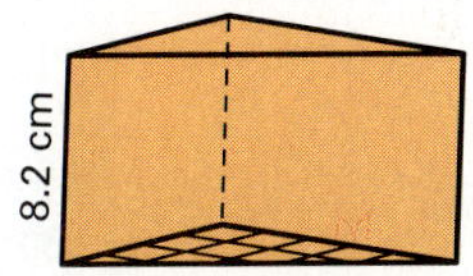

13. Use the dimensions shown for the cereal boxes. Estimate the volume of each box. Then calculate the volume of each box. Which box holds more? How much more?

a.

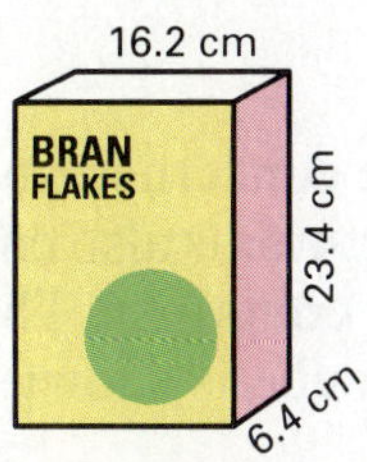

b.

Copy and complete the chart for rectangular prisms.

	Length	Width	Height	Volume
14.	25 cm	15 cm	?	7500 cm^3
15.	1.8 m	?	3.0 m	60.75 m^3

Indicate the most appropriate unit of measure for the following.

16. mass of a cat

17. amount in a milkshake

18. liquid in a cup

19. your weight

SECTION 9–8

MEASURING SURFACE AREA

WARM-UP ACTIVITY

A cube has six faces. Which of the following patterns can be folded to make a cube?

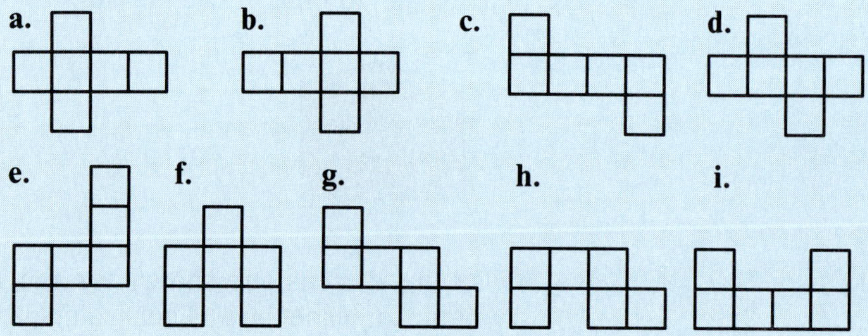

INTRODUCTION

As a consumer you are concerned about how much a container holds. However, the manufacturer is also concerned with how much material is needed to make the container. The material needed to construct a box in the shape of a cube is shown at the top of the page.

SKILLS DEVELOPMENT

In this chapter, you studied figures from a one-dimensional point-of-view. For example, when you found the perimeter of a rectangle, you were finding the sum of segments. This length could be measured in feet or meters.

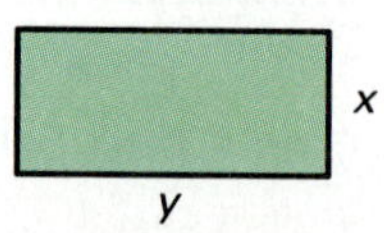

$$\text{perimeter} = 2y + 2x$$

You studied figures from a two-dimensional view when you considered the area of a figure. The area is the sum of square units needed to cover a region.

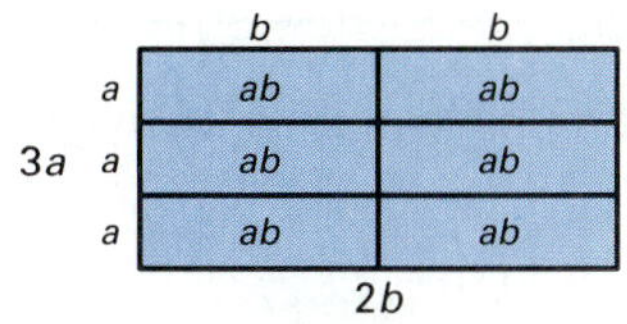

area = $6ab$ square units

You studied figures from a three-dimensional view when you found the volume of a solid. The volume could be measured as the number of cubic units needed to fill the solid. It could also be described in terms of capacity.

Are there other ways you can think about a three-dimensional figure besides volume?

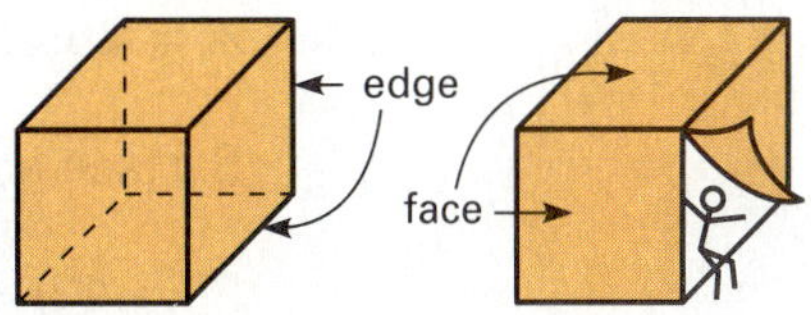

A **skeleton** of a figure shows its framework or edges.

When the skeleton is covered with material, the amount of material needed is called the **surface area.**

To calculate the surface area of a cube or a prism, you need to calculate the total area of the **faces,** or surfaces. Since you are finding an area, the measuring unit is a square unit.

Note: Think of a diagram to help you find areas of rectangles.

Example

Find the surface area of this rectangular solid.

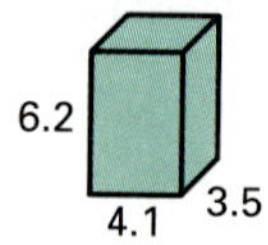

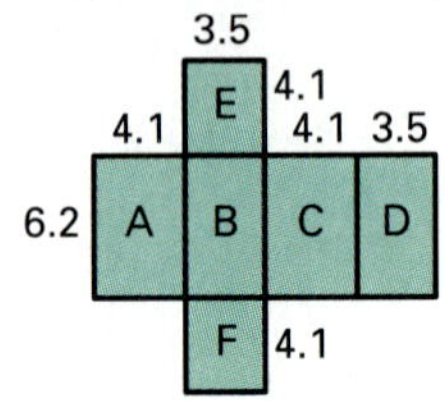

To calculate the surface area, calculate the area of each face or surface. Decide which faces have the same area. The unit of measure is centimeters.

Solution

faces A, C	faces B, D	faces E, F
$A = \ell w$	$A = \ell w$	$A = \ell w$
$A = 6.2 \times 4.1$	$A = 6.2 \times 3.5$	$A = 4.1 \times 3.5$
$A = 25.42$	$A = 21.7$	$A = 14.35$

Surface area is given by the calculations,

$$\begin{aligned} 2(25.42) + 2(21.7) + 2(14.35) &= 50.84 + 43.4 + 28.7 \\ &= 122.94 \\ &= 122.9 \text{ (to 1 decimal place)} \end{aligned}$$

The surface area is 122.9 cm^2.

In general, the formula for the surface area of a rectangle solid is

$$SA = 2\ell w \times 2wh \times 2\ell h$$

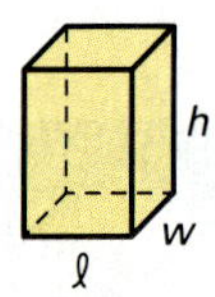

EXERCISES

1. A cube has six faces. Using the diagram, write a formula to calculate the surface area of the cube.

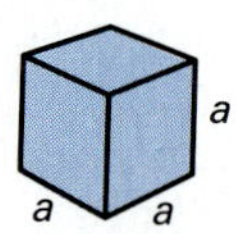

2. Calculate the surface area of a cube if its dimensions are 20 cm by 20 cm by 20 cm.

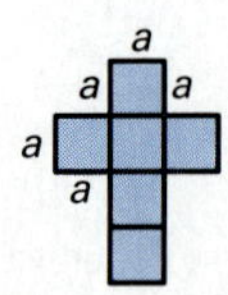

Find the surface area and volume of each solid.

3. 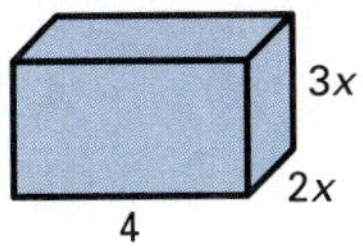

4. 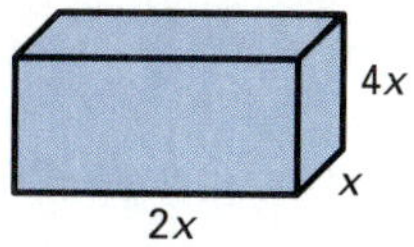

5. 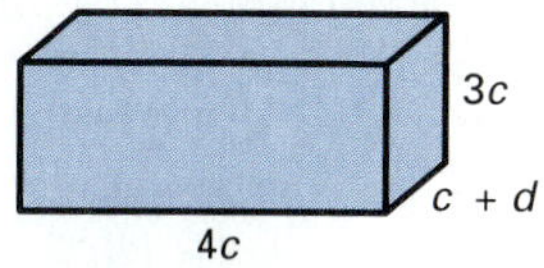

Use the figure for Exercises 6 and 7.

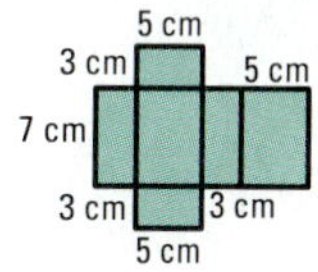

6. Think of folding the pattern to form a solid. What is the volume?

7. Find the surface area of the figure.

Use the figure for Exercises 6–8.

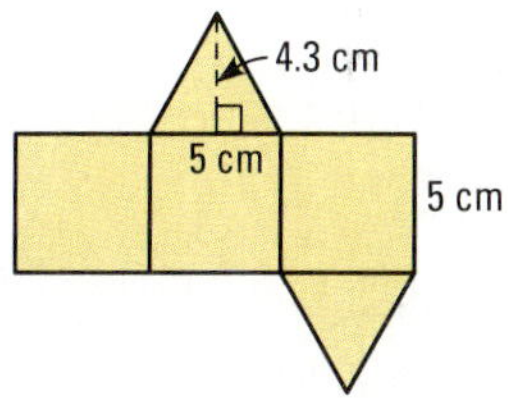

8. Draw a solid that corresponds to this pattern if it were folded. What is the name of the solid?

9. Find the surface area of the solid.

10. Find the volume of the solid.

11. Suppose the sum of all of the edges of a box is 24 cm and that all the dimensions are whole numbers. Given this information and using cubes, make all the possible boxes that you can. Then complete the chart below.

Length	Width	Height	Volume in Cubic Units	Surface Area in Square Units
1	1	4	4	18

12. For Exercise 11, which box had the largest volume? Which box had the smallest surface area?

13. Collect various boxes from the store. Then take the necessary measurements and calculate the volume and the total surface area of each box. Now, compare the volumes. Which type of box is most commonly found?

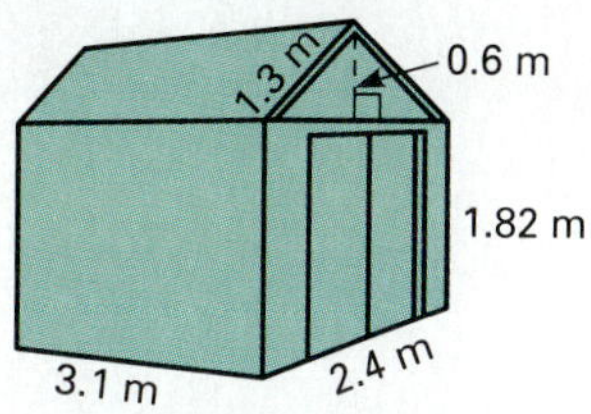

SECTION 9–9

MEASUREMENT TOOLS

WARM-UP ACTIVITY

Copy the following chart. Place the common symbol for various types of measurement in the appropriate column.

Length	Area	Volume	Mass	Temperature	Time

day, C, m^3, kg, mL, F, mg, m, cm^2, ft, in., c, hr, kL, dm^2, km, min, A.M., ha, sec, cm, g, P.M., yd, qt, lb, oz, mi, ′, ″, °

INTRODUCTION

During winter break, Perra plans to leave her hometown in Maine and vacation in Southern California. How will the time and temperature be different?

SKILLS DEVELOPMENT

The basic unit of time is the second. The table shows the units of time and relationships.

Unit	Abbreviation	Relationships
second	sec	60 sec = 1 min
minute	min	60 min = 1 hr
hour	hr	24 hr = 1 day
day	day	1 day = 24 hr
year	yr	1 yr = 365 days
leap year		1 leap year = 366 days

Example 1

How many hours is 1000 minutes?

Solution $\overset{100}{\cancel{1000}}\text{ min} \times \frac{1\text{ hr}}{\underset{6}{\cancel{60}}\text{ min}} = 16.7\text{ hours}$

Since there are 24 hours in a day but only 12 hours on a clock, you need to have a method to tell whether the time you state is in the morning or afternoon; A.M. indicates time from midnight until noon and P.M. indicates time from noon until midnight.

To have a somewhat uniform sunrise time across the world, the earth is divided into time zones. The time zones in the United States are given below.

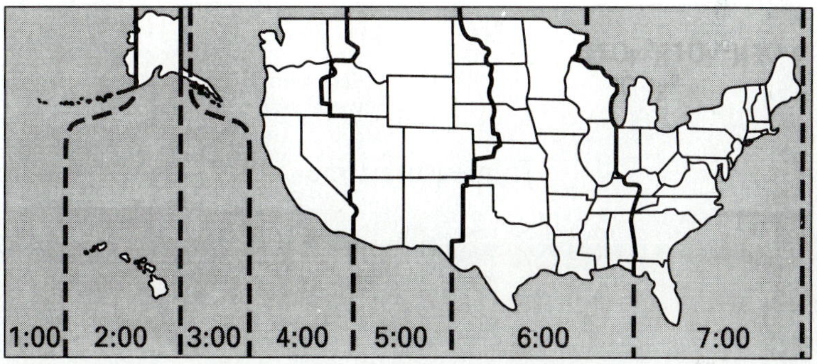

Temperature is measured in degrees. Temperature can be stated using the metric system (degrees Celsius) or the customary system (degrees Fahrenheit). The following chart lists some common temperatures.

Description	Temperature Range	
	Celsius	Fahrenheit
Boiling water	100°C	212°F
Normal human body temperature	37°C	98.6°F
Hot	30°C to 35°C	86°F to 95°F
Warm (room temperature)	20°C to 25°C	68°F to 77°F
Cold	0°C to 10°C	32°F to 50°F
Freezing	0°C	32°F
Below freezing (water)	Below 0°C	Below 32°F

Example 2

Refer to the problem in the introduction: How will the time and temperature be different in Maine and California?

Solution There will be a 3-hour time difference. If it is 2:00 P.M. in Maine, it will be 11:00 A.M. in California. Perra should expect it to be warmer in California than in Maine.

When you measure, you also have a variety of devices from which to choose.

To measure length

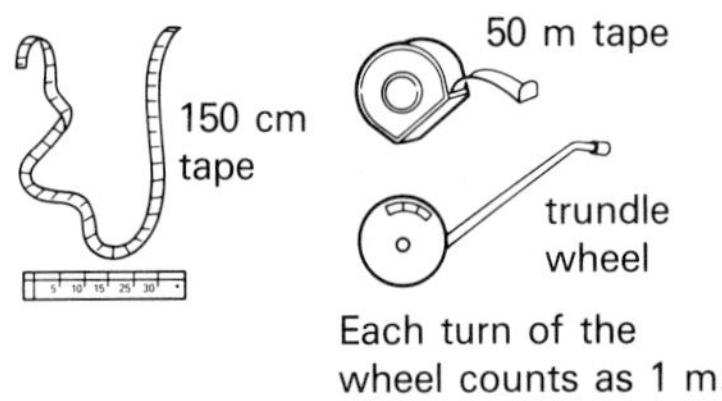

To measure mass

To measure volume

To measure time

To measure temperature

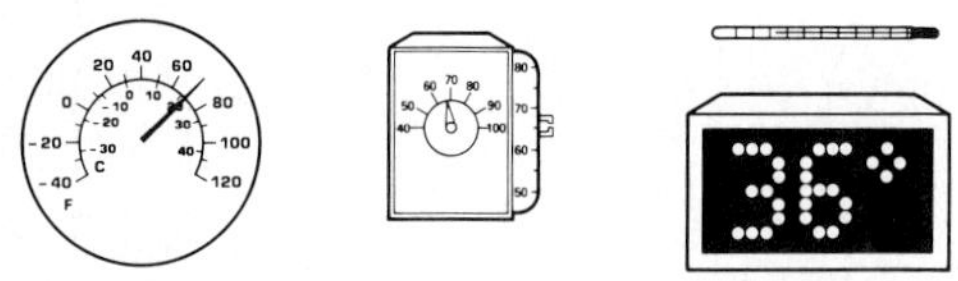

Example 3

For each of the following, which unit of measure would you most likely need to use?

a. the length of a roller coaster ride
b. the amount of air in your lungs
c. the distance from Maine to California
d. the area of postage stamps
e. the thickness of this paper
f. the time needed to make a pizza
g. the volume of water in a glass

Solutions **a.** meters or feet
b. cubic meters or cubic inches
c. kilometers or miles
d. square centimeters or square inches
e. millimeters or inches
f. minutes
g. milliliters or cups

EXERCISES

For each of these activities, measurement is involved in some way. List which measurements are most useful. Then decide on a suitable unit of measurement. For example, when painting a floor you need to know the area of the floor (m^2), the width of the brush (cm), the volume of paint needed (L).

1. baking a pie

2. washing a car

3. pouring a glass of water

4. delivering papers

5. playing basketball

6. taking a hike

Estimate how long you think each activity takes. Then find out by doing each activity. How long does it take to

7. wash your hands before you eat?

8. drink a glass of milk?

9. read 10,000 words?

Refer to the chart of temperature ranges in Skills Development. Estimate a suitable temperature for each of the following weather conditions. Give reasons for your answer.

10. an autumn afternoon

11. a short summer shower

How much is a million dollars? How rich is a billionaire? How big a debt is 3 trillion dollars? Very large numbers are hard to imagine. Consider the following situation. Suppose you spend $1 every second around the clock; no time off to eat or sleep. Also assume you are not making or paying any interest on your money.

12. How much money would you be spending each hour?

13. How long would it take you to spend 1 million dollars?

14. How long would it take you to spend 1 billion dollars? (1 billion = 1,000,000,000)

Give the correct time in each of the following situations.

15. It is noon in Chicago. What time is it in New York?

16. It is 2:00 A.M. in Arizona. What time is it in Hawaii?

CHAPTER SUMMARY

This chapter helps you develop the laws of exponents. By patterning, you learn that $x^0 = 1$ and $x^{-n} = \frac{1}{x^n}$ as long as $x \neq 0$. Using exponents and the decimal system of numbers, you studied a way to write very large or small numbers. This way of writing numbers is called scientific notation. You looked at how your calculator and a computer handles scientific notation.

As you study mathematics, you are probably surprised to see how the separate ideas you studied as a child are connected to each other. This chapter used those connections to help you learn about measurement.

The chapter concludes with the measurement of time and temperature and investigating the tools used to measure the world around you.

VOCABULARY

area
base
capacity
dimensional analysis
exponent
faces
linear measurement
perimeter
scientific notation
skeleton
standard unit
surface area
tesseract
volume

CHAPTER REVIEW

Simplify and express your answer in exponential form.

1. $(3^3)(3^5)$ **2.** $(x^5)(x^2)$

3. $\frac{y^6}{y^4}$ **4.** $(-x)^{20}$

If $x = -2$ and $y = 4$, find the value of

5. $2xy^2$ **6.** $(2xy)^2$

Each number is to be converted to scientific notation on a computer. What will be displayed?

7. 72,000,000 **8.** 0.0000085

Simplify.

9. $3_0 + 2^{-1}$ **10.** $(4_2)_0$

List the customary unit and the metric unit you would use to measure the length of each.

11. width of fingernail

12. length of a desk top

13. thickness of a penny

Find the perimeter for each figure.

14. 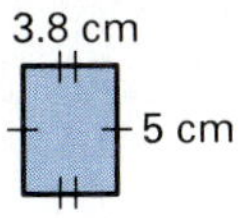

15. 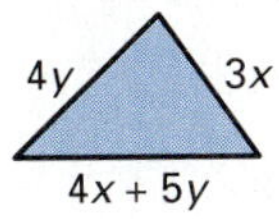

16. The largest postage stamp has dimensions 10.8 cm by 15.8 cm. Find the area to the nearest tenth of a square centimeter.

Complete the following statements.

17. 1 square foot = □ square inches

18. 1 square meter = □ square centimeters

19. 1 square yard = □ square feet

20. Draw a diagram to find the product of $(x + 3)(2x + 1)$.

21. Expand $(2x - 1)^2$.

Calculate the volume of each prism.

22. 

23. 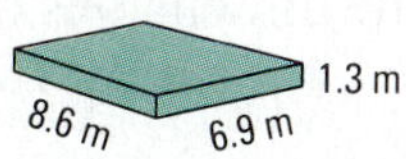

24. The average person drinks about 115 L of soft drinks in a year. If there is an average of 500 mL in a container of soft drinks, how many containers would a person use in a year?

Which has the greater mass, the object in column A or column B?

	A	B
25.	15 kg	150 g
26.	10 lb	100 oz

Find the surface area and volume of each solid.

27. 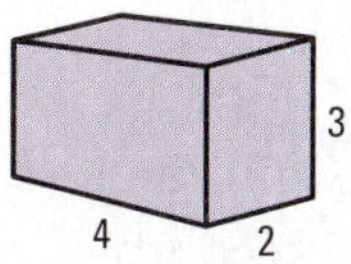

28. 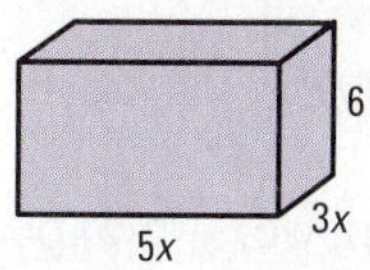

List the measurements needed to complete the work for each of the following activities.

29. building a pole barn

30. planning a class barbecue

CHAPTER 9 TEST

Fill in □ in each of the following.

1. 4.5 m = □ cm

2. 1.3 km = □ m

3. 758 mg = □ g

4. 1 sq yd = □ sq ft

Match the suitable measurement unit with the item to be measured.

5. the length and width of a stamp

6. area of a baseball diamond

7. volume of water in a reservoir

8. mass of a nickel

9. time to run a 50-m dash

a. square meters
b. kiloliters
c. grams
d. millimeters
e. seconds

Simplify. Express your answers in exponent form.

10. $(2^3)(2^2)$ **11.** $(n^4)(n)$

12. $(3^6) \div (3^2)$ **13.** $(q^7) \div (q^2)$

14. $(3^2)^3$ **15.** $(m^3)^4$

Evaluate.

16. 5^0 **17.** 4^{-2}

Write each number in scientific notation.

18. 4,500,000 **19.** 0.000029

Find the perimeter and the area for each figure.

20.

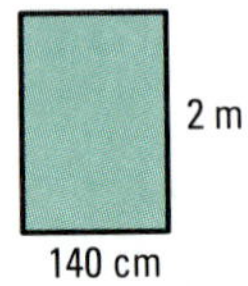

21.

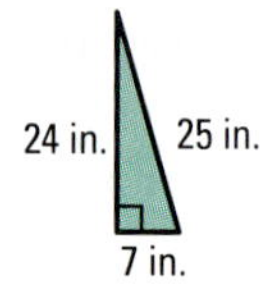

Simplify.

22. $(m + 5)(m - 2)$ **23.** $(2h - 4)^2$

Find the volume and the surface area of each solid.

24.

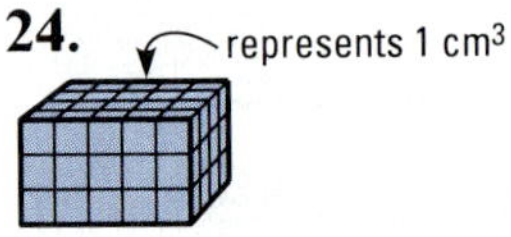

25.

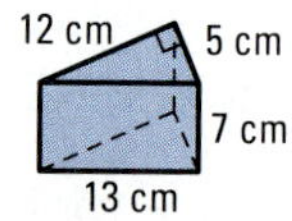

A tennis court has dimensions 23.8 m by 11.0 m.

26. How much fencing will be needed to completely enclose the courts?

27. How many square meters of playing surface will need to be poured?

The dimensions of an aquarium are 42.5 cm by 24.5 cm by 18.6 cm.

28. What is the volume?

29. Which two dimensions are most likely the base of the aquarium?

30. Using your answer from Exercise 29, how many square centimeters of glass are in the aquarium?

SKILLS PREVIEW

List two things from the world around you that illustrate each word.

1. point

2. line

3. angle

4. ray

5. line segment

6. circle

7. parallel lines

8. intersecting lines

9. perpendicular lines

Calculate.

10. $4^2 + 3^2$

11. $7^2 - 5^2$

12. $10^2 + 2^2$

13. $4^2 - 3^2$

14. $7^2 + 5^2$

15. $9^2 - 1^2$

16. $11^2 - 4^2$

17. $12^2 + 3^2$

18. $8^2 - 6^2$

Find each product.

19. $2.3 \times 4 \times 4.82$

20. $3.14 \times 6 \times 2$

21. $6.28 \times 5 \times 3 \times 1.9$

22. $1.98 \times 8 \times 1 \times 3.14$

Estimate and then measure these line segments to the nearest centimeter. Write a ratio to compare the lengths listed in Exercises 23–26.

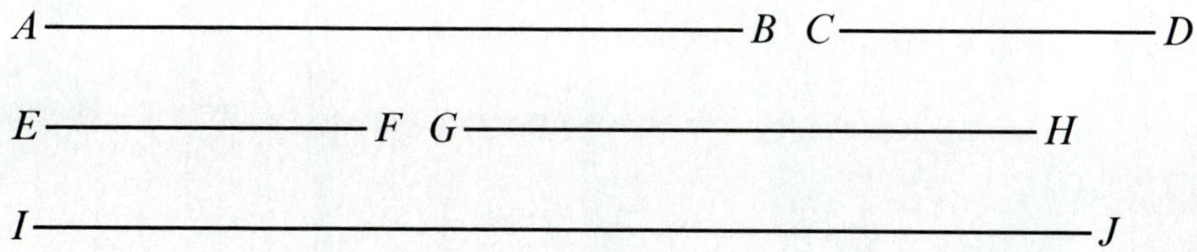

23. *GH* to *CD*

24. *AB* to *IJ*

25. *EF* to *GH*

26. *EF* to *IJ*

Find the missing term by solving the proportion.

27. $\frac{x}{5} = \frac{20}{10}$

28. $\frac{4}{a} = \frac{12}{36}$

29. $\frac{25}{y} = \frac{15}{9}$

30. $\frac{9}{10} = \frac{b}{5}$

Suppose you are challenged to find the height of a tree without actually measuring the tree itself. With the help of two friends, a tape measure, and a mirror, you can employ the concept of similar triangles to find the tree's height.

Suppose you place the mirror on level ground away from the tree. One friend measures the distance from the tree to the center of the mirror. You stand where you can see the top of the tree reflected in the mirror. Then your other friend measures the distance from the mirror's center to the spot on which you stand and then measures your height as you stand erect.

Since you know that light bounces off a mirror at the same angle at which it strikes the mirror, you can make a few quick calculations and confidently meet the challenge.

Whether finding the height of a tree, traveling on the highway, or even planning or building the highway, geometry plays an important part in real life.

Ask Yourself

- *What ideas of geometry do you see being used to find the height of the tree?*
- *How do you think the height of the tree was found?*
- *How is geometry used in the construction of highways?*

CHAPTER 10
GEOMETRY

SECTION 10–1

LANGUAGE OF GEOMETRY

WARM-UP ACTIVITY

Copy the diagram at the left below. Four segments are drawn, and equal segments are marked. Use your copy. Draw a line segment to join A to 6. Join B to 5. Join C to 4, and so on. Continue the pattern.

Can you reproduce the design shown at the right below?

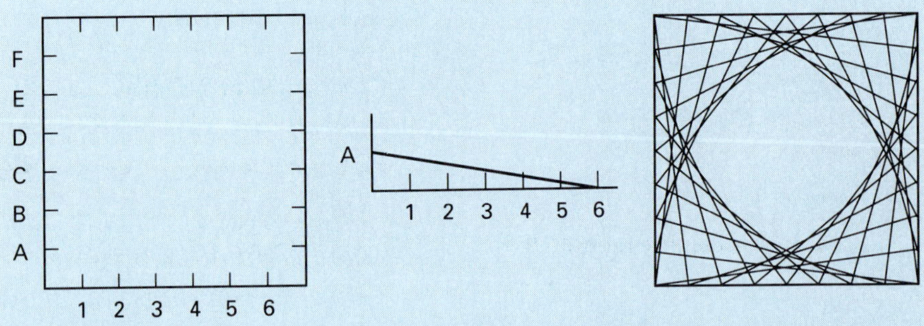

INTRODUCTION

The word **geometry** comes from the Greek *geo* (earth) and *metria* (measurement) and means the measurement of the earth. About 2000 years ago, Euclid organized geometric knowledge into a system of propositions, beginning with some basic axioms.

- Plane geometry is the study of the properties and relationships between figures.
- Coordinate geometry is the study of the methods of locating the position of figures.
- Transformational geometry is the study of the properties of slides, flips, and turns.

SKILLS DEVELOPMENT

To study geometry you need to learn the precise meaning of special words. In geometry some basic ideas such as point, line, plane, and space are difficult to define. They are said to be **undefined.**

A **point** shows position. Capital letters are used to name points. Two points, *A* and *B*, name a **straight line.** The arrows show that the line extends indefinitely in each direction.

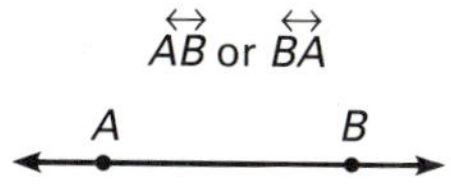

Line segment *AB* is part of a line. It has end points *A* and *B*. Its length can be measured. Two line segments having the same length are said to be **congruent.**

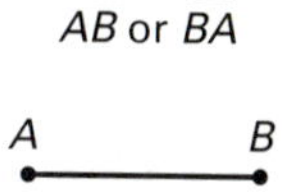

AB is also part of a line. It is called a **ray.** It starts at an end point *A* and extends indefinitely in one direction.

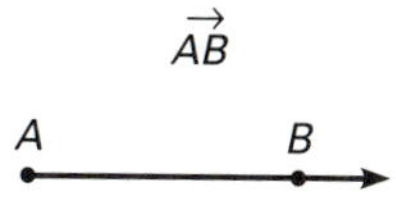

An **angle** *ABC* is constructed from two rays as shown. These rays are often referred to as the **sides** of the angle. When naming an angle, the symbol ∠ may be used for the word angle. The letter at the vertex of the angle is always written in the middle. Two angles are **congruent** if they have equal measures. A standard unit, the **degree,** is used for measuring angles. In this book, the symbol ∠*A* will be used to refer to either an angle or its measure.

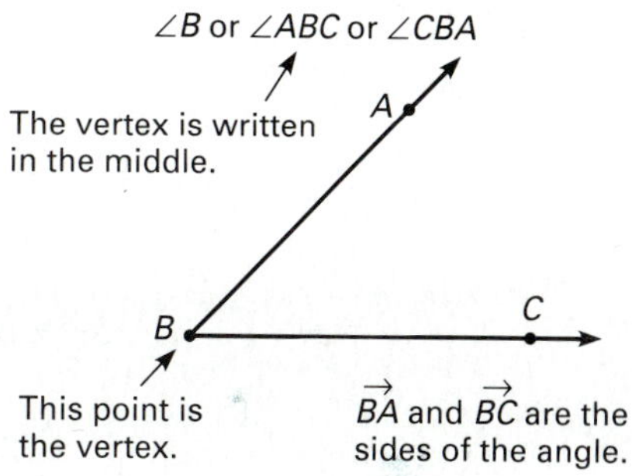

Special vocabulary is also used to describe the way in which pairs of lines are related.

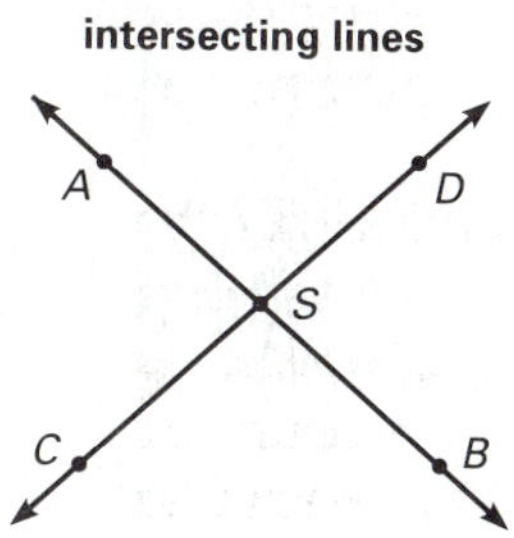

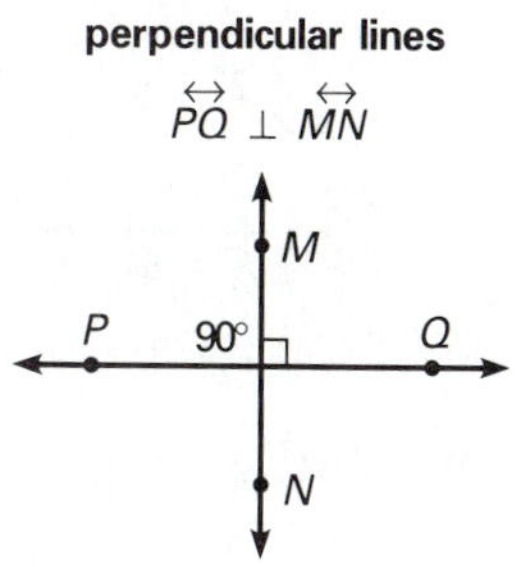

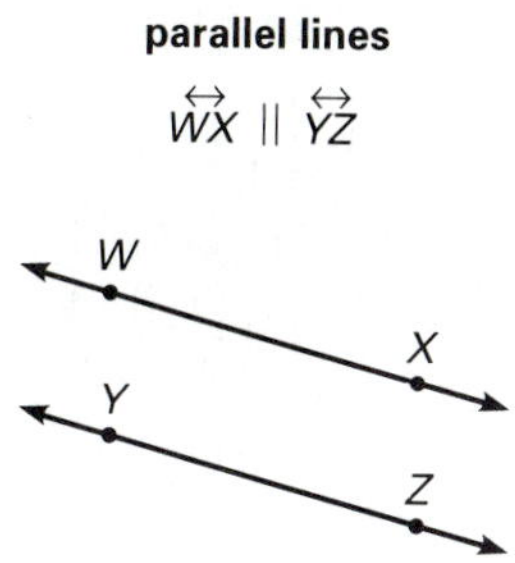

In the left drawing above, the lines *AB* and *CD* intersect at one point *S*. ∠*CSB* and ∠*ASD* are called **vertical angles.**

EXERCISES

From the diagram, name:

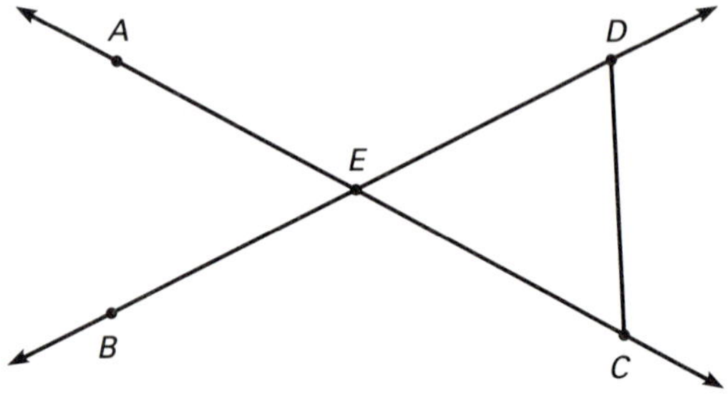

1. two lines.

2. two rays.

3. two angles.

4. two line segments.

For Exercises 5–6, refer to the diagram below.

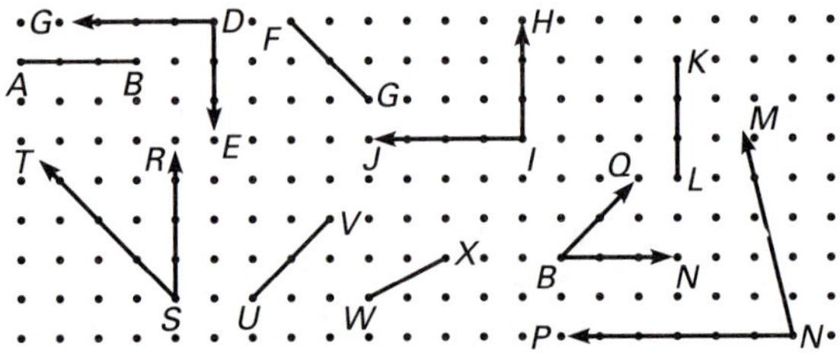

5. Name two pairs of congruent line segments.

6. Name two pairs of congruent angles.

Angles can be classified by size. For example, a right angle measures 90°. The measure of each angle is shown. Write a definition for each angle.

7. Acute angle

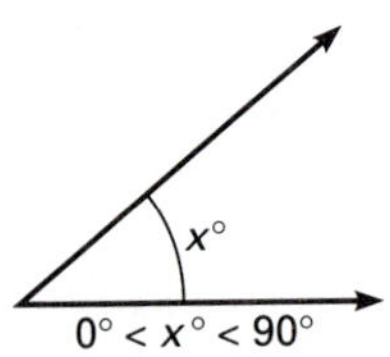

8. Right angle

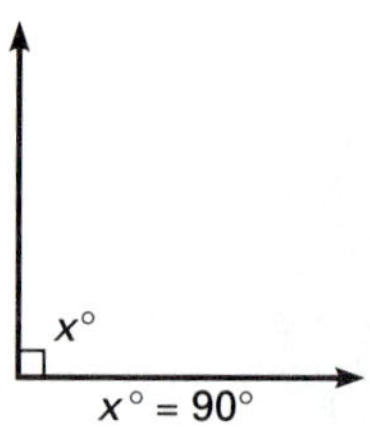

9. Obtuse angle

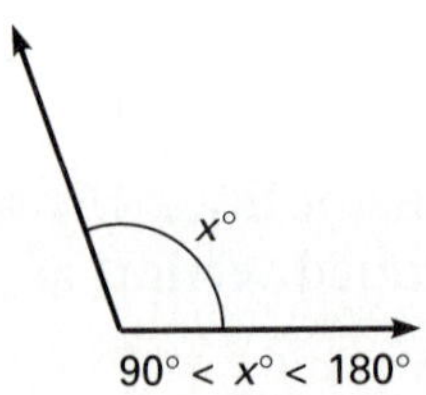

10. Straight angle

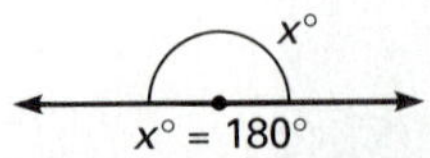

For Exercises 11–16, refer to the figure at the right.

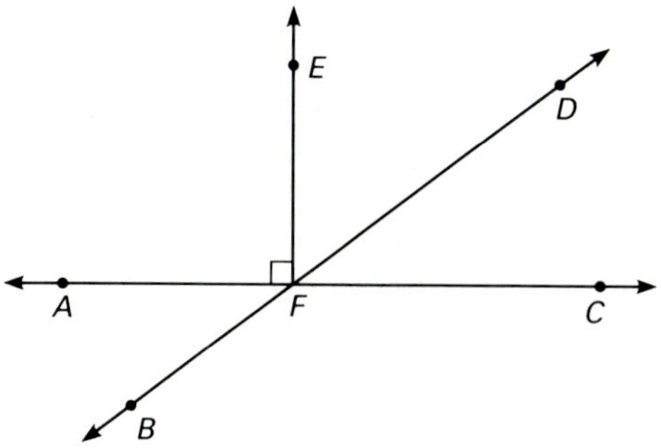

11. Name two acute angles

12. Name two right angles

13. Name one straight angle

14. Name one obtuse angle

15. Name one pair of intersecting lines

16. Name two pairs of perpendicular line segments

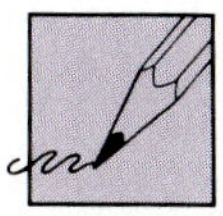

Each diagram illustrates angles with special names. Write a definition for each type of angle.

17. $\angle ABC$ and $\angle CBD$ are adjacent angles.

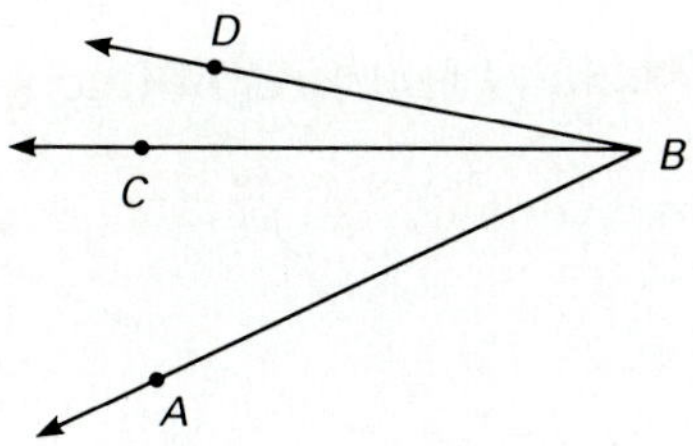

18. $\angle ABD$ and $\angle DBC$ are supplementary angles.

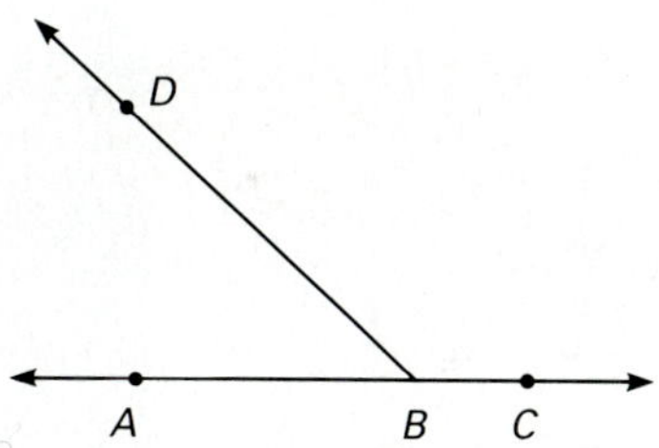

19. $\angle ABC$ and $\angle CBD$ are complementary angles.

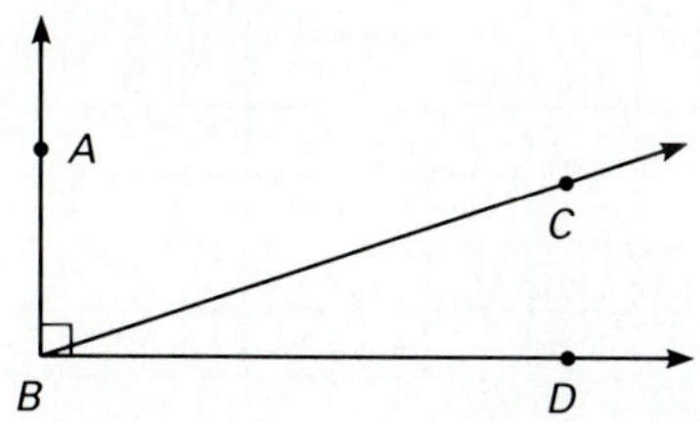

20. $\angle AOD$ and $\angle COB$ are vertical angles.

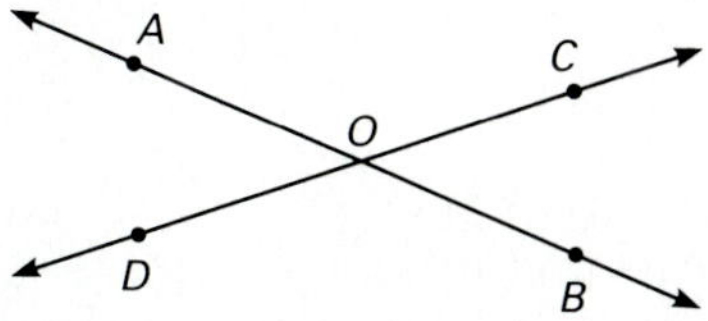

21. The design at the right is similar to that in the Warm-Up Activity. Make a copy of the design. Then construct a similar design of your own.

SECTION 10–2

TYPES OF THINKING

WARM-UP ACTIVITY

Four different triangles are drawn.

Measure the angles.

Copy and complete the table.

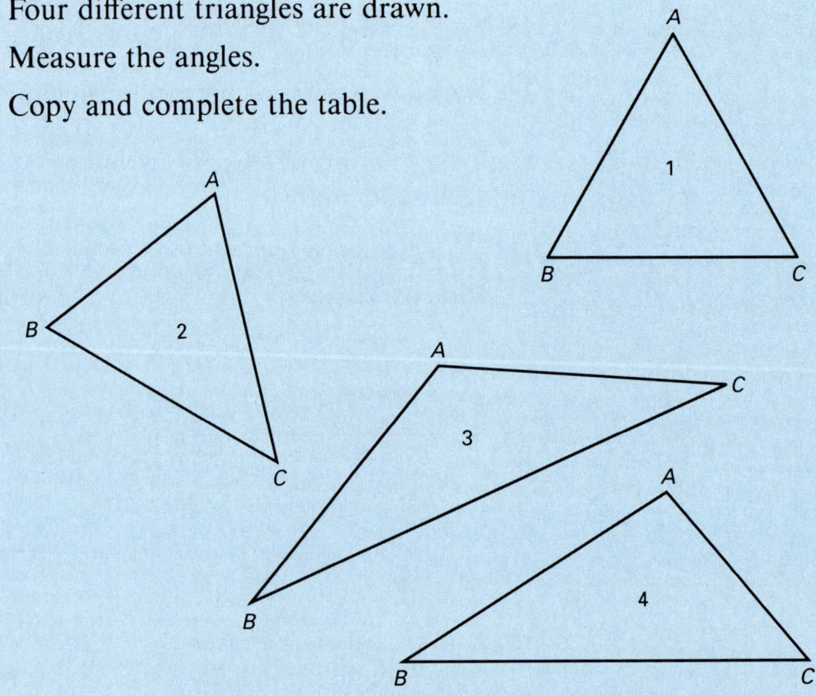

Triangle	Measure ∠A	∠B	∠C	Sum
1	?	?	?	?
2	?	?	?	?
3	?	?	?	?
4	?	?	?	?

Use the results from the table.
What do you notice about the sum of the angle measures?

Round the sums on the chart to the nearest 10. What do you notice about your answers?

INTRODUCTION

In data analysis, which you studied in Chapter 4, you carried out experiments in order to make discoveries. Scientists also perform experiments and have made many discoveries which have important applications in your everyday living.

SKILLS DEVELOPMENT

Using particular cases to arrive at a probable conclusion is known as **inductive reasoning.** The probable conclusion is known as a **conjecture** or **hypothesis.**

Your work in the Warm-Up Activity was inductive. For example, by measuring the angles of a triangle you found that the sum of the angles was 180°. Your conjecture was that the sum of the angles of any triangle is 180°. You can show the same result by folding the angles as shown below.

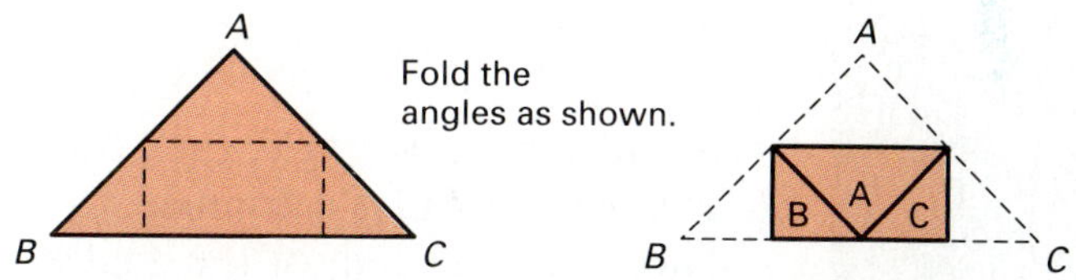

By experiment you learned that the sum of the angles of any triangle is 180°. You can use this probable conclusion to learn some new information. For example, a quadrilateral is created from two triangles, and you are able to **deduce** new information. This method of reasoning is called **deductive reasoning.**

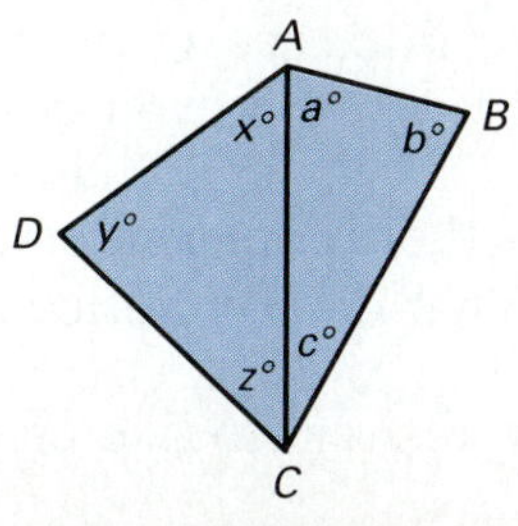

$$x° + y° + z° = 180°$$
$$a° + b° + c° = 180°$$
$$360°$$

Thus the sum of all measures of the angles in quadrilateral *ABCD* is 360°.

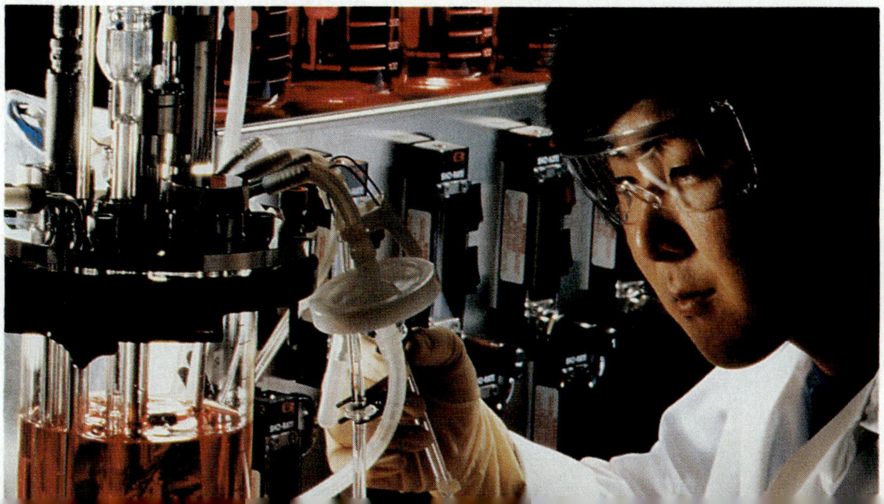

Example 1

Find the measure of $\angle C$.

Solution

$$\begin{aligned} \angle A + \angle B + \angle C &= 180 \\ 100 + 50 + x &= 180 \\ x &= 180 - 150 \\ x &= 30 \end{aligned}$$

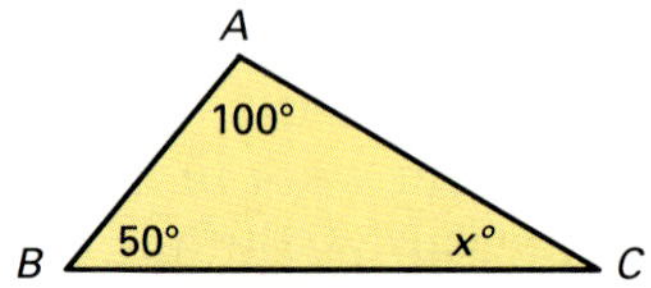

Use the fact that the angle sum of a triangle is 180°

A common form in which to write deductive reasoning is the if-then statement. This conditional statement states our accepted conclusion.

If a figure is a triangle, then the sum of its angles is 180°.
The figure *ABC* is a triangle.
Therefore, the sum of the angles of $\triangle ABC$ is 180°.

EXERCISES

Exercises 1–9 explore the properties of geometric figures.

1. Draw two intersecting lines. Measure the vertical angles with a protractor. What do you notice about your results?

2. Draw another pair of intersecting lines and measure the vertical angles. What do you notice about your results?

3. Use your answers to Exercises 1 and 2 to write a conjecture.

An **isosceles triangle** is a triangle with two congruent sides.

4. Measure the angles of the isosceles triangle *ABC*. What do you notice about your results?

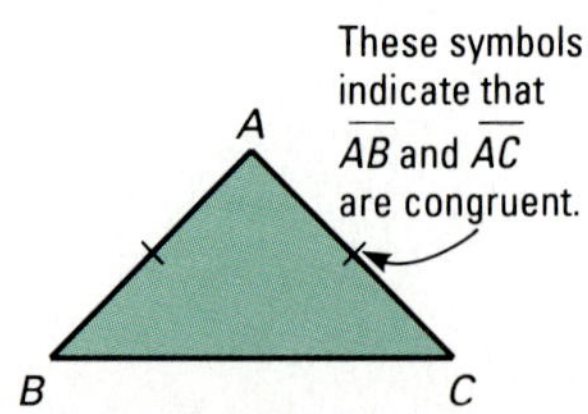

5. Draw another isosceles triangle and measure the angles. What do you notice about your results?

6. Use your answers to Exercises 4 and 5 to write a conjecture.

An **equilateral triangle** has three congruent sides.

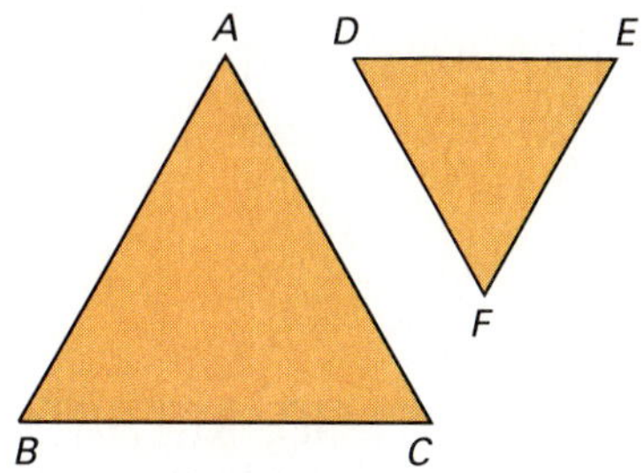

7. Measure the angles of the equilateral $\triangle ABC$ and $\triangle DEF$.

8. Draw another equilateral triangle. Measure the angles. What do you notice about your results?

9. Use your answers to Exercises 7 and 8 to write a conjecture.

Find the missing measures.

10.

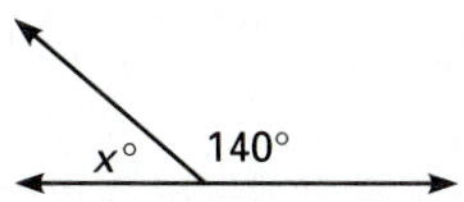

11.

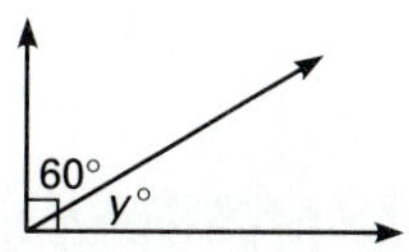

12.

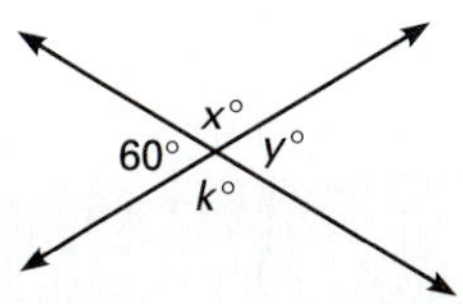

Find the missing measures of each triangle.

13.

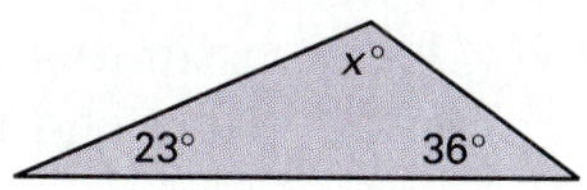

14.

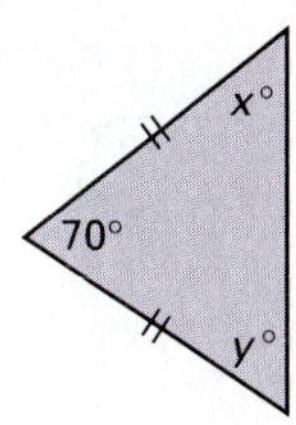

15.

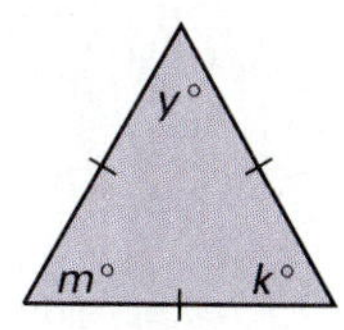

Find the missing measures in each figure. Be able to justify your answer.

16.

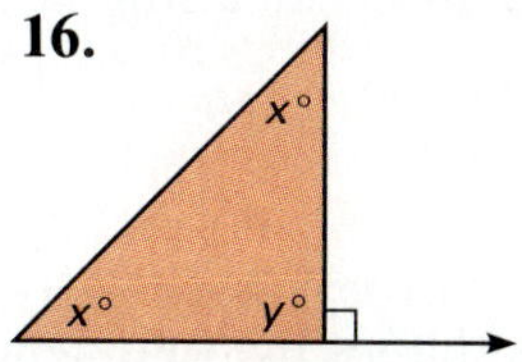

17.

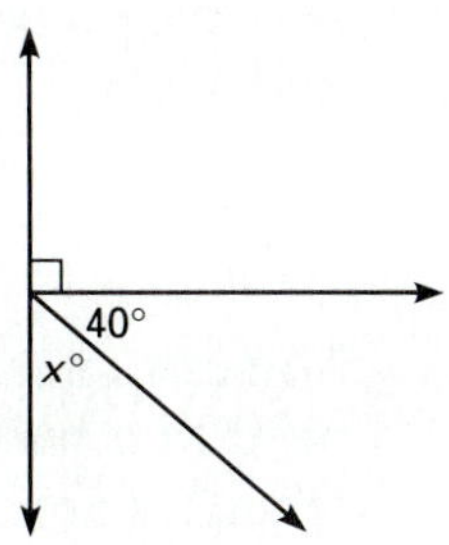

18.

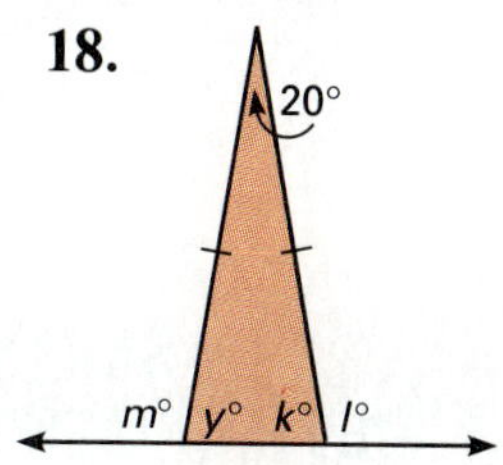

SECTION 10–3

POLYGONS

WARM-UP ACTIVITY

The six dots represent the vertices of a hexagon. Two players, each using a different color, draw either a side or a diagonal of the hexagon. The first player who completes a triangle completely in his color loses.

Is the game fair? That is, assuming each player makes the best possible moves, does the game end in a tie?

INTRODUCTION

Often in the study of mathematics special names are used to describe figures. A **polygon** is a simple closed figure that is constructed from line segments. The word polygon is formed from two Greek words, *poly* meaning many, and *gonia* meaning angle. There are many sizes and shapes of polygons.

SKILLS DEVELOPMENT

You can refer to all these shapes by their general name—polygons. To refer to a particular polygon, you use its special name. For example, triangles are polygons with three sides.

Some other members of the polygon family are shown. A polygon is said to be a regular polygon if:
all sides have equal measures.
all angles have equal measures.

Pentagon	**Hexagon**	**Heptagon**	**Octagon**
Five sides	Six sides	Seven sides	Eight sides

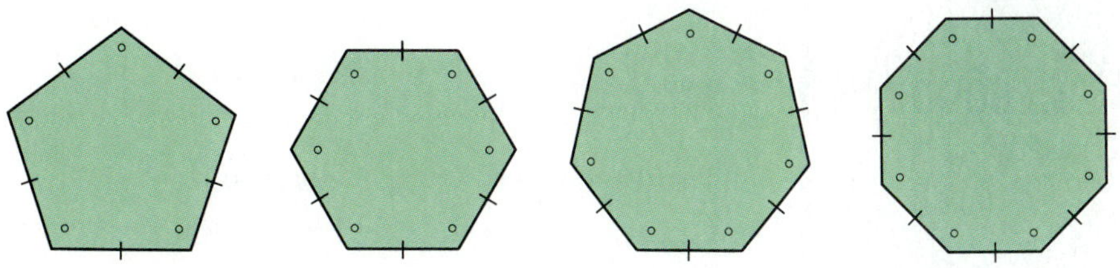

Triangles can be named according to the measures of their sides.

A **scalene triangle** has no equal sides.
An **isosceles triangle** has two equal sides.
An **equilateral triangle** has three equal sides.

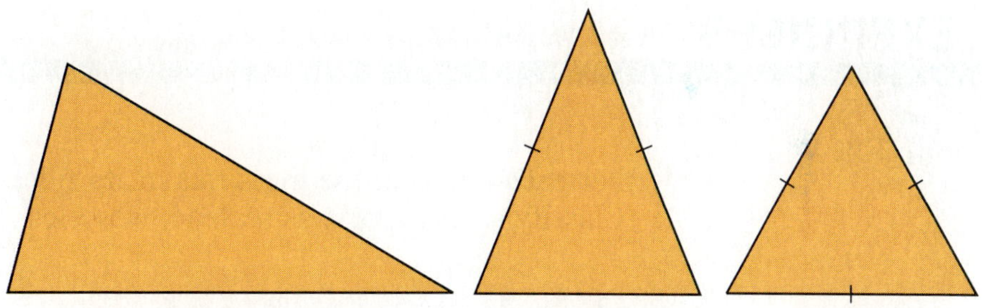

Triangles can be named according to the measures of their angles.

An **acute triangle** has three acute angles.
A **right triangle** has one right angle.
An **obtuse triangle** has one obtuse angle.
An **equiangular triangle** has three congruent angles.

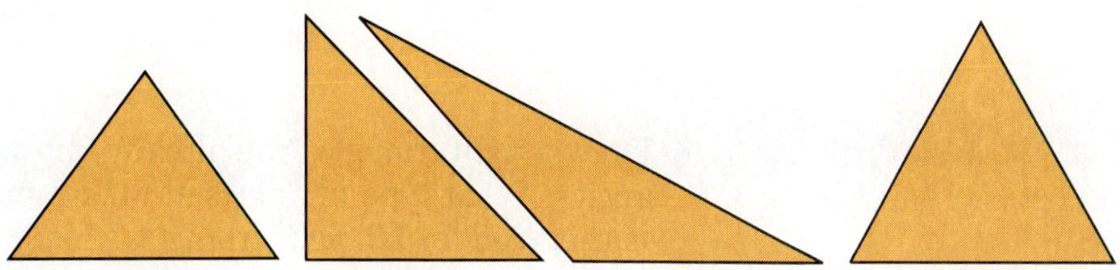

Polygons with four sides are called **quadrilaterals.** Quadrilaterals are named according to their properties.

Square
Equal sides, all angles equal to 90°

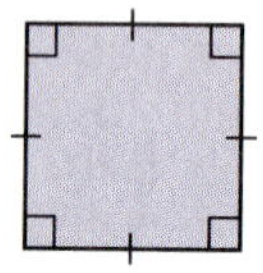

Rhombus
Equal sides

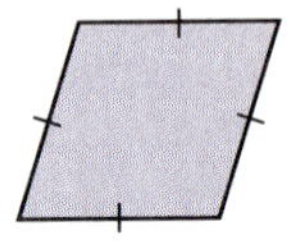

Rectangle
All angles equal to 90°, opposite sides equal

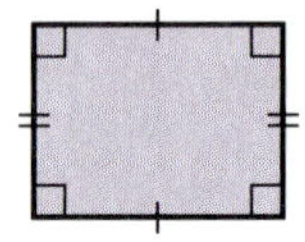

Parallelogram
Opposite sides are parallel.

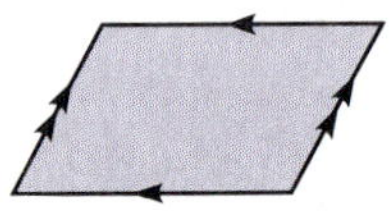

Trapezoid
Exactly one pair of opposite sides are parallel.

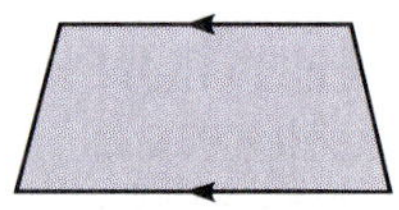

Kite
Two pairs of adjacent sides are equal.

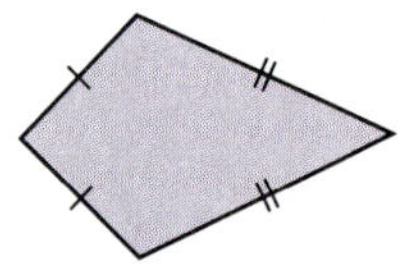

EXERCISES

For each triangle, the measures of its sides are given in centimeters. Classify each triangle as scalene, isosceles, or equilateral.

1. 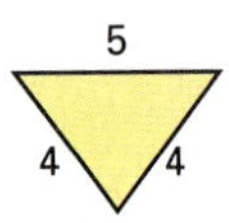

2. 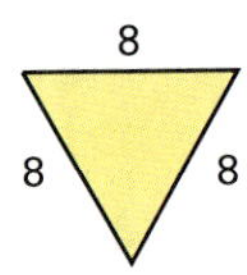

3. 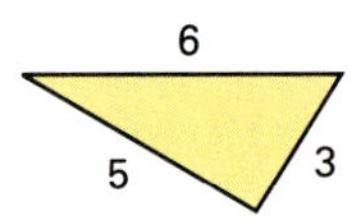

For each triangle, the measures of the angles are shown. Classify each triangle as acute, right, obtuse, or equiangular.

4. 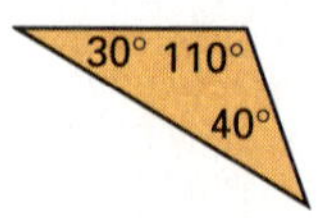

5. 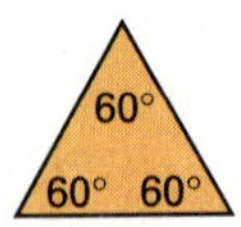

6. 

7. 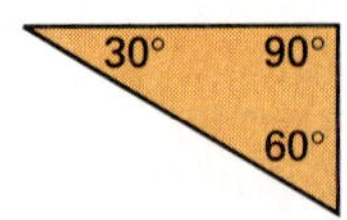

8. Draw a right triangle. You already know the measure of one of the angles. What type of angles are the remaining ones? Is this true for all right triangles? Explain.

Which of the following statements are true (T)? Which are false (F)? Give reasons for your answers.

"I can draw a triangle that is:

9. scalene and obtuse."

10. isosceles and right."

11. equilateral and right."

12. obtuse and isosceles."

Identify each polygon.

13. 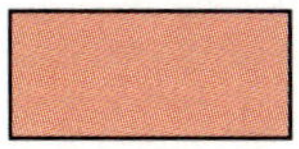

14. 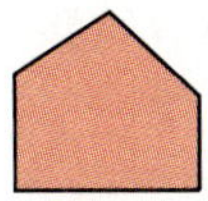

15. 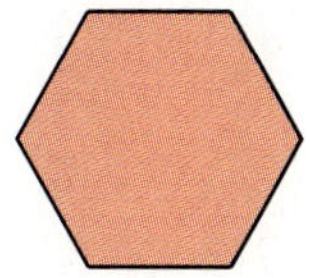

16. 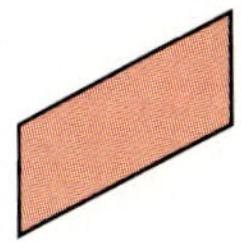

The area formulas for a parallelogram, a triangle, and a trapezoid are given below. For these exercises, find the area of the figure.

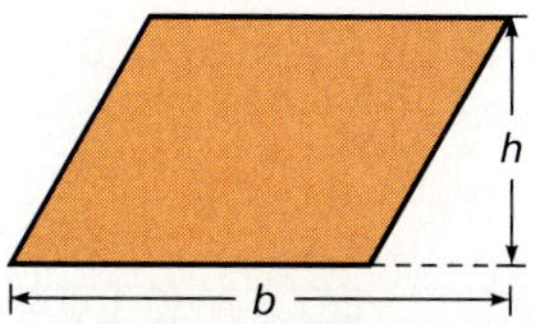

$A = b \times h$

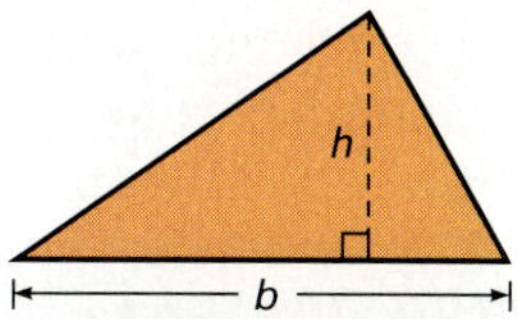

$A = \frac{1}{2} \times b \times h$

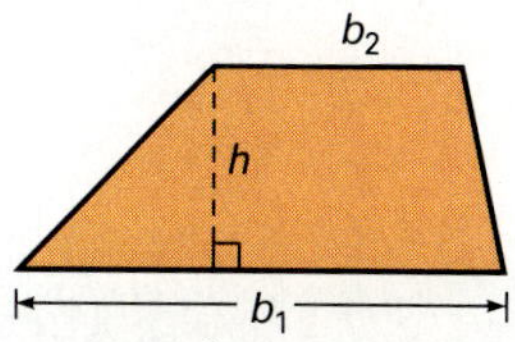

$A = \frac{1}{2}\, h\,(b_1 + b_2)$

17. A parallelogram has a base of 8 m and height of 5 m.

18. A triangle has a base of 12 ft and a height of 7 ft.

19. A trapezoid has bases of 3 cm and 6 cm and a height of 4 cm.

SECTION 10–4

PYTHAGOREAN THEOREM

WARM-UP ACTIVITY

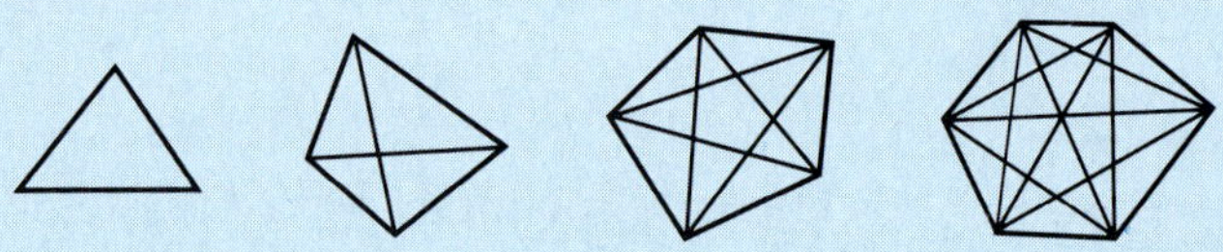

Copy and complete the chart. Continue the chart until the number of sides equals 12.

Name of polygon	Number of sides	Number of diagonals
Triangle	3	0
Quadrilateral	?	?

How many diagonals would be in a decagon?

How many diagonals would be in a dodecagon?

Write the general rule for this pattern.

INTRODUCTION

Because triangular structures provide strength, you often see them used in building and bridge construction. Many of these triangles are right triangles.

One example of a property of the right triangle that is useful in solving problems was studied by Pythagoras, an ancient Greek mathematician. By working with many right triangles he proved an important relation among the three sides.

SKILLS DEVELOPMENT

Pythagoras drew a square on each side of a right triangle as shown.

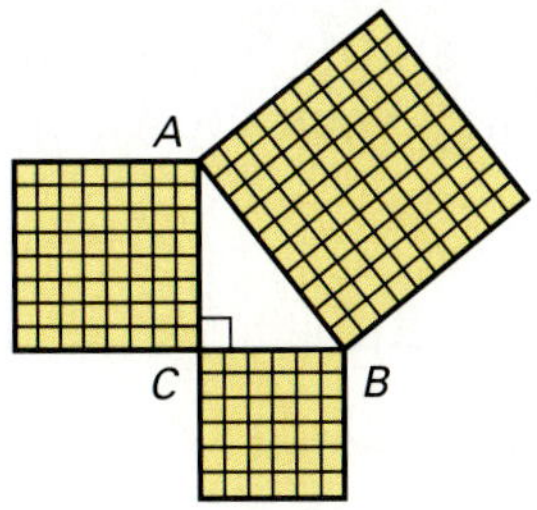

AB is the **hypotenuse** and faces the right angle.

a. Calculate the area of the square drawn on the hypotenuse AB.

b. Calculate the areas of the squares drawn on the other two sides.

c. How does the area of the square drawn on AB relate to the areas of the squares drawn on AC and BC?

From the activity above, you can make the following conjecture.

In any right triangle, the square of the length of the hypotenuse is equal to the sum of the squares of the other two sides, or $a^2 + b^2 = c^2$. This relationship is called the **Pythagorean Theorem.**

To use the Pythagorean Theorem you must be able to take the **square root** of a number. To square a number, you multiply it by itself. To take the square root of a number, ask yourself what number multiplied by itself will give the original number. Thus,

$4^2 = 16$ so $\sqrt{16} = 4$. Read: "The square root ($\sqrt{\ }$) of 16 is 4."

You can use a scientific calculator to find square roots. First, input the number. Then, depending on your calculator, press either the $\boxed{\sqrt{\ }}$ key or the $\boxed{\text{INV}}$ and $\boxed{x^2}$ keys in that order. The square root is displayed. You may want to round the square root to the nearest hundredth.

Example 1

Find the square root of 13. Round the answer to the nearest hundredth.

Solution Using the calculator, $\sqrt{13} = 3.6055513$ or 3.61.

Example 2

Find the length of the diagonal of the square. Round the answer to the nearest hundredth.

Solution Using the Pythagorean Theorem,

$$\begin{aligned} a^2 + b^2 &= c^2 \\ 8^2 + 8^2 &= c^2 \\ 64 + 64 &= c^2 \\ \sqrt{128} &= c \\ 11.31 &= c \end{aligned}$$

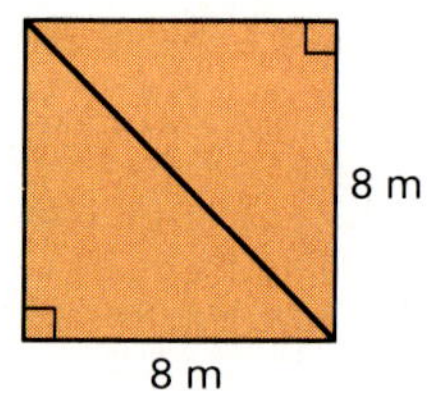

The length of the diagonal is 11.31 m.

EXERCISES

Write the square of each number.

1. 3 **2.** 6 **3.** 8 **4.** 9

Find the square root of each number.

5. 16 **6.** 25 **7.** 100 **8.** 64

Find the square root of each number. Use a calculator. Round each answer to the nearest hundredth.

9. 7 **10.** 12 **11.** 24 **12.** 95

Find the missing measure in each triangle.

13.

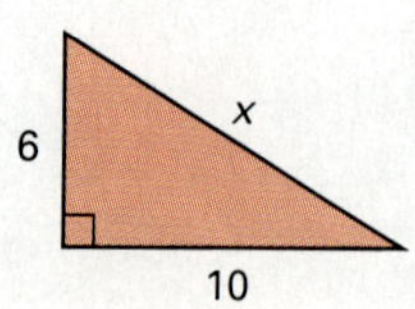

14.

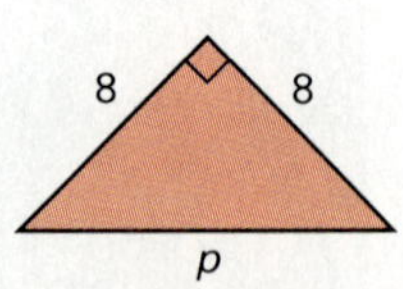

15.

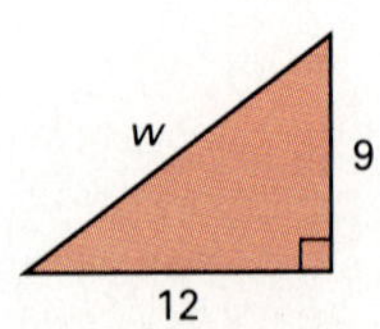

16. Calculate the length of the wire AB. Estimate first. Then calculate. Round the answer to the nearest tenth meter.

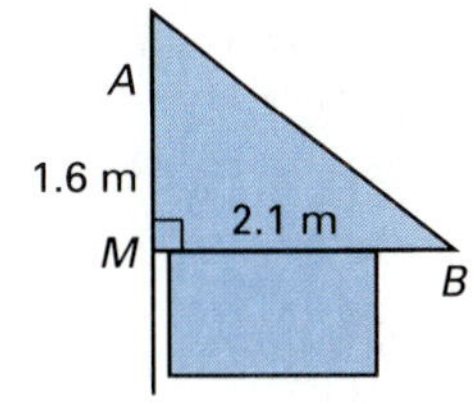

17. To calculate the height of the kite, the base of the triangle is measured. John is standing vertically below the kite. Use the data in the diagram. Calculate the height of the kite.

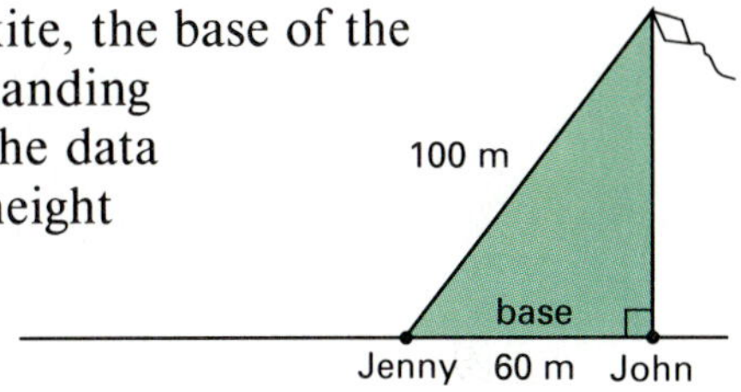

18. Use the measures of this baseball diamond to calculate the length of the throw from home to second base.

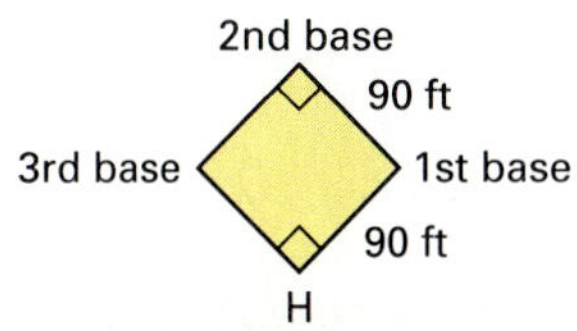

Galileo used to drop stones off the top of the leaning tower of Pisa to study the motion of falling bodies. To find the distance, d, in meters a body falls, use the formula $d = 5 \times t^2$. Time of fall, t, is given in seconds.

	1 s	2 s
Time fallen	1 s	2 s
Distance fallen	$d = 5 \times t^2$	$d = 5 \times t^2$
	$= 5 \times 1^2$	$= 5 \times 2^2$
	$= 5$ m	$= 20$ m

From the information calculated you can obtain other information.

$$\begin{pmatrix}\text{distance fallen}\\ \text{in the 2nd second}\end{pmatrix} = \begin{pmatrix}\text{distance fallen}\\ \text{in 2 seconds}\end{pmatrix} - \begin{pmatrix}\text{distance fallen}\\ \text{in 1 second}\end{pmatrix}$$
$$= 20 \text{ m} - 5 \text{ m}$$
$$= 15 \text{ m}$$

A suitcase is dropped from a bridge.

19. How far will it fall in 4 seconds? in 3 seconds?

20. How far did the suitcase fall in the 4th second?

SECTION 10–5

REAL NUMBERS

WARM-UP ACTIVITY

From your earlier work with integers and rational numbers, you can also obtain negative square roots. For example, 25 is the product of the factors shown.

$$(5)(5) = 25 \qquad (-5)(-5) = 25$$

The square roots of 25 are 5 and −5.

When you are working real-world problems involving the Pythagorean Theorem, you find the positive or principal square root. The symbol $\sqrt{\ }$ always means the principal square root.

Find the square roots of each of the following.
Circle the principal square root.

400 36

INTRODUCTION

In your earlier work you found that terminating and repeating decimals represent rational numbers.

$\frac{1}{8} = 0.125$ $\frac{2}{3} = 0.6666\ldots$ $1\frac{45}{99} = 1.454545\ldots$

There is another type of decimal that is neither terminating nor repeating, such as the following,

0.131131113111131 1 . . . 0.123456789101112 . . .

Decimals such as those above which cannot be written in rational form are called **irrational numbers.** An irrational number you used when you worked with the circle is π. The first 100 decimal places of π are shown below.

3.141 592 653 589 793 238 462 643 383 279 502 884 197 169 399 375 105 820 974 944 592 307 816 406 286 208 998 628 034 825 342 117 067 9

SKILLS DEVELOPMENT

In the previous section, you found the square roots of numbers. The decimals for numbers such as $\sqrt{2}$ or $\sqrt{45}$ are non-repeating and non-terminating. Thus, $\sqrt{2}$ and $\sqrt{45}$ are irrational numbers. Together, the rationals and irrationals form the set of **real numbers,** denoted by the symbol R.

Example 1

Identify each number as rational or irrational.

a. $-\sqrt{25}$ **b.** $\sqrt{45}$ **c.** $-\frac{3}{5}$

Solutions a. $-\sqrt{25} = -5$ since $\sqrt{25} = 5$. Thus, $-\sqrt{25}$ is rational.

b. $\sqrt{45} = 6.70820393\ldots$ Since $\sqrt{45}$ is not a perfect square, it is irrational.

c. All fractions including $-\frac{3}{5}$ are rational numbers.

A number line graph is often used to describe the answers to an inequality. The set of numbers from which the **domain,** or available answers, is chosen determines how the solution of the inequality is graphed.

Example 2

Hans bought apples at the store. He bought between 1 and 3 apples. Graph the number of apples he could have bought.

Solution Apples come in whole number amounts.

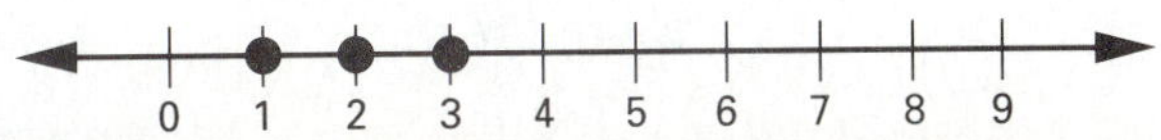

Example 3

Mike decided to get the bulk fruit trail mix and buy between 1 and 3 ounces.

Solution
When buying by weight, Mike can purchase any real number amount between 1 and 3 ounces inclusive.

A solid line is used if the domain is real numbers.

EXERCISES

In Exercises 2–10, identify each number as rational or irrational.

1. $-\sqrt{16}$ [$-\sqrt{16} = -4$ since $\sqrt{16} = 4$.] Thus $-\sqrt{16}$ is rational.

2. 6 **3.** $\sqrt{4}$ **4.** $-\frac{2}{3}$

5. $-\sqrt{9}$ **6.** $\sqrt{2}$ **7.** $2\frac{3}{5}$

8. $-\sqrt{36}$ **9.** $\sqrt{6}$ **10.** 0.136

The rational number $\frac{355}{113}$ was used by Tsu Ch'ung-chih to give an approximate value for π.

11. For which decimal places do the decimals for $\frac{355}{113}$ and π match?

12. Which is a better approximation to π, $\frac{22}{7}$ or $\frac{355}{113}$? Why?

To check your calculator, it is important to develop a method of estimating square roots. To estimate $\sqrt{20}$, ask yourself what two integers $\sqrt{20}$ would be between. That is, $4^2 = 16$ and $5^2 = 25$. You know the answer must be between 4 and 5.

Estimate each of the following.

13. $\sqrt{40}$ **14.** $\sqrt{84}$ **15.** $-\sqrt{75}$

16. A chart can be used to show the relationships of the sets of numbers. Any set of numbers contains all of the numbers of the set below it. Complete the chart with the sets of irrationals, integers, real numbers, whole numbers, natural numbers, and rationals.

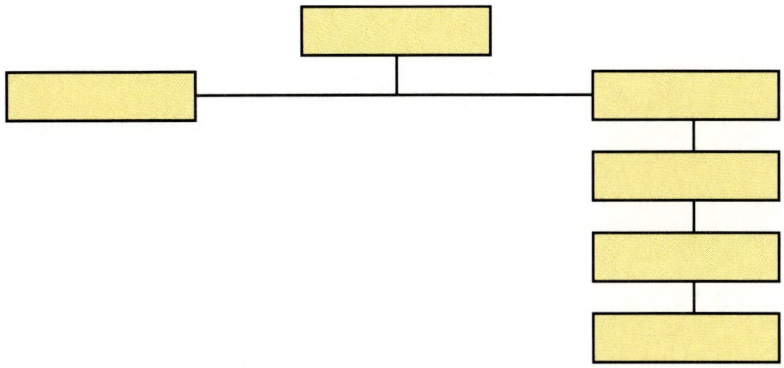

The graphs of two sets are shown.

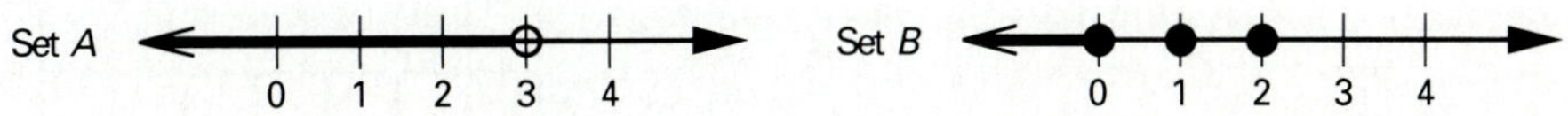

17. Name the domain for A.

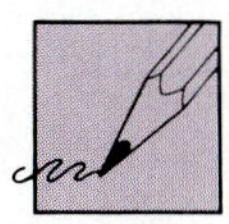

18. Name the domain for B.

19. How are the graphs alike? How do they differ?

Draw the graph that describes the answer to each of the following problems.

20. Buddy ate between 2 and 5 prunes inclusive each day. How many prunes could he have eaten each day?

21. Carla drank between 4 and 10 pints of water each day for a month. How many pints could she have drunk each day?

The number properties you studied previously may or may not apply to different sets within the set of real numbers.

22. Does the Commutative Property hold for addition of integers? Give an example.

23. Does the Associative Property hold for subtraction of rational numbers? Give an example.

SECTION 10–6

CONGRUENCE AND SIMILARITY

WARM-UP ACTIVITY

Complete each sentence. Then draw and complete the crossword puzzle.

Across

2 A triangle with one angle measuring 90° is called a __?__ triangle.

6 A triangle with one angle greater than 90° is a __?__ triangle.

7 A triangle with all sides of equal length is a __?__ triangle.

8 A triangle whose sides are different lengths is a __?__ triangle.

Down

1 The point where two sides of a triangle meet is a __?__.

3 A triangle with only two sides of equal length is an __?__ triangle.

4 A triangle whose angles measure less than 90° is an __?__ triangle.

5 A triangle whose angles all have equal measures is an __?__ triangle.

INTRODUCTION

If two figures are **congruent,** they can be placed one on top of each other and they match exactly. Congruent figures have the same size and shape.

To test whether two figures are congruent, you can make a tracing of one figure and see whether it fits exactly on top of the other.

The figures match.
The figures are congruent.

The figures do not match.
The figures are not congruent.

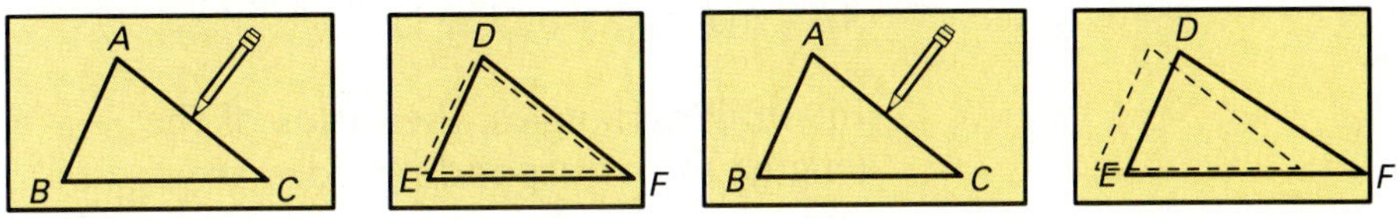

SKILLS DEVELOPMENT

When triangles are congruent, they are aligned so the equal sides and equal angles are in the same positions. Congruent sides are said to be **corresponding sides.** Similarly, congruent angles are said to be **corresponding angles.**

If $\triangle ABC \cong \triangle DEF$, then corresponding sides and angles are congruent.

Sides	Angles
$\overline{AB} \cong \overline{DE}$	$\angle A \cong \angle D$
$\overline{AC} \cong \overline{DF}$	$\angle B \cong \angle E$
$\overline{BC} \cong \overline{EF}$	$\angle C \cong \angle F$

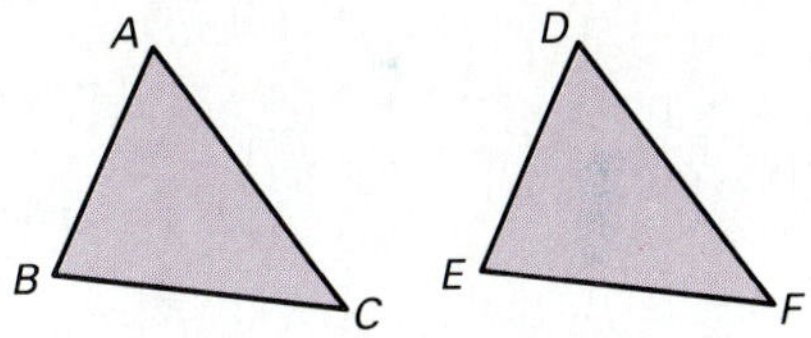

Similar figures have the same shape but are different in size.

Just as you enlarge a photograph, you can enlarge figures to obtain similar figures.

The ratio below compares the length of $\overline{AB}$ in the original figure to the length of $\overline{FG}$ in the enlargement.
$AB : FG = 2:4$ or $1:2$

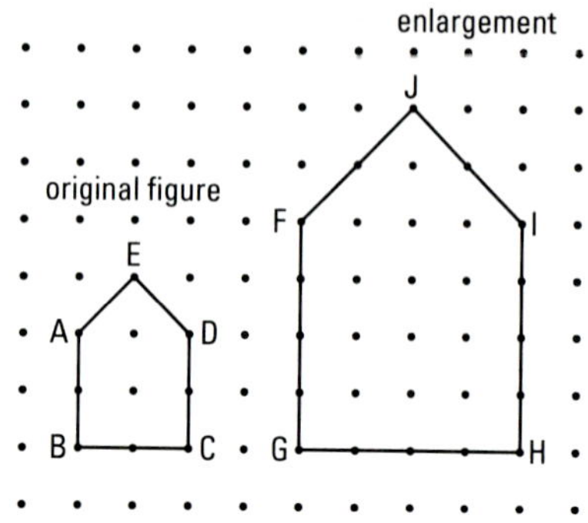

In similar triangles, the ratios of the lengths of corresponding sides are equal.

Example

Find the lengths of the sides of $\triangle DEF$.

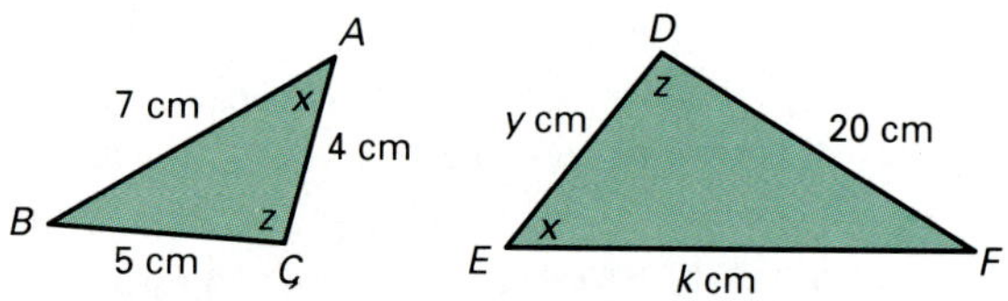

Solution Since $\angle A \cong \angle E$ and $\angle C \cong \angle D$, then $\angle B \cong \angle F$ (angle sum of triangle is 180°). Thus $\triangle ABC \sim \triangle EFD$ (symbol for similarity is $\sim$).

Thus $\frac{DE}{CA} = \frac{DF}{CB}$ and $\frac{EF}{AB} = \frac{DF}{CB}$

$\frac{y}{4} = \frac{20}{5}$ $\qquad$ $\frac{k}{7} = \frac{20}{5}$

$y = 16$ $\qquad$ $k = 28$

Thus $DE = 16$ cm and $EF = 28$ cm.

EXERCISES

For each pair of similar triangles, find the missing measures.

1.

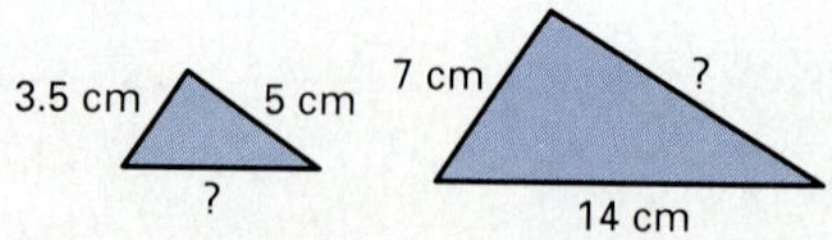

2.

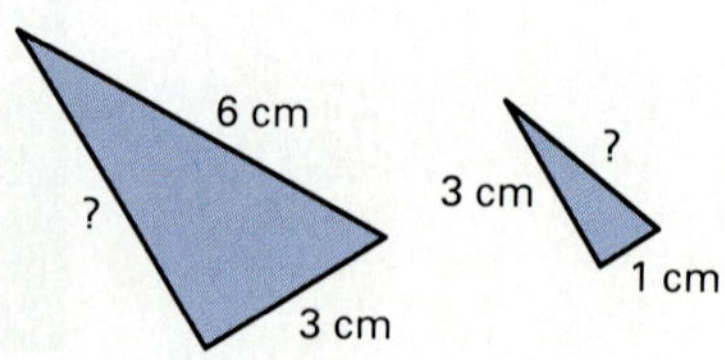

3. Which pairs of figures appear to be similar? Which appear to be congruent? Give reasons for your answer.

To measure the distance across the river, Francis, Michael, Helen, and Bob placed stakes as shown and marked off triangles. Michael located the point E lining up the rock at C and the tree at A.

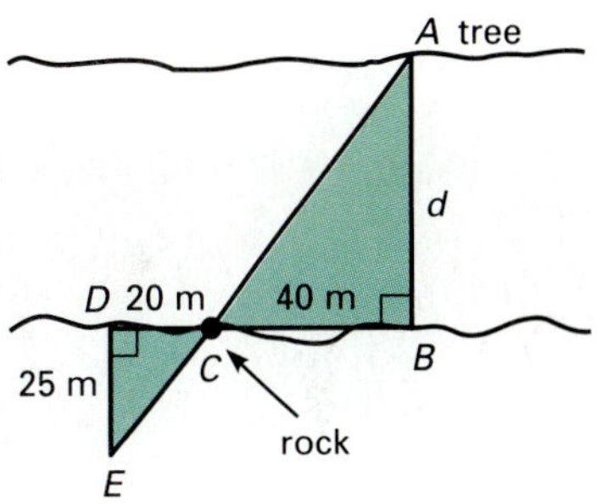

4. Use the diagram to explain why $\angle DCE = \angle ACB$.

5. Why are $\triangle ABC$ and $\triangle EDC$ similar?

6. Why is $\frac{AB}{ED} = \frac{BC}{DC}$?

7. Use the proportion in Exercise 6 to find the width of the river.

During takeoff, a jet climbs (rises) 10 m for a horizontal distance of 60 m.

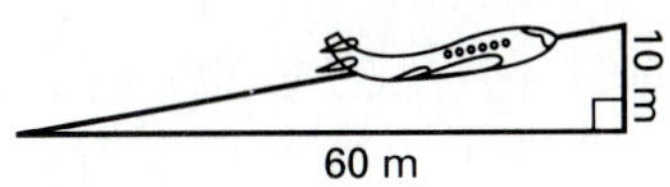

8. Calculate the height of the jet after it has traveled 1 km horizontally.

9. What assumption do you make in Exercise 8 to obtain your answer?

SECTION 10–7

CIRCLES

WARM-UP ACTIVITY

Copy and complete the chart.

Shape	Base	Height	Area
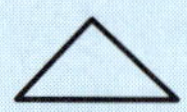	10 cm	9 cm	
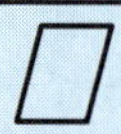	15 m	20 m	
	20 km		150 km^2
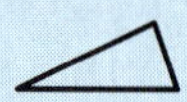		13 m	169 m^2
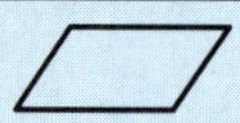	3 mm		15 mm^2

INTRODUCTION

A **circle** is a closed curve whose points are the same distance from a given point within the curve called the **center.** A circle is named by its center.

Certain characteristics of a circle have special names.

A **radius** joins the center to any point on the circle. $\overline{OC}$ is a radius.

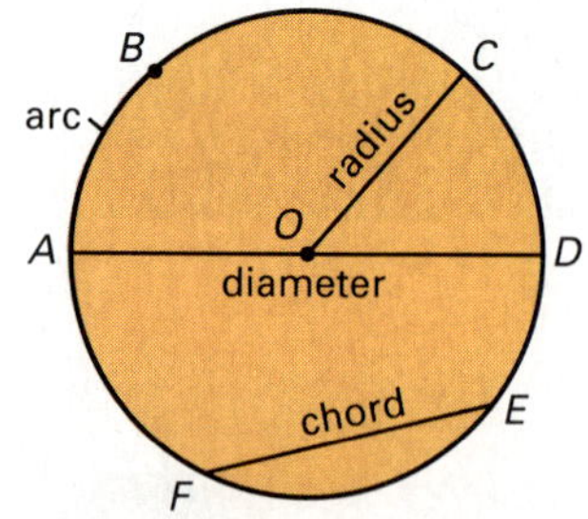

A **chord** joins any two points on the circle. $\overline{FE}$ is a chord.

A **diameter** is a chord through the center. $\overline{AD}$ is a diameter.

An **arc** is a portion of the circle itself. $\overline{AB}$ is an arc.

SKILLS DEVELOPMENT

Since the ratio of the circumference of a circle to its diameter, C/d, is π or about 3.14, a formula can be written for calculating the circumference, or perimeter, of a circle.

$$C = \pi \times d$$

The radius, r, and the diameter, d, of a circle are related by $d = 2r$ or $d = 2 \times r$. Therefore, the formula can also be written as $C = \pi \times 2 \times r$ or $C = 2 \times \pi \times r$.

Example 1

Calculate the circumference of each circle. Round your answer to the nearest tenth.

a.

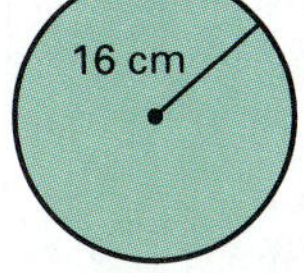

b.

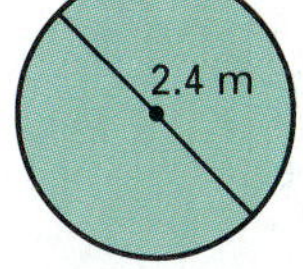

Solutions a. Use $C = 2 \times \pi \times r$ — The radius, r, is 16 cm.

$C \approx 2 \times 3.14 \times 16$

≈ 100.48 — The symbol $\approx$ means "is approximately equal to."

The circumference is about 100.5 cm.

b. Use $C = \pi \times d$ — The diameter, d, is 2.4 m.

$C \approx 3.14 \times 2.4$

≈ 7.536

The circumference is about 7.5 m.

A formula for the area of a circle can be developed from the formula for the area of a parallelogram. Suppose a circle is cut into equal sections and is used to form a rough parallelogram.

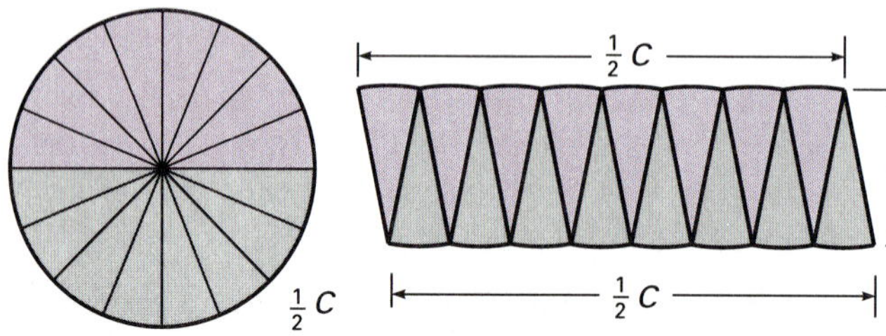

$A = b \times h$

$A = \frac{1}{2} \times C \times r$

$A = \frac{1}{2} \times 2 \times \pi \times r \times r$

$A = \pi \times r \times r$

$A = \pi \times r^2$

Example 2

Find the area of each circle. Use $\pi \approx 3.14$. Round each answer to the nearest tenth.

a. 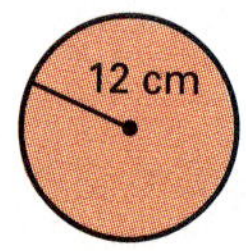

b. 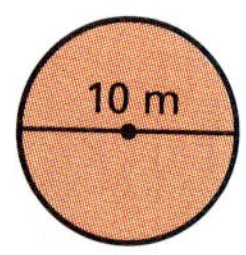

Solutions a. Use the formula:

$A = \pi \times r \times r$ $\qquad r = 12$ cm

$A \approx 3.14 \times 12 \times 12$

≈ 452.16

The area of the circle is about 452.2 cm^2.

b. Use the formula:

$A = \pi r^2$ $\qquad d = 10$m

$A \approx 3.14(5)^2$ $\qquad r = 5$m

$\approx 3.14(25)$

≈ 78.5

The area of the circle is about 78.5 m^2.

EXERCISES

1. An ordinary basketball has a diameter of about 10 in. What would be the measure of the circumference?

2. Telescopes let us see far into the sky. One of the larger telescopes in the world is in Russia. The circular mirror used in the telescope has a diameter of 600 cm. Calculate the circumference of the mirror. Would this mirror fit into your classroom?

3. How far would you walk to travel around the track?

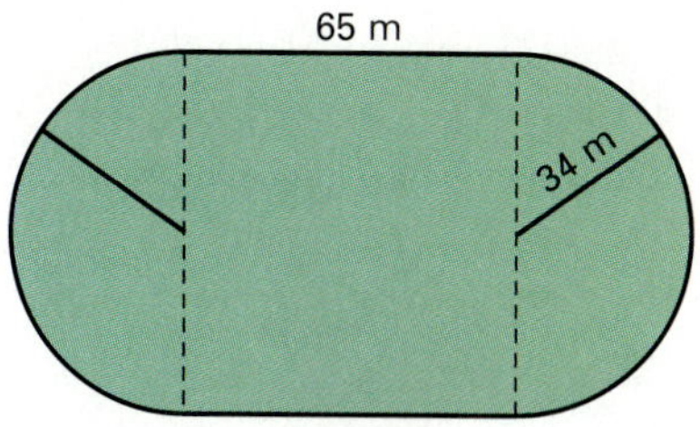

4. An equilateral triangle is inscribed (drawn) in a circle with center O as shown. $\angle COB$ is called a **central angle.** Name two other central angles in this diagram.

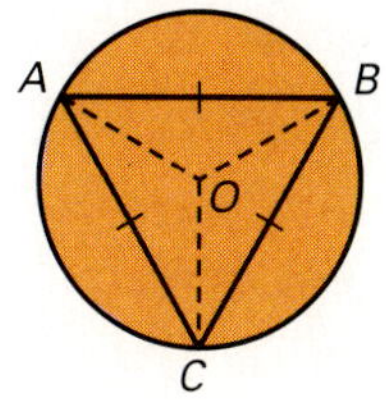

5. Measure the central angles $\angle AOB$, $\angle AOC$, and $\angle BOC$. What do you notice?

6. Explain why $\angle AOB + \angle AOC + \angle BOC = 360°$.

A **circle** or **pie graph** can be used to show data that can be expressed as a percent of the total. The percents are then used to calculate the size of the **central angle** for each sector of the circle graph.

The monthly sales for an appliance firm are shown as a circle graph.

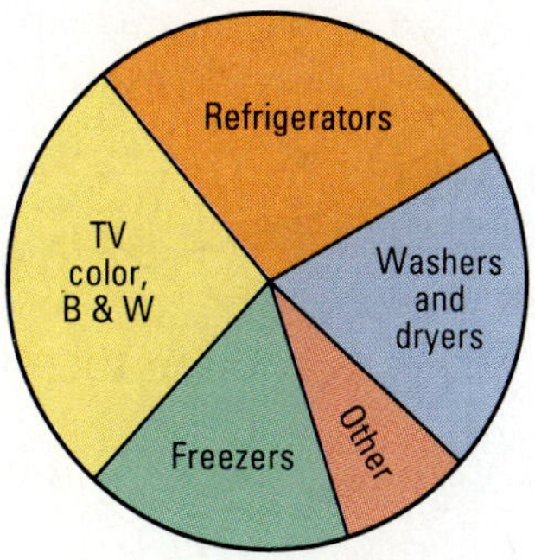

Monthly Sales

7. The central angle for the washers and dryers is 72°. How can you tell they account for 20% of total sales?

8. Measure the central angle for the refrigerators.

9. What percent of total sales do refrigerators account for?

SECTION 10–8

WORKING IN THREE DIMENSIONS

WARM-UP ACTIVITY

In a triangle, if the sum of the squares of the two shorter sides equals the square of the longest side, then the triangle is a right triangle. Which of the following triangles are right triangles?

a	b	c
4 yd	4 yd	6 yd
9 ft	12 ft	15 ft
5 m	12 m	13 m
6 cm	8 cm	15 cm

INTRODUCTION

A **polyhedron** is a three-dimensional object with flat surfaces. Pyramids and prisms are polyhedra.

Pyramid

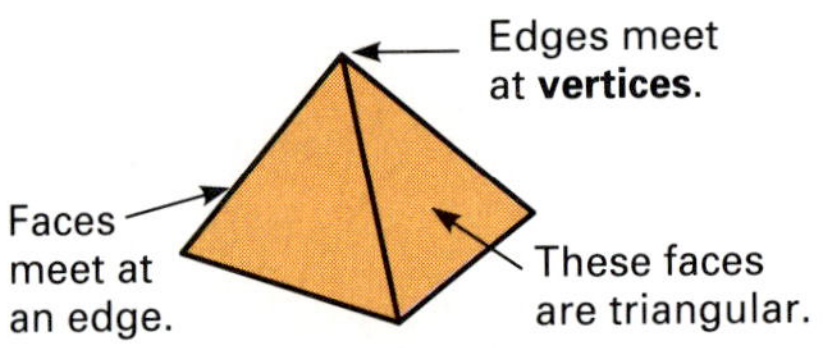

Prism

A10.

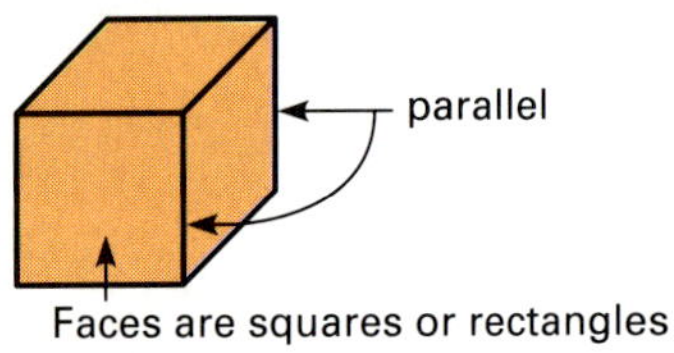

These common three-dimensional objects are not polyhedra.

Cylinder

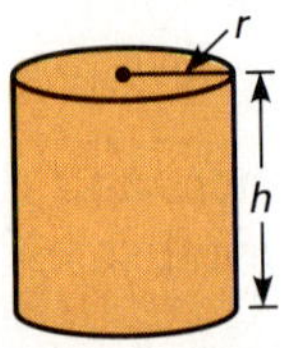

Cone

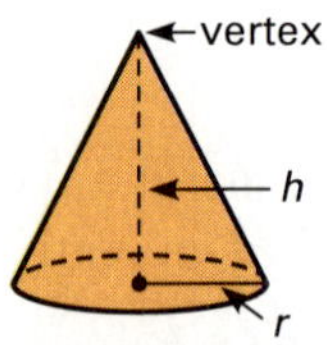

Sphere

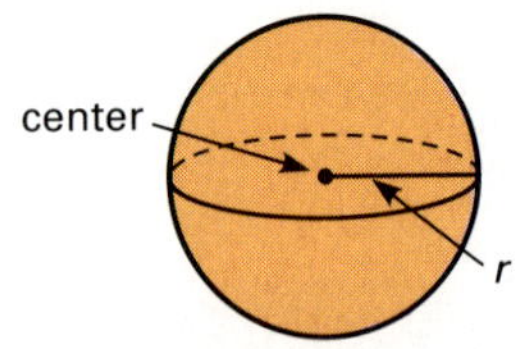

To calculate the volume of a prism or a cylinder, use $V = B \times h$ where B is the area of the base and h is the height.

SKILLS DEVELOPMENT

Recall that for a rectangular prism, volume = area of base × height, or $V = B \times h$. Since the base of the prism is rectangular, B is written as $l \times w$. Thus, $V = l \times w \times h$.

For a cylinder, volume = area of base × height, or $V = B \times h$. Since the base of a cylinder is a circle, B is written as $\pi \times r \times r$. Thus, $V = \pi \times r \times r \times h$.

Example 1

How much does the can hold? Round the answer to the nearest whole centimeter.

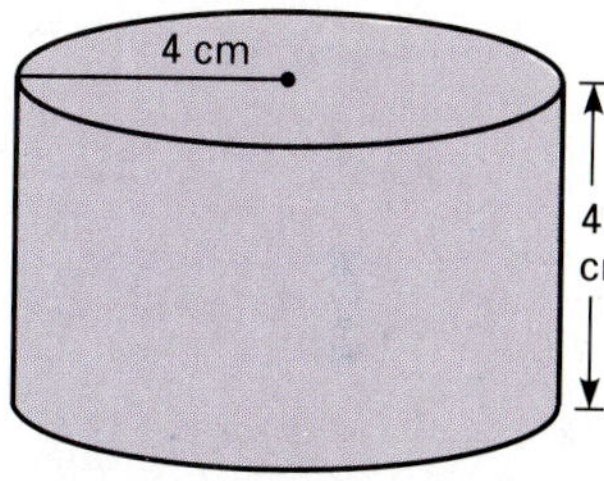

Solution Use the formula:

$$V = \pi \times r \times r \times h$$
$$\approx 3.14 \times 4 \times 4 \times 4.5$$
$$\approx 226.08$$

The can holds about 226 cm^3.

Pyramids and cones have special relationships to prisms and cylinders, respectively. Given a prism and a pyramid with the same bases and heights, the volume of the pyramid is one-third the volume of the prism. Thus, the formula for the volume of a pyramid is $V = \frac{1}{3}Bh$.

Given a cylinder and a cone with the same bases and heights, the volume of the cone is one-third the volume of the cylinder. Thus, the formula for the volume of a cone is $V = \frac{1}{3}Bh$ or $V = \frac{1}{3}\pi r^2 h$.

Example 2

Find the volume of each container. Round the answers to the nearest cubic centimeter.

Solutions a. Use the formula:

$$V = \frac{1}{3}Bh$$
$$= \frac{1}{3} \times l \times w \times h$$
$$= \frac{1}{3} \times 6 \times 3 \times 5$$
$$= 30$$

The volume is 30 cm^3.

b. Use the formula:

$$V = \frac{1}{3}\pi r^2 h$$
$$\approx \frac{1}{3} \times 3.14 \times (2.1)^2 \times 6.2$$
$$\approx 28.61796$$

The volume is about 29 cm^3.

EXERCISES

1. What is meant by the term regular polygon? What do you think is meant by the term regular polyhedron?

2. There are only five regular polyhedra. The surface area patterns of four of them are shown. Make a larger copy of each one. Construct each regular polyhedron. Use construction paper. Each prefix is in Greek. What do you think is its meaning?

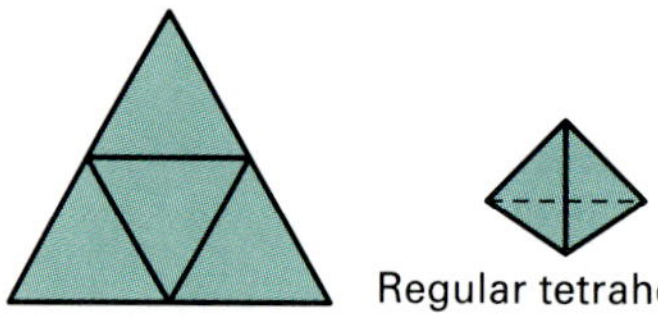

Regular tetrahedron

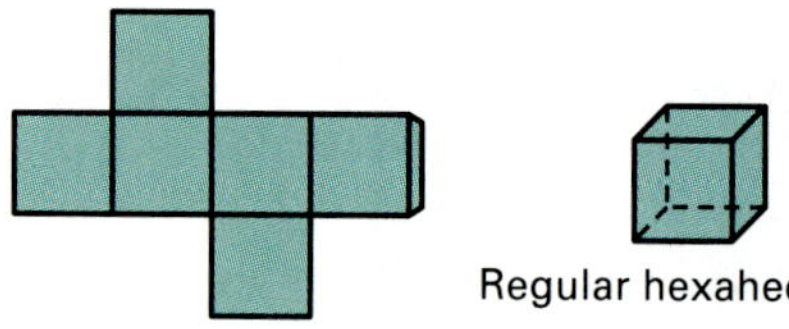

Regular hexahedron

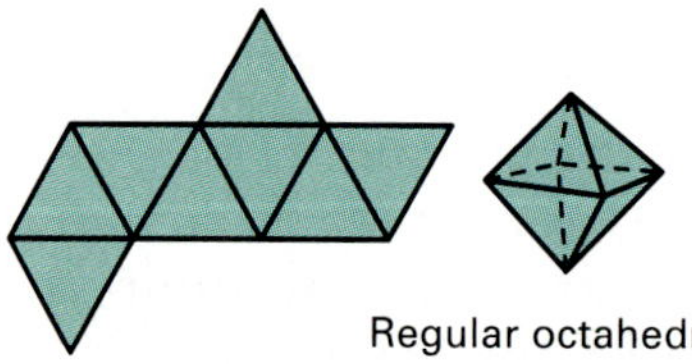

Regular octahedron

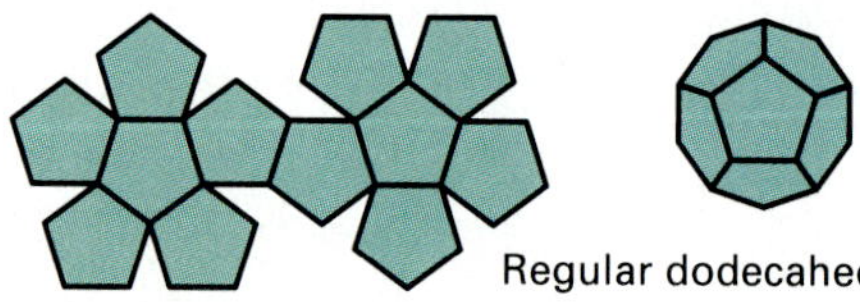

Regular dodecahedron

Find the volume.

3.

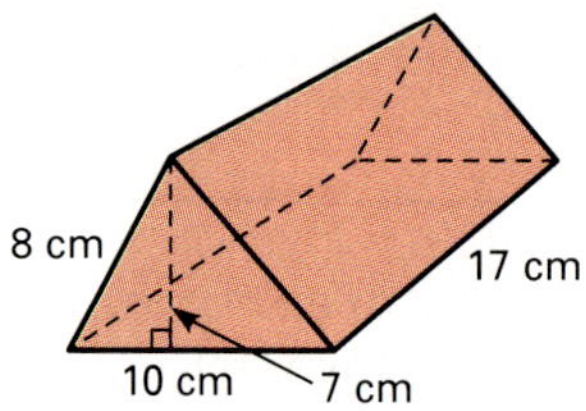

4.

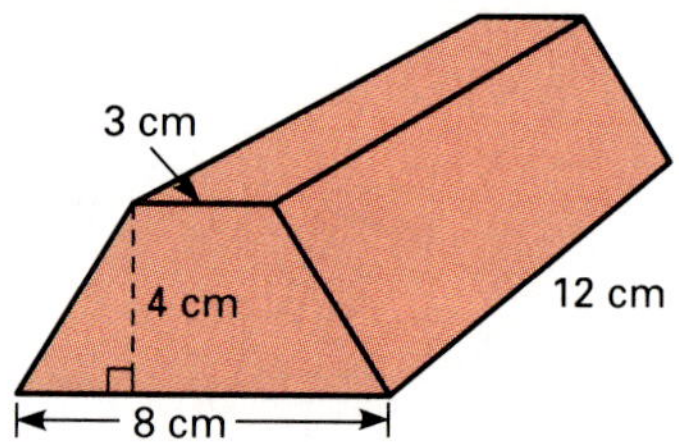

5. Which container holds most? Arrange the containers in order from the one that holds the most to the one that holds the least.

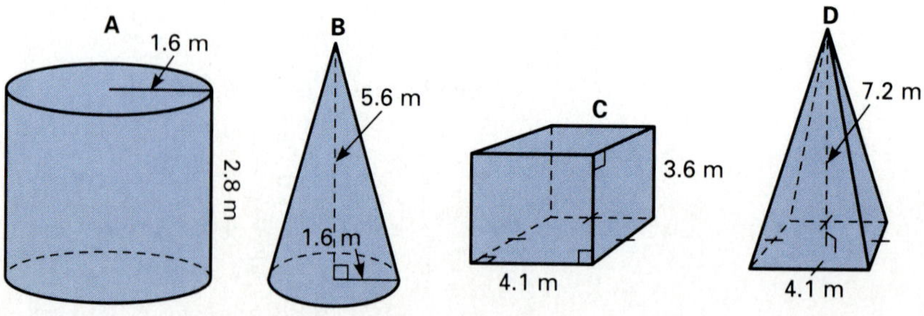

Lisa drinks six glasses of water a day. The dimensions of her glass are shown.

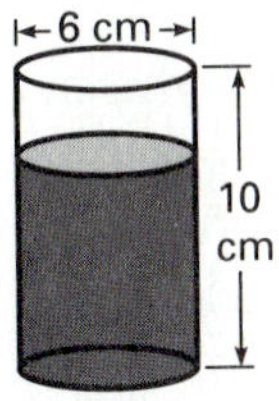

6. Calculate the volume of water she drinks in a day.

7. Calculate the volume of water she drinks in a year.

8. What assumptions did you make to find your answers?

9. How much of your classroom would this volume of water fill?

The formula for the volume of a sphere is $V = \frac{4}{3}\pi r^3$.

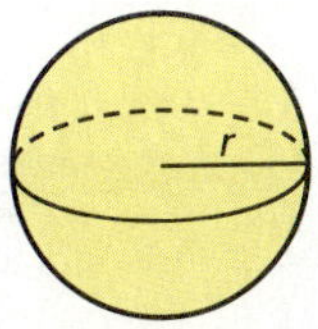

10. Calculate the volume of a sphere with radius 6 cm.

11. Calculate the volume of gas in a balloon with radius 2.6 m.

12. Calculate the volume of a spherical tank with radius 3.8 m.

CHAPTER 10 SUMMARY

To communicate with precision you need to know the vocabulary and reasoning skills employed in geometry. You learned the fundamental terms and definitions of geometric figures and studied the use of conjecture in inductive reasoning.

In the previous chapter you investigated perimeter, area, and volume. In this chapter you extended those concepts to develop formulas for the figures and shapes you studied. As you looked at figures, you noticed some were the same size and others were the same shape but not the same size. These ideas of congruence and similarity were explored. As you investigated triangles, you discovered a relationship between the sides of a right triangle known as the Pythagorean Theorem. This led to working with square roots and irrational numbers to solve problems.

VOCABULARY

arc
center
central angle
chord
circle
circle graph
congruence
conjecture
deductive reasoning
dense
diameter
hypotenuse
inductive reasoning
irrational numbers
line segment
point
polygon
polyhedron
Pythagorean Theorem
radius
ray
real numbers
similar
square root
straight line
undefined

CHAPTER REVIEW

Measure the angles in the diagram. Use the diagram to answer the true and false questions.

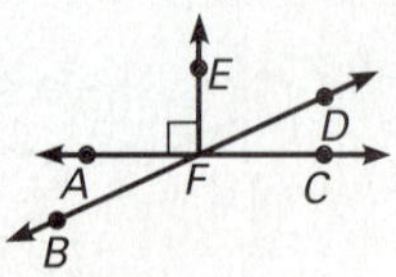

1. $\overline{EF}$ is perpendicular to $\overline{FC}$.
2. $\overline{BD}$ is a ray.
3. $\overline{AC}$ is a line.
4. $\angle EFC = 90°$.
5. $\angle DFC$ is adjacent to $\angle DFE$.
6. $\angle BFC$ is an acute angle.

7. Find the missing measure.

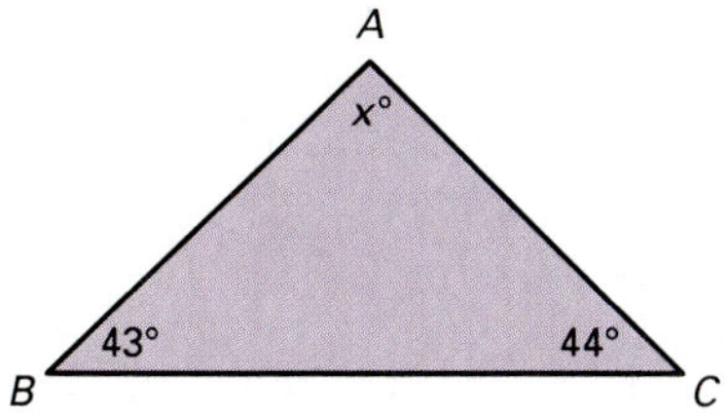

Classify each of the following triangles according to the measures of their angles.

8.

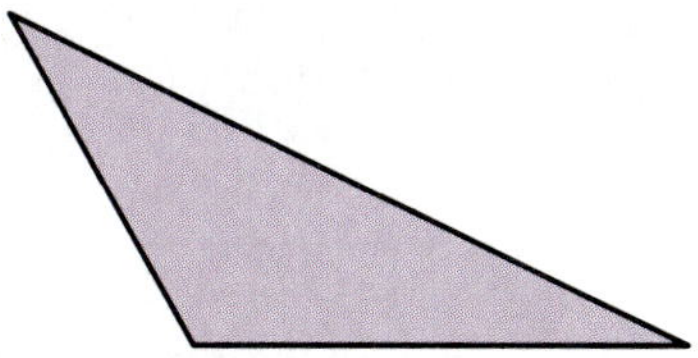

9.

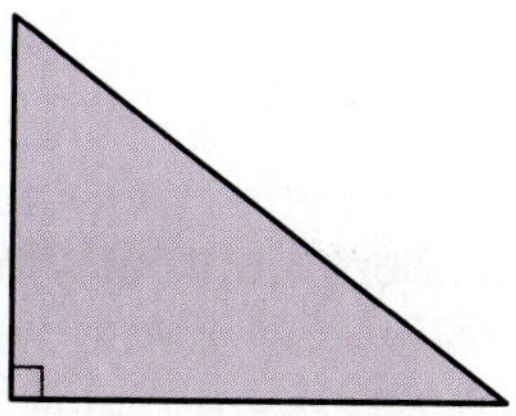

10.

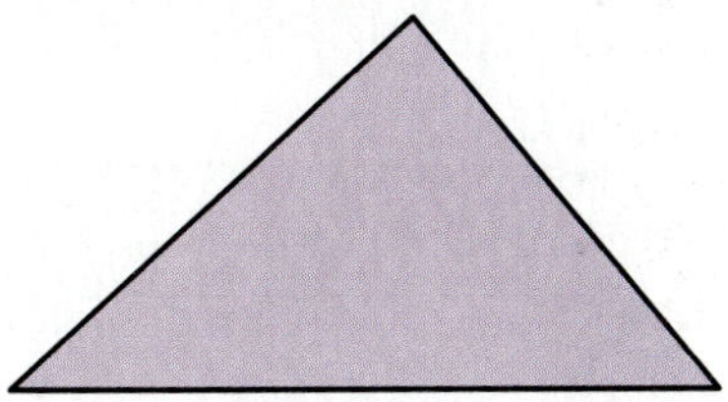

Use the Pythagorean relation to find the missing measures.

11.

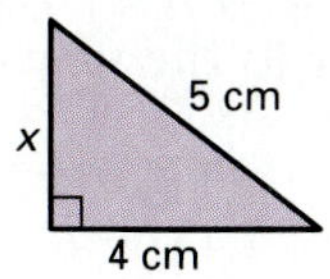

12.

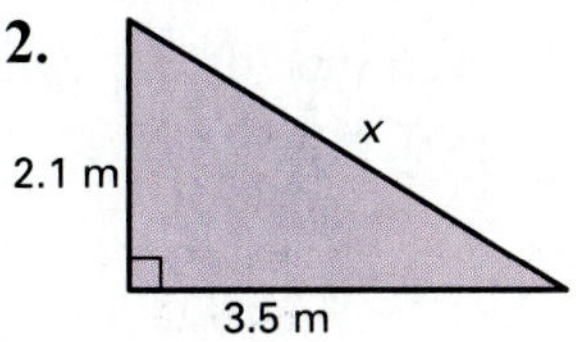

Tell if the following numbers are rational (R) or irrational (I).

13. 3.45464748 . . . **14.** 3.4534 **15.** 85

16. Graph on a number line the real numbers greater than -2.

17. Graph on a number line the integers greater than -2.

18. Find the missing measures. $\triangle ABC$ is similar to $\triangle DEC$.

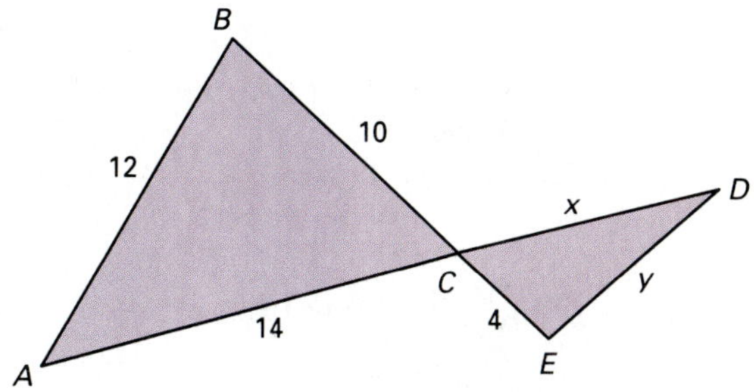

19. Calculate the circumference and the area of the circle.

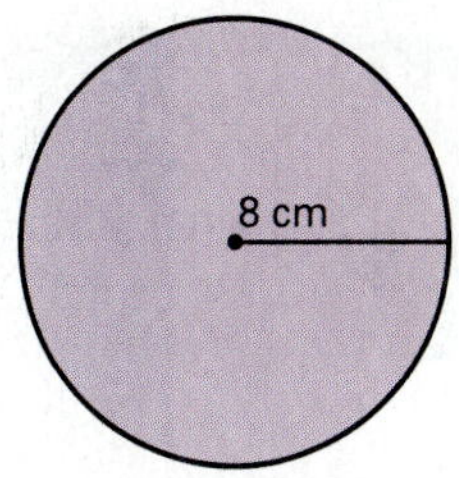

Identify each shape.

20

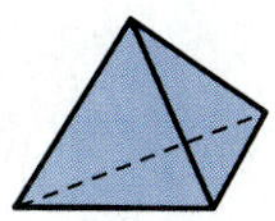

21.

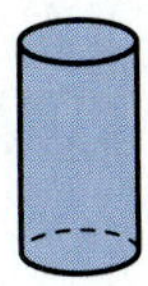

22.

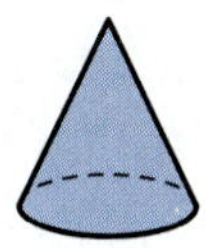

CHAPTER 10 TEST

Match the figure in column A with its description in column B and its area in column C.

	Column A	Column B	Column C
1.	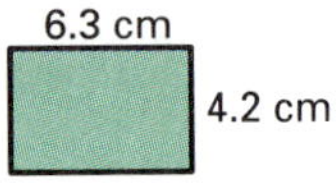	trapezoid	113 cm^2
2.	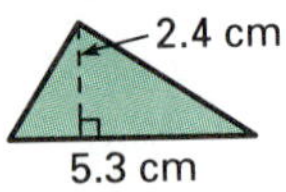	rectangle	51.5 cm^2
3.		triangle	26.5 cm^2
4.	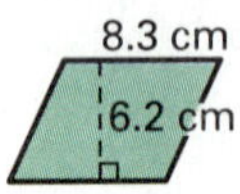	rhombus	6.4 cm^2
5.	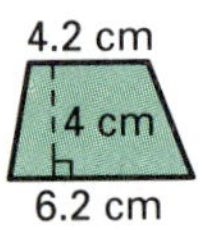	circle	21 cm^2
6.	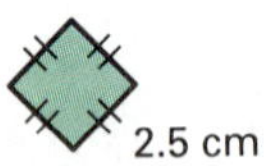	parallelogram	6.3 cm^2

7. Find the measure of y.

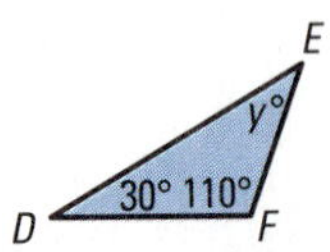

8. Calculate the length of the guywire. Round the answer to the nearest tenth.

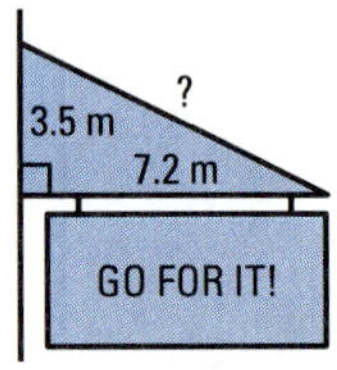

9. True or false: Square roots are always irrational numbers.

10. To measure the width of a river, the triangles shown were constructed on the banks. Find the width of the river.

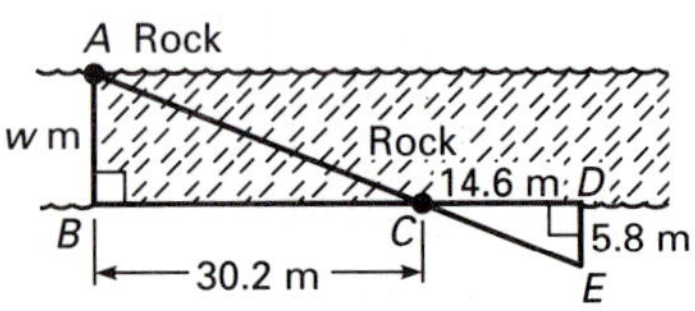

11. The formula for the volume of a cylinder is $V = \pi r^2 h$. Find the volume of the cylinder to the nearest tenth.

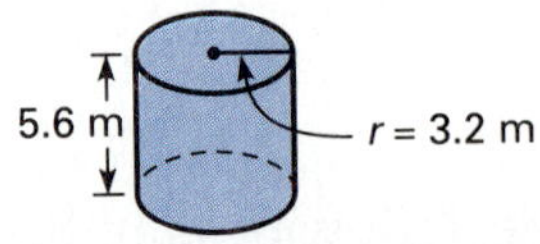

CUMULATIVE REVIEW

1. Jill starts from the center of a garden maze and finds the way out by walking 3 meters forward, 2 meters right, 7 meters right, 1 meter left, 3 meters left, 2 meters right, 8 meters left, 5 meters left, and 2 meters right. How far from the center is the outside opening of the maze?

2. Heinz and Martha live on a road that leads straight to the beach. The sum of their distances to the beach is 15 miles. Martha lives 4 miles closer to the beach than Heinz. How many miles from the beach do Heinz and Martha live?

Solve each percent problem.

3. What percent of 36 is 20?

4. 3 is 8% of what number?

5. What number is 2% of 72?

Solve each proportion.

6. $\frac{x}{8} = \frac{3}{24}$

7. $\frac{2}{5} = \frac{x}{45}$

8. $\frac{7}{5} = \frac{6}{x}$

Write each number in scientific notation.

9. 540

10. 3800

11. 567,000,000

12. Measure in millimeters to find the perimeter and the area of the figure.

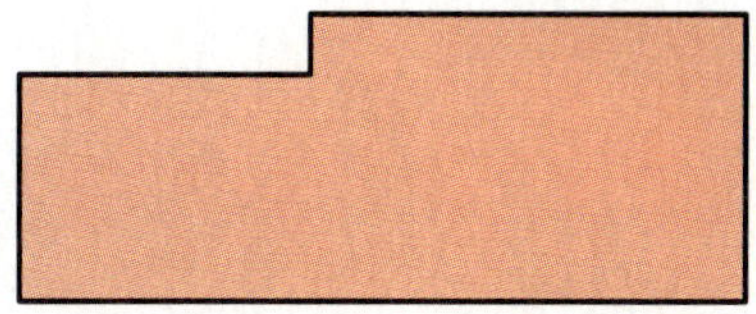

13. Name the three-dimensional shape having the given surface area pattern.

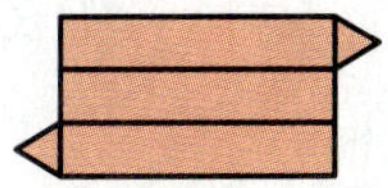

14. Find the perimeter of the triangle.

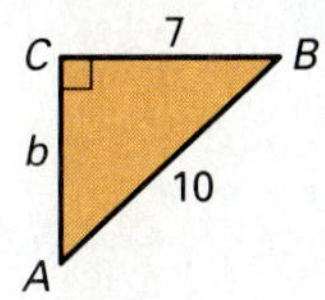

15. Identify which figures are polygons.

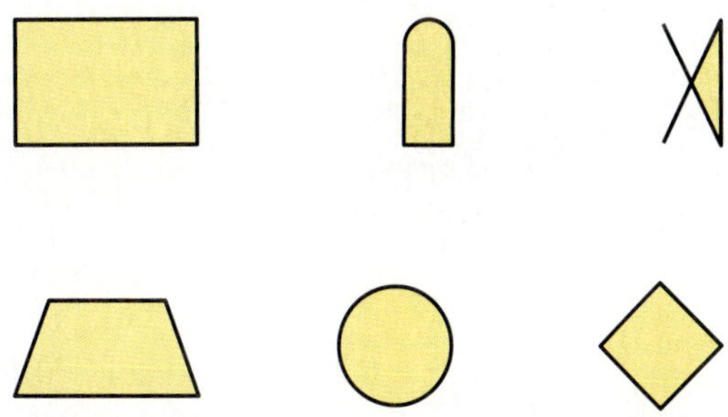

The study of relationships is important in fields such as business, psychology, mathematics, and in everyday life. Relationships suggest patterns that can be observed and analyzed. Relationships can be shown in several ways.

Traveling by Car	
Time (hours)	Distance (miles)
0	0
2	100
4	200
6	300

Amount of gasoline used for a trip

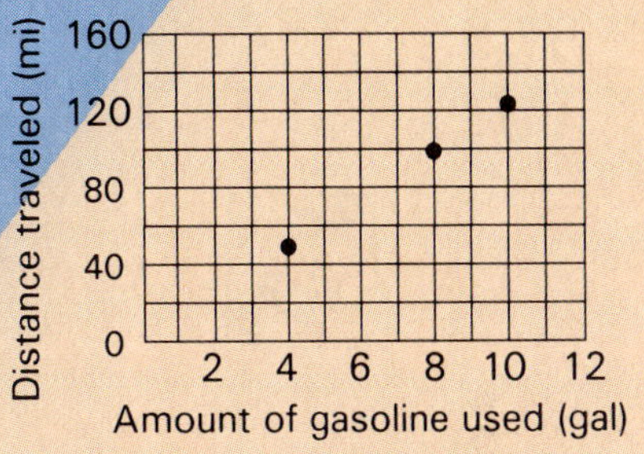

Ask Yourself

- *What relationship do you see between the elapsed time and the distance traveled?*
- *What relationship do you see between the amount of gasoline used and the distance traveled?*
- *What relationship exists between the number of people attending a sporting event and the profit?*

CHAPTER 11

RELATIONS AND FUNCTIONS

SECTION 11–1

ORDERED PAIRS

WARM-UP ACTIVITY

To solve this problem you have to ask some important questions. How are the sums of digits related to each other? There are only 10 digits: 0, 1, 2, 3, 4, 5, 6, 7, 8, and 9. Use them each once to complete the addition problem.

INTRODUCTION

In Chapter 10 you were introduced to some of the types of geometry you will study. In this section you are introduced to yet another aspect of geometry. Geometry with coordinates enables you to describe the position of geometric figures much like a map describes the location of places.

The grid with a horizontal axis and a vertical axis is often called the **Cartesian plane** after the French mathematician René Descartes. This geometry is referred to as coordinate geometry because it uses coordinates to describe location.

Notice the numbering of the horizontal axis. It is exactly like the number lines you have already used. The vertical axis is also a number line. The positive numbers go up and the negative, down. The point of intersection of the axes is the **origin.**

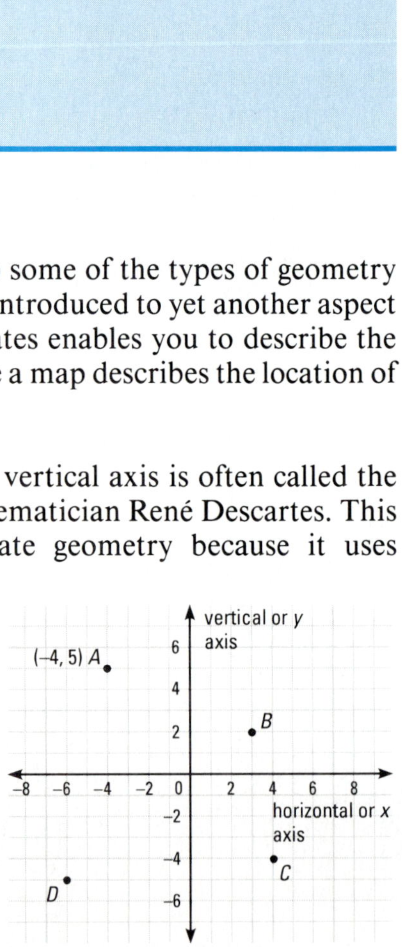

SKILLS DEVELOPMENT

Coordinates are numbers in an ordered pair, such as $(-4, 5)$. These coordinates locate the position of a point on the Cartesian plane. The point is represented by a dot and labeled with a capital letter.

Let us examine the ordered pair $A(-4, 5)$. The coordinate -4 is called the x-coordinate (or abscissa) and determines the direction and the horizontal distance of the point from the origin. The coordinate 5 is called the y-coordinate (or ordinate) and determines the direction and the vertical distance of the point from the origin. To plot or graph point A, begin at the origin, move 4 units to the left to -4, and 5 units up to 5. Mark the point and label it with a capital A.

Example 1

Refer to the coordinate plane on the previous page.

a. Which point has coordinates $(4, -4)$?

b. What are the coordinates of B? of D?

Solutions a. C has coordinates $(4, -4)$.

b. The coordinates of the points are:

$B(3, 2)$: 3 right, 2 up

$D(-6, -5)$: 6 left, 5 down

Example 2

For the ordered pair $(-1, 3)$ give:

a. the x-coordinate.

b. the y-coordinate.

Solutions a. The x-coordinate is -1. It is written first in an ordered pair.

b. The y-coordinate is 3. It is written second in an ordered pair.

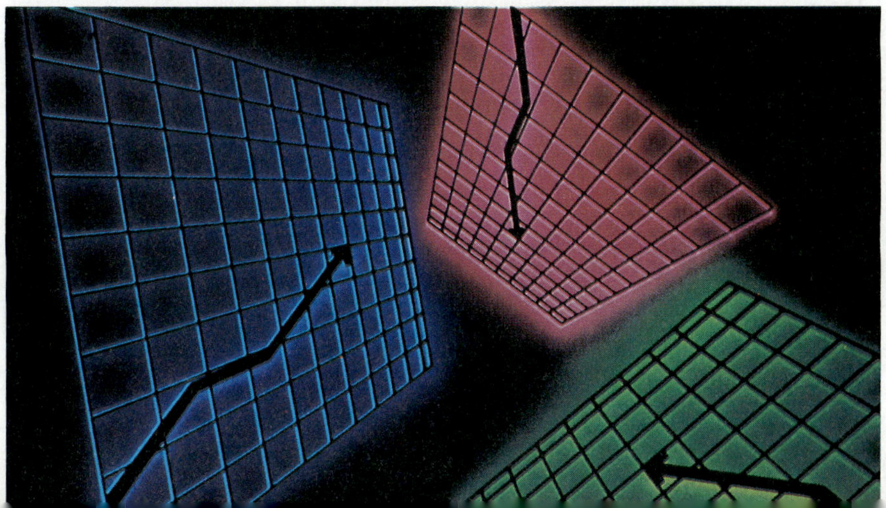

The word **quadrant** is used to refer to each of the four parts of the Cartesian plane.

EXERCISES

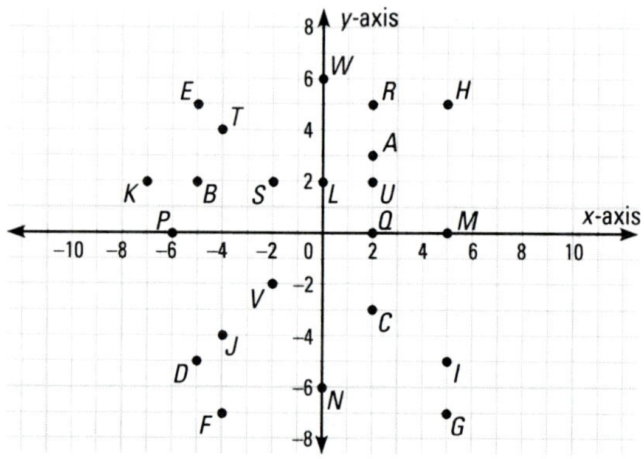

Name the coordinates for each of the following points.

1. *U* **2.** *B* **3.** *C* **4.** *V*

5. *G* **6.** *K* **7.** *E* **8.** *N*

Each ordered pair names a point. Write the corresponding letter for each ordered pair.

9. (2, 3) **10.** (−2, 2) **11.** (−4, −4)

12. (5, −5) **13.** (2, 0) **14.** (0, 2)

15. (2, 5) **16.** (−5, −5) **17.** (5, 5)

18. Name the coordinates of points *P, M,* and *Q*. What property do these points have in common?

19. Name the coordinates of points *W, L,* and *N*. What property do these points have in common?

20. Name the points that have both coordinates negative.

21. Name the points that have equal coordinates.

22. Name the coordinates of the origin.

23. Why do the ordered pairs (5, −5) and (−5, 5) name different points?

Draw a Cartesian plane and plot the following points.

24. $A(-2, 0)$	25. $B(0, 2)$	26. $C(1, 3)$
27. $D(-3, -1)$	28. $E(5, 7)$	29. $F(-4, -2)$
30. $G(2, 4)$	31. $H(-1, 1)$	32. $I(3, 5)$

33. What do you notice about your graph?

These ordered pairs represent a geometric figure.

$A(0, 4)$ $B(4, 0)$ $C(0, -4)$ $D(-4, 0)$

34. Plot the points. Then join them in order. What geometric figure have you drawn?

35. The points $A(3, -3)$, $B(3, 7)$, and $C(-7, 7)$ are three vertices of a square. Find the coordinates of the remaining vertex.

36. Follow these instructions:
Join (−8, −7) to (−8, 2).
Join (−8, 2) to (−3, −7).
Join (−3, −7) to (2, 2).
Join (2, 2) to (2, −7).
What letter of the alphabet have you made?

A message is written in code based on ordered pairs. The letters of each word are scrambled. Unscramble the letters to find the message.

37. How many wings does a bee have? Do you think it's two, four, six, or eight? Use the coded message to find the answer.

Join (−5, 3) to (−7, 3) to (−7, −1). Join (−5, 1) to (−7, 1).

Join (−2, 3) to (−1, 3) to (0, 2) to (0, 0) to (−1, −1) to (−2, −1) to (−3, 0) to (−3, 2) to (−2, 3).

Join (1, 3) to (1, 0) to (2, −1) to (3, −1) to (4, 0). Join (4, 3) to (4, −1).

Join (5, −1) to (5, 3) to (7, 3) to (8, 2) to (7, 1) to (8, −1). Join (5, 1) to (7, 1).

SECTION 11–2

RELATIONS AND THEIR RULES

WARM-UP ACTIVITY

Play the game of Mathematical Steeplechase. The track has different questions on computation. After you have completed the race, give yourself one point for each correct answer and deduct one point for each incorrect answer.

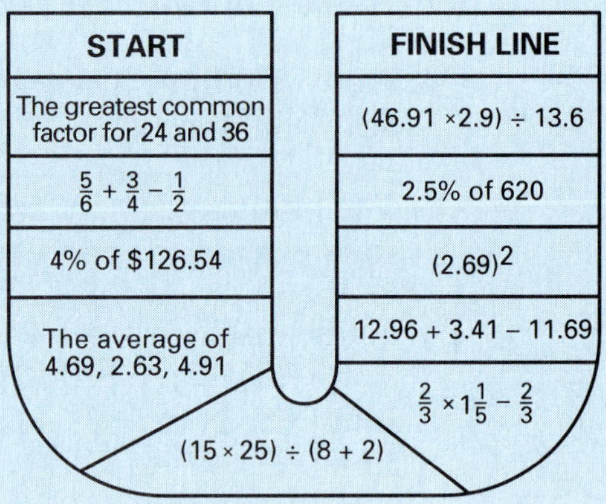

INTRODUCTION

In the game of steeplechase above you performed a variety of operations on numbers. Each time you perform an operation on a number, that number is in relation to the result. An everyday example would be the operation of an automobile engine. The gasoline injected into the cylinders is ignited by a spark and transformed into energy to push the piston. There is a relation between the amount of gasoline and the amount of energy produced.

SKILLS DEVELOPMENT

A set of pairs of numbers that have been changed in the same manner is called a **relation.** A relation rule describes how the members of one set, the domain, are related to the members of another set, the range. A relation rule can be given in many different ways. It can be shown as a mapping diagram, a word description, a table of ordered pairs, a list of ordered pairs, an equation, or relation boxes.

Mapping Diagram

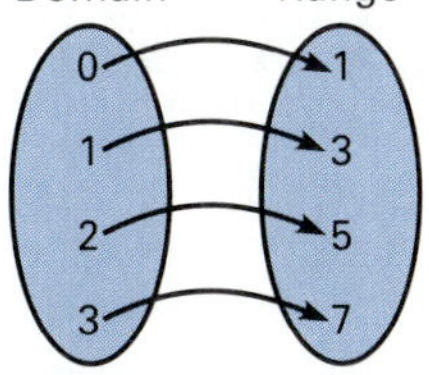

Word Description

This is read as:
0 is mapped onto 1,
1 is mapped onto 3,
2 is mapped onto 5,
3 is mapped onto 7.

Table

k	$2k + 1$
0	1
1	3
2	5
3	7

↑ domain ↑ range

Ordered Pairs

(0, 1), (1, 3), (2, 5), (3, 7)

The first numbers are the domain.
The second numbers are the range.

Equation

$y = 2x + 1, x \in S = \{0, 1, 2, 3\}$
S is the domain set.

Relation Boxes

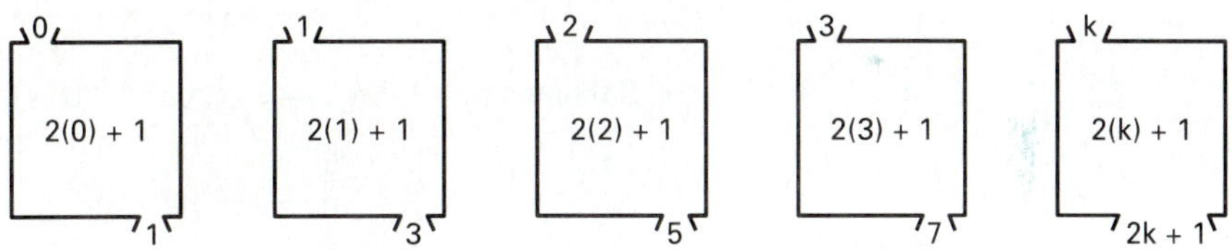

Example 1

Use the relation rule $y = x + 5$ and a domain of D = {1, 2, 3, 4} to find the range.

Solution Substitute the values for x and solve for y.

$y = 1 + 5$ $y = 2 + 5$ $y = 3 + 5$ $y = 4 + 5$

$y = 6$ $y = 7$ $y = 8$ $y = 9$

The range is R = {6, 7, 8, 9}.

Example 2

Use the table to write the rule as

a. a mapping diagram.

b. a word description.

c. a list of ordered pairs.

d. an equation.

x	$2x - 2$
1	0
2	2
3	4
4	6

Solutions a.

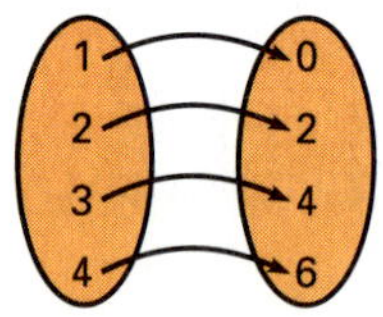

b. The range is two less than twice the domain.

c. (1, 0) (2, 2) (3, 4) (4, 6)

d. $y = 2x - 2$

EXERCISES

A relation is shown by a table of values.

n	$n - 1$
1	0
2	?
3	?
4	?

1. Find the missing values.

2. List the domain.

3. List the range.

A relation is shown as ordered pairs.

(1, 3)
(2, 6)
(3, ?)
(4, ?)

4. Find the missing values.

5. List the domain.

6. List the range.

7. A relation is shown by the equation $y = x^2 + 2$. Using D = {0, 1, 2, 3} draw a mapping diagram.

The mapping diagram shows a relation.

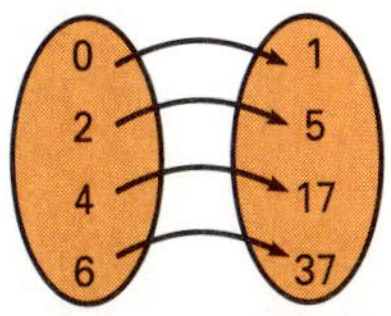

8. Show the relation as a table of values.

9. Use ordered pairs to show the relation.

A relation rule is expressed in words, "The second number is one less than twice the first." Find the missing values.

10. (3, ?) **11.** (9, ?) **12.** (12, ?)

13. (128, ?) **14.** (4.8, ?) **15.** (16.5, ?)

Write each relation rule as a word description.

16. $y = 2x$ **17.** $y = 2x + 1$

18. $y = x^2$ **19.** $y = 2(x + 1)$

In the Zecchino family, there were Mr. and Mrs. Zecchino, Mr. Zecchino's brother Harry, and Mrs. Zecchino's sister Martha. Four sons, David, Januk, Herb, and Bert, and three daughters, Cassandra, Karen, and Lina, lived at home with Mr. and Mrs. Zecchino.

20. How would you interpret the relation rule for (Bert, Januk)? Write two more ordered pairs for this relation rule.

21. How would you interpret the relation rule for (Karen, Herb)? Write two more ordered pairs for this relation rule.

22. How would you interpret the relation rule for (Harry, David)? Write two more ordered pairs for this relation rule.

SECTION 11–3

ORDERED PAIRS AND RELATIONS

WARM-UP ACTIVITY

What did one test say to another? To find out, locate the letter that corresponds to each set of ordered pairs. Then place the letter that names that point above the corresponding question number.

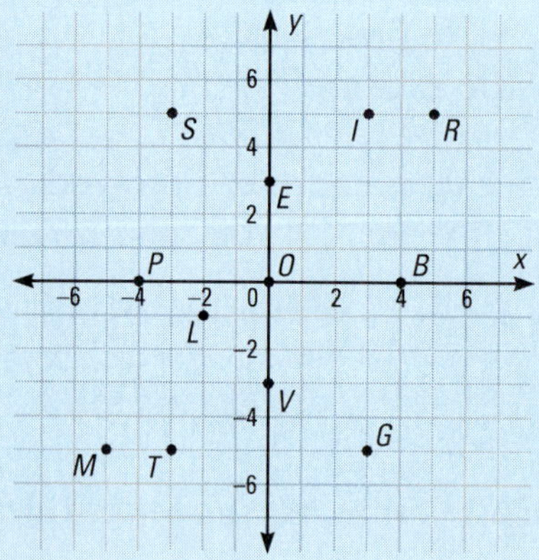

1	(0,3)
2	(–3,–5)
3	(–3,5)
4	(3,5)
5	(3,–5)
6	(4,0)
7	(–4,0)
8	(0,–3)
9	(0,0)
10	(5,5)
11	(–5,–5)
12	(–2,–1)

_ _ _   _ _ _   _ _ _ _ _ _ _ _
4 8 1   5 9 2   7 10 9 6 12 1 11 3

INTRODUCTION

Mrs. Edwards decided to find out how many minutes her students were spending on homework and how many minutes they talked on the phone on a given night. All 14 of Mrs. Edwards' students recorded in a table the number of minutes spent on the phone and the number of minutes doing homework for one night.

Student	Minutes Phone	Minutes Homework	Student	Minutes Phone	Minutes Homework
Alice	90	90	Ian	0	20
Bettye	100	45	Juan	30	0
Cho Chen	60	90	Maria	150	30
Domonic	10	105	Nicholas	30	30
Elsa	50	60	Ruth	15	180
Fred	130	60	Sheketha	60	80
Gabriel	0	150	Velvet	180	30

SKILLS DEVELOPMENT

The class plotted the information on a coordinate plane. In order to graph the relation, the students used (phone time, homework time) for ordered pairs. Since neither one of the coordinates could have negative values, they only drew one quadrant. Their graph is shown below. Notice the labeling of the axes. The horizontal axis corresponds to the x-coordinate, minutes spent on the phone. The vertical axis corresponds to the y-coordinate, minutes spent on homework.

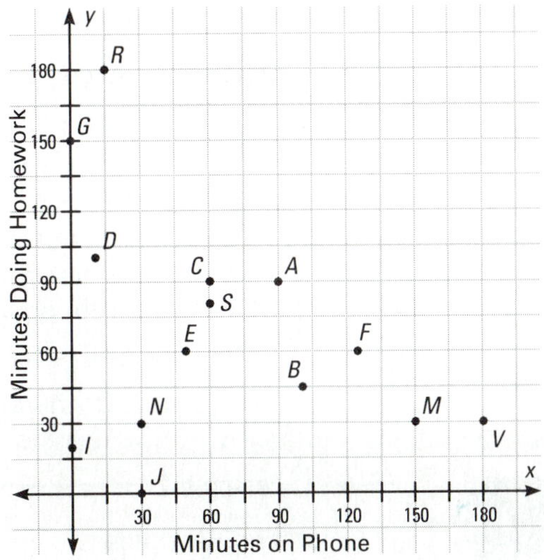

Example 1

Use the students' table and graph to answer these questions.

a. Which student(s) spent the same amount of time on the phone as doing homework?

b. How many students spent more than one hour on the phone?

c. How many students spent one hour doing homework?

d. Which student is represented by (60, 80)?

e. How many numbers will be listed in the range?

Solutions a. Alice: (90, 90); Nicholas: (30, 30)

b. There are 5 points to the right of the vertical line for 60 minutes. Thus, five students spent more than one hour on the phone.

c. Count the number of points along the horizontal line for 60 minutes. Two students, Elsa and Fred, spent one hour doing homework.

d. Sheketha is represented by the point (60, 80).

e. 10; there are 10 different *y*-coordinates.

Example 2

The Riverside Swimming Pool's income for the five months that it is open is in the table at the right.

Month	Income in Dollars
May	−500
June	250
July	500
August	750
September	100

a. List the domain.

b. List the range.

Solutions a. D = {May, June, July, August, September}

b. R = {−\$500, \$100, \$250, \$500, \$750}

EXERCISES

For Exercises 1–4, refer to Example 1.

1. List the domain.

2. Which student spent the most time talking on the phone?

3. How many students spent less than one hour doing homework?

4. What do you think this graph shows about Mrs. Edwards' students' grades?

For Exercises 5–7, refer to Example 2.

5. Why do you think the income in May was negative?

6. In which month was the income greatest? Why?

7. Write the ordered pairs from the table.

The graph below shows the speed of a train during the seven hours it takes to travel from San Antonio to Dallas.

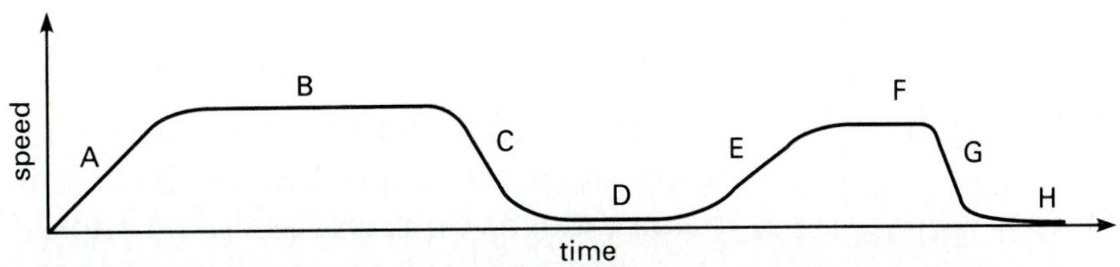

8. What parts of the graph show the train speeding up?

9. What parts of the graph show the train stopped?

10. What parts of the graph show the train slowing down?

The table at the right shows the closing price of a share of stock for the MATH Company during a one-week period.

Day	Closing Price in Dollars
Mon.	$30\frac{1}{4}$
Tue.	30
Wed.	31
Thur.	$30\frac{1}{2}$
Fri.	$30\frac{3}{4}$

11. Plot the information on a graph using the ordered pairs (day, closing price).

12. List the domain.

13. List the range.

SECTION 11–4

EQUATIONS AS RELATIONS

WARM-UP ACTIVITY

Copy the nine dots at the right. How can you draw four line segments to cross each dot without lifting your pencil off the paper?

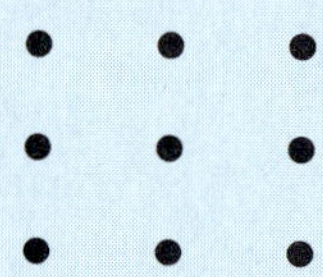

INTRODUCTION

Lucinda earns \$4.00 per hour working at Movie Mania. That is, her wages equal \$4 multiplied by the number of hours worked. This relation can be expressed as the equation $W = 4h$, where W is the wage in dollars and h is hours worked. You can make a table of values from the equation.

h	W
0	0
1	4
2	8
3	12
4	16

The table of values can be expressed as ordered pairs (h, W).

$$(0, 0), (1, 4), (2, 8), (3, 12), (4, 16)$$

Graphing the ordered pairs gives this result.

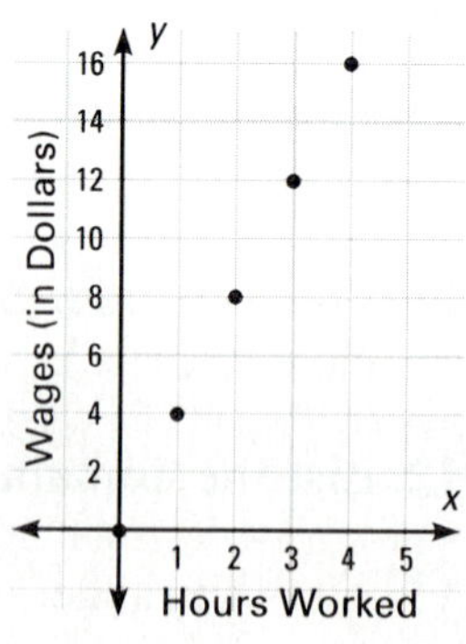

SKILLS DEVELOPMENT

Relation rules can be expressed as a table of values. Ordered pairs can be written from a table of values and then plotted on a coordinate plane. When given a replacement set for the domain, substitute to find the range.

Example 1

Find three ordered pairs for the relation $y = x + 4$ where $x \in \{-2, 0, 2\}$.

Solution Make a table.

x	y
−2	2
0	4
2	6

Write ordered pairs (x, y).
(−2, 2)
(0, 4)
(2, 6)

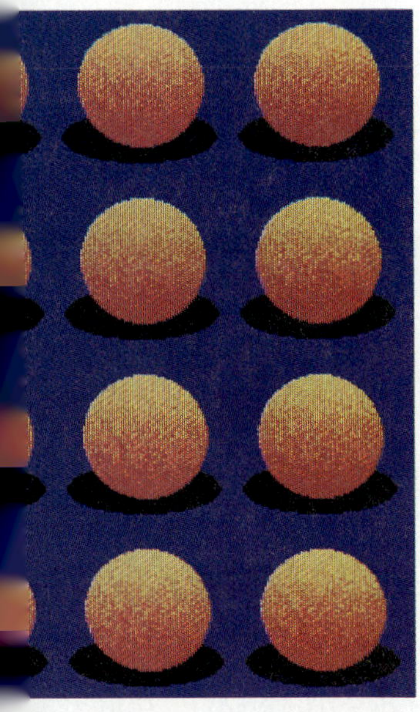

Once a relation has been graphed, it is possible to make predictions about other ordered pairs by observing the pattern.

Example 2

Graph the relation $y = 3x - 1$ for $x \in \{-2, -1, 1, 2\}$ to predict:

a. The value of y when $x = 0$.

b. The value of y when $x = 3$.

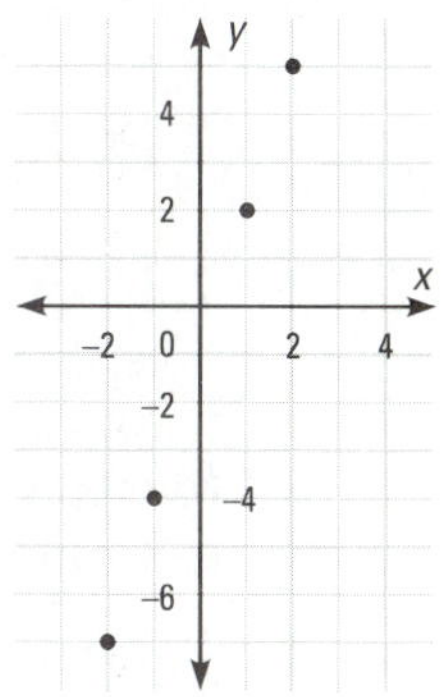

Solutions

x	y	Ordered pairs
−2	−7	(−2, −7)
−1	−4	(−1, −4)
1	2	(1, 2)
2	5	(2, 5)

a. According to the pattern, when $x = 0$, $y = -1$.

b. According to the pattern when $x = 3$, $y = 8$.

EXERCISES

1. For the relation $y = 2x + 1$ for $x \in \text{I}$, find the missing values in the table.

x	y
0	?
-1	?
1	?
-2	?
2	?

2. Write the ordered pairs from the table in Exercise 1.

For each of the following relations, find the missing values.

3. $y = x^2 + 1$
   $(-3, ?)$
   $(0, ?)$
   $(3, ?)$

4. $y = 2x^2 - 1$
   $(-1, ?)$
   $(0, ?)$
   $(1, ?)$

5. $y = 2x - 3$
   $(-2, ?)$
   $(-1, ?)$
   $(0, ?)$

6. $y = -x + 3$
   $(-1, ?)$
   $(-3, ?)$
   $(5, ?)$

7. Draw a graph of the following ordered pairs.
   $A(0, 1)$ $B(-1, 3)$ $C(1, -1)$ $D(-3, 7)$

8. Use your graph in Exercise 7 to predict another ordered pair that would seem to fit the pattern.

These ordered pairs are obtained from a relation.

$A(-2, -5)$ $B(-1, -2)$ $C(0, 1)$ $D(1, 4)$

9. Draw a graph of the ordered pairs.

10. Use your graph in Exercise 9 to predict two other ordered pairs that belong to the graph.

A relation is given by $y = x - 3$ for $x \in \{-1, 0, 1, 2, 3\}$.

11. Write five ordered pairs for the relation.

12. Draw a graph of the relation.

13. Draw a graph of the ordered pairs for the relation $y = 2x - 1$ for $x \in \{-2, -1, 0, 1, 2\}$.

14. Use the same Cartesian plane as in Exercise 13 and draw a graph of $y = 2x$ for $x \in \{-2, -1, 0, 1, 2\}$.

15. Use the same Cartesian plane as in Exercise 13 and draw a graph of $y = 2x + 1$ for $x \in \{-2, -1, 0, 1, 2\}$.

16. Compare the graphs you drew in Exercises 13, 14, and 15.

The graphs of various relations are shown. For each relation:
Write three additional ordered pairs.
Write the relation rule in the form $y = \square x + \square$ for $x \in \mathrm{I}$.

17.

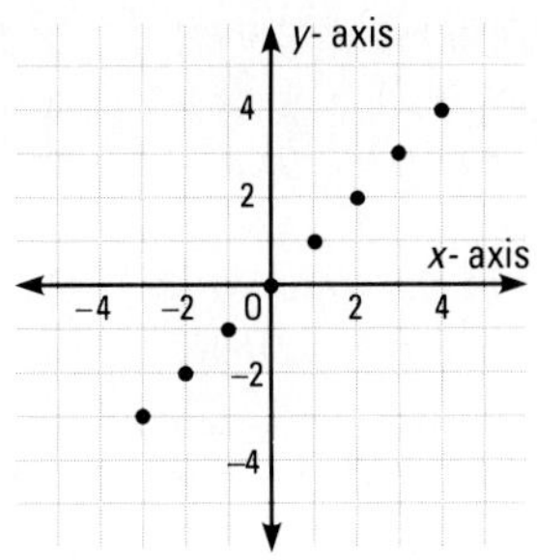

18.

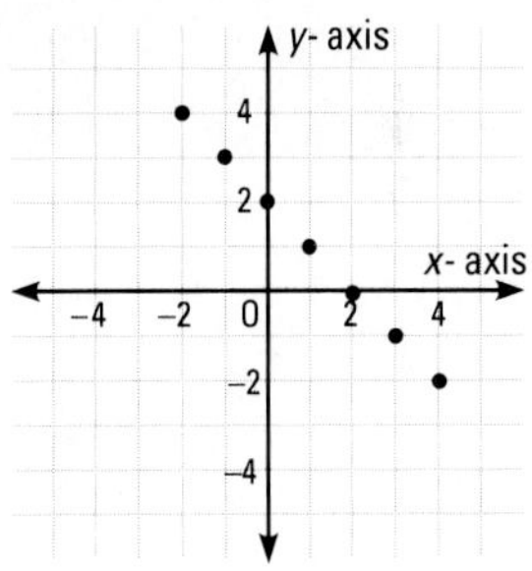

19.

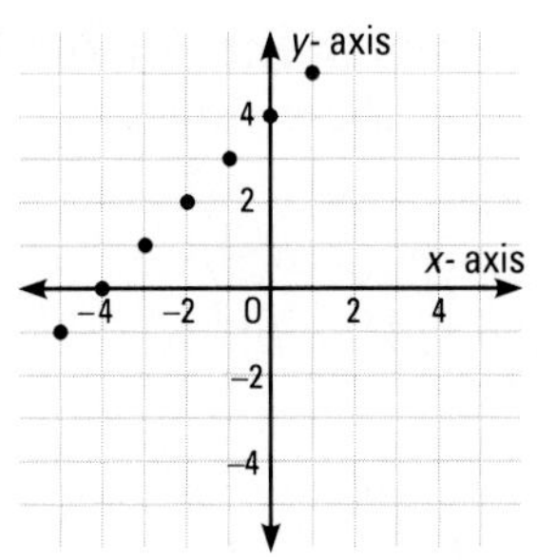

20.

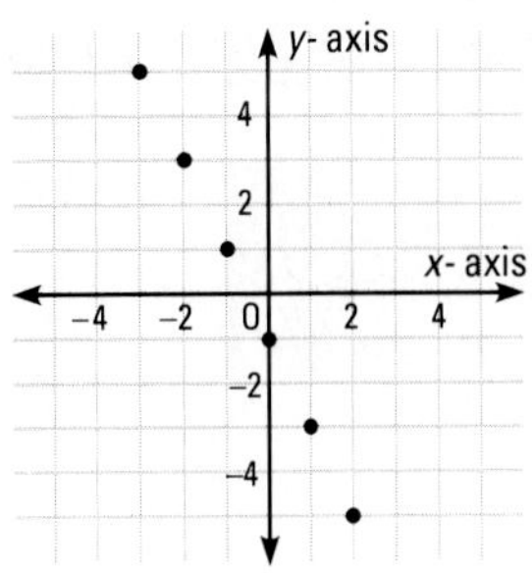

SECTION 11–5

FUNCTIONS

WARM-UP ACTIVITY

Answer the questions to complete the cross number puzzle.

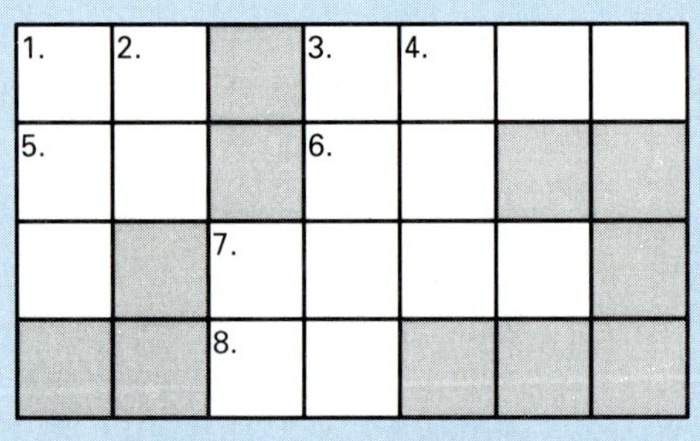

Across

1 18 is 25% of what number?

3 2040 is 85% of what number?

5 1% of this number is 0.65.

6 What percent is 410 of 500?

7 What number is 30% of 5190?

8 279 is what percent of 465?

Down

1 35% of this number is 266.

2 40% of this number is 10.

3 1140 is 40% of what number?

4 8% of what number is 34?

7 25% of 64 is what number?

INTRODUCTION

Observe the two curves in the graphs. How are they similar? How are they different? Describe some real-world objects that have these shapes.

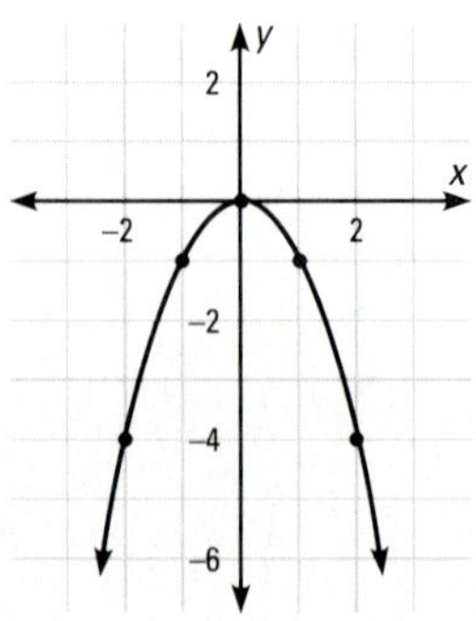

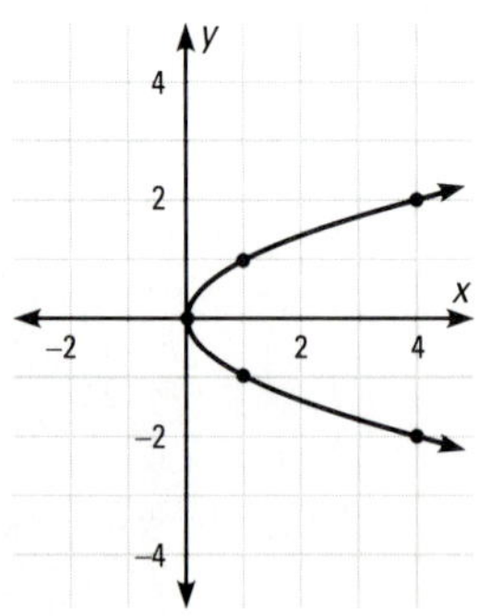

SKILLS DEVELOPMENT

The graphs on the previous page are both those of relations. One has an equation of $y = -x^2$ and the other $x = y^2$. One of the graphs is that of a special type of relation, called a function.

A **function** is a relation in which each value in the domain has one and only one value in the range.

This means that each ordered pair in a function must have different x-coordinates. Which one of the graphs is that of a function?

Vertical Line Test

If the graph of a relation touches a vertical line at more than one point, the relation is not a function.

Example

Use the Vertical Line Test to determine whether the graphs are those of a function.

a. b. c.

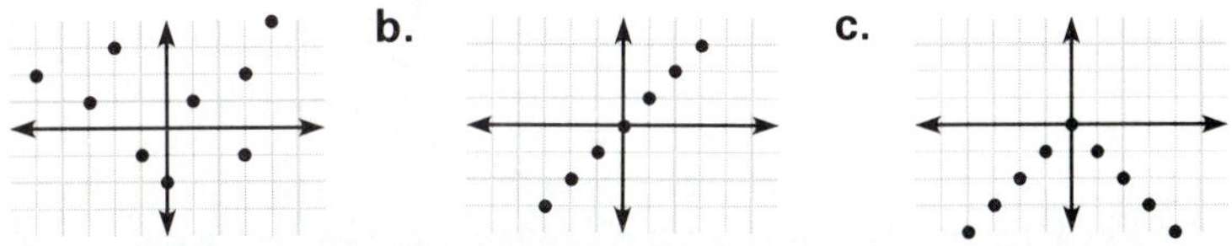

Solutions a. No, because a vertical line touches the graph at (3, 2) and (3, −1).

b. Yes, because a vertical line touches only one point anywhere on the graph.

c. Yes, because a vertical line touches only one point anywhere on the graph.

EXERCISES

Tell if the set of ordered pairs represents a function. If it is not a function, give a reason why.

1. {(1, 2), (3, 1), (−2, 5), (−4, −3)}

2. {(1, 1), (0, 2), (−1, 3), (−1, 1), (1, 3)}

3. {(3, −1), (3, 0), (3, 1), (3, 2)}

4. {(−1, 5), (0, 5), (1, 5), (2, 5)}

Tell if the mapping diagram represents a function. If it is not a function, give a reason why.

5.
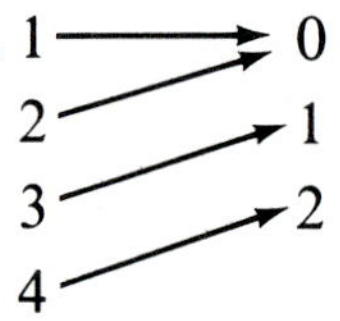

6.
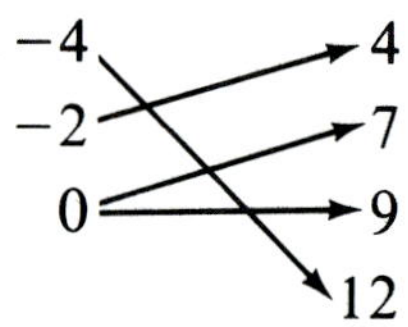

7.
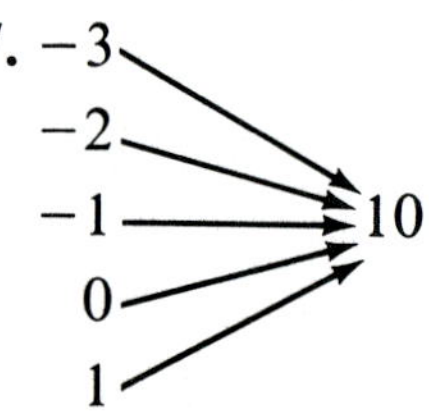

Tell if the table represents a function. If it is not a function, give a reason why.

8.

x	y
−3	−3
−2	1
2	−1
3	−1
2	3

9.

x	y
1	1
2	2
0	0
−1	−1
−2	−2

Tell if the graph represents a function. If it is not a function, give a reason why.

10.
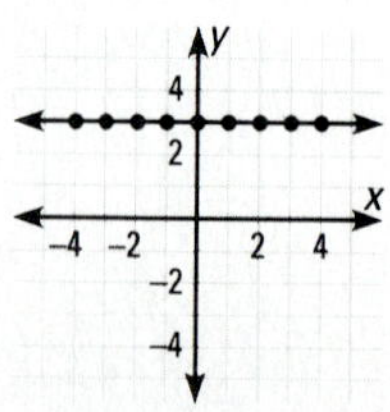

11.
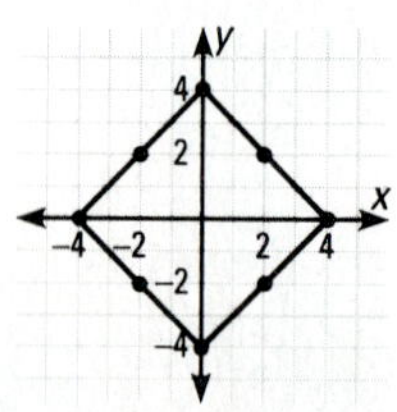

12.
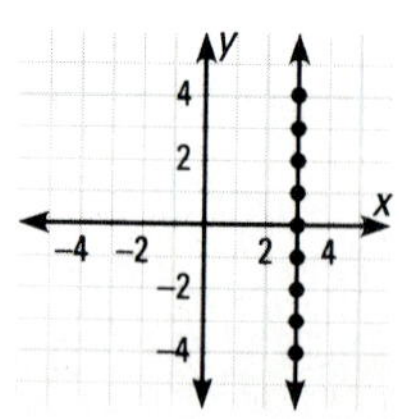

Write the ordered pairs for the relation rules for $x = \{-2, -1, 0, 1, 2\}$. Tell whether the rules represent functions.

13. $y = 2x - 3$

14. $y = \frac{1}{2}x$

15. $y = x^2$

16. $y = -x + 1$

17. Is it possible to determine if the equations in Exercises 13–16 represent functions without writing ordered pairs? Give reasons for your answer.

18. A special function is the Absolute Value Function. Its equation is $y = |x|$. A partial graph of this function is shown. What is the domain and the range for the graph?

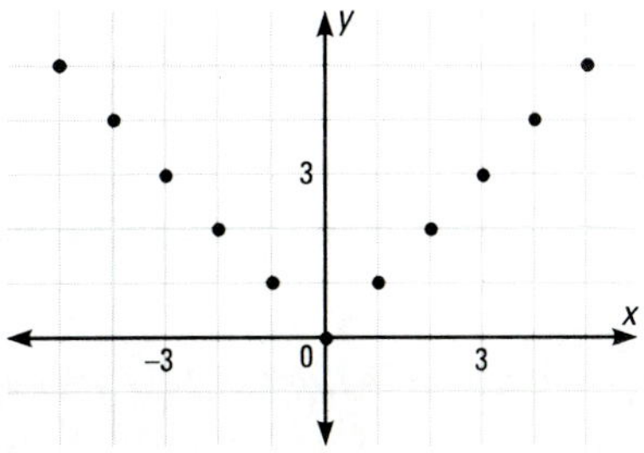

19. Is the relation (x, y) where x is every student in the freshman class and y is that person's birthday a function? Give a reason why.

20. Is the relation (x, y) where x is a member of the tennis team and y is that person's racquet a function? Give a reason why.

21. Is the relation (x, y) where x is the students in your Pre-Algebra class and y is each student's grade point average a function? Give a reason why.

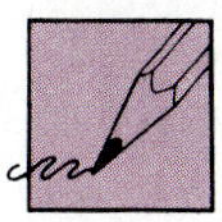

22. Is the relation (x, y) where x is a teenage female and y is her boyfriend a function? Give a reason why.

23. Describe a situation in which a relation is a function.

24. Describe a situation in which a relation is not a function.

CHAPTER 11 SUMMARY

In this chapter you have learned to identify relations, tell if the relation is a function, state its domain and range, and represent it as a mapping diagram, a table, a set of ordered pairs, a graph, and sometimes as an equation. You know that a relation is a set of ordered pairs of numbers. The ordered pairs were plotted on a coordinate plane. Remember that the first number in the ordered pair is the x-coordinate and the second number is the y-coordinate. The domain is a set that is made up of x-coordinates and the range is a set of the y-coordinates.

Ordered pairs can be found from mapping diagrams that map one number onto another, from tables that list the domain and range, and from substituting a known value into an equation. When you observed a pattern in the points on a graph, you extended the pattern to make predictions of other points on that graph.

VOCABULARY

Cartesian plane
function
origin
quadrant
relation

CHAPTER REVIEW

For Exercises 1–4, use the graph at the right.

1. Give the coordinates for points *B, C, H, I,* and *L.*

2. What is the location of *D* called?

3. Name two points in Quadrant II.

4. Identify the points with coordinates (2, −3), (0, −4), and (−4, 5).

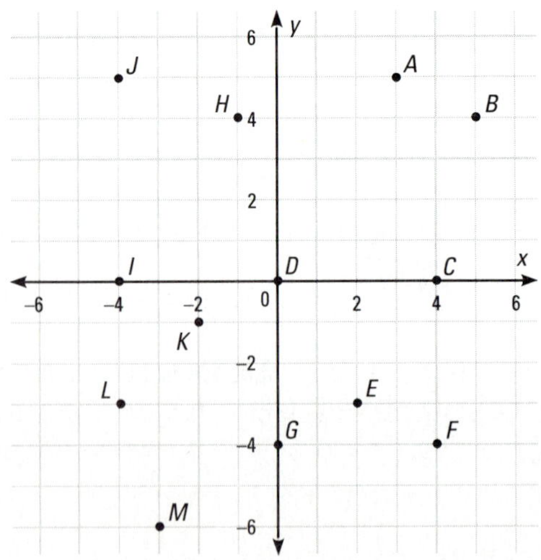

A relation is shown in the table.

x	$x - 2$
5	?
6	?
7	?
8	?

5. Find the missing values.

6. List the domain.

7. List the range.

Use the rule $y = x^2 + 1$ and a domain of D = {0, 1, 2, 3}.

8. Make a table to represent the relation.

9. Make a mapping diagram to represent the relation.

10. Write the ordered pairs.

11. Graph the relation.

12. Which statement best describes the graph.

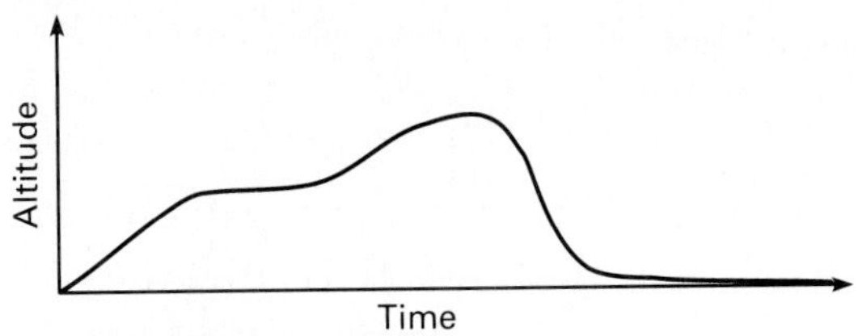

An airplane takeoff.
An airplane landing.
An airplane trip from takeoff to landing.
An airplane at the terminal.

13. Draw a graph that represents the speed of a cross-country runner as she runs up a hill and down the other side.

14. Draw a graph of the following ordered pairs.
$A(-1, 0)$ $B(1, 2)$ $C(0, 1)$ $D(3, 4)$

15. Use your graph in Exercise 14 to predict another ordered pair that would seem to fit the pattern.

A relation is represented by $y = x - 3$ for $x \in \{-1, 0, 1, 2, 3\}$.

16. Write the ordered pairs for the relation.

17. Graph the relation.

Tell if the relation is a function. If not, give a reason why.

18. (1, 2), (2, 2), (2, 1), (1, 1)

19.

x	x^2
2	4
1	1
0	0
−1	1
−2	4

20. 1 → 5
2 → 4
3 → 3
4 → 2
5 → 1

21.

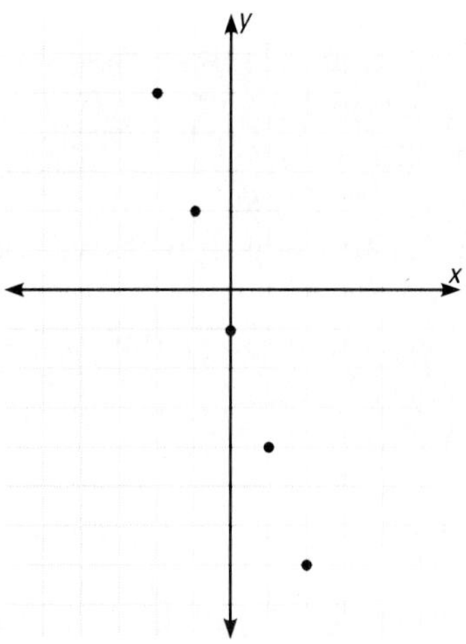

CHAPTER 11 TEST

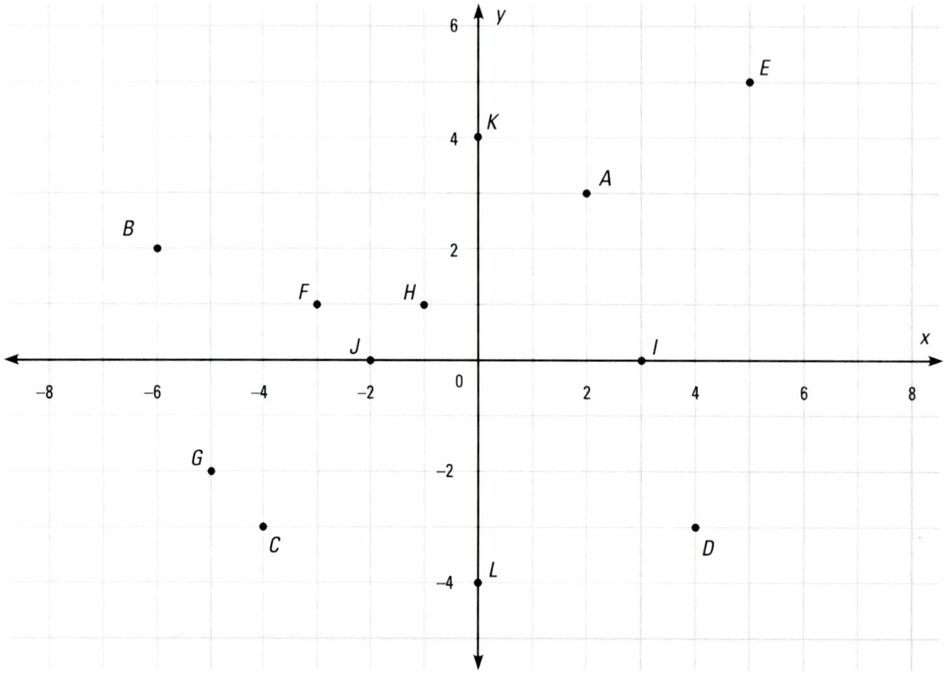

Refer to the graph above for Exercises 1–7.

Give the coordinates of each point as an ordered pair.

1. C **2.** H

3. J **4.** L

Name the point on the graph that corresponds to the ordered pair.

5. (2, 3) **6.** (−5, −2)

7. (0, 4)

A relation is shown in the table.

x	$3x$
−1	?
0	?
1	?

8. Find the missing values.

9. List the domain.

10. List the range.

Refer to the set of ordered pairs below.
{(−2, 1), (−1, 2), (0, 3), (1, 4), (2, 5), (3, 6)}

11. Show the relation as a mapping diagram.

12. Write an equation that defines the rule.

13. Graph the relation.

A relation is represented by $y = |x|$. Use $x \in \{-4, -2, 0, 2, 4\}$.

14. Make a table to represent the relation.

15. Graph the relation.

16. Is the relation a function?

SKILLS PREVIEW

One relation is represented by $y = x + 2$. Another relation is represented by $y = 2x + 1$. Use $x \in \{-2, -1, 0, 1, 2\}$ for both equations.

1. Make a table of values for each relation.

2. Graph each relation on the same coordinate plane. Draw a straight line through each set of points.

3. Moving from left to right along the coordinate plane, which graph appears to be steeper?

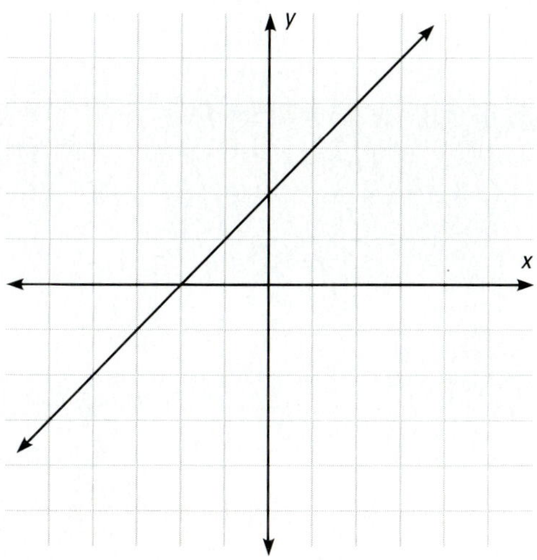

Choose any two points on the graph of $y = x + 2$. Draw a right triangle using the segment between the two chosen points as the hypotenuse.

4. Use the grid to count the vertical change or rise between the two points. HINT: This is the length of the vertical leg of the right triangle.

5. Use the grid to count the horizontal change or run between the two points. HINT: This is the length of the horizontal leg of the right triangle.

6. Write the ratio of the vertical change to the horizontal change.

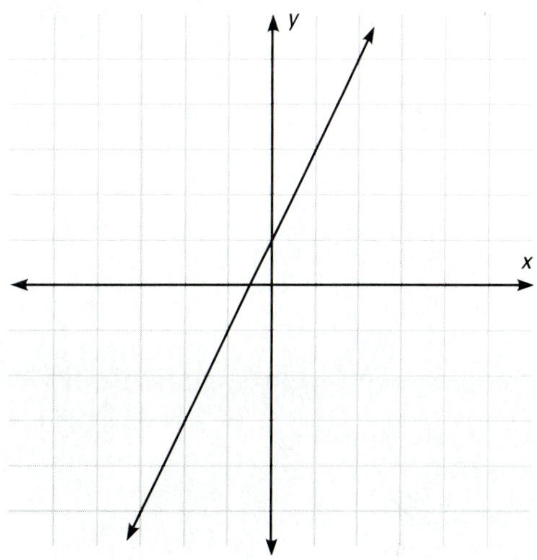

Choose any two points on the graph of $y = 2x + 1$. Draw a right triangle using the segment between the two chosen points as the hypotenuse.

7. Use the grid to count the vertical change between the two points.

8. Use the grid to count the horizontal change between the two points.

9. Write the ratio of the vertical change to the horizontal change.

10. Compare the ratios in Exercises 6 and 9. Does the greater ratio represent the steeper line?

You will find lines all around you. Look at a corner where two walls meet. Those two walls intersect in a line. Look at the photograph of the house. You can see many different lines. Can you see any parallel lines? What about perpendicular lines? What other geometric shapes can you see in the photographs?

In mathematics you learn various skills and, by applying those skills to different situations, you find that similarities exist. In Chapter 11 you developed skills with coordinates. You can now apply these skills to explore patterns that are described by relationships that use coordinates.

Ask Yourself

- *How are lines used in the building pictured on this page?*
- *Do the roof and windows shown here have the same steepness? Why do you think it was designed this way?*
- *What other items have you seen that have a slanted line for a particular reason?*
- *Do you know how to use numbers to describe the steepness or slope of a line?*

CHAPTER 12

GRAPHING RELATIONS AND FUNCTIONS

SECTION 12–1

GRAPHING LINEAR RELATIONS AND FUNCTIONS

WARM-UP ACTIVITY

Work in cooperative learning groups to solve these puzzles. You will need 12 toothpicks.

Arrange 12 toothpicks to form 5 squares. (Hint: Each square does not have to be the same size.)

INTRODUCTION

David is responsible for bringing two dozen pieces of fruit to the Student Council Leadership Conference. When he got to the store, he decided to get apples and bananas.

Since there are 24 pieces of fruit in two dozen, David had to make a decision on how many of each to get. This purchase could be described by this equation: apples + bananas = 24 pieces of fruit. For convenience, let's use $A + B = 24$.

SKILLS DEVELOPMENT

The table shows the choices David could make. The ordered pairs from the table can then be graphed. David has 25 different choices. Why?

A	B
0	24
1	23
2	22
3	21
.	.
.	.
.	.
21	3
22	2
23	1
24	0

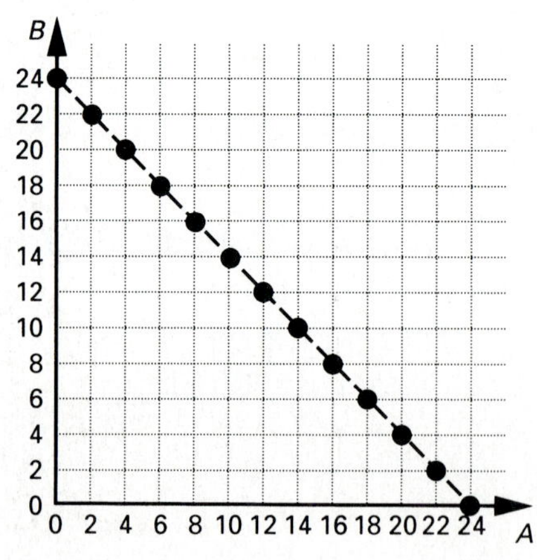

David's graph could not include negative values because it is not possible to select a negative number of pieces of fruit. Nor could it include fractional values, since David could not buy a part of a piece of fruit.

There are situations in mathematics, however, when all possible values for two numbers whose sum is 24 need to be graphed. The equations for such situations would be $x + y = 24$. The table could be expanded to include ordered pairs such as: (12.5, 11.5), (8.5, 15.5), (26, −2), and (28, −4). The graph shows all possible values.

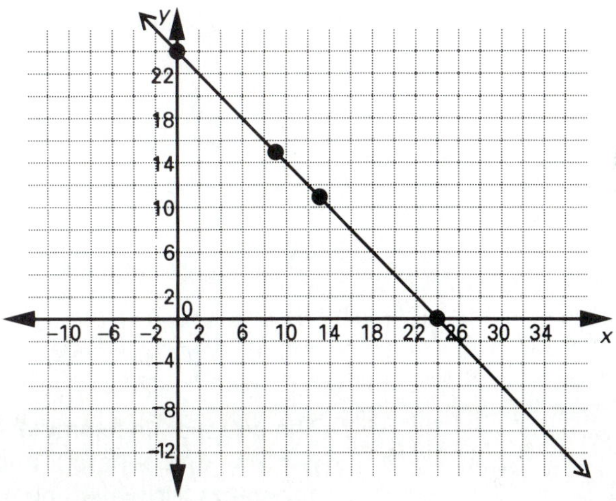

Equations whose graphs are lines are called **linear equations.** When the domain is not given as specific values, it is understood that the domain is all real numbers.

Example 1

Graph the equation $y = 2x + 1$.

Solution Make a table showing x and y values.

x	y	
-2	-3	$2(-2) + 1$
-1	-1	$2(-1) + 1$
0	1	$2(0) + 1$
1	3	$2(1) + 1$
2	5	$2(2) + 1$

Plot the ordered pairs. Draw a line.

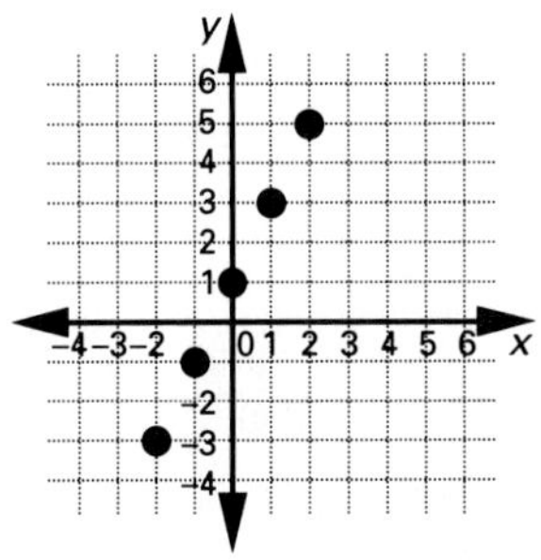

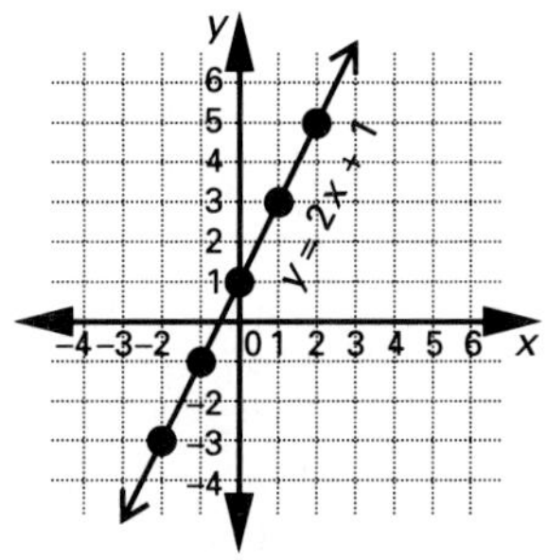

The equation in Example 1, $y = 2x + 1$, is written in the form $y = mx + b$, where x and y are variables and m and b are constants. However, not all equations are written in the form $y = mx + b$. In these cases it may be more efficient to solve the equation for y before selecting values for x.

Example 2

Graph the equation $x + 2y = 6$.

Solution First, solve the equation for y.

$$2y = -x + 6$$
$$y = -\frac{1}{2}x + 3$$

Then, make a table, plot the points, and draw a line.

x	y
-2	4
-1	$3\frac{1}{2}$
0	3
1	$2\frac{1}{2}$
2	2

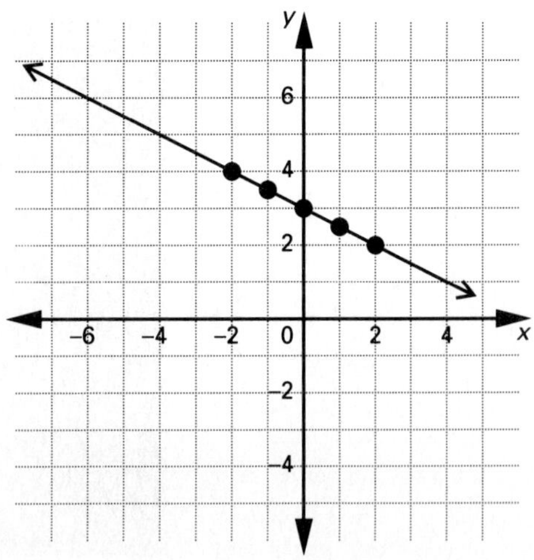

EXERCISES

Use the graph to find the missing values.

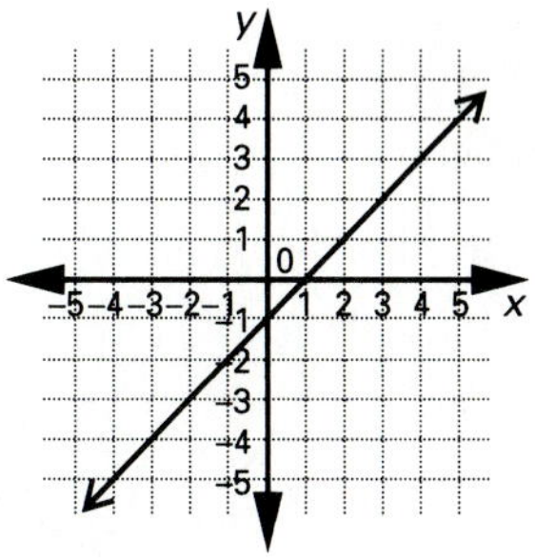

1. (2, ?)

2. (−3, ?)

3. (?, 2)

4. (?, 0)

For the relation defined by $y = 2x + 1$, find the missing values.

5. (0, ?) **6.** (?, 3) **7.** (−1, ?) **8.** (?, −3) **9.** (2, ?)

10. Plot the points in Exercises 5–9. What do you notice?

Use these ordered pairs for Exercises 11–12.
A (0, −2) B (5, −1) C (2, 4)
D (−1, 5) E (1, 1) F (5, 0)

11. Which ordered pairs satisfy the equation $y = 3x - 2$? Which do not?

12. Plot the ordered pairs. Which ordered pairs are on the graph of $y = 3x - 2$?

For each equation, copy and complete the table. Then plot the ordered pairs and draw the graph of the equation.

13. $y = 2x + 3$

x	y
−1	?
0	?
1	?

14. $3x - y = 1$

x	y
0	?
1	?
2	?

For the following relation rule, choose values for the variables and draw a graph. Describe the graph. Then create 2 problems based on your graph. Write a solution for each problem you create.

15. The temperature, T, of the pool in degrees Celsius rose according to the relation rule, $T = 2.5t$, where t is the time in hours.

SECTION 12–2

FINDING SLOPE

WARM-UP ACTIVITY

Find a path through the pond so that the sum of the answers for the rocks stepped on is 69.865. Write the letters of the rocks along your path on your paper. Do not write in your textbook.

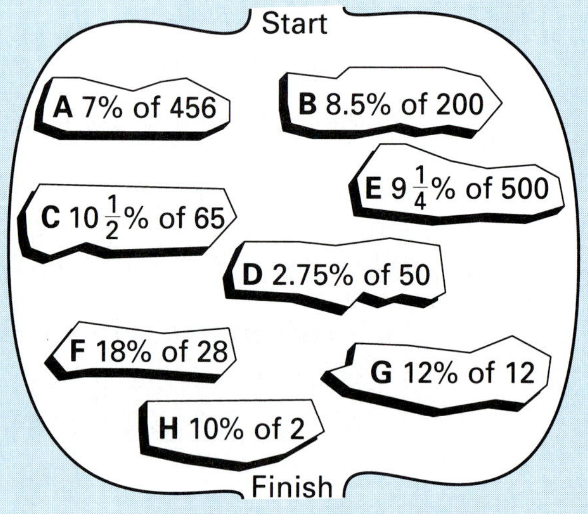

INTRODUCTION

Often the meaning of a word in mathematics is related to its meaning in everyday activities. For example, when you think of slope in your everyday activities, you might think of steepness. The **slope** of a line or line segment is the measure of its steepness. The slope of a roof or a mountain is also the measure of its steepness.

SKILLS DEVELOPMENT

The slope of a line or line segment is the ratio of its rise (vertical distance) compared to its run (horizontal distance).

$$\text{slope} = \frac{\text{rise}}{\text{run}}$$

You can find the slope of a line from its graph by choosing two points on the line. Then calculate the ratio of the rise to the run from one point to the other as shown for each line.

slope $= \frac{\text{rise}}{\text{run}} = \frac{2}{5}$

The slope of line ℓ is $\frac{2}{5}$.

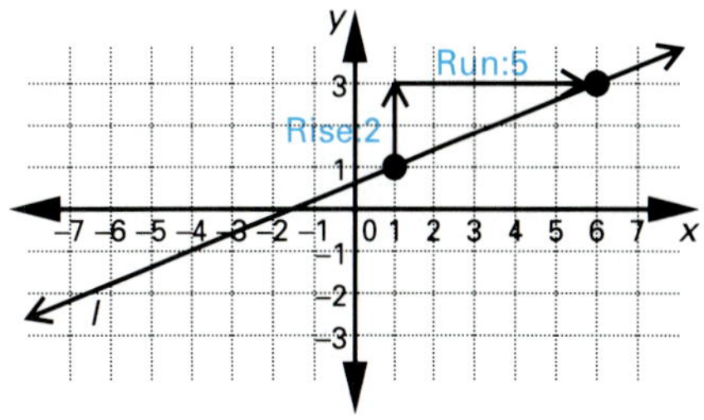

slope $= \frac{\text{rise}}{\text{run}} = \frac{-2}{4} = -\frac{1}{2}$

The slope of line p is $-\frac{1}{2}$.

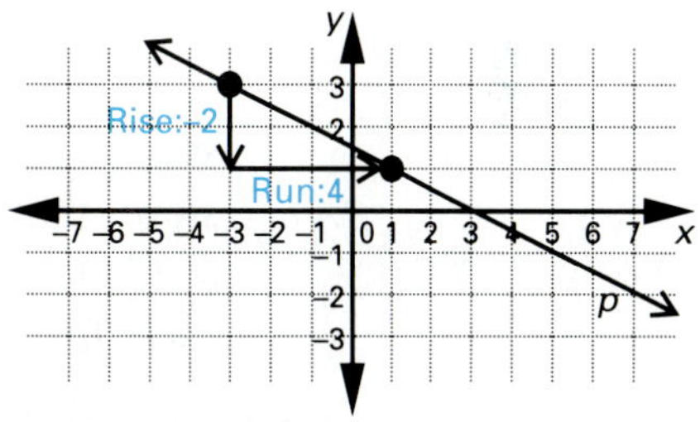

Look at the directions in which these two lines slope from left to right. Now look at the signs of their slopes. Can you predict a relationship between the sign of the slope of a line and the direction that it slants from left to right? Test your prediction with example on the next page.

Example

Find the slope of each line.

Solutions line *a:* slope $= \frac{\text{rise}}{\text{run}} = \frac{2}{2} = 1$ line *b:* slope $= \frac{\text{rise}}{\text{run}} = \frac{-4}{2} = -2$

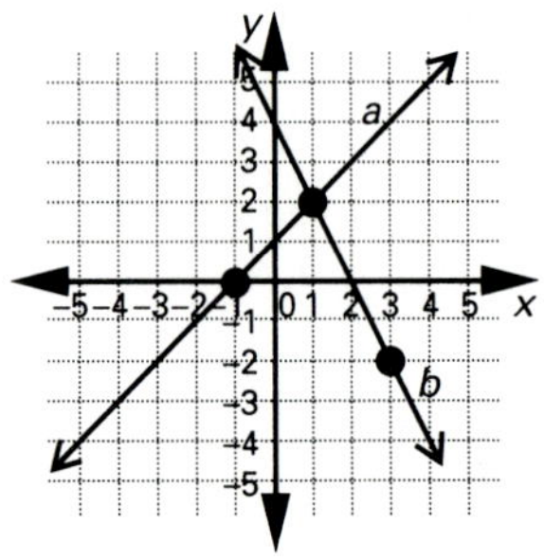

Was your prediction correct? If you said that a line with a negative slope slants downward to the right and a line with a positive slope slants upward to the right, you predicted correctly! A vertical line has no slope, and a horizontal line has a slope of zero.

EXERCISES

Draw each line segment using the coordinates given. Then find the slope of each line segment.

1. $\overline{AB}$: $A(-6, -3)$; $B(2, -1)$
2. $\overline{CD}$: $C(-3, 4)$; $D(7, 0)$
3. $\overline{EF}$: $E(3, 11)$; $F(2, 0)$
4. $\overline{GH}$: $G(-2, -1)$; $H(-5, -4)$
5. Choose two points on the line that is graphed. Record the coordinates of each point. Use the points you chose to calculate the slope of the line.

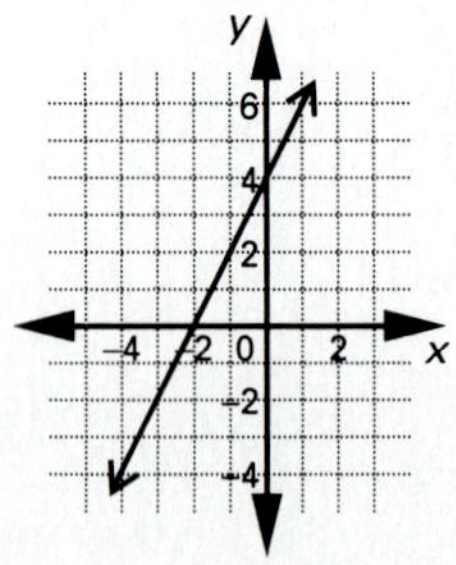

Calculate the rise, run, and slope of each line.

6.

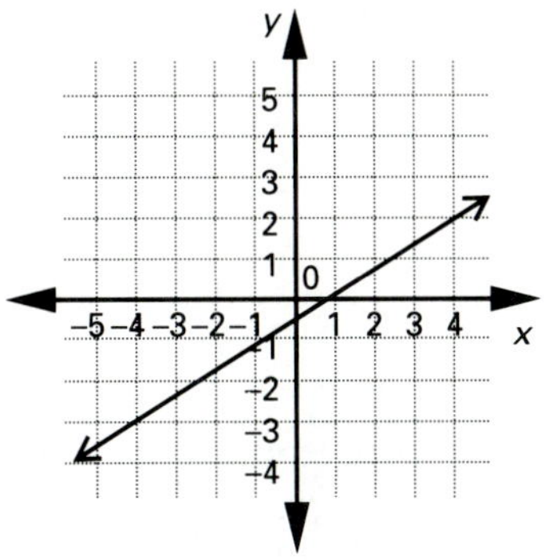

7.

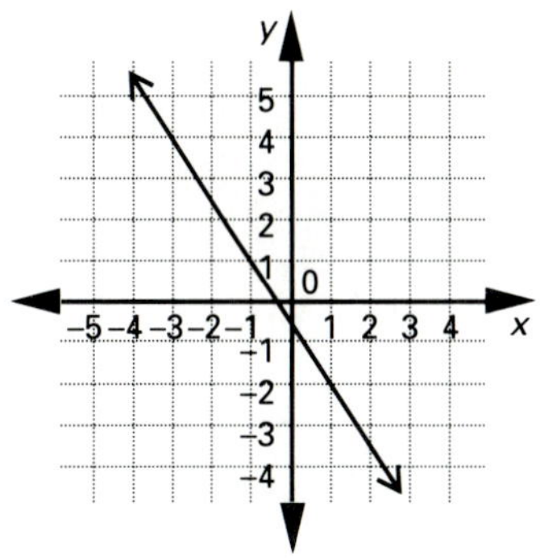

Without graphing, tell if the line that passes through each pair of points would have a positive slope, a negative slope, no slope, or a slope of zero.

8. $A(5, 4)$, $B(-1, 3)$

9. $C(2, 4)$, $D(2, -6)$

10. $E(0, 0)$, $F(-3, -5)$

11. $G(-1, 8)$, $H(4, 8)$

Three points are said to be **collinear** if they lie on the same line. $A(-4, 2)$, $B(-2, 4)$, $C(0, 6)$ are three points.

12. Find the slope of $\overline{AB}$ and $\overline{BC}$. What do you notice about the slopes?

13. Based on your findings, predict whether the points A, B, and C lie on the same line. Then graph the points to check your prediction.

Determine which sets of points are collinear.

14. $A(-6, -4)$, $B(-2, -6)$, $C(2, -8)$

15. $D(-5, 1)$, $E(0, 3)$, $F(7, 6)$

16. $J(-5, 3)$, $K(1, 1)$, $L(7, -1)$

SECTION 12–3

PROPERTIES OF SLOPE

WARM-UP ACTIVITY

Use a calculator to find the first three products. Predict the other products. Check using your calculator.

$15{,}873 \times 7 = ?$ $15{,}873 \times 28 = ?$ $15{,}873 \times 49 = ?$
$15{,}873 \times 14 = ?$ $15{,}873 \times 35 = ?$ $15{,}873 \times 56 = ?$
$15{,}873 \times 21 = ?$ $15{,}873 \times 42 = ?$ $15{,}873 \times 63 = ?$

INTRODUCTION

Just as the steepness of a roller coaster varies, the slopes of lines vary.

A linear equation written as $y = mx + b$ is said to be in **slope-intercept form.** The coefficient of x, m, represents the slope.

Study the sets of lines graphed on this page. Particularly observe the steepness of the lines and the change in the value for m in each equation.

When the slope is positive, the greater the value of the slope, the steeper the line is.

When the slope is negative, the smaller the value of the slope, the steeper the line is.

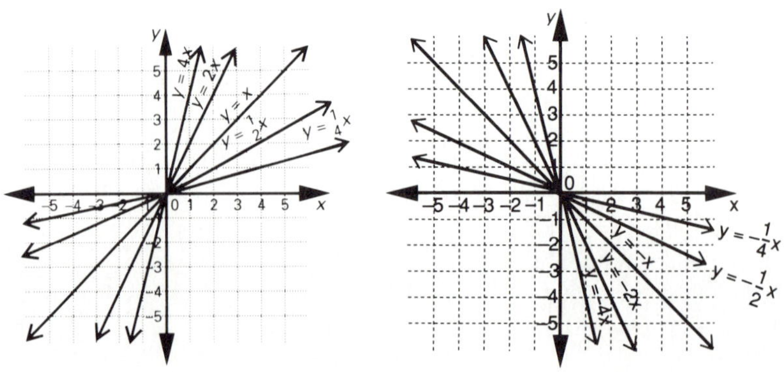

SKILLS DEVELOPMENT

To determine the steeper of two lines, given their equations, look at the absolute value of each slope. Then apply this property: the greater the absolute value of the slope, the steeper the line.

Example 1

Rank in order from least steep to steepest.

A slope of $\frac{1}{3}$ **B** slope of $\frac{3}{5}$ **C** slope of $-\frac{1}{4}$
D slope of -3 **E** slope of 2 **F** slope of -1

Solution Find the absolute value of each slope.

A $\frac{1}{3}$ **B** $\frac{3}{5}$ **C** $\frac{1}{4}$ **D** 3 **E** 2 **F** 1

Rank the absolute values from least to greatest.

$$\frac{1}{4}, \frac{1}{3}, \frac{3}{5}, 1, 2, 3$$

Now rank the slopes, using the original values.

C slope of $-\frac{1}{4}$, **A** slope of $\frac{1}{3}$, **B** slope of $\frac{3}{5}$,
F slope of -1, **E** slope of 2, **D** slope of -3

Compare the slopes of each pair of lines.

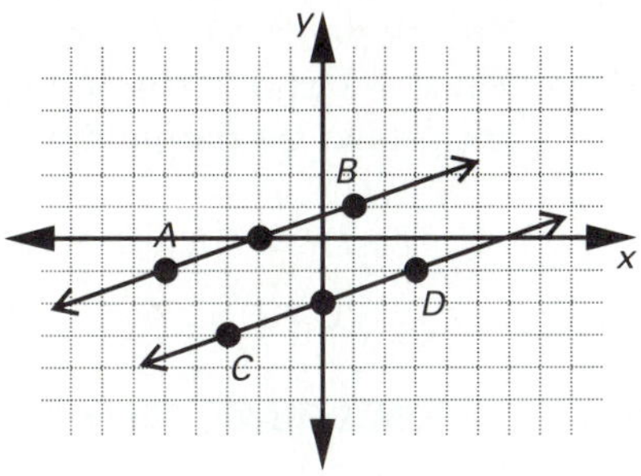

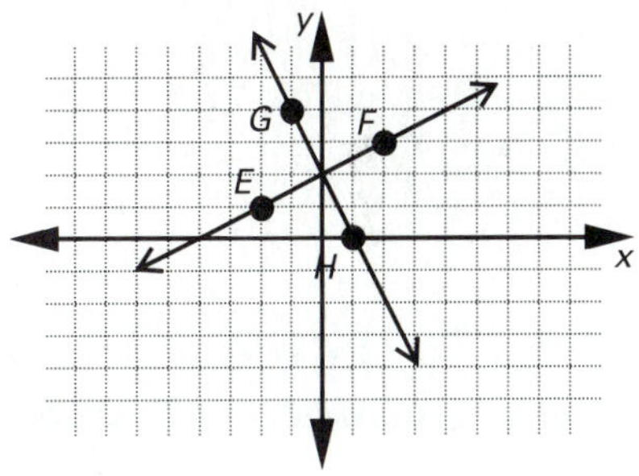

The slope of $\overleftrightarrow{AB}$ is $\frac{2}{6}$ or $\frac{1}{3}$.
The slope of $\overleftrightarrow{CD}$ is $\frac{2}{6}$ or $\frac{1}{3}$.
The lines appear to be parallel, and their slopes are equal.

The slope of $\overleftrightarrow{EF}$ is $\frac{2}{4}$ or $\frac{1}{2}$.
The slope of $\overleftrightarrow{GH}$ is $-\frac{4}{2}$ or -2.
The lines appear to be perpendicular, and their slopes are opposite reciprocals.

These findings lead us to three other properties of slope:

- The slopes of non-vertical parallel lines are equal.
- The slopes of oblique perpendicular lines are opposite reciprocals.
- If one line has slope 0 and the other has no slope, the lines are perpendicular.

Example 2

Determine whether each pair of lines are parallel, perpendicular, or neither.

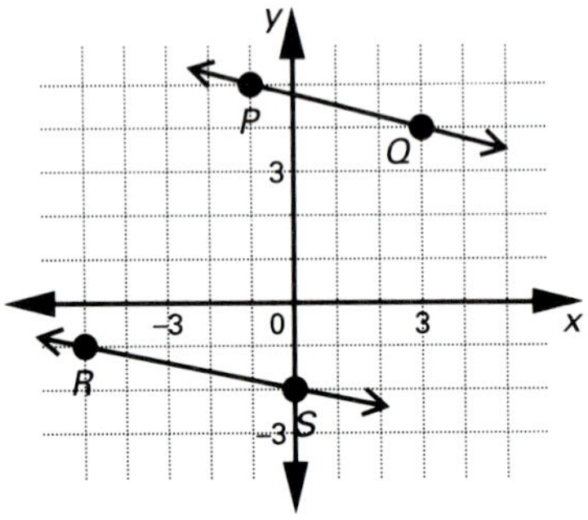

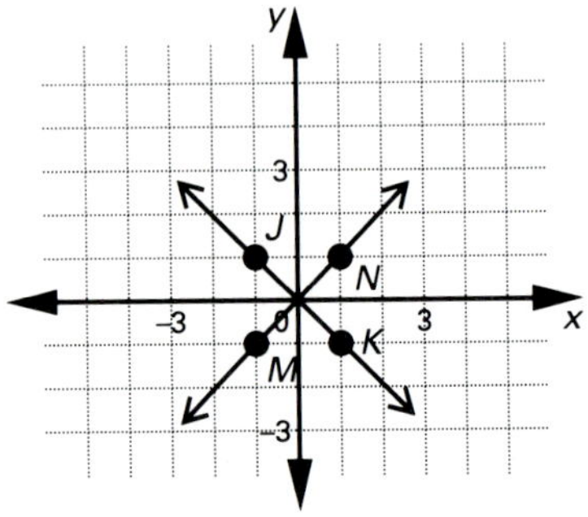

Solutions a. The slope of $\overleftrightarrow{PQ}$ is $-\frac{1}{4}$.
The slope of $\overleftrightarrow{RS}$ is $-\frac{1}{5}$.
These slopes are not equal nor are they opposite reciprocals, so lines are neither parallel nor perpendicular.

b. The slope of $\overleftrightarrow{JK}$ is $-\frac{2}{2}$ or -1.
The slope of $\overleftrightarrow{MN}$ is $\frac{2}{2}$ or 1.
Since -1 and 1 are opposite reciprocals, these lines must be perpendicular.

EXERCISES

1. List the lines in order from least steep to steepest.

a: slope of $\frac{1}{5}$ **b:** slope of 0 **c:** slope of $-\frac{2}{3}$
d: slope of $\frac{3}{2}$ **e:** slope of -2 **f:** slope of $\frac{4}{5}$

Give the slope of the line defined by each equation.

2. $y = -x$

3. $y = \frac{3}{5}x$

4. $y = 2x$

5. $y = -\frac{1}{6}x$

6. Using their equations, list the lines in order from steepest to least steep.

a: $y = x$ **b:** $y = -\frac{5}{4}x$ **c:** $y = 4x$

Sometimes the equation of a line must be rewritten into its slope-intercept form in order to identify the slope of that line from its equation. Rewrite each equation into the form $y = mx + b$. Then identify the slope of the line.

7. $3x + y = 0$

8. $2x + 2y = 0$

9. $2x + y = -1$

10. $5x - 3y = 6$

Graph each pair of lines using the coordinates given. Then identify each pair of lines as parallel, perpendicular, or neither.

11. $\overleftrightarrow{AB}: A(1, 2), B(3, -1)$
$\overleftrightarrow{CD}: C(0, 0), D(-3, 2)$

12. $\overleftrightarrow{DE}: D(3, 0), E(3, -5)$
$\overleftrightarrow{FG}: F(4, 3), G(-1, 3)$

13. $\overleftrightarrow{HJ}: H(-4, 2), J(0, 3)$
$\overleftrightarrow{KL}: K(-3, -4), L(5, -2)$

14. $\overleftrightarrow{MN}: M(1, 1), N(2, 5)$
$\overleftrightarrow{QR}: Q(-4, -4), R(-3, -1)$

Tell if the lines defined by each pair of equations are parallel, perpendicular, or neither.

15. $y = \frac{1}{3}x - 2$
$y = -3x - 2$

16. $2x + y = 0$
$x + 2y = 4$

17. $4x - 2y = 8$
$2x - y = -4$

18. The rate of interest is found by comparing the amount of interest to the amount of principal. Graph the information given in the table to determine the interest rate (slope) on these accounts. Then rename the slope as a percent. (Hint: Use the horizontal axis marked in \$50 units to represent principal. Use the vertical axis marked in \$5 units to represent interest.)

Account Number	Principal	Interest
101	\$200	\$30.00
202	325	48.75
303	75	11.25
404	400	60.00
505	150	22.50

19. Find an example of a line graph in your social studies or science textbook. Describe the trend(s) shown by this graph in terms of the slope(s) of the line.

GRAPHING USING THE SLOPE

WARM-UP ACTIVITY

Plot the following points for each part on a set of co-ordinate axes labelled from 0 to 10. Join the points in each part in the order given. Place the points for all the parts on the same grid.

A $(1,8) \rightarrow (9,8) \rightarrow (9,3) \rightarrow (7,1) \rightarrow (3,1) \rightarrow (1,3) \rightarrow (1,8)$
B $(1,8) \rightarrow (2,10) \rightarrow (3,8)$
C $(2,6) \rightarrow (3,7) \rightarrow (4,6) \rightarrow (2,6)$
D $(4,4) \rightarrow (5,5) \rightarrow (6,4) \rightarrow (4,4)$
E $(6,6) \rightarrow (7,7) \rightarrow (8,6) \rightarrow (6,6)$
F $(7,8) \rightarrow (8,10) \rightarrow (9,8)$
G $(3,3) \rightarrow (4,2) \rightarrow (6,2) \rightarrow (7,3)$

INTRODUCTION

In the last section you learned that, if the equation of a line is written in the form $y = mx + b$, you can identify the slope of that line by looking at the equation. The slope is the coefficient of x, the value of m. This equation also provides other information about the graph of that line. b represents the *y-intercept,* the point at which the line intersects the y-axis.

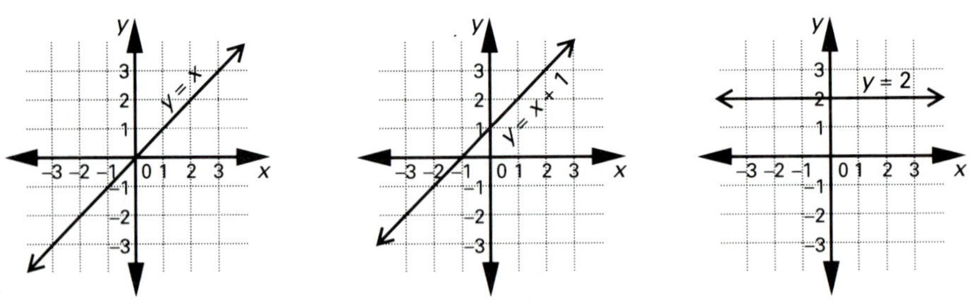

b is zero, so the y-intercept is $(0,0)$. The line intersects the y-axis at $(0,0)$.

b is 1, so the y-intercept is $(0,1)$. The line intersects the y-axis at $(0,1)$.

b is 2, so the y-intercept is $(0,2)$. The line intersects the y-axis at $(0,2)$.

SKILLS DEVELOPMENT

When you know the slope and the y-intercept, a line can easily be graphed.

Example 1

Graph the line defined by the equation $y = 2x + -3$.

Solution From the equation, we know that the slope is 2 and the y-intercept is $(0, -3)$.

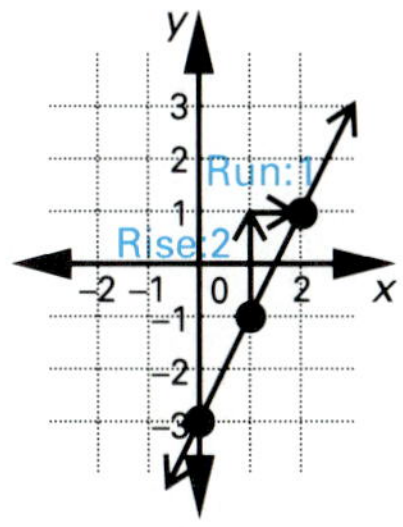

Since the y-intercept is $(0, -3)$, plot the ordered pair $(0, -3)$.

Using the slope of 2 (rise of 2 and run of 1), plot another point.

From that point, use the slope again to plot the third point. Then draw a line through the points. Since the slope is positive, the graph of the line should slant upward and to the right. It does.

If you know any point on the line, you can use the slope to draw the graph of that line.

Example 2

Draw the graph of a line with slope $\frac{2}{3}$ that passes through $(1, 3)$.

Solution Plot the point $(1, 3)$. Use the slope to locate the other points.

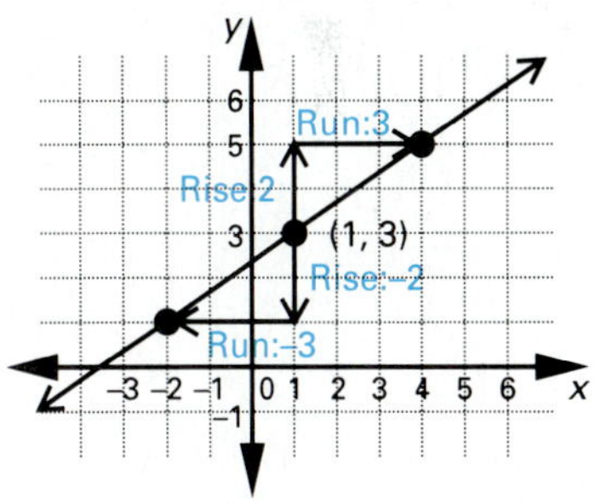

When graphing lines, remember all the facts you learned about slope in this chapter.

- Lines with positive slopes slant upward to the right.
- Lines with negative slopes slant downward to the right.
- The greater the absolute value of the slope, the steeper the line.
- The slope of a vertical line does not exist. It is said to have no slope.
- The slope of a horizontal line is zero.
- The slopes of parallel lines are equal.
- The slopes of perpendicular lines are opposite reciprocals.

EXERCISES

Name the slope and y-intercept of the line defined by each equation.

1. $y = 2x$

2. $y = 2x + 3$

3. $y = -3x - 1$

4. $3x - y = 2$

5. $2x + 3y = 0$

6. $y = -3$

Graph each line given its slope and y-intercept.

7. slope of -1 and y-intercept of $(0, 3)$

8. slope of $\frac{4}{5}$ and y-intercept of $(0, 1)$

9. slope of $-\frac{3}{2}$ and y-intercept of $(0, 0)$

10. slope of 0 and y-intercept of $(0, -2)$

Graph each line given its slope and a point, P, on the line.

11. slope of 3 and $P(2, -4)$

12. no slope and $P(-3, -1)$

13. slope of $-\frac{1}{2}$ and $P(0, 0)$

14. slope of 0 and $P(5, 1)$

Graph the line defined by each equation.

15. $y = -\frac{1}{4}x - 4$ **16.** $y = -2$

17. $x = -2$ **18.** $y = x + 4$

19. $2x + 3y = 6$ **20.** $x - 2y = 4$

21. Graph the line that passes through the origin and is parallel to a line that has a slope of -2.

22. Draw the graph of the line that contains $P(3, 5)$ and is perpendicular to a line that has a slope of 4.

Another point that can be useful in graphing a line is the ***x*-intercept,** the point at which the line crosses the x-axis. Any point on the x-axis must have a y-coordinate of 0, so to calculate the x-intercept from an equation, substitute 0 for y and solve for x.

Example:

$y = 3x - 9 \rightarrow 0 = 3x - 9$

$0 + 9 = 3x - 9 + 9$

$9 = 3x$

$\frac{9}{3} = \frac{3x}{3}$

$3 = x$, so the x-intercept is $(3, 0)$.

Find the x-intercept of each line.

23. $y = 2x + 4$ **24.** $5x + y = 20$

25. $x + 3y = -7$ **26.** $y = 12$

Graph each line by plotting its x- and y-intercepts.

27. $y = -4x - 2$ **28.** $y = -2x + 1$

29. $3x - 2y = 6$ **30.** $4x + 3y = 12$

You have been directed to graph the equation $2x + 5y = 10$.

31. Describe the methods you could use to carry out this direction.

32. Which of the methods that you described in Exercise 31 would you prefer to use? Explain your answer.

SECTION 12–5

GRAPHING SYSTEMS OF LINEAR EQUATIONS

WARM-UP ACTIVITY

Use each of the digits from 1 to 7 in place of each ■ to give a sum of 100.

$$\begin{array}{r} ■■ \\ ■■ \\ ■■ \\ +\ \ ■ \\ \hline 100 \end{array}$$

INTRODUCTION

The decoration committee for the Good-bye School–Hello Summer Dance decided to decorate using balloons. Two companies were contacted about the cost of balloons. Party Supply, Inc. charges \$0.25 per balloon with a \$5 packing charge, and Balloon Crazy charges \$0.50 per balloon with no packing charge. Let y = cost per company and x = number of balloons ordered. This situation can then be represented by these two equations:

Party Supply, Inc.: $y = 0.25x + 5$
Balloon Crazy: $y = 0.50x$

The graph of these equations is shown below.

The graph of these equations shows that the cost, \$10, is the same for both companies if 20 balloons are ordered.

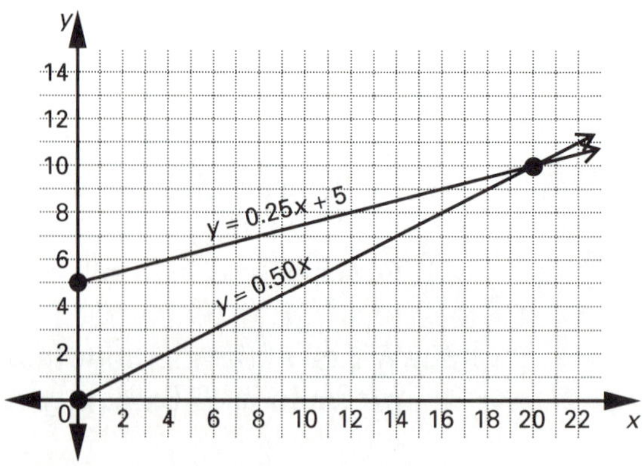

SKILLS DEVELOPMENT

Graphing two or more lines on the same coordinate plane in order to determine a point of intersection is a method of solving a **system of linear equations.** The point of intersection is the solution. When the coordinates of this point are substituted for x and y in the equations for these lines, true statements should be obtained.

Example 1

Solve this system of linear equations.

$y = x + 4$
$2x + 3y = -3$

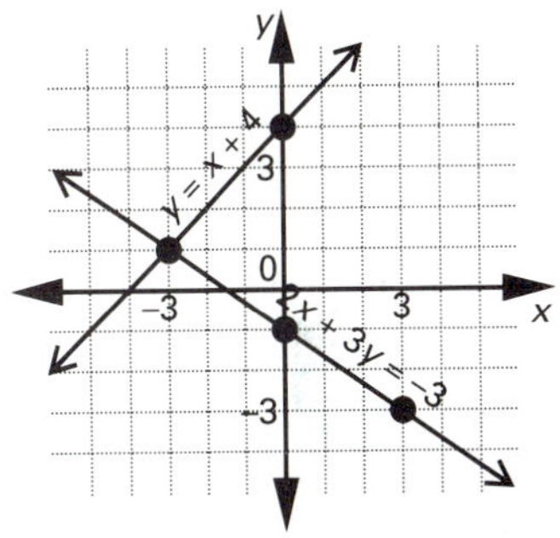

Solution Rewrite $2x + 3y = -3$ in slope-intercept form.

$2x + 3y = -3$

$y = -\frac{2}{3}x + -1$

Then graph the lines.

The lines intersect at $(-3, 1)$.

Check: $y = x + 4$ | $2x + 3y = -3$
$1 \stackrel{?}{=} -3 + 4$ | $2(-3) + 3(1) \stackrel{?}{=} -3$
$1 = 1$ | $-3 \stackrel{?}{=} -3$

Example 2

A plane is flying along the line defined by $y = x + 1$. A ship is sailing along a path defined by the line $y = -2x + 4$. Find the coordinates of the point where the path of the plane crosses the path of the ship.

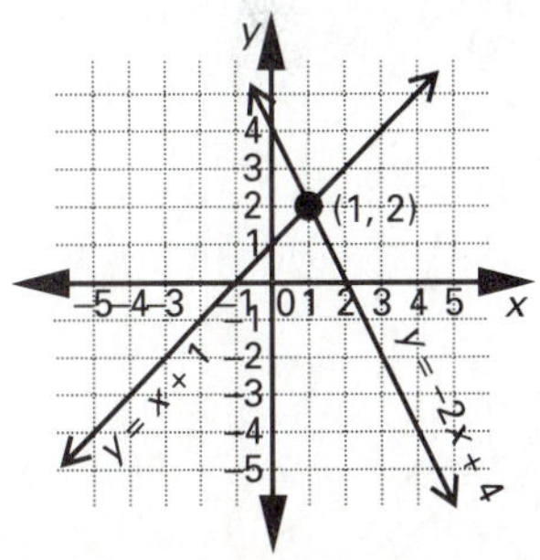

Solution Graph the lines.

The solution is $(1, 2)$.

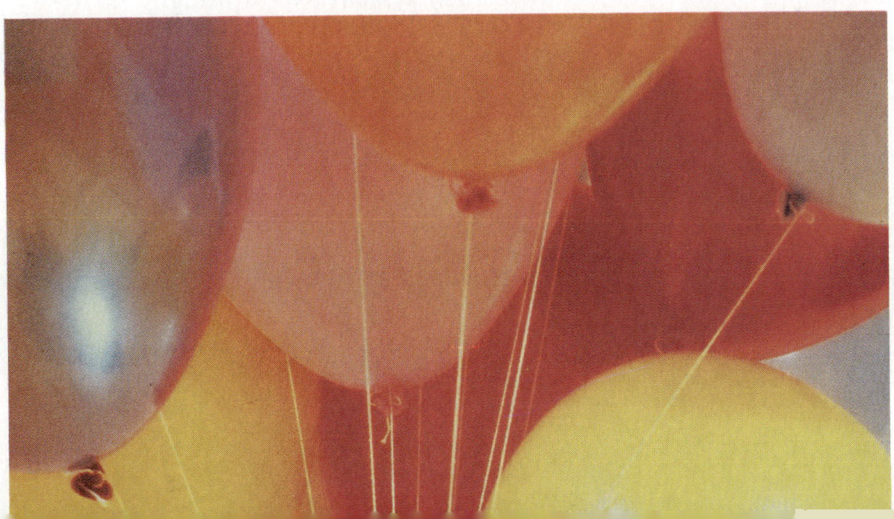

Example 3

The sum of two numbers, x and y, is 5. The first number, x, is 3 less than the second number, y.

a. Write a system representing the problem.

b. Find the two numbers by graphing the system.

Solutions a. The sum of x and y is 5. $x + y = 5$
x is less than y. $x = y - 3$

b. Write each equation in slope-intercept form.

$x + y = 5 \rightarrow y = -x + 5$

$x = y - 3 \rightarrow y = x + 3$

Graph the lines.

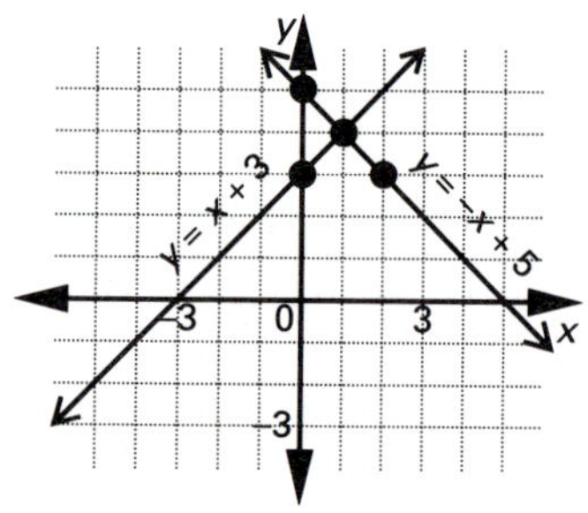

The intersection is (1, 4).
Check: Is 1 + 4 = 5? Yes
Is 1, 3 less than 4? Yes
Since both statements are true, the two numbers are 1 and 4.

EXERCISES

1. Graph $y = 2x - 1$ and $y = 3x - 3$ on the same coordinate plane. At what point do these two lines intersect?

Determine if the given point is the solution of the system.

2. $y = -x + 7$
$y = x - 3$
(5, 2)

3. $x - y = 0$
$3x + 2y = -10$
(2, −2)

4. $y = -\frac{1}{2}x + 1$
$y = 2$
(−2, 2)

Solve each system by graphing.

5. $y = x$
$y = -x + 2$

6. $y = x - 2$
$y = \frac{1}{3}x + 2$

7. $y = -x$
$x - y = 2$

8. $y = x + 2$
$x + y = 2$

9. $2x + y = -2$
$2x + 3y = 6$

10. $x - y = -4$
$-3x + y = 0$

Rewrite each system into the slope-intercept form. Do not solve.

11. $4x + 2y = 6$
$3x - 4y = -12$

12. $2x + 3y = -9$
$4x + 6y = 12$

13. $x + y = 1$
$2x + 2y = 2$

Use the systems in Exercises 11–13 for Exercises 14–16.

14. What information do the equations tell about the graph of each system?

15. Make a prediction about the solution of each system.

16. Solve each system by graphing. How well did you predict the solution?

17. A ship is sailing along a path defined by $y = 2x + 4$. A motor boat is traveling along the line defined by $y = -x + 1$. Find the point of intersection of the two paths.

18. Two planes are flying along the paths defined by $y = x - 7$ and $y = 2x - 10$. An airport is located at the point of intersection. Describe the location of the airport with an ordered pair.

19. A helicopter is flying on a path defined by $3x - 2y = -6$. A bird is flying on a path defined by $y = 3x$. Find the coordinates of the point at which their paths cross.

Rhonda has a total of 14 apples and oranges. She has 4 more oranges than apples.

20. Write a system representing this information.

21. Find the number of apples and oranges Rhonda has by graphing the system.

These two relation rules show the distance, *d*, in miles traveled after time, *t*, in hours.

$d = 4t$

$d = 2t + 4$

22. Draw the graph of each relation rule on the same coordinate plane. Use time, *t*, for the horizontal axis.

23. What is the value of *t* at the point of intersection? the value of *d*?

24. Create a problem based on your graph. Then find its solution.

CHAPTER 12 SUMMARY

In this chapter you expanded the skills you developed in Chapter 11. The domain was no longer limited—it contained all real numbers, and you learned to graph linear equations by plotting points from tables of x and y values. You found that the slope of a line is the ratio of its rise (vertical change) to its run (horizontal change). You then learned to graph lines defined by equations written in the slope-intercept form ($y = mx + b$) by using the y-intercept, b, and the slope, m. Finally you found that the solution of a system of linear equations can be found by graphing the two equations on the same coordinate plane and finding their point of intersection. This method is especially useful in business and industry to solve current problems and predict answers to future problems.

VOCABULARY

collinear
linear equation
slope
slope-intercept form
system of linear equations
x-intercept
y-intercept

CHAPTER REVIEW

Tell whether each ordered pair would lie on a line defined by the equation $y = x + 2$.

1. (0, 2) **2.** (−2, 1)
3. (−1, 1) **4.** (2, 4)

Rewrite each equation in the form $y = mx + b$.

5. $2x + y = 6$ **6.** $-3x - 4y = 8$

7. $-x + 3y = 0$ **8.** $x + 2y = -2$

Find the slope of the line.

9.

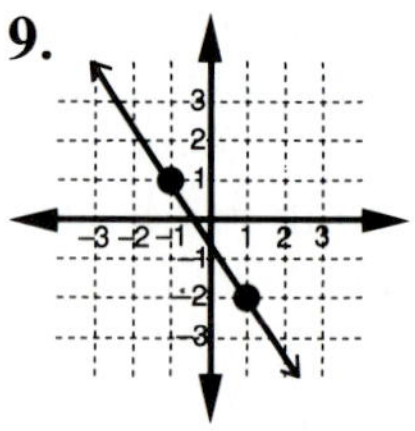

10.

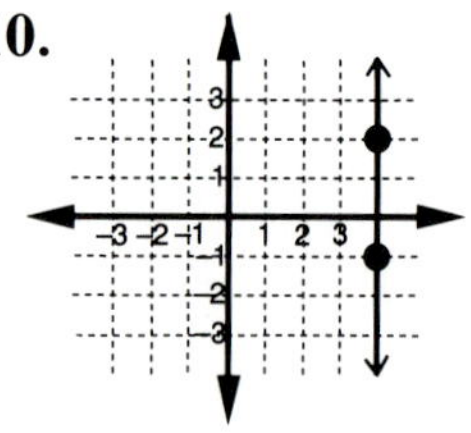

11.

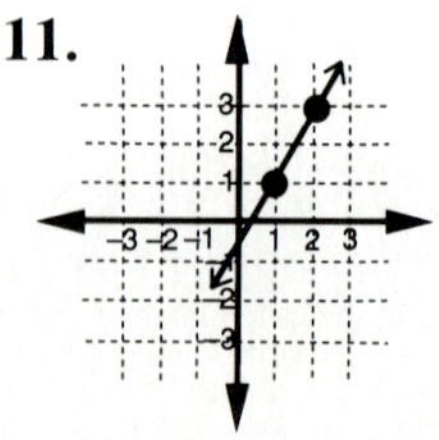

Draw a line for each given slope.

12. slope $\frac{1}{2}$ **13.** slope -3

14. slope 0 **15.** no slope

16. slope $-\frac{2}{3}$ **17.** slope 1

18. How can you determine if three points are collinear?

19. List the lines in order from least steep to steepest.
a: slope of 2 **b:** slope of $-\frac{1}{2}$
c: slope of $\frac{1}{4}$ **d:** slope of -3

Give the slope of the line defined by each equation.

20. $y = 4x$ **21.** $y = -\frac{4}{5}x - 1$

22. $x + y = 0$ **23.** $3x - 2y = 10$

Tell if the lines defined by each pair of equations are parallel, perpendicular, or neither.

24. $y = 3x + 6$, $y = -3x + 6$ **25.** $y = x + 6$, $y = x - 6$

26. $2x + y = 8$, $x - 2y = 12$ **27.** $3x - 2y = 6$, $2x - 3y = 6$

Name the slope and y-intercept of the line defined by each equation.

28. $y = 3x - 5$ **29.** $y = 6$

30. $2x + 3y = 9$ **31.** $4x - 8y = 24$

Graph the line defined by each equation.

32. $y = -\frac{1}{2}x + 4$ **33.** $2x - 3y = 0$

34. $y = x - 5$ **35.** $x + 4y = 8$

Graph each line given its slope and a point, P, on the line.

36. slope of 0 and $P(1, 5)$

37. no slope and $P(-2, 3)$

38. slope of $\frac{4}{5}$ and $P(0, 0)$

39. slope of -3 and $P(1, -4)$

Find the x-intercept of each line.

40. $y = 2x - 8$ **41.** $2x - y = -10$

42. $x + 5y = 15$ **43.** $-3x - 3y = 12$

Solve each system by graphing.

44. $y = \frac{2}{3}x + 1$, $x + y = 6$ **45.** $y = x - 3$, $y = 2x + 2$

46. $y = -x + 2$, $2x - y = 4$

The sum of two numbers, x and y, is 0. The first number, x, is 4 less than the second number, y.

47. Write a system representing this information.

48. Find the numbers by graphing the system.

CHAPTER 12 TEST

Tell if the ordered pair $(3, -1)$ will lie on the line defined by each equation.

1. $y = x - 4$ **2.** $y = 2x - 5$

3. $3x - 4y = 5$

Rewrite the equation in the slope-intercept form. Then name the slope and y-intercept of the line defined by each equation.

4. $y = \frac{1}{2}x - 5$ **5.** $x + y = 1$

6. $3x - 2y = 6$

7. Determine whether the points $(5, 6)$, $(4, 5)$, and $(-2, -1)$ are collinear.

8. List the lines in order from steepest to least steep.

a: slope of -3
b: slope of 0
c: slope of $\frac{1}{2}$
d: slope of $-\frac{1}{4}$

Tell if the lines defined by each pair of equations are parallel, perpendicular, or neither.

9. $y = -\frac{1}{2}x + 3$
$-2y = x + 4$

10. $6x + y = -4$
$6y = x + 12$

11. $2x + 3y = 3$
$2x - 3y = -3$

Graph the line defined by each equation.

12. $y = 2x + 1$ **13.** $x + 3y = -6$

14. $x = 4$

Graph the line that

15. has a slope of -1 and a y-intercept of $(0, 4)$.

16. has a slope of 0 and passes through $P(1, 1)$.

17. passes through $Q(3, -2)$ and is parallel to the line defined by $y = \frac{1}{3}x$.

18. is perpendicular to the line defined by $y = 4x - 2$ and passes through $R(-3, 4)$.

Find the x-intercept of the line defined by each equation.

19. $x + y = 9$ **20.** $y = -3x + 6$

Solve each system by graphing.

21. $y = 2x + 1$
$y = x + 3$

22. $x - y = -2$
$y = 3x$

23. $y = 3$
$x = -2$

24. A jeep is crossing the desert on a path defined by $y = -\frac{3}{2}x + 1$. A camel is traveling on a path defined by $y = -2x$. Find the coordinates of the point where their paths cross.

Jonel has a total of 15 coins. She has twice as many dimes as nickels.

25. Write a system representing this information.

26. Find the number of dimes and nickels Jonel has by graphing the system.

SKILLS PREVIEW

List all the ways each event could occur.

1. A die is rolled, and one number lands on top.

2. A bag contains the letters A, B, C, D, and E.
   - **a.** One letter is pulled out of the bag.
   - **b.** One letter is pulled out of the bag and not put back. Another letter is pulled out of the bag.

3. A bag contains the letters X, X, Y, Y, Z.
   - **a.** One letter is pulled out of the bag.
   - **b.** One letter is pulled out of the bag and not put back. Another letter is pulled out of the bag.

4. Gina has blue, red, green, and yellow markers.
   - **a.** She chooses three colors to use on a poster.
   - **b.** She chooses one color for the top line of a sign, another color for the middle line, and another for the bottom line.

The heights in centimeters of 15 teens are listed below:
156, 169, 154, 170, 174, 165, 159, 172, 158, 167, 159, 169, 175, 155, 159

5. Place these heights in order. Which height is in the middle of your list?

6. What is the difference between the shortest and tallest height?

7. Which height is listed most often?

You received the following grades on math tests:
90 95 90 92 93 62

8. To find your average test score, your teacher added all of scores together and divided by 6? What was your average test score?

9. Suppose your teacher only uses your 5 best scores. What would your average be then?

10. How does one very low score affect your average test score?

Tony is getting ready for a long distance race next Saturday. Here is his plan for the number of miles he will run each day this week.

Sat	Sun	Mon	Tues	Wed	Thurs	Fri
15	20	15	20	15	5	5

11. How many miles is he planning to run in all?

12. Divide the total miles by 7 to find the average number of miles per day Tony plans to run.

13. What is the average for Saturday through Wednesday?

14. Give a reason why Tony plans to run only 5 miles on Thursday and Friday.

LOUISIANA
CARTER MONDALE
NTANA
DEM. AB

Politicians, baseball players, and musicians use statistics and probability. Politicians need to know the chances of winning an election. Baseball managers use statistics when deciding whether to use a right- or left-handed pinch hitter. Musicians need to know the probability of making a profit from a music video or musical tour.

You also use probabilities and statistics everyday. The chance of rain reported in a weather report may have helped you decide whether to take an umbrella today. Daily you make decisions that involve predictions. In order for you to be able to make those decisions, you have to be able to process information, or data.

In Chapter 4 you learned how to gather, collect, organize, and interpret data. In this chapter you will study probability and statistics so you can do even more with data.

Ask Yourself

- *How does a meteorologist determine the chance of rain?*
- *What decision have you made recently based on the chance of something occurring?*
- *Do you know how to predict the likelihood of something occurring?*

CHAPTER 13

PROBABILITY AND STATISTICS

SECTION 13–1

LANGUAGE OF PROBABILITY

WARM-UP ACTIVITY

Work in cooperative learning groups to conduct an experiment. You will need these materials:

a die   a tally chart like the one below.

Number rolled	Tally	Frequency
1		
2		
3		
4		
5		
6		

If you roll the die 30 times, how many times do you think you will roll a 4?

Roll the die 30 times. Record your results in the tally chart.

Are your results what you expected? Explain your answer.

Save your results.

INTRODUCTION

In the experiment you just conducted, the result of rolling the die is called an **outcome** or an **event**. There are 6 **equally likely** outcomes: 1, 2, 3, 4, 5, and 6. We say that the **sample space** of this experiment is {1, 2, 3, 4, 5, 6}. The measure of the likelihood of a certain outcome or event is called the **probability** of the event.

SKILLS DEVELOPMENT

When you roll a die, you expect 1 of 6 outcomes. So, to find the **expected** or **theoretical** probability, the following is used:

$$P(\text{event}) = \frac{\text{number of favorable outcomes}}{\text{total number of possible outcomes}} \qquad P(4) = \frac{1}{6} \begin{array}{l} \leftarrow \text{one 4 on die} \\ \leftarrow \text{total possible outcomes} \end{array}$$

Probability can be written as a ratio between 0 and 1 or as a percent.

Example 1

A die is rolled. What is the probability of rolling a

a. 7? **b.** 2? **c.** 1, 2, 3, 4, 5, or 6?

Solutions

0 $\frac{1}{6}$ 1

a. There is no 7. **b.** There is one 2. **c.** There are 6 favorable outcomes.

$P(7) = \frac{0}{6} = 0$, or 0% $P(2) = \frac{1}{6}$, or $16\frac{2}{3}\%$ $P(1, 2, 3, 4, 5, 6) = \frac{6}{6} = 1$, or 100%

This is a **certainty**.

When you conducted the experiment of rolling a die, 4 may not have come up in exactly $\frac{1}{6}$ of your rolls. The probability of an outcome based on the results of an experiment is known as the **experimental probability**.

Example 2

A die is rolled 30 times. A 4 occurred 6 times. What is the experimental probability of rolling a 4?

Solution $P(4) = \frac{\text{number of times 4 occurred}}{\text{total number of rolls}} = \frac{6}{30} = \frac{1}{5}$, or 20%

EXERCISES

Assign an approximate value from 0 to 1 to show the probability of each event.

1. A glass bottle breaking after being dropped on concrete

2. You swimming outdoors on January 1 of this coming year

3. Your buying lunch in the school cafeteria tomorrow

Express each probability as a percent.

4. The probability is $\frac{4}{5}$ that the team will score a touchdown from that position.

5. For 30 people, the probability is 0.7 that two people will have the same birthday.

Suppose you experiment by tossing a coin.

6. What is the probability it will come up heads? tails?

7. In 100 tosses of a coin, how many times would you expect a head to come up? a tail?

8. Perform this experiment 100 times. Record the number of heads and tails.

9. Compare your answers to Exercise 7 and 8. What do you notice?

A bag contains 6 green balls, 9 white balls, and 12 blue balls. You are to draw a ball. Find the probability of each event.

10. drawing a white ball

11. drawing a blue ball

12. drawing a green ball

13. drawing a black ball

Use the spinner at the right. Find each probability.

14. P(black)

15. P(red)

16. P(white)

17. Suppose you tossed a thumbtack 50 times and it landed point up 34 times. What is the probability that the next time it is tossed the thumbtack will land point up?

A deck of cards has 52 cards in all, including 13 diamonds, 13 clubs, 13 hearts, and 13 spades. A card is drawn from a well-shuffled deck. Find each probability.

18. P(ace) **19.** P(king) **20.** P(5 of clubs)

21. P(black ace) **22.** P(face card) **23.** P(even-numbered card)

Suppose you experiment by tossing a nickel, a dime, and a penny.

24. List the sample space. Use H for head and T for tail.

25. Study this tree diagram. How does this diagram help you find all the different possible outcomes of this experiment?

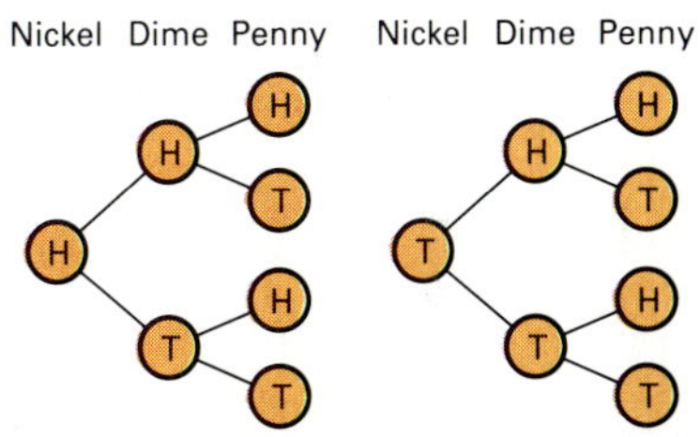

26. What is the probability of 3 heads? What other outcome is equally likely?

27. What other probabilities can you write for this experiment?

28. Copy and complete the chart below. Use your results from the experiment you conducted at the beginning of the lesson to complete the third column. Conduct the experiment 30 more times and then another 30 more times to complete the other columns.

Number rolled	Theoretical probability	Experimental probability after . . . 30 rolls	60 rolls	90 rolls
1				
2				
3				
4				
5				
6				

29. Look at your chart from Problem 28. Compare the experimental probabilities to the theoretical probabilities. Describe your findings.

SECTION 13–2
COUNTING

WARM-UP ACTIVITY

What kind of muffin is an unhappy muffin? To find out, write the letters as indicated by the clues. Put the letters together to form a word.

1. The first $\frac{2}{7}$ of BLENDER
2. The last $\frac{1}{2}$ of TRUE
3. The last $\frac{1}{3}$ of SEPTEMBER
4. The last $\frac{2}{5}$ of STORY

Make up your own fraction word puzzle and give it to a friend to solve.

INTRODUCTION

Kara is upset as she looks into her closet to find something to wear. Her wardrobe is not as extensive as she would like. What Kara does not realize is that the 5 T-shirts she has can be worn with any of her 3 pairs of jeans. How would this information and a little math help Kara?

Let's examine Kara's situation more closely. Kara has 5 T-shirts that can be worn with 3 pairs of jeans. The first T-shirt can be matched will all 3 pairs of jeans, so she actually has 3 different outfits for the first T-shirt. We can illustrate all the possible outfits by making a tree diagram.

SKILLS DEVELOPMENT

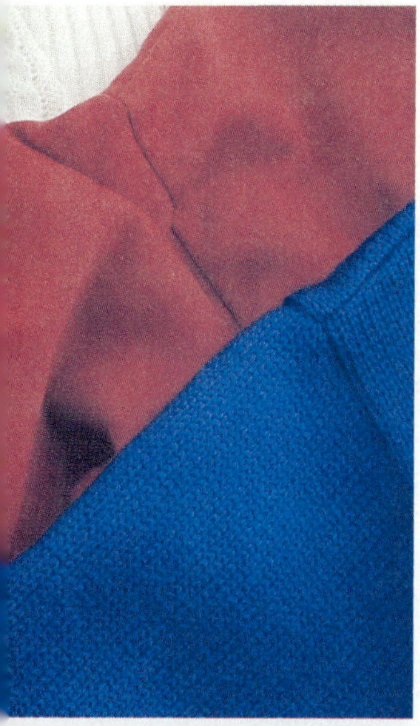

The tree diagram shows 15 possible outfits. We can also compute this number by multiplying 5 times 3. Consider why this works.

$\underset{\text{T-shirts}}{\underline{\ \ ?\ \ }}\quad \underset{\text{jeans}}{\underline{\ \ ?\ \ }}$

There are $\underline{\ 5\ }$ (T-shirt choices) and $\underline{\ 3\ }$ (jeans choices), so in all there are $5 \cdot 3$, or 15, different outfits possible.

This example uses what we call the **Fundamental Counting Principle**.

Fundamental Counting Principle

If there are a ways of doing one task and b ways of doing a second task, then there are ab different ways in which these two tasks can be completed.

The Fundamental Counting Principle can be extended to more than two tasks. The total possibilities are always obtained in the same way, multiplying the possibilities for each task.

Example 1

How many different ways can you answer a 5-question true/false test if no answer may be left blank?

Solution Use short lines to represent the questions.

___ ___ ___ ___ ___ (There are 5 lines, since there are 5 questions.)

Since there are two possible answers for each question, we write 2 on each line. Using the Fundamental Counting Principle, all we do now is multiply the 2s.

$$\underline{\ 2\ } \cdot \underline{\ 2\ } \cdot \underline{\ 2\ } \cdot \underline{\ 2\ } \cdot \underline{\ 2\ } = 2^5 = 32$$

There are 32 different ways to answer this test.
Be careful of very subtle differences in the wording of problems.

Example 2

a. How many different ways can you choose sandwich **and** juice from 20 different sandwiches and 5 flavors of juice?

b. How many different ways can you choose sandwich **or** juice from 20 different sandwiches and 5 flavors of juice?

Solutions The only difference in the two problems involves the use of *and* and *or*.

a. With *and*, use the Fundamental Counting Principle.
$20 \cdot 5 = 100$ ways

b. With *or*, add the possibilities.
$20 + 5 = 25$ ways

Have you ever noticed that license plates differ in the quantity and arrangement of letters and numbers? Each state seems to use a different format. The Fundamental Counting Principle can be used to determine how many possible license plates can be generated using letters and numbers.

Example 3

How many different license plates can be made using 3 digits, 1 letter, and 3 more digits?

Solution Use 7 short lines (3 digits + 1 letter + 3 digits).

Since there are 10 digits (0, 1, 2, 3, 4, 5, 6, 7, 8, 9), we record 10 choices on each of the first 3 lines. There are 26 letters in the alphabet, so we record 26 choices on the fourth line.

The 10 choices for the digits are recorded on the remaining 3 lines.

10	10	10	26	10	10	10	D = digit
D	D	D	L	D	D	D	L = letter

Then use the Fundamental Counting Principle.

$10 \cdot 10 \cdot 10 \cdot 26 \cdot 10 \cdot 10 \cdot 10 = 26{,}000{,}000$ possible license plates

Example 4

Certain license plates will have 3 digits, 1 letter, and 3 more digits. However, no two digits can be the same. How many different license plates can be made?

Solution Use 7 short lines, as in Example 3.

We record 10 choices on the first line. However, since no digit can be repeated, once we choose a digit, only 9 choices are left, so 9 is recorded on the second line. Once we choose another digit, only 8 choices are left. We continue in this manner until we have recorded the number of choices on each of the six lines marked D. There are still 26 choices for letters. Then we multiply. This time, a calculator is helpful.

10	9	8	26	7	6	5	= 3,931,200 possible
D	D	D	L	D	D	D	license plates

EXERCISES

1. How many different ways can a 10-question multiple-choice test be answered if the answers must be **a**, **b**, **c**, or **d**?

2. There are 12 boys and 15 girls in a class. How many different dancing partners can be made consisting of a boy and a girl?

3. Tran had 5 ties and 4 shirts that match. How many different outfits are possible?

A code name consists of 4 letters. How many different code names are possible

4. using any letters of the alphabet?

5. if repetition of a letter within a code name is not allowed?

6. if the code name must end in a vowel?

7. if only consonants can be used and repetition of a letter within a code name is not allowed?

8. A basketball coach has 7 players who can play forward, 5 who can play guard, and 3 who can play center. How many different starting teams made up of 2 forwards, 2 guards, and 1 center are possible?

Do some research on telephone numbers. The first 3 digits are the area code. Are they any restrictions on which digits can be used? What about the other 7 digits? Use this information to solve these problems.

9. How many different area codes are possible?

10. How many different local numbers (last 7 digits) are possible?

11. Yogurt Delight has 5 equally popular flavors of frozen yogurt and 20 different equally popular toppings. What is the probability that the next person who orders frozen yogurt with topping will choose banana yogurt and peanuts?

12. Write your own counting problem and solve it.

SECTION 13–3

ARRANGEMENTS

WARM-UP ACTIVITY

Use a 5-by-5 dotted grid as shown.

Draw the shape 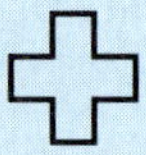 by connecting 12 dots in such a way that there are 8 dots outside the shape and 5 dots inside.

INTRODUCTION

Have you ever had your picture taken with your family or a group of people? Sometimes it seems that no one knows how the people should line up. This can take a very long time. Think about how many ways we could place just 5 people in a straight line.

SKILLS DEVELOPMENT

Just as in the last lesson, we can use short lines and the Fundamental Counting Principle to help us solve the photograph problem. We will need 5 lines. Since there are 5 people, we record 5 for the first position. But what about the second position? There are only 4 choices, because one person can occupy only one space at a time. After recording the number of choices for each position, we multiply.

$5 \cdot 4 \cdot 3 \cdot 2 \cdot 1 = 120$ different possible ways

Now you can appreciate the photographer's dilemma!

The arrangement of the elements or part of the elements of a set in a specific order is called a **permutation**.

Example 1

Suppose that, instead of placing 5 people in 1 straight line, the photographer decides to place them in 2 rows. There will be 3 people in the back row. In how many ways can this be done?

Solution Two cases must be considered.

Case 1: The back row is filled first.

$\underline{\;5\;} \cdot \underline{\;4\;} \cdot \underline{\;3\;} \cdot \underline{\;2\;} \cdot \underline{\;1\;} = 120$ ways

back-row choices front-row choices

Case 2: The front row is filled first

$\underline{\;5\;} \cdot \underline{\;4\;} \quad \cdot \quad \underline{\;3\;} \cdot \underline{\;2\;} \cdot \underline{\;1\;} = 120$ ways

front-row choices back-row choices

For each case the number of ways is the same as the number of ways using the straight-line arrangement. However, since there are 2 possible cases for the 2-row arrangement, we must find the total of the ways for both cases.

$$\begin{matrix}\text{ways for}\\ \text{Case 1}\end{matrix} + \begin{matrix}\text{ways for}\\ \text{Case 2}\end{matrix} = 120 + 120 = 240 \text{ ways}$$

Notice the pattern in the multiplication of the numbers.

$5 \cdot 4 \cdot 3 \cdot 2 \cdot 1$

The factors count down from 5 to 1. This can be written as 5!. The expression 5! is known as a **factorial**.

Example 2

Compute 7! (Read as "seven factorial.")

Solution $7! = 7 \cdot 6 \cdot 5 \cdot 4 \cdot 3 \cdot 2 \cdot 1 = 5040$

For convenience, 0! is defined to be equal to 1.

The formula $n!$ represents the number of arrangements or permutations of n objects.

Example 3

How many different ways can 10 flags be arranged on a flagpole?

Solution $10! = 3{,}628{,}800$ ways

${}_nP_r$ designates taking r of n objects at a time and placing them in a definite order.

$$ {}_nP_r = \frac{n!}{(n-r)!} $$

Example 4

Compute ${}_7P_3$.

Solution ${}_7P_3 = \frac{7!}{(7-3)!} = \frac{7!}{4!} = \frac{7 \cdot 6 \cdot 5 \cdot 4 \cdot 3 \cdot 2 \cdot 1}{4 \cdot 3 \cdot 2 \cdot 1} = 7 \cdot 6 \cdot 5$
$= 210$

Example 5

How many different ways can 3 prizes be awarded to 10 people?

Solution ${}_{10}P_3 = \frac{10!}{(10-3)!} = 10 \cdot 9 \cdot 8 = 720$ ways

EXERCISES

Evaluate.

1. ${}_5P_2$ **2.** ${}_9P_4$ **3.** $\frac{8!}{2!4!}$ **4.** $\frac{9!}{7!}$

5. In how many different ways can a pitcher and a catcher be chosen from 8 players who can play either position?

6. In how many ways can 6 different books be arranged on a shelf?

7. Given triangle ABC, how many different ways can the triangle be named?

8. In how many different ways can a president, a vice president, and a secretary be selected from a class of 30 students?

Use the word PROBLEM.

9. In how many different ways can a 3-letter arrangement be selected?

10. How many different ways can a 3-letter arrangement be made that begins and ends with a vowel?

11. How many different ways can all the letters be arranged?

The number of arrangements of the letters in a word such as MINIMUM will be less than expected since some of the letters repeat. Rearranging M, M, and M will result in a permutation that doesn't look any different. For problems like this, we divide the expected number of permutations by the number of repeats written with the factorial symbols.

MINIMUM has 7 letters; 3 M's and 2 I's repeat. Therefore, the number of permutations of all the letters is found as follows:

$$\frac{7!}{3!2!} = \frac{7 \cdot 6 \cdot 5 \cdot 4 \cdot 3 \cdot 2 \cdot 1}{3 \cdot 2 \cdot 1 \cdot 2 \cdot 1} = 420$$

Find the number of permutations of the letters in these words.

12. MISSISSIPPI

13. ATTENTION

14. BASEBALL

15. LITTLE

For items arranged in a circle, there is another consideration. If the letters A, B, C, D are placed in a straight line, a new arrangement is made when the A is moved to the other end of the line: B, C, D, A. However, nothing happens when A, B, C, D are written in a circle and A is moved to the end. The arrangement stays the same. The formula for the number of permutations of n objects arranged in a circle is $(n - 1)!$.

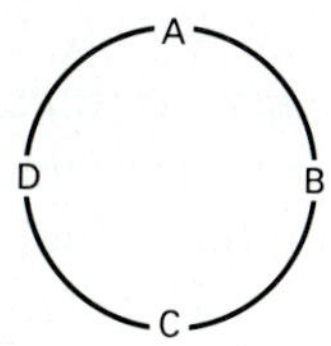

16. How many different ways can 6 people be seated at a round table?

17. How many different ways can 7 Scouts be seated around a campfire?

SECTION 13–4

COMBINATIONS

WARM-UP ACTIVITY

Copy this cross-number puzzle. Then use the clues to complete it.

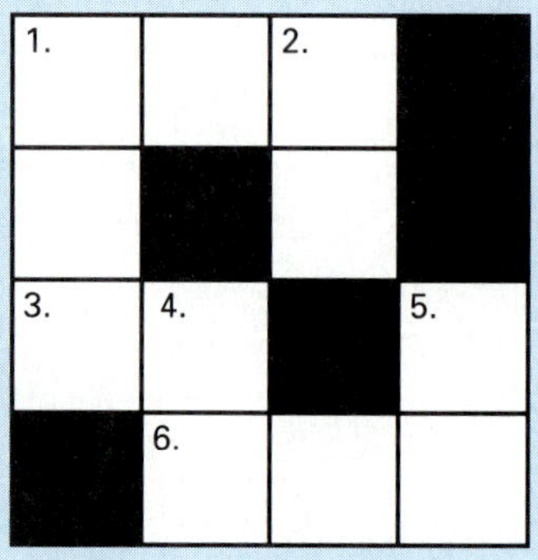

Across

1. 195 is 60% of me.

3. I am 75% of 60.

5. 100% of me is 3.

6. I am 40% of 1530.

Down

1. 15% of me is 57.6.

2. I am 200% of 25.

4. 7% of 800 gives me.

5. I am 80% of 40.

INTRODUCTION

To stay competitive, pizza companies make decisions about specials that affect their profit margins. Deciding how many toppings to offer is a major decision that impacts the total operation of a business.

Week's Special Choice of 2 toppings from 10 selections		
	orig	pan
small	4.99	5.99
medium	6.99	7.99
large	8.99	9.99

SKILLS DEVELOPMENT

Notice that this week's Pizza Special involves choosing 2 out of 10 toppings. This type of problem is different from the kind we encountered in the previous lesson. If we select pepperoni and green pepper, the order in which we make our selections is not important. The choice of part of a set without regard to order is called a **combination**.

${}_nC_r$ designates taking r of n objects at a time when order is not important.

$${}_nC_r = \frac{n!}{(n-r)!r!}$$

The number of different ways in which 2 toppings can be chosen out of 10 can be represented as ${}_{10}C_2$.

Example 1

Compute ${}_{10}C_2$.

Solution $${}_{10}C_2 = \frac{10!}{(10-2)!2!}$$

$$= \frac{10 \cdot 9 \cdot 8 \cdot 7 \cdot 6 \cdot 5 \cdot 4 \cdot 3 \cdot 2 \cdot 1}{8 \cdot 7 \cdot 6 \cdot 5 \cdot 4 \cdot 3 \cdot 2 \cdot 1 \cdot 2 \cdot 1}$$

$$= \frac{10 \cdot 9}{2 \cdot 1} = \frac{90}{2} = 45$$

Example 2

In how many different ways can:

a. 3 officers be selected from a class of 15 students?

b. a president, a vice president, and a secretary be selected from a class of 15 students?

Solutions

a. This problem involves combinations because only the 3 people selected, not the order of their selection, is being considered. Selecting Bill, Donya, and Meketa is the same as selecting Donya, Bill, and Meketa.

$$_{15}C_3 = \frac{15!}{(15-3)!3!} = \frac{15 \cdot 14 \cdot 13}{3 \cdot 2 \cdot 1} = 455 \text{ ways}$$

b. This problem involves arrangements or permutations, since choosing 3 people is not enough. The choices must then be arranged into the different officer positions. Bill can be president, Donya can be vice president, and Meketa can be secretary. Or, Meketa can be president, Bill can be vice president, and Donya can be secretary. Or, Donya can be president, Meketa can be vice president, and Bill can be secretary.

$$_{15}P_3 = \frac{15!}{(15-3)!} = 15 \cdot 14 \cdot 13 = 2730 \text{ ways}$$

Example 3

How many combinations of 3-letter codes can be made from the letters in the word STAND? List the combinations.

Solution $_5C_3 = \frac{5!}{(5-3)!3!} = \frac{5 \cdot 4 \cdot 3}{3 \cdot 2 \cdot 1} = 10$ combinations

STA SAN TAN AND STN
STD SND TAD SAD TND

EXERCISES

Evaluate.

1. $_6C_4$ **2.** $_{10}C_1$ **3.** $_{12}C_3$ **4.** $_{11}C_7$

5. As an assignment students must solve 6 out of 10 problems. In how many different ways can these problems be selected?

6. In how many different ways can you form a committee of 4 people from a group of 7 people?

Trina wants to order a baked potato. She can choose from butter, sour cream, chives, bacon bits, cheese, and onion bits. How many different ways could she order a potato with

7. two toppings?

8. three toppings?

9. The cast of a school play requires 4 girls and 3 boys to be selected from 7 girls and 10 boys. In how many different ways can a cast be selected?

10. Hamburger Heaven offers a choice of 9 toppings for their hamburgers. If the price of the Cloud Nine Burger includes a choice of 3 toppings, how many different Cloud Nine Burgers are possible? with a choice of 4 toppings? with a choice of 5 toppings?

11. In how many different ways can a subcommittee of 3 Democrats and 4 Republicans be chosen from a Senate committee of 8 Democrats and 10 Republicans?

12. A chord is defined as a line segment connecting any 2 distinct points on a circle. If you have 10 points to choose from on a circle, how many chords are possible?

13. The principal asked a teacher to send 2 of her 30 students to help out in the office. How many different ways could the teacher do this?

14. A test contains 2 sections. Section A has 5 questions of which a student must answer 3. Section B has 6 questions of which a student must answer 4. How many different ways can a student answer the test?

15. Think about what is called a combination lock. Is this an accurate name for this lock? Explain your answer.

SECTION 13–5

PROBABILITY

WARM-UP ACTIVITY

Estimate each of the following. Then use a calculator to find the actual answers. Round answers to the nearest tenth. How close were your estimates to the actual answers?

1. 7% of 130

2. 19% of 360

3. 8.2% of 600

4. $35\frac{3}{4}\%$ of 62

5. What percent is 24 of 50?

6. What percent is 24 of 80?

7. What percent is 19 of 38?

8. What percent is 35 of 42?

9. 40% of what number is 80?

10. 35 is 35% of what number?

11. 85% of what number is 33?

12. What is 250% of 13?

INTRODUCTION

Benjamin is not exactly the neatest person in the world. His drawer contains 10 black socks and 8 blue ones. If he reaches in his drawer early in the morning before it's light and removes 2 socks, what is the probability that they will match?

SKILLS DEVELOPMENT

The sock problem actually puts together several of the ideas we have learned about in the previous sections. Remember how probability is defined.

$$\text{P(event)} = \frac{\text{N(favorable outcomes)}}{\text{N(total possible outcomes)}}$$

Let's examine this problem in parts.

Example 1

What is the probability that both socks will be black?

Solution 1

$$P(\text{2 black socks}) = \frac{\text{number of ways to remove 2 black socks}}{\text{number of ways to remove 2 socks}}$$

Since order is not important, we will use ${}_nC_r$.

$$P(\text{2 black socks}) = \frac{{}_{10}C_2}{{}_{18}C_2} = \frac{\frac{10!}{(10-2)!2!}}{\frac{18!}{(18-2)!2!}}$$

$$= \frac{\frac{10 \cdot 9}{2 \cdot 1}}{\frac{18 \cdot 17}{2 \cdot 1}} = \frac{10 \cdot 9}{18 \cdot 17} \text{ or about } 0.29$$

Solution 2

Removing a black sock first and removing a black sock second are **dependent** events because the outcome of removing the first sock affects the probability of removing a black sock second.

$P(\text{first sock is black}) = \frac{10}{18}$ $\qquad$ $P(\text{second sock black}) = \frac{9}{17}$

$$P(\text{black and black}) = \frac{10}{18} \cdot \frac{9}{17} = \frac{5}{17} = 0.29$$

Example 2

What is the probability that both socks will be blue?

Solution 1

$$P(\text{2 blue socks}) = \frac{{}_{8}C_2}{{}_{18}C_2} = \frac{\frac{8!}{(8-2)!2!}}{\frac{18!}{(18-2)!2!}}$$

$$= \frac{\frac{8 \cdot 7}{2 \cdot 1}}{\frac{18 \cdot 17}{2 \cdot 1}} = \frac{8 \cdot 7}{18 \cdot 17} \text{ or about } 0.18$$

Solution 2

$$P(\text{blue and blue}) = \frac{8}{18} \cdot \frac{7}{17} = \frac{28}{153} = 0.18$$

Now the original problem asked for the probability that the 2 socks will match. This implies that they will both be black or both be blue.

Example 3

What is the probability that both socks will be black or both will be blue?

Solution
P(2 black socks) + P(2 blue socks) $= \frac{5}{17} + \frac{28}{153} =$

$\frac{765 + 476}{2601} = \frac{1241}{2601}$, or $0.29 + 0.18 = 0.47$

Therefore, Benjamin has a 47% chance of removing 2 socks that match.

Events are **independent** if the outcome of one does not affect the outcome of the other.

Example 4

A coin is tossed and a die rolled. What is the probability of obtaining a head and a 2?

Solution Tossing a coin and rolling a die are independent events,
so $P(H) = \frac{1}{2}$ $\qquad P(2) = \frac{1}{6}$
$P(H \text{ and } 2) = \frac{1}{2} \cdot \frac{1}{6} = \frac{1}{12}$

Examining the sample space shows the same result.
{H1, H2, H3, H4, H5, H6, T1, T2, T3, T4, T5, T6}

EXERCISES

A coin is tossed, a die is rolled. Find each probability.

1. P(a head and 3)

2. P(a tail and 6)

3. P(a tail and an even number)

4. P(neither a head nor an even number)

A coin is tossed twice. Find each probability.

5. P(2 heads)

6. P(a head and then a tail)

7. P(not 2 tails)

A card is drawn from a deck of 52 cards, and a coin is tossed. Find each probability.

8. P(a king and a head)

9. P(a spade and a tail)

10. A deck of 52 cards is shuffled. A card is drawn, replaced in the deck, and another card drawn. What is the probability of drawing 2 aces?

There are 4 nickels and 6 dimes in a bag. A coin is drawn from the bag and not replaced. Then another coin is drawn. Find each probability.

11. P(2 nickels)

12. P(nickel, then dime)

A bag contains 5 black and 9 blue chips. Once a chip is drawn, it is not replaced. Find each probability.

13. P(black, then blue)

14. P(3 blue)

In a box of 100 plugs, 5 plugs are defective. What is the probability

15. of not choosing a defective plug?

16. that the first 2 plugs chosen will be defective?

A bag contains 25 marbles. Ten are red, 7 green, 5 yellow, and 3 pink. Once a marble is drawn, it is not replaced. What is the probability that

17. if one is drawn, it is not red?

18. if one is drawn, it is pink?

19. if two are drawn, both are yellow?

20. if two are drawn, the first is green and the second is red?

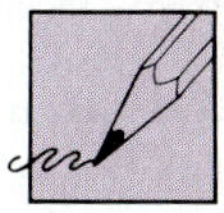

TRY THIS

Design a probability experiment that consists of two events. Then do the following:

21. Describe the experiment.

22. Tell whether the events are independent or dependent.

23. List at least 3 possible outcomes of the experiment and tell the probability of each.

SECTION 13–6

DATA AND PREDICTIONS

WARM-UP ACTIVITY

Copy this grid. Then do the calculations and place your answers in the boxes. Check your work by using the Check Statements.

Check Statements

1. The difference between E and J is 150.
2. The total of A, B, and C is 82.
3. The total of D and E is 1012.
4. The total of F, G, and H is 82.
5. The total of I and J is 886.

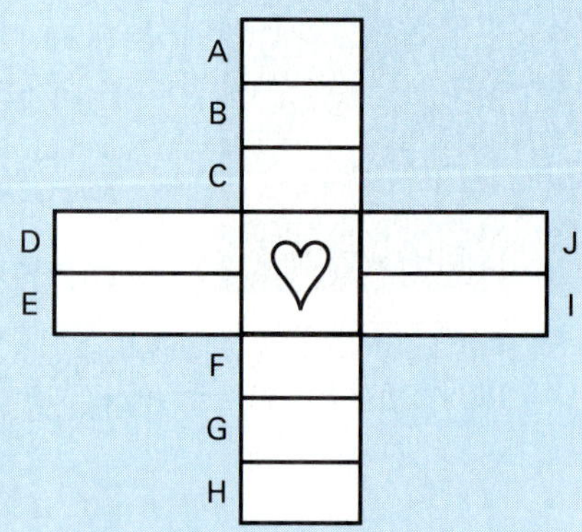

A $14 + 3 \times 4$

B $4^2 + 5 \times 6$

C $(2^2 + 9) - 3$

D $(4^2 + 3^2 + 2^3) \times 4$

E $10 \times (4^3 - 1^3 + 5^2)$

F $(4^2 + 3^2 + 50) \div 5$

G $\frac{6 \times 9}{27 - 18} + 3^3$

H $2^3 + 5^2 + 1^2$

I $12 + 12 \times 12$

J $(9^2 - 2^3) \times 10$

INTRODUCTION

A major factor in deciding to sail is the weather. Often, a sailing race will be cancelled because of bad weather. To make this decision, the weather forecast is considered. Forecasts are based on as much information as is available. The more data a meteorologist has access to, the better the forecast will be.

In this section, we will apply the skills we have developed in this chapter to problems involving data and predictions.

SKILLS DEVELOPMENT

You have learned that probabilities can help you predict what will happen. You also know that the experimental probability for a particular event is not always the same as the theoretical probability for that event.

Example

A die is rolled 10 times. The results are recorded in this table.

Number	1	2	3	4	5	6
Occurrence	1	1	4	2	1	1

a. In the next 10 rolls, how many 3s would you expect to roll?
b. Roll a die 10 times and record the results.
c. Was your prediction accurate? Why or why not?
d. How would you improve your prediction?

Solutions a. Based on only the data in the table, you would predict 4 out of 10 rolls would be a 3.
b. Try the experiment, record your results.
c. Answer the question. If your prediction was not accurate, it is probably because of the limited number of rolls recorded in the table.
d. The prediction would be more accurate if it was based on more trials.

EXERCISES

As a coin was tossed 500 times, the data in the box were gathered. Based on this data, if the coin is tossed 1000 times, estimate how many times the outcome will be

Heads	239
Tails	261

1. a head.

2. a tail.

3. Based on this data, what is the probability of getting a head in 1 toss?

Results from 3 sets of rolling a die 500 times are shown below.

Number	1	2	3	4	5	6
Trial 1	93	93	82	80	93	59
Trial 2	87	81	92	73	74	93
Trail 3	88	77	84	93	80	78
Total	268	?	?	?	?	?
Average per trial	?	?	?	?	?	?

4. Copy and complete the last two rows of the table. Round averages to the nearest tenth.

5. Use the data completed in Exercise 4 to predict the number of 3s that would be obtained in another trial of 500 rolls.

Data are given for the range of scores on a test. You select a student's test paper at random. What is the probability that, if passing is above 60,

6. the student passed?

7. the student scored above 80?

8. the student did not have a score in the 61–70 range?

Class Limits	Frequency
1–10	0
11–20	0
21–30	3
31–40	4
41–50	5
51–60	18
61–70	20
71–80	15
81–90	9
91–100	2

A thumbtack can land in one of 2 ways, point up or point down.

9. Drop a thumbtack 50 times and record the number of times it lands in each position.

10. Based on your results, which would be more likely: a thumbtack to land point up or point down?

11. Based on your results, what is the probability of a thumbtack landing point up?

This histogram shows the frequency distribution of the normal pulse rates of students.

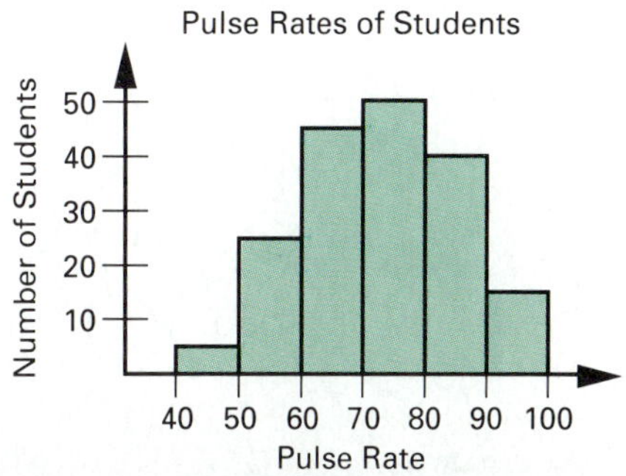

If a student is selected at random, what is the probability that the student has a pulse rate

12. less than 60?

13. greater than 80?

14. between 60 and 80?

15. To estimate the redfish population, 25 redfish were tagged and released. Out of 124 fish sampled, 8 fish were tagged. Estimate the redfish population.

16. It is necessary to predict the alligator population before the Wildlife and Fisheries Commission will allow any to be killed in the Louisiana swamps. Thirty-five alligators were tagged and released in the Maurepas area. Out of the 250 alligators sighted, 6 were tagged. Estimate the Maurepas alligator population.

17. Interview a person who is in a career you are interested in entering. Find out what work-related decisions this person has had to make that were based on predictions. What data were used to make the predictions and how were these data collected? Summarize your interview in a few paragraphs.

SECTION 13–7

STATISTICS

WARM-UP ACTIVITY

Find the decimals, using the clues for their location, and add.

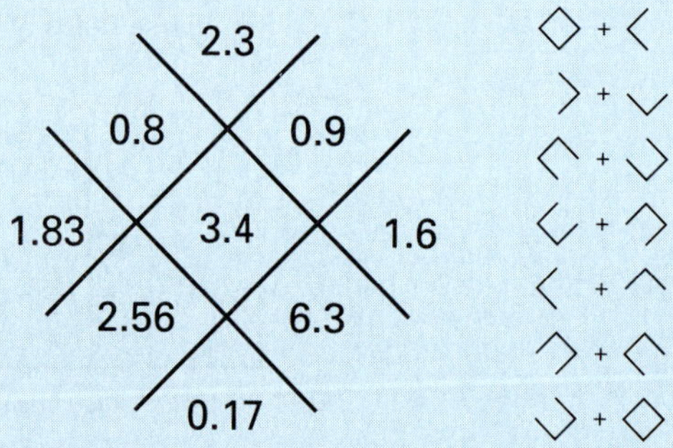

INTRODUCTION

Yoko is not particularly thrilled with her average in history because of two rather low scores. She decides to try to convince her history teacher to use another method to calculate her average. In this section, we will learn why Yoko thinks this would be helpful.

Statistics has grown into an important branch of mathematics in which data are used to solve problems. Each piece of data is often referred to as a statistic. Yoko's test scores are statistics about her performance.

SKILLS DEVELOPMENT

Example 1

These are Yoko's grades: 63, 85, 71, 88, 95, 97, 98, 97, 100.

Calculate the **a.** mean, **b.** median, and **c.** mode of Yoko's scores.

Solutions a. Another name for the average is the **mean**.

$$\text{mean} = \frac{\text{sum of all the data}}{\text{the number of pieces of data}}$$

$$= \frac{794}{9} = 88.2 \text{ (to the nearest tenth)}$$

b. To find the **median**, arrange the data in order from least to greatest. The median is the middle number.

63, 85, 71, 88, 95, 97, 98, 97, 100

The middle number is 95, so Yoko would consider it "better" for her teacher to calculate the median.

c. The **mode** of a set of data is the number that occurs most frequently. The mode of Yoko's scores is 97, so the "best" choice for Yoko would be the mode.

Of course, deciding how fair using the median and mode are to reflect Yoko's work will be left to Yoko and her history teacher.

When you determine the mean, median, or mode of a set of numbers, you obtain a single value which is an indicator of the data. The data numbers "cluster" around this single value. For this reason, these indicators are often referred to as the **measures of central tendency.**

You often have to decide which measure of central tendency is most representative of the data.

- The mean is the most appropriate when there are no extreme values within the data.
- The median is more appropriate if there are a few extreme values, as it is least affected by these.
- The mode is useful when the measure of the most characteristic value of a group of data is required.

Another characteristic of data which provides additional information is the **range**. The range is the difference between the greatest and smallest pieces of data.

Example 2

What is the range of Yoko's scores?

Solution Highest score: 100 Lowest score: 63
Since 100 − 63 = 37, the range of Yoko's scores is 37.

EXERCISES

Use the data 4, 6, 8, 10, 12.

1. What is the median?

2. What is the mean?

3. Why is there no mode?

For each of the following sets of data, find the mean, median, and mode. Round your answers to the nearest tenth.

4. 12, 16, 8, 23, 20, 21, 10, 12, 11

5. 38, 40, 42, 38, 41, 38, 42

6. 21, 20, 30, 24, 28, 22, 16, 18, 31, 17

7. 6.8, 8.5, 5.7, 9.6, 10.2, 9.8, 4.3, 5.7

8. The median for a set of data is 50. Does 50 need to be one of the data? Use an example to illustrate your answer.

The number of accidents at an intersection during 8 consecutive months is shown in this table.

A	M	J	J	A	S	O	N
13	15	18	8	8	18	10	11

9. Find the mean, median, and mode of the data. Round your answers to the nearest tenth.

10. Which of the measures of central tendency best represents the data? Why?

11. Predict the number of accidents for the missing months.

12. Find the mode of the following data: 0, 1, 3, 4, 6, 7, 135, 135.

13. Why is the mode not a useful piece of information for the data in Exercise 12?

The points scored by the school swimming team during the past 8 swim meets are shown.

210, 186, 170, 73, 175, 180, 73, 196

14. Find the mean, median, and mode of the scores. Round your answers to the nearest tenth.

15. Which measure of central tendency best represents the data?

16. Predict the number of points the team will score in the next swim meet.

Use the following histogram for Exercises 17–23.

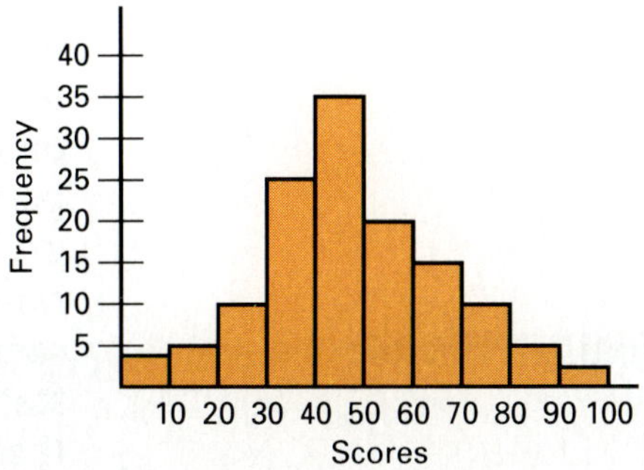

17. In which range do the most data occur?

18. In which range do the fewest data occur?

19. Predict in which one range the mean might occur.

20. Predict in which ranges the mean might occur.

21. Which prediction is more likely to be correct? Why?

22. Describe how the histogram can be used to determine in which one range the median occurs.

23. Can the histogram be used to determine in which range the mode occurs? Explain why or why not.

24. Although Yoko would have preferred otherwise, it seems that the mean is usually an appropriate measure of central tendency to use as an indicator of a student's grades. Give other examples of when the mean, median, mode, and range would be the most appropriate indicators to use.

CHAPTER 13 SUMMARY

In this chapter you examined how much our world relies on data. You learned to calculate the probability that something will occur. You used the Fundamental Counting Principle to find the number of permutations and related permutations to compute the number of ways of doing different things.

In the last sections you learned how to apply the concepts of probability and statistics to real-world data in order to make more accurate predictions.

VOCABULARY

combination
dependent events
equally likely
event
experimental probability
factorial
Fundamental Counting Principle
independent events
mean
measures of central tendency
median
mode
outcome
permutation
probability
range
sample space
statistics
theoretical probability

CHAPTER REVIEW

A box contains 10 pennies, 5 dimes, and 5 nickels. A coin is chosen at random. Find each probability.

1. P(nickel)

2. P(dime)

3. P(penny)

4. Conduct the experiment described above 20 times. Record your results. Find the experimental probability of choosing each type of coin.

5. Are the probabilities you found in Exercise 4 different from those you found in Exercises 1–3? Explain.

6. A library just received 6 new mysteries and 5 new science fiction books. If you wanted to read one of each, how many different ways could you do this?

Evaluate.

7. ${}_5P_3$

8. $6!$

9. $\frac{6!}{4!}$

How many different ways can the letters in each word be arranged?

10. VIDEO **11.** FOOTBALL

Evaluate.

12. ${}_{10}C_2$ **13.** ${}_{11}C_8$

14. The Yogurt Parlor has a choice of 6 delicious toppings for their banana splits. In how many ways could you order 4 of the toppings?

A bag contains 5 yellow, 8 red, 7 green, and 5 black marbles. Once a marble is drawn, it is not replaced. What is the probability that

15. if one is drawn, it is yellow?

16. if one is drawn, it is not green?

17. if two are drawn, both are red?

As a coin was tossed 500 times, the following data were gathered.

Heads	242
Tails	258

Based on this data, if the coin is tossed 1000 times, estimate how many times the outcome will be

18. a head.

19. a tail.

20. Based on these data, what is the probability of getting a head in one toss?

The points won by the school swimming team during the past eight competitions are shown.

205, 172, 189, 85, 182, 177, 85, 203

21. Find the mean, median, and mode of the scores. Round your answers to the nearest tenth.

22. Which measure of central tendency best represents the data?

23. Predict the number of points the team will score in the next swim meet.

24. Predict the total number of points the team will score in the next 4 meets.

25. Which prediction would you expect to be more accurate? Why?

CHAPTER 13 TEST

1. In how many different ways can a 6-question true/false test be answered if no answer may be left blank?

2. In how many ways can 7 different books be arranged on a shelf?

3. In how many different ways can 4 students be chosen from a class of 25 to serve on the dance committee?

4. In how many ways can a pitcher and a catcher be chosen from 10 eligible players?

Experiment: A box contains 12 tennis balls, 15 golf balls, 9 Ping-Pong balls, and 6 baseballs. Once a ball is chosen at random, it is not replaced. What is the probability that

5. if one is chosen, it is a Ping-Pong ball?

6. if one is chosen, it is a golf ball or a baseball?

7. if one is chosen, it is not a tennis ball?

8. if one is chosen, it is both a golf ball and a Ping-Pong ball?

9. if two are chosen, both are golf balls?

10. if two are chosen, one is a tennis ball and one is a baseball?

11. In a sample of 100 batteries, 3 are defective. What is the probability of buying a good battery?

This table shows the number of tackle boxes sold by a sporting goods store during four months.

Month	Total number of tackle boxes sold	Number of Old Pal tackle boxes sold
April	58	35
May	47	28
June	40	22
July	35	19

12. If 32 tackle boxes are sold in August, how many would you expect to be Old Pal tackle boxes?

13. Predict the total number of tackle boxes that will be sold in September.

Find each for the given data: 15, 18, 13, 15, 16, 13, 14, 15, 15.

14. the mean (the nearest tenth)

15. the median

16. the mode

17. the range

The number of people attending a local golf tournament each day is shown.

Day	1	2	3	4	5
Attendance	965	836	753	921	862
Day	6	7	8	9	
Attendance	903	1048	2012	6012	

18. Calculate the mean attendance per day. Round your answer to the nearest whole person.

SKILLS PREVIEW

Copy each figure. Then draw a line to show where each figure could be folded in half so that one half matches the other. Some figures could be folded in more than one way.

1. 

2. 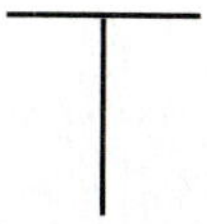

3. 

4. 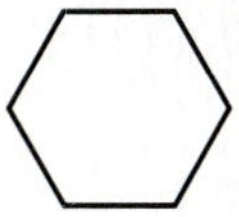

For each pair of figures, tell whether the second figure was obtained by sliding, flipping, or turning the first figure.

5. 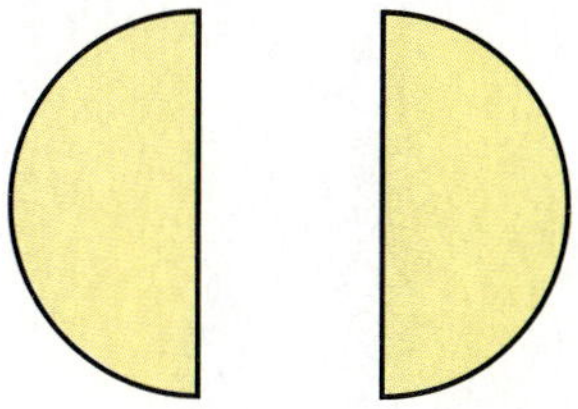

6. 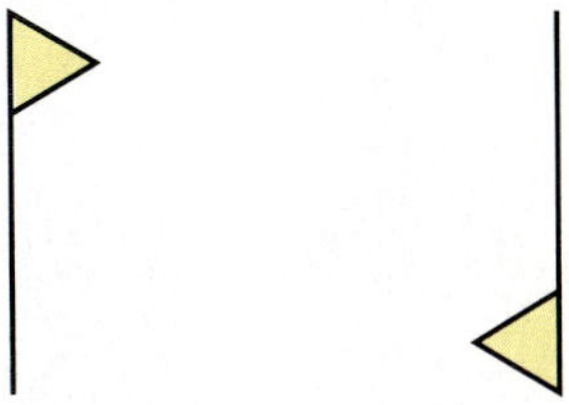

7. 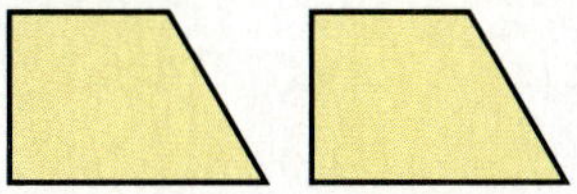

8. 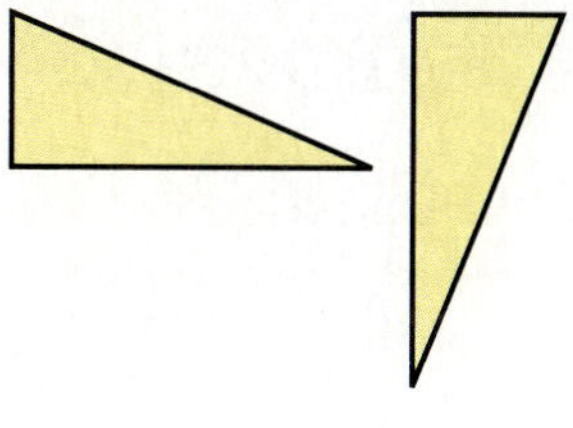

Everyday you see transformations in patterns around you. Look at the photograph. The pattern is made with fruit sections. You could use slides and dilations to draw a similar pattern on paper.

The butterfly shows another geometric relationship: symmetry. If you could draw a line through the body of the butterfly, the left side of the butterfly is the mirror image of the right side. We say that the butterfly has one line of symmetry. Do the sections of fruit have a line of symmetry? Does a section of fruit have more than one line of symmetry?

In this chapter, you will investigate the properties of translations, reflections, rotations, and dilations. You will investigate ways these movements can be expressed using coordinates and mappings. You will study the answers to these questions and others.

Ask Yourself

- *What is a tessellation?*
- *Which transformations result in a congruent image?*
- *What is a glide reflection?*
- *What image results from combining transformations?*

CHAPTER 14

TRANSFORMATIONAL GEOMETRY

SECTION 14–1

SLIDES

WARM-UP ACTIVITY

To solve some problems, begin by thinking of all possible answers. Then throw out the ones that are not correct.

Write the date for a year (in the last 200 years) that reads the same

backwards
or
forwards
or
upside down.

INTRODUCTION

There are many examples around you that suggest geometry and motion. Each example that follows suggests a **slide**. Often designs are based on slides or **translations**.

SKILLS DEVELOPMENT

All the points of square ABCD have been translated 5 units to the right. The image of square ABCD is the square A′B′C′D′.

A translation is shown by a **slide arrow** or **translation arrow**. It shows the direction and distance of the translation. A translation puts square *ABCD* in a new position called its image position.

A is the **pre-image** point which corresponds to *A′*, which is called the **image** point.

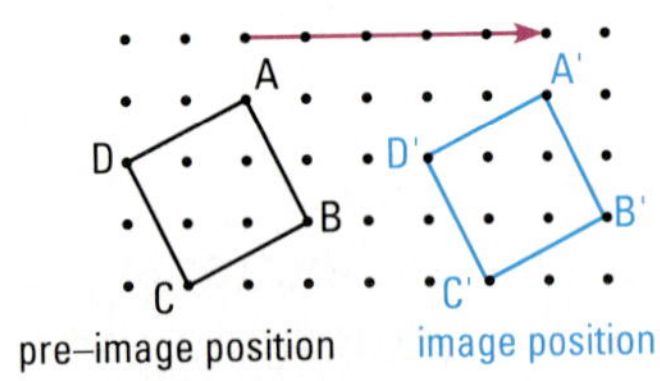

A′ is read as "*A* prime."

$A \rightarrow A'$ $\quad B \rightarrow B'$

$C \rightarrow C'$ $\quad D \rightarrow D'$

Example 1

Find the image of the figure for the slide arrow.

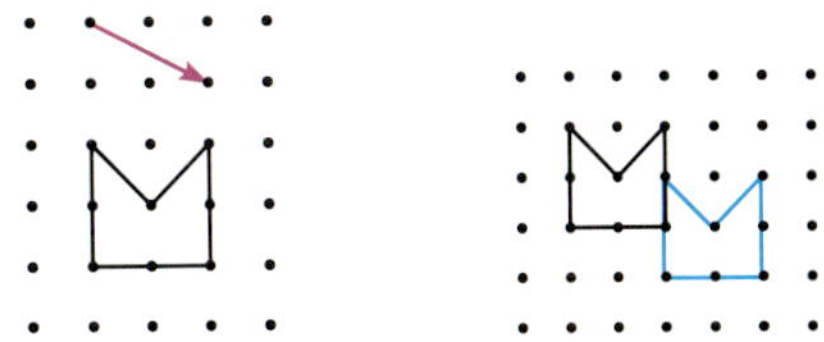

Solution Copy the figure onto dot paper and move it 2 units to the right and 1 unit down.

Example 2

You can use your skills with coordinates to show translations. $\triangle XYZ$ and its translation image $\triangle X'Y'Z'$ are shown. The coordinates of the corresponding vertices are related.

> 6 has been added to each *x*-coordinate.
>
> The *y*-coordinate remains the same.

Solution

orginal or pre-image points	image points
$X(-4, 3) \rightarrow$	$X'(2, 3)$
$Y(-5, -2) \rightarrow$	$Y'(1, -2)$
$Z(-2, -2) \rightarrow$	$Z'(4, -2)$

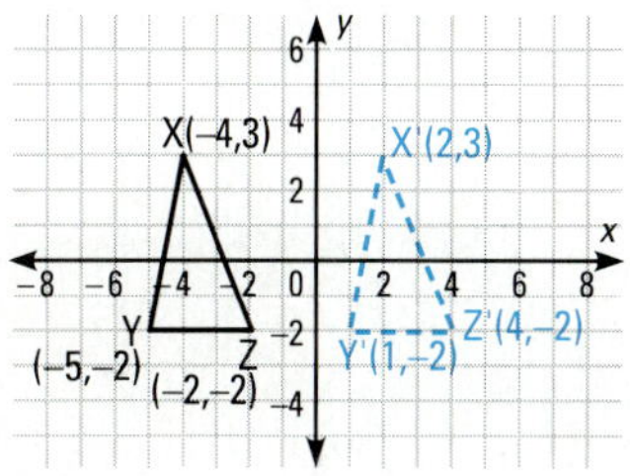

Mapping notation shows the translation as $(x, y) \rightarrow (x + 6, y)$.

EXERCISES

For each figure, the pre-image and image are shown. Describe each slide in words.

1.

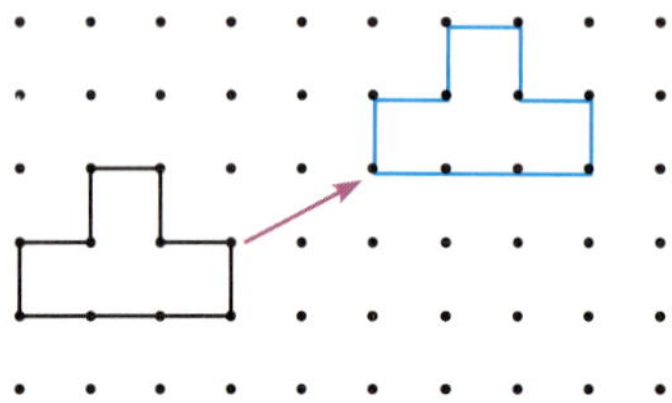

2.

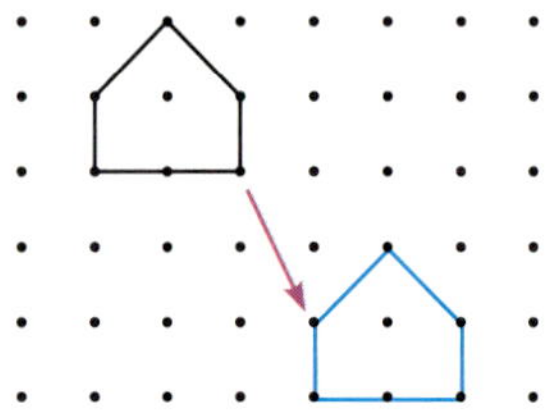

These diagrams do not show slides. Why not?

3.

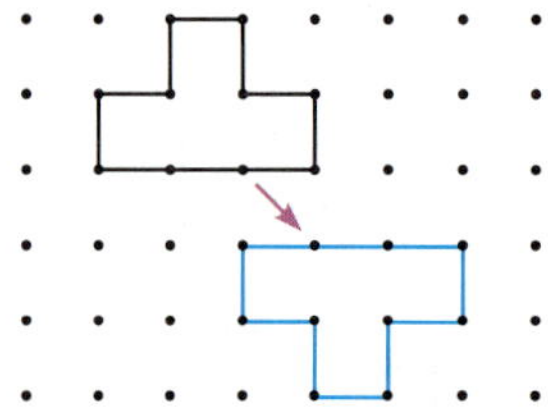

4.

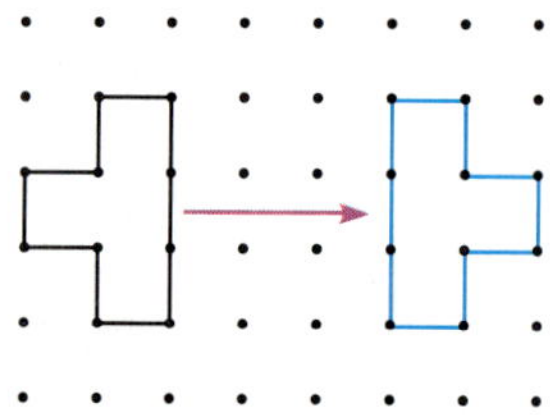

A shape is shown. Find its image for the slide arrow shown.

5.

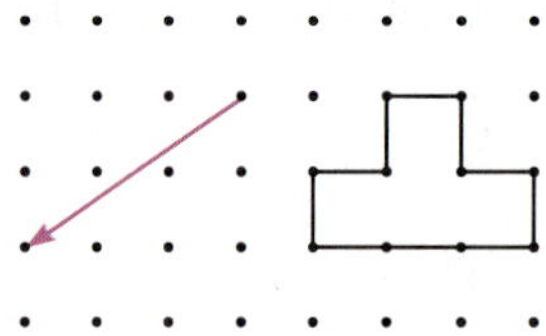

6.

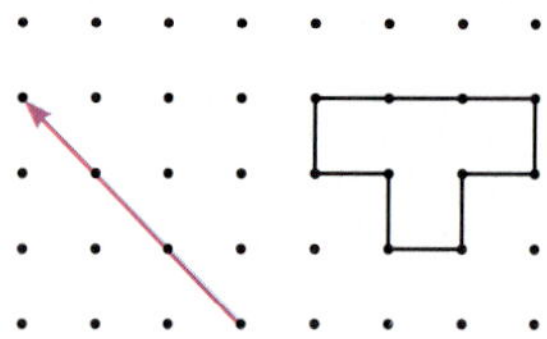

A translation is applied to obtain the image of figure *PQRST*. Use the diagram.

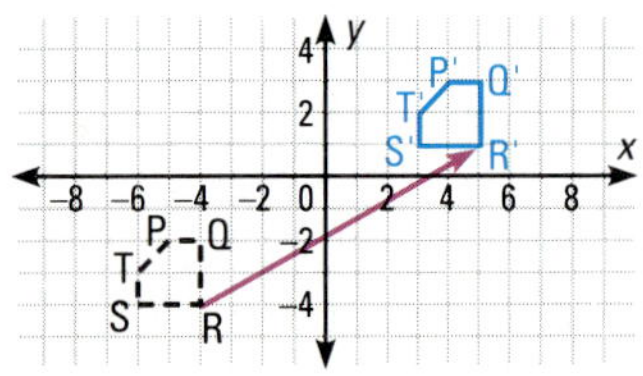

7. Write the coordinates of corresponding vertex points.

8. How are the x coordinates related?

9. How are the y coordinates related?

10. Write the mapping for this translation as $(x, y) \rightarrow (?, ?)$.

11. A mapping is given by $(x, y) \rightarrow (x, y + 3)$. Find the image of $STUV$, with vertex points $S(1, 2)$, $T(-2, -1)$, $U(1, -4)$, and $V(4, -1)$.

12. Why does the mapping show a translation?

The mapping for a translation is given by $(x, y) \rightarrow (x + 1, y - 2)$. Find the coordinates of the image of each figure.

13.

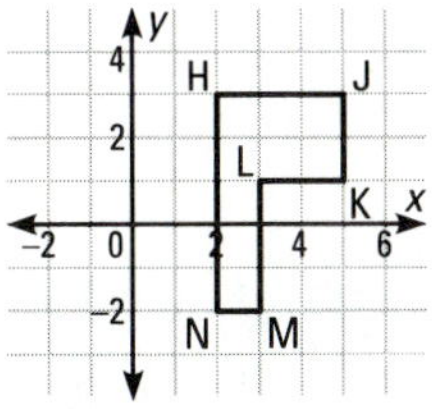

14.

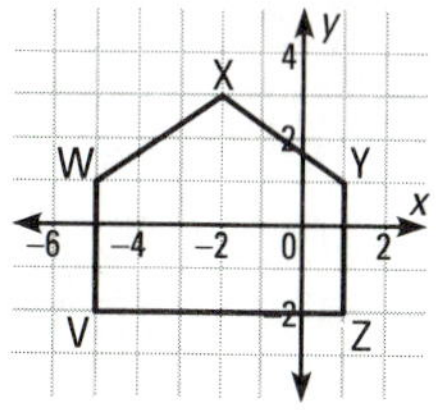

In learning mathematics, you also look for relationships. How are the facts or properties the same? How are the facts or properties different?

The following exercises investigate the fundamental properties of translations using coordinates of $\triangle ABC$.

A translation, given by $(x, y) \rightarrow (x + 1, y - 2)$ is applied to $\triangle ABC$ with coordinates $A(-3, 6)$, $B(-7, 4)$, $C(-2, 1)$.

15. Find the coordinates of A', B', C', and graph $\triangle ABC$ and its image.

16. Find and compare the slopes of AC and $A'C'$; AB and $A'B'$; BC and $B'C'$.

Measure and compare each of the following distances in $\triangle ABC$ and $\triangle A'B'C'$.

17. AC and $A'C'$ **18.** AB and $A'B'$ **19.** BC and $B'C'$

Compare the measures of each pair of corresponding angles in $\triangle ABC$ and $\triangle A'B'C'$.

20. What do you notice about the angle measures?

A translation, given $(x, y) \rightarrow (x - 3, y + 4)$, is applied to the square with coordinates $A(5, -3)$, $B(9, -1)$, $C(7, 3)$, and $D(3, 1)$.

21. Find the coordinates of A', B', C', D'.

22. Test the conclusions that you found in Questions 16–20 for this translation.

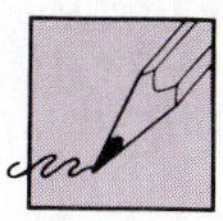

23. Test any of your previous conclusions. Choose a polygon and apply a translation, $(x, y) \rightarrow (x + ?, y + ?)$, of your own choice. Repeat the steps of the previous questions. List your probable conclusions.

SECTION 14–2

REFLECTIONS

WARM-UP ACTIVITY

- Plot the points $A(-3, 3)$, $B(2, -2)$, $C(-6, -3)$. The sense of points A, B, C, written in order, is **clockwise**. The sense of A, C, B, written in order, is **counterclockwise**.
- Find the coordinates of A', B', and C' for the translation $(x, y) \rightarrow (x - 3, y + 2)$. Plot these points.
- Are the points A', B', and C' clockwise or counterclockwise? How about A', C', and B'?

INTRODUCTION

Another type of transformation is a **flip** or **reflection**. You see examples of reflection in the world around you.

SKILLS DEVELOPMENT

To find the **reflection** or **flip image** of a figure like $\triangle ABC$, a **line of reflection** is drawn. The **reflection image**, $\triangle A'B'C'$, is what you would see if you place a mirror along the **reflection line**, ℓ.

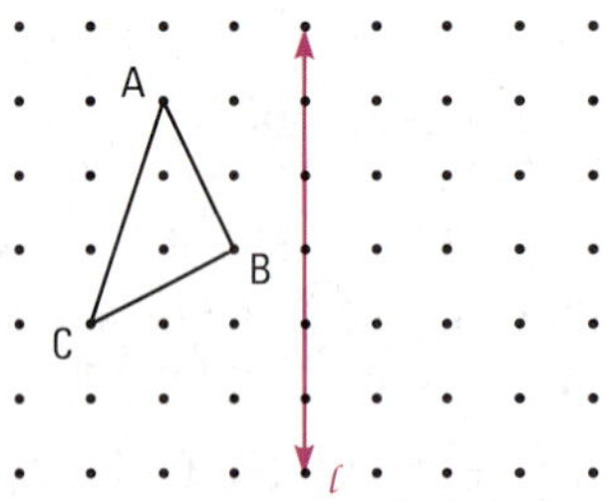

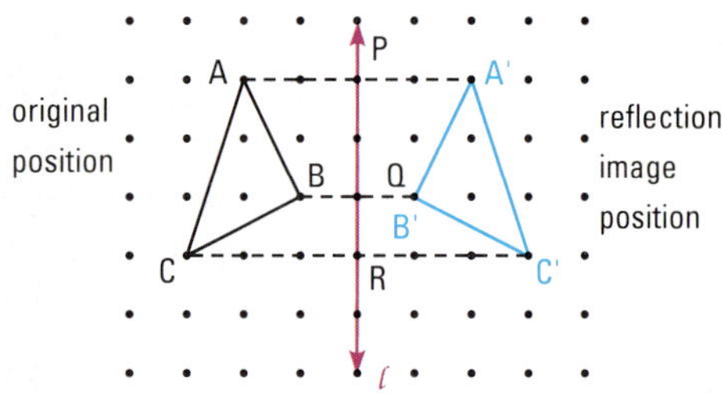

If you measure angles and line segments, you will find the following relations between the original position and the reflection image.

$$AP = A'P \qquad BQ = B'Q \qquad CR = C'R$$

AA', BB', CC' are perpendicular to the reflection line.

To construct a reflection image, use the properties listed above.

Example 1

Find the reflection image of $\triangle ABC$ in line ℓ.

Solution Step 1: Find the reflection image of each vertex of $\triangle ABC$.

Step 2: Draw $\triangle A'B'C'$.

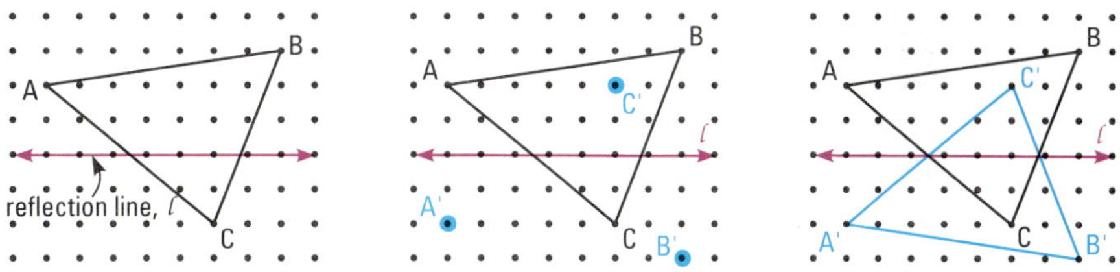

In this section, you will explore the properties that apply to reflections. Again you will see how your skills with coordinate geometry apply to reflections and their properties.

$\triangle A'B'C'$ is the reflection image in the *y*-axis of $\triangle ABC$. Compare the coordinates of the corresponding vertices.

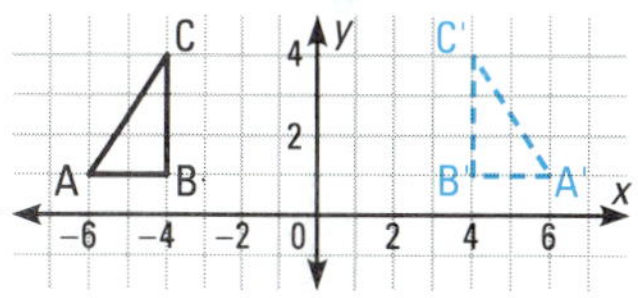

$A(-6, 1)$	$A'(6, 1)$
$B(-4, 1)$	$B'(4, 1)$
$C(-4, 4)$	$C'(4, 4)$

The reflection in the *y*-axis can be shown using mapping notation.
$(x, y) \rightarrow (-x, y)$

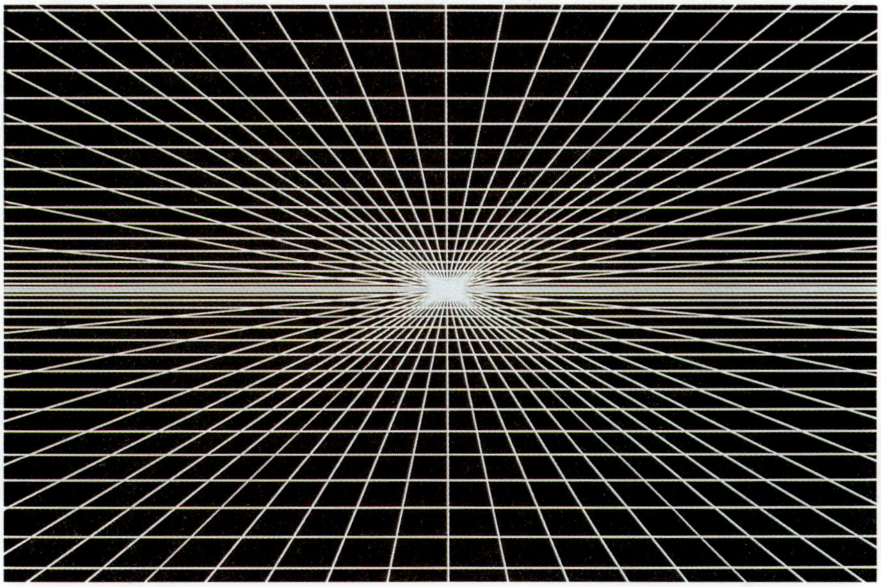

EXERCISES

Read each of the following reflections. (Do you need a mirror?)

1. ?slwarc dna ,depirts ,der si tahW
2. ?1 ni noitseuq eht ot rewsna eht wonk uoy oD
3. .relwarc depirts der a si 1 ni rewsna ehT

Use a reflecting image to tell the time.

4. 
5. 
6. 

Use grid paper. Draw the reflection image of each figure.

7. 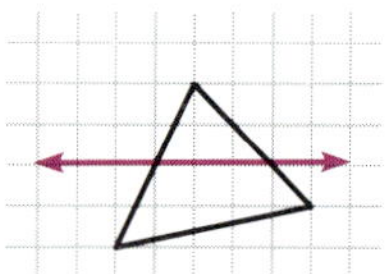
8. 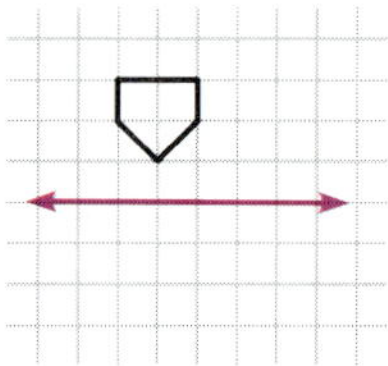
9. 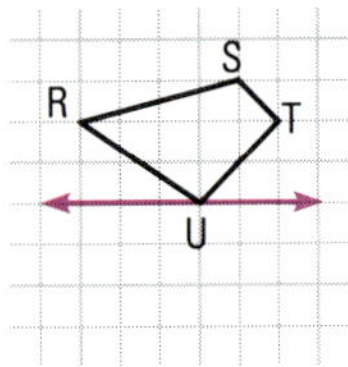

10. In Exercise 9, name the line segments of the reflection and of the original figure that are equal.
11. What point is its own reflection?

Use the properties of a reflection. Copy each figure and find the reflection image.

12. 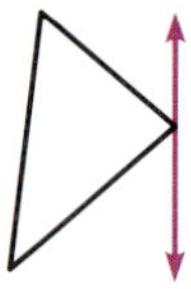
13. 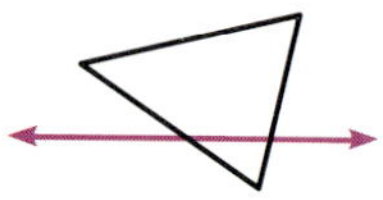

Use grid paper. A flip or reflection line is shown for each design. Find the reflection of each design in relation to the flip line.

14. 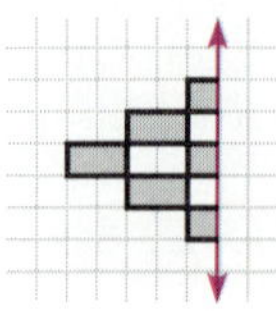
15. 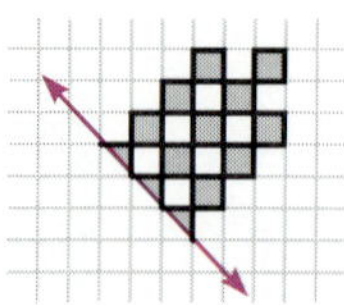

Copy and complete the chart for each reflection.

	Original Point	Image Point	Line of Reflection
16.	(4, ?)	(?, 5)	x-axis
17.	(?, 5)	(3, ?)	y-axis
18.	(−6, −3)	(−6, 3)	?
19.	(7, 2)	(−7, 2)	?

$\triangle ABC$ is reflected in the y-axis. Its coordinates are $A(2, 7)$, $B(3, 1)$, and $C(-4, 4)$.

20. Find the coordinates of $\triangle A'B'C'$.

21. Draw the original figure and the image on the Cartesian plane. What do you notice?

22. Compare the slopes of AB and $A'B'$; BC and $B'C'$; CA and $C'A'$. What do you notice?

23. Compare the lengths of AB and $A'B'$; BC and $B'C'$; CA and $C'A'$. What do you notice?

24. Reflect $\triangle$ABC in the x-axis. Then repeat your investigations as in Questions 20–23.

Based on the results above, write a probable conclusion about

25. the slopes of corresponding line segments of a figure and its image after a reflection through an axis.

26. the length of corresponding line segments of a figure and its image after a reflection through an axis.

A quadrilateral $PQRS$ has coordinates given by $P(-6, 4)$, $Q(-4, 5)$, $R(2, 2)$, $S(-7, -3)$. Reflect quadrilateral $PQRS$ in the y-axis.

27. Draw the figure and its image.

28. Compare the sense of the vertices of the original figure and the image. (Are they clockwise or counterclockwise?)

Make a summary of all the properties you found about reflections.

29. In a reflection, what changes? What is invariant (does not change)?

30. Compare the reflection properties with those for translations.

SECTION 14–3

ROTATIONS

WARM-UP ACTIVITY

A piece of copper has the shape shown at right.

How can it be divided into three congruent pieces?

INTRODUCTION

In the previous sections, you worked with the transformations of translations and reflections. Another transformation is a **rotation**. Skills and concepts related to rotations are important in the study of geometry.

Imagine a spinning wheel, the whirling blades of a helicopter, or a circular ride in an amusement park. Each suggests a turn or rotation about a point.

A turn or rotation about a point can be applied to a figure to obtain its **turn** or **rotation image**.

SKILLS DEVELOPMENT

Point O is called the **rotation** or **turn center**. The amount of rotation is 90° clockwise. A rotation can be clockwise or counterclockwise and any number of degrees.

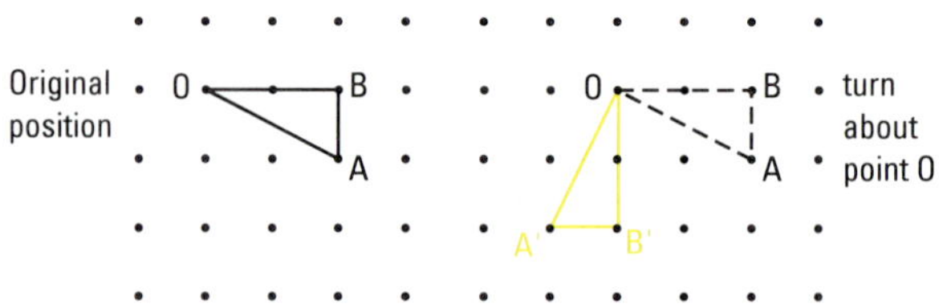

Example

A figure is shown. Find its turn image about the rotation center O for a rotation angle of 90° counterclockwise.

Solution

Step 1: Trace the figure.

Step 2: Turn the figure to obtain the position of the image.

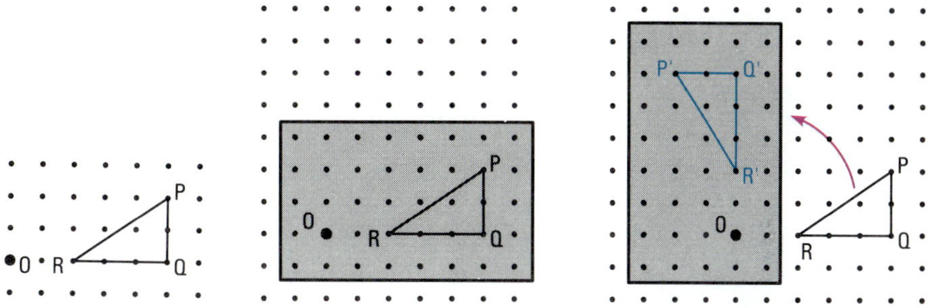

Use your coordinate geometry skills to explore properties of rotations.

$\triangle ABC$ is shown with coordinates.

$\triangle ABC$ is rotated and the image triangle, $\triangle A'B'C'$ is shown.

$\angle C'OC$ is referred to as the angle of rotation.

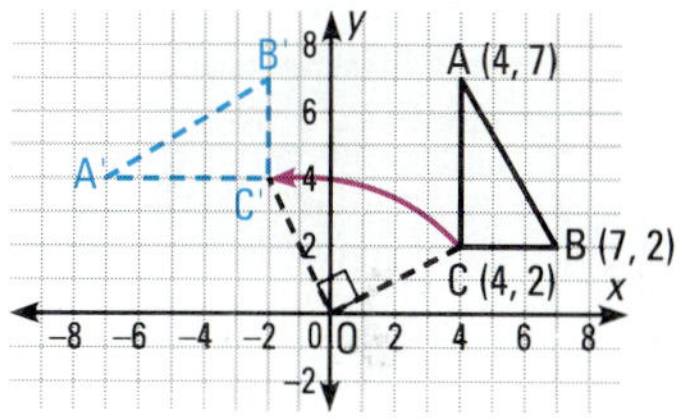

Remember, indicate clearly

- amount of rotation 90°
- direction of rotation counterclockwise
- rotation center origin

As you explore rotations in the exercises, remember to ask yourself:

- What has changed?
- What has not changed?

EXERCISES

A figure *ABCD* is rotated about the turn center *O*. What is the image of each point about the turn center?

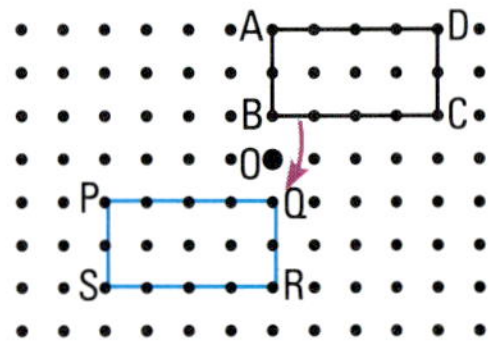

1. *A* **2.** *B* **3.** *C* **4.** *D*

For each figure, the original figure and the turn image are shown. What is the amount of each turn in degrees?

5.

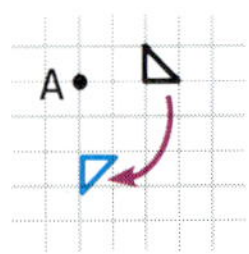

6.

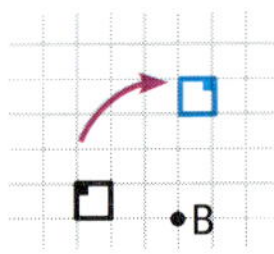

7.

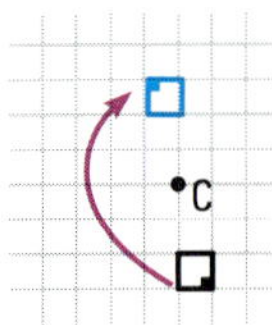

8.

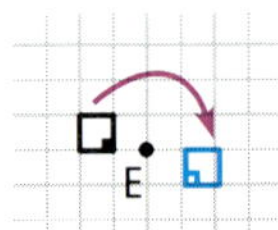

Find the image about each turn center, *O*, for the turn angle given.

9.

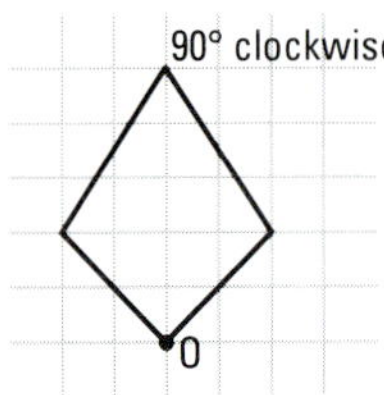

10.

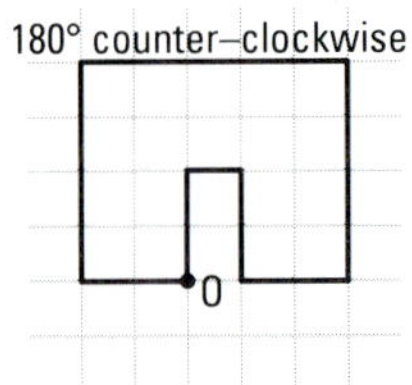

11.

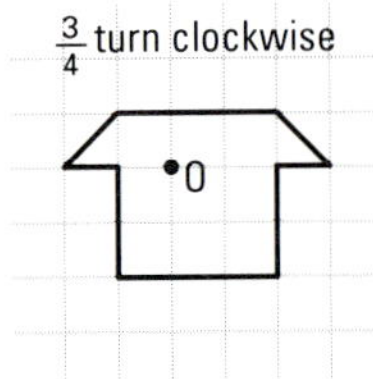

Use dot or square paper to answer the following.

12. A right triangle is rotated $\frac{1}{2}$ turn about the point *O*. What figure is obtained? *O* is the midpoint of *AC*.

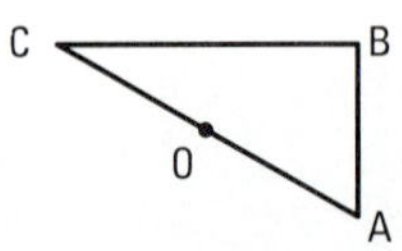

O is the midpoint of AC.

13. An isosceles right triangle is rotated $\frac{1}{2}$ turn about *O*. What figure is obtained? *O* is the midpoint of *AC*.

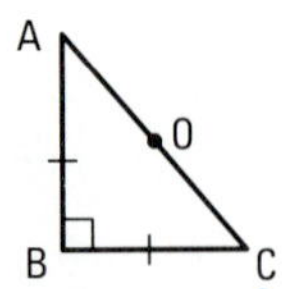

O is the midpoint of AC.

Often the angle of rotation is expressed in another way, as you have seen. A half turn ($\frac{1}{2}$ turn) is 180°. A quarter turn ($\frac{1}{4}$ turn) is 90°.

Often an integer is used to show a rotation.

+90° means 90° clockwise
−180° means 180° counterclockwise

Copy each figure. Draw the image for each rotation around the origin.

14. +90°

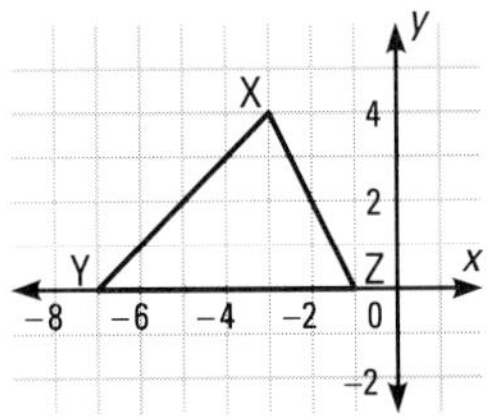

15. −180°

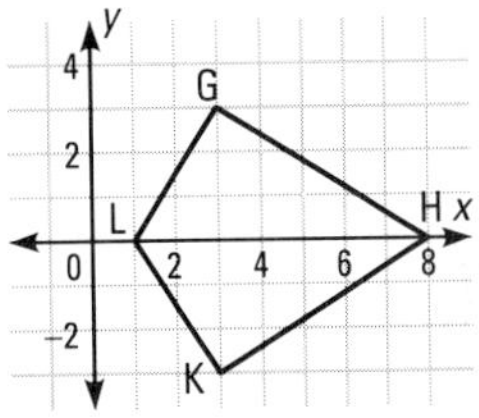

16. Write a mapping to show each rotation above.

Use the previous figures.

17. Compare the measure of corresponding angles and sides of the original and image figure.

18. Compare the order of the vertices of the original and image figure. (Are they clockwise or counterclockwise?)

19. Compare the slopes of the corresponding sides of the original figure and the image figure.

20. Based on your results, when $\triangle ABC$ is rotated to $\triangle A'B'C'$ what changes? What does not change?

21. Use other figures and rotate them to test your answers.

SECTION 14–4
SYMMETRY

WARM-UP ACTIVITY

A REVIEW

1. Identify each of the following as a translation, a reflection, or a rotation.

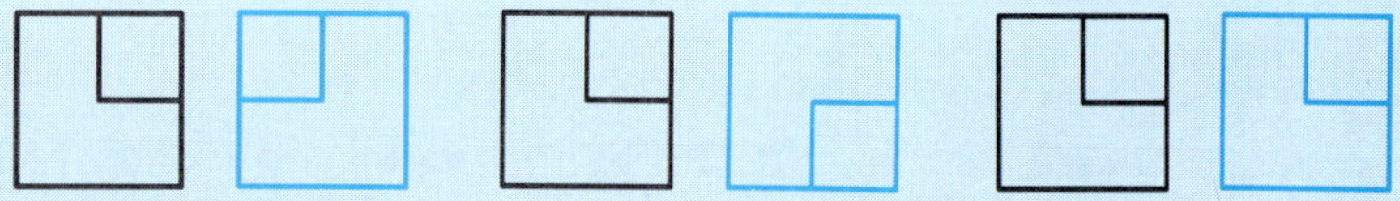

2. Construct the image of the figure for each translation.

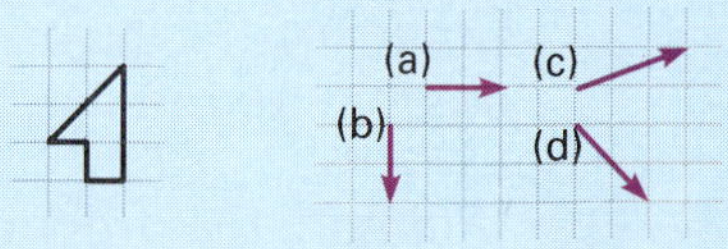

INTRODUCTION

How are the clown and snowflake pictures related?

For each figure, a reflection line can be drawn. One part of the figure can be reflected on the other part so that they match. The matching parts are **congruent**. Each figure is said to have **reflectional symmetry** or **line symmetry**. The **line of symmetry** is *ST*.

SKILLS DEVELOPMENT

A figure may have more than one line of symmetry.

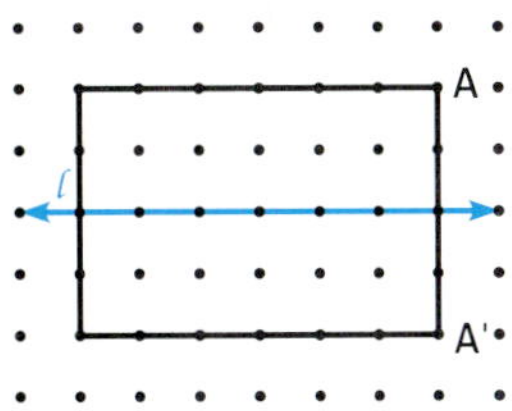

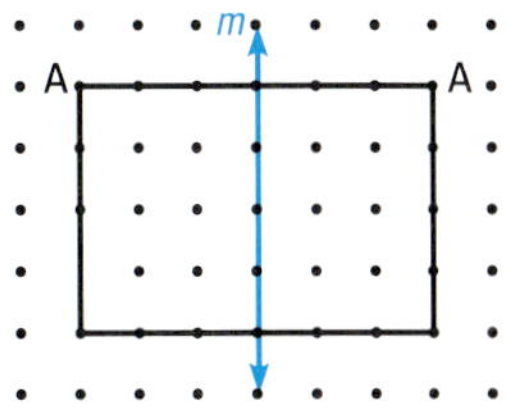

The rectangle is reflected onto itself in the line of reflection ℓ.

The rectangle is also reflected onto itself in the line of reflection m.

For the reflection, the point A is mapped onto the point A'. The lines ℓ and m are the lines of symmetry.

Example 1

Find the lines of symmetry for the given figure. How many are there?

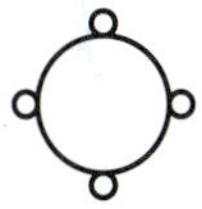

Solution The lines of symmetry for a figure are the lines in which the figure can be reflected onto itself. For example, the given figure has four lines of symmetry.

To find the lines of symmetry, you can use a mirror. If you place a mirror on a line of symmetry, you will see the figure again.

Each of the following symmetrical designs has been constructed by using rotation.

When any of the above designs are turned about a point, the design matches again. If a figure can be mapped onto itself during a turn about a point, the figure is said to have **rotational symmetry** or **turn symmetry**.

For example, the figure on the left above can be turned about its center to match its original position. In a complete rotation, the figure maps onto itself four times. The figure is said to have rotational symmetry of **order four**. Its center point is called the **turn center** or **center of rotation**.

To find the **order** of rotational symmetry of a figure,

- decide on a center of rotation.
- find the number of times that the figure maps onto itself in a complete rotation.

Example 2

What is the order of rotational symmetry for $\triangle ABC$?

Solution

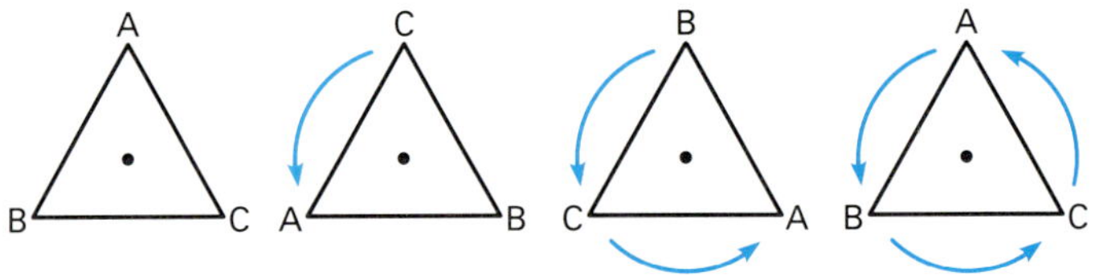

For each turn, you can think of a mapping of points. For one turn, *A* maps onto *B*, *B* maps onto *C*, and *C* maps onto *A*.

The number of times $\triangle ABC$ maps onto itself for one complete rotation is three. $\triangle ABC$ has rotational symmetry of order three.

EXERCISES

Draw each kind of triangle. How many lines of symmetry does each triangle have?

1. isosceles **2.** scalene **3.** equilateral

How many lines of symmetry does each figure have?

4. A **5.** H **6.** 8 **7.** 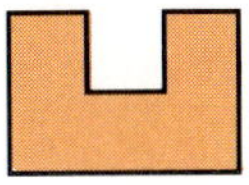 **8.** 

9. Construct a figure with four lines of symmetry.

A **regular** polygon is one that has all angles equal and all sides equal. How many lines of symmetry does each regular polygon have if that polygon has

10. four sides? **11.** five sides? **12.** six sides? **13.** eight sides?

For each of the following figures, decide how many lines of symmetry there are. Which of the figures have only rotational symmetry (order greater than one)?

14. 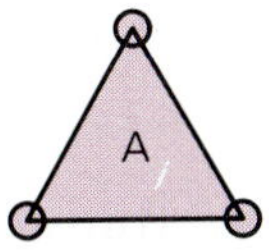

15. 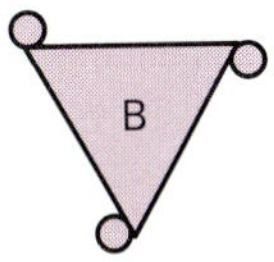

16. 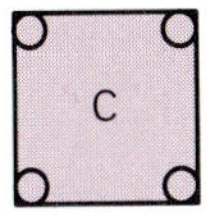

17. 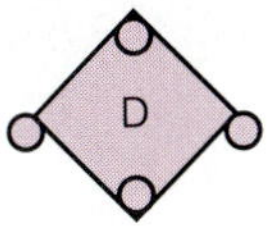

18. 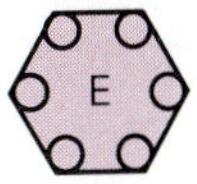

19. 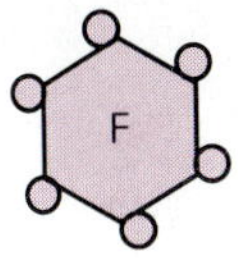

A solid figure has a plane of symmetry as shown. Each part, A and B, has the same shape and size.

What if you cut each of the following solids? Describe the planes of symmetry. How many planes of symmetry does each have?

20. 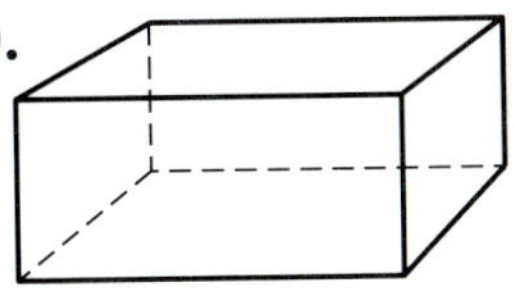 **21.** 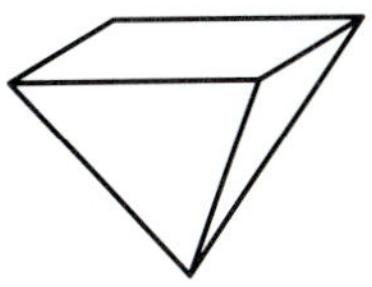 **22.** 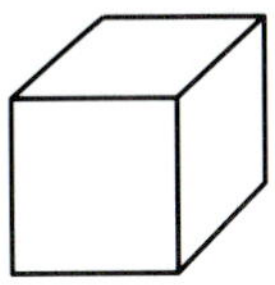

SECTION 14–5

COMBINING TRANSFORMATIONS

WARM-UP ACTIVITY

Some letters have a vertical line of symmetry: A, H, I, M are a few. List all such letters.

What is the longest sentence you can make with the letters that have a vertical line of symmetry? Sample: Timothy hit my Toyota with a hot wax tomato.

INTRODUCTION

So far you have studied three transformations: translations, reflections, and rotations. In all three, the original figure and image figure are congruent, so they are called **congruence transformations**. You have explored the properties of these transformations when they have been applied to a figure one at a time. What if you apply more than one transformation to a figure?

SKILLS DEVELOPMENT

For example, a translation is applied to $\triangle XYZ$. Then a reflection is applied. The resulting image is the **product of the mappings**.

Step 1
A translation given by $(x, y) \rightarrow (x + 3, y + 2)$ is applied to $\triangle XYZ$.

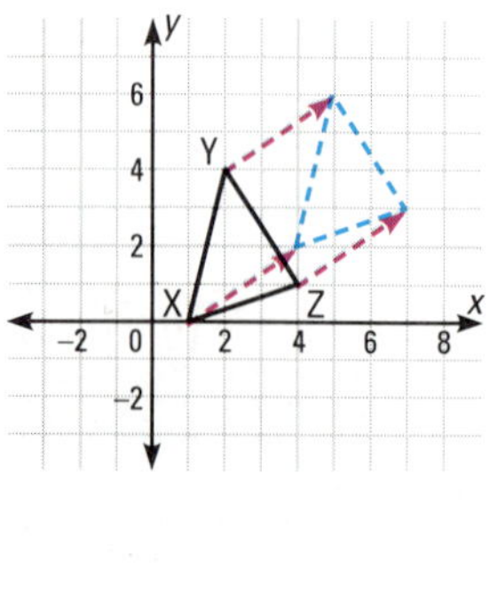

Step 2
The translation is followed by a reflection in the x-axis to obtain $\triangle X'Y'Z'$.

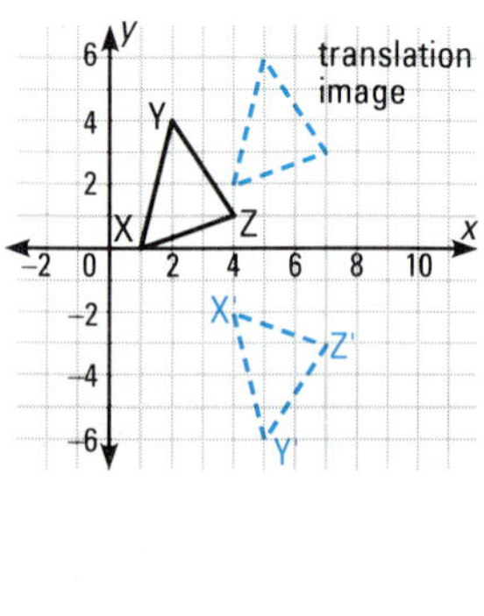

Example 1

A product of two mappings is described by the translation $(x, y) \rightarrow (x + 2, y - 1)$ and a reflection about line k. Draw the image of $\triangle ABC$.

Solution

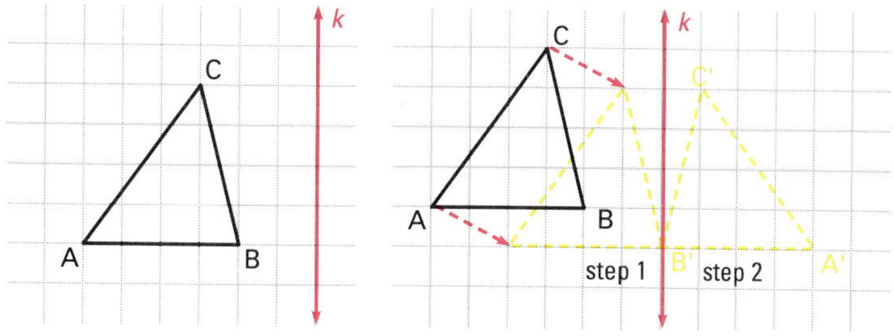

Example 2

Reflect the shape about line k. Then reflect this image about line m.

Solution

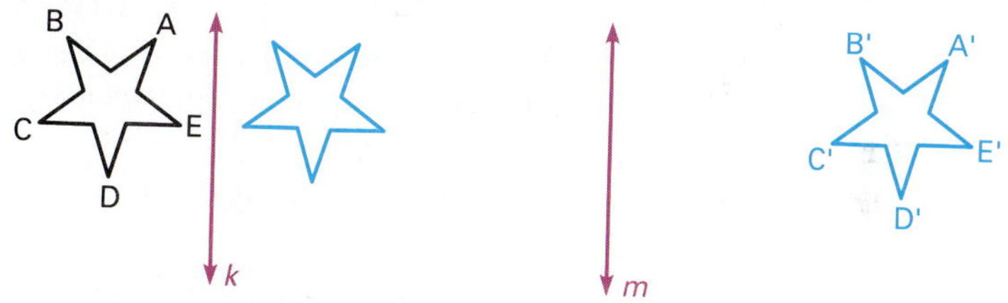

EXERCISES

Find the image for each combination of transformations.

1.

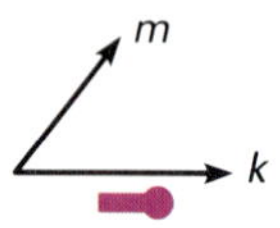

- reflection about line k
- reflection about line m

2.

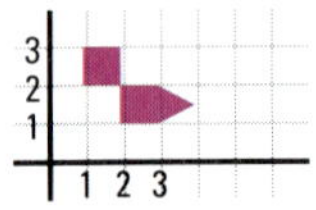

- translation $(x, y) \rightarrow (-x, y + 1)$
- translation $(x, y) \rightarrow (x + 1, -y)$

Two reflection lines, m_1 and m_2, are parallel as shown.

3. Find the image of each figure reflected in line m_1.

a.

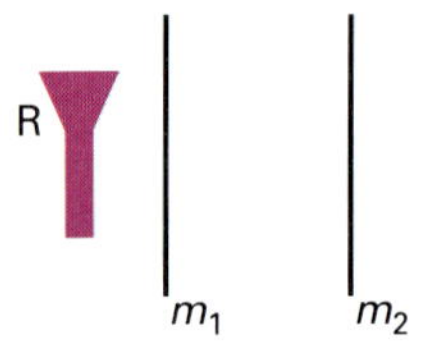

b.

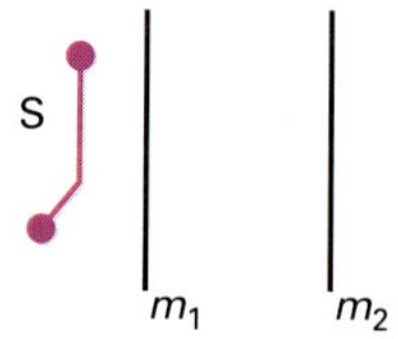

4. Construct the images of the answers for Exercise 5 about line m_2.

5. What single transformation could be used to obtain the images?

Two reflection lines, p_1 and p_2, are perpendicular as shown.

6. Find the image of each figure reflected in line p_1.

a.

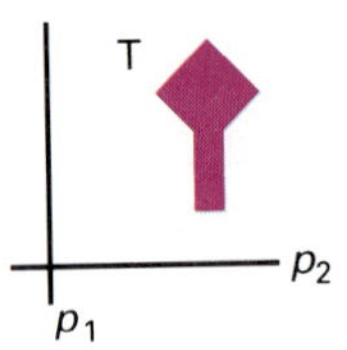

b.

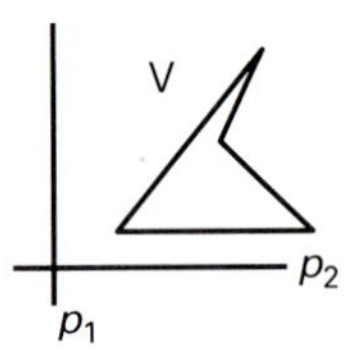

7. Construct the images of the answer for Exercise 8 reflected about line p_2.

8. What single transformation could be used to obtain the images?

Given $\triangle ABC$ as shown.

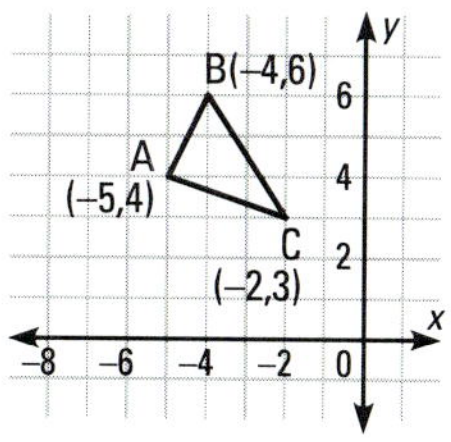

9. Find the image of $\triangle ABC$ translated by the mapping $(x, y) \rightarrow (x, y - 3)$.

10. Reflect the image of $\triangle ABC$ in the y-axis. Write the vertices of the final triangle.

$\triangle PQR$ has vertices $P(-4, -6)$, $Q(-2, -2)$, and $R(-3, 5)$. For a glide reflection, the translation must be parallel to the line of reflection. A glide reflection is described by
a translation $(x, y) \rightarrow (x, y + 2)$, and
then a reflection in the y-axis.

11. Draw the glide reflection of $\triangle PQR$.

12. Find the length of corresponding sides of the triangles and its image. What do you notice about your results?

13. Find the measure of corresponding angles of the triangle and its image. What do you notice about your results?

14. Find the slopes of corresponding sides of the triangle and its image. What do you notice about your results?

What if you reverse the steps in the previous question? Try it. A combination of transformations is described by

- a reflection in the y-axis.
- a translation $(x, y) \rightarrow (x, y + 2)$.

15. Draw the image of the two transformations of $\triangle PQR$.

16. Is your result the same as for Question 11?

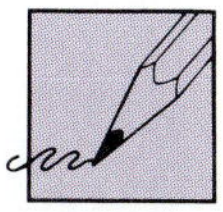

17. Use your results to explore glide reflections. When a glide reflection is applied to a figure

- what changes?
- what does not change?

SECTION 14–6

TILING AND TRANSFORMATIONS

WARM-UP ACTIVITY

How many colors?

Copy the map and color the different "countries" so that no two countries that share a common boundary have the same color. Make another map and try it again. What is the smallest number of colors you need?

INTRODUCTION

A design in which one or more polygons are used to cover a plane is called a **tiling pattern.**

In this design, the basic shape is an isosceles triangle. Translation and reflection are used to create the tiling pattern.

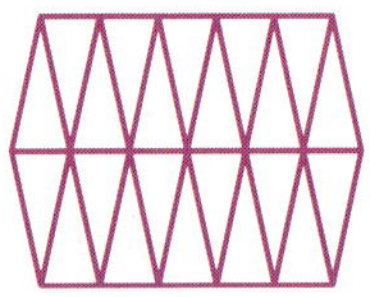

In this design, a rhombus and an equilateral triangle are the basic shapes. Reflections are used to create the tiling pattern.

SKILLS DEVELOPMENT

A tiling of a flat surface is often called a **tessellation**. In a tessellation, the shapes do not overlap. There are no spaces left between the shapes.

Example

A design is based on one tile. Transformations are used to construct the design. Can a tessellation be made using only

a. equilateral triangles? **b.** regular pentagons?

Solutions

a. Begin by cutting out several equilateral triangles and fitting them together. The equilateral triangles will tessellate because each angle equals 60° and the six triangles meet to form a 360° angle.

b. Cut out several regular pentagons and try to tile a surface. The regular pentagons will not tessellate. Each angle of a regular pentagon is 108°. Three pentagons form an angle of 324°—but 360° is needed.

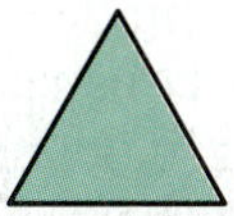

One or more tile shapes can then be used to cover a flat surface. To tile the surface, you repeat the pattern.

One tile is used. Two tiles are used. Three tiles are used.

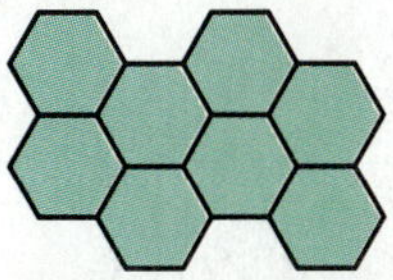

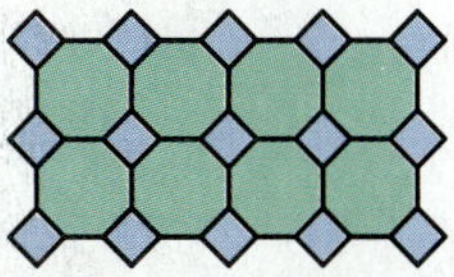

To create tiling patterns, follow these steps.

1. Begin with a figure used for tiling a surface.

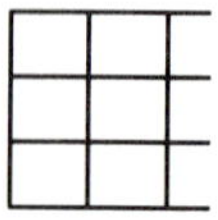

2. Do the same construction on each figure.

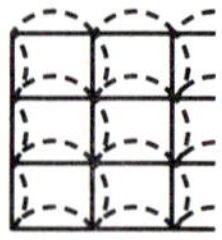

3. Add a design of your own.

EXERCISES

A design can be made from one tile. Flips are used to construct the design. Construct a design based on each of the following tiles.

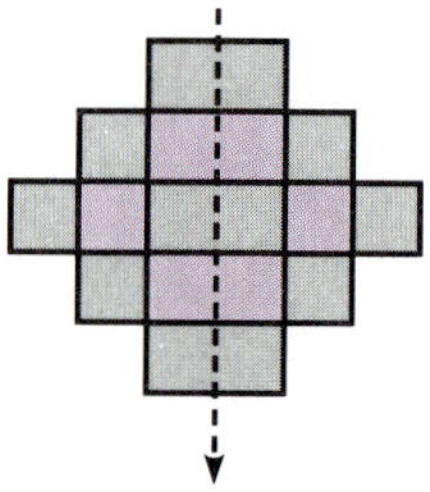

1. 

2. 

3. 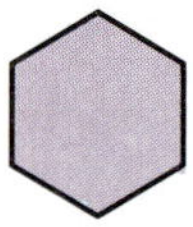

A design can be based on two different tiles. Construct a design for each pair of tiles.

4. 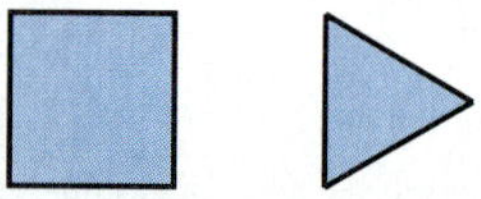

5. 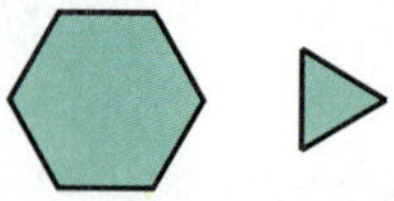

Different tiles can be combined to tile a floor (a plane).

6. For the following tiles, which pairs can be used to tile a floor?

A tetromino is a shape made using four squares sharing at least one common side.

7. How many different tetronimos can you make.

8. Make a design using one tetronimo.

Use dot paper or graph paper to draw the following. Tell what kind of figure is formed.

9. Reflect an isoscles triangle over its base.

10. Reflect a scalene triangle over its longest side.

11. Reflect a square over one side.

12. Reflect a trapezoid over one of its parallel sides.

Of these patterns, which have reflectional symmetry? Which have rotational symmetry?

13.

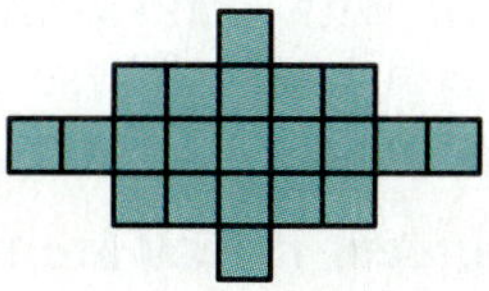

14.

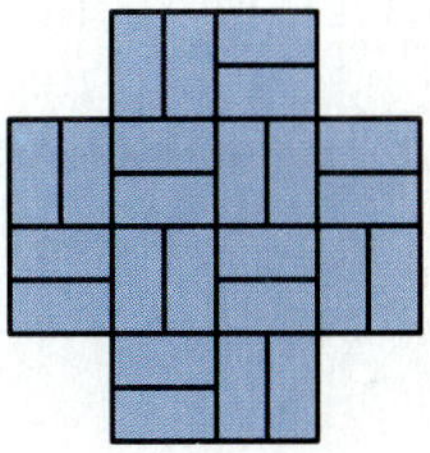

Use each of the following to create a tiling pattern. Use shading to create a design.

15.

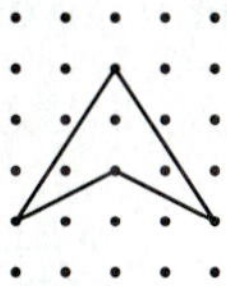

16.

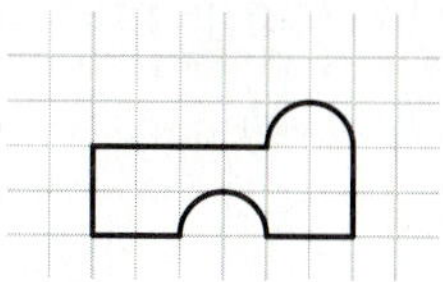

17.

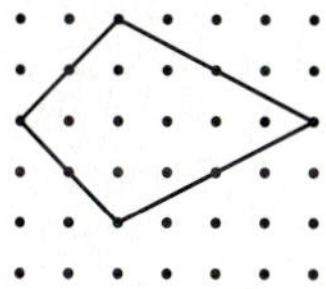

18. Create a tiling pattern of your own. Which transformation did you use?

SECTION 14–7

DILATIONS

WARM-UP ACTIVITY

Two people were playing cards. They played seven games and each won the same number of games. How can this be?

INTRODUCTION

In your earlier work with transformations, the original figure and image figures were congruent. These transformations are called congruence transformations.

- Figures are **congruent** if they are the same shape and size.
- Figures are **similar** if they have the same shape. If you enlarge a picture, the figure in the picture and the enlarged picture are similar.

You can use grid or dot paper to make an enlargement. In the diagram, the enlargement image has a **scale factor** of 2. Namely, since AD measures 1 unit, $A'D'$ measures 2 units.

The enlargement or **scale ratio** is given by $AD{:}A'D' = 1{:}2$.

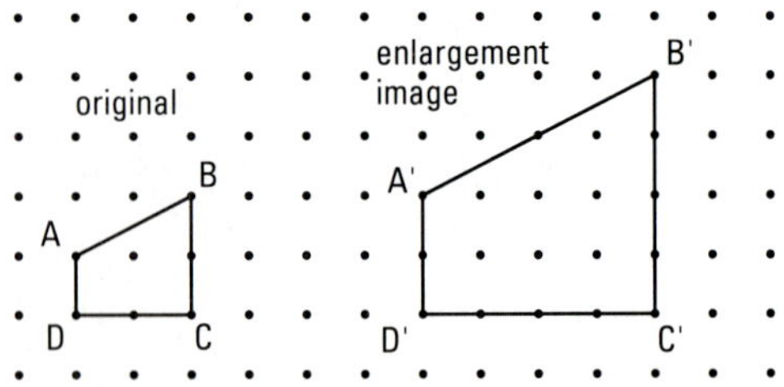

SKILLS DEVELOPMENT

You can enlarge a figure to twice its size by applying an **enlargement factor** of 2. The construction lines show three ways to do this. The enlargement can be written as the mapping $(x, y) \rightarrow (2x, 2y)$.

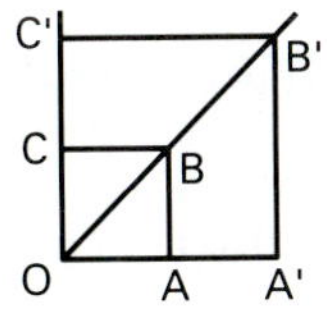

The **center of the enlargement**, O, is a vertex of the figure.
$OA{:}OA' = 1{:}2$

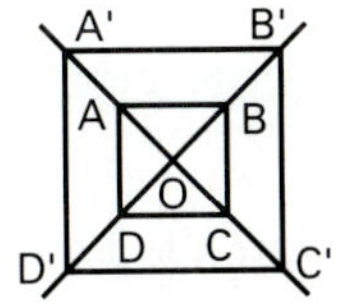

The center of the enlargement, O, is in the interior of the figure.
$OA{:}OA' = 1{:}2$

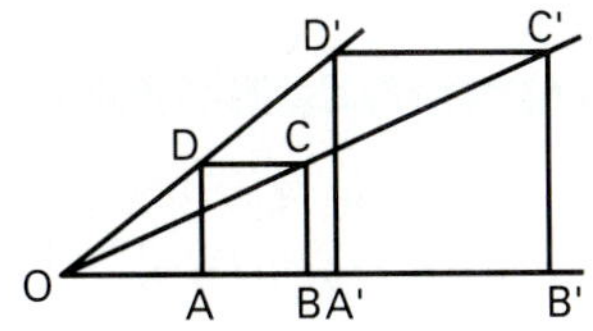

The center of the enlargement, O, is outside the figure.
$OA{:}OA' = 1{:}2$

In this reduction, compare corresponding sides like AF and $A'F'$. The scale ratio is $AF{:}A'F' = 4{:}2$ or $2{:}1$, and the scale factor is $\frac{1}{2}$.

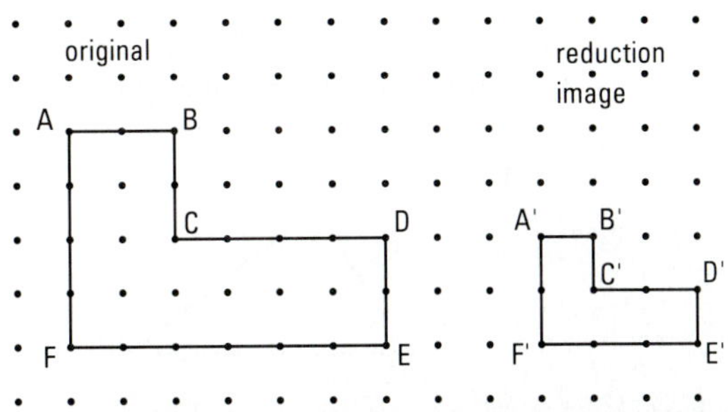

For the above enlargements and reductions, the shape of the figures remains the same, but the size changes. This type of transformation is often called a **size transformation** or **dilation**.

EXERCISES

A square *ABCD* is enlarged as shown.

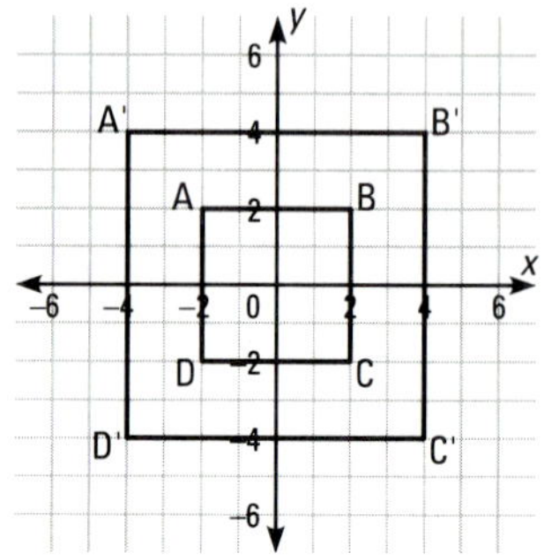

1. Which points correspond?

2. Write a relation rule that relates the coordinates.

3. Compare the lengths of the corresponding sides.

Copy each figure and construct the following enlargement for each figure. Use an enlargement factor of 2 and center of enlargement *E*.

4.

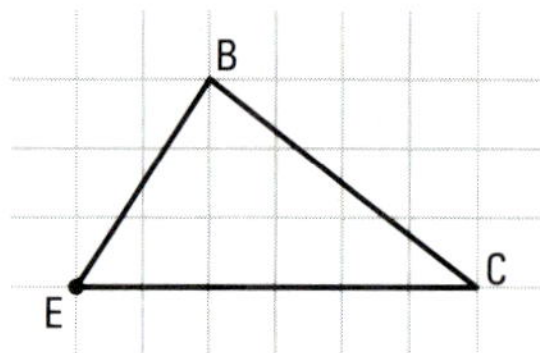

5.

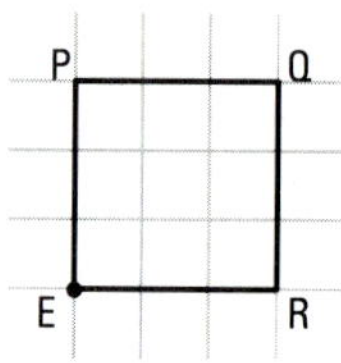

6.

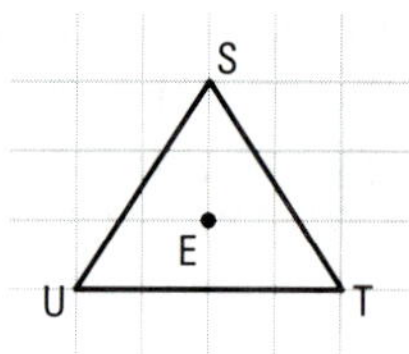

7.

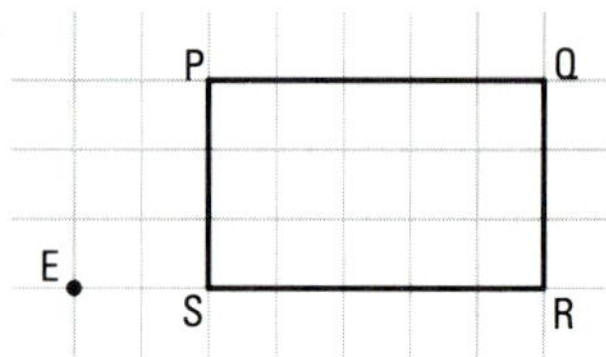

8. Find the perimeter and the area of each figure above.

9. Based on your results, what observations can you make?

Draw $\triangle EFG$ on graph paper. Points *E*, *F*, and *G* are shown clockwise: $E(-2, 2)$, $F(-5, 2)$, $G(-4, 4)$.

10. Find the image of $\triangle EFG$ for the relation rule: double the value of each coordinate.

11. Are the vertices of the image labeled in a clockwise or counterclockwise direction?

12. Compare the original and image figures. Which corresponding sides are parallel?

13. Is the image figure congruent to the original figure? Why?

Use the results from any preceding question.

14. List the properties of size transformation or dilation.

15. When a size transformation is applied to a figure, what remains unchanged? What changes?

To measure the distance across the river, Francis, Michael, Helen, and Bob placed stakes as shown and marked off triangles. Helen located the point E lining up the rock at C and the tree at A.

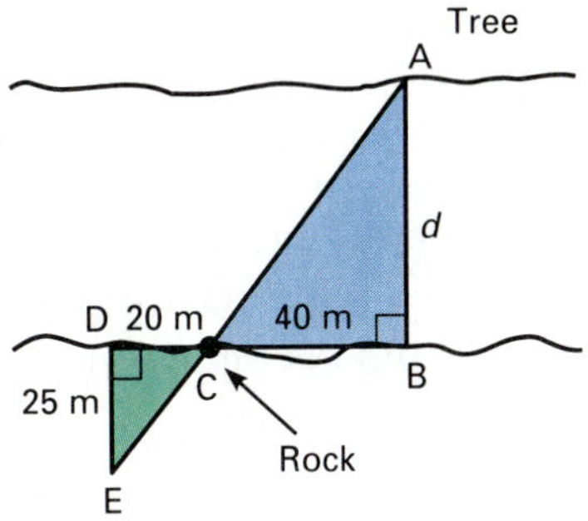

16. Use the diagram to explain why $\angle DCE \cong \angle ACB$.

17. Why are $\triangle ABC$ and $\triangle EDC$ similar?

18. Why is $\frac{AB}{ED} = \frac{BC}{DC}$?

19. Use the proportion to find the width of the river.

Each of the following are sets of points in the original figure and the image for a dilation. Find the missing coordinates.

20. $(1, 3) \rightarrow (2, 6)$
$(-2, 5) \rightarrow (-4, 10)$
$(-2, 1) \rightarrow (?, 2)$
$(x, y) \rightarrow (?, ?)$

21. $(-2, -4) \rightarrow (1, 2)$
$(6, -2), \rightarrow (-3, ?)$
$(?, -4), \rightarrow (1, ?)$
$(x, y), \rightarrow (?, ?)$

22. What is the dilation factor for each of the above dilations?

Draw a square for each diagram and place the rotation center at O as shown.

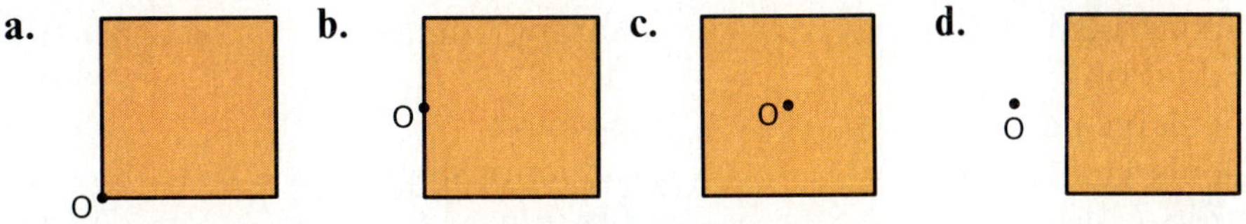

23. For each diagram, find the dilation image for a scale factor of 3.

24. For each diagram, find the dilation image for a scale factor of $\frac{1}{2}$.

CHAPTER 14 SUMMARY

Take a shape; slide it, flip it, rotate it, and enlarge it to create a pattern. In this chapter, you learned about transformational geometry. You explored the properties of transformations using the Cartesian plane.

There are three congruence transformations; translation, reflection, and rotation. Under these transformations, the original figure and its image have the same shape and the same size.

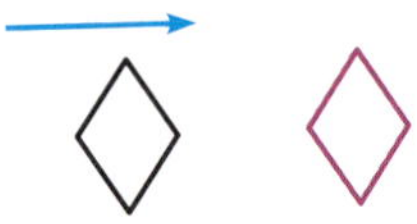

Translation

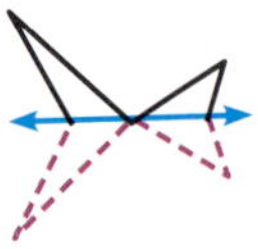

Reflection

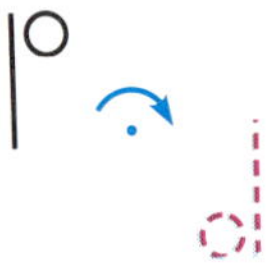

Rotation

You also considered symmetry, both reflectional and rotational. Below are pictures of these symmetries.

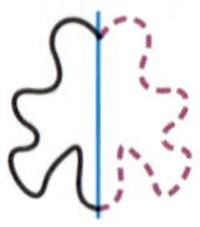

Reflectional symmetry

Rotational symmetry

Having considered these transformations, you learned to put them together. So you explored combining transformations. A translation parallel to the line of reflection followed by a reflection is called a glide reflection. When using a glide reflection, the order in which you combine the transformations does not matter. However, for most transformations you have to be careful to perform them in the order given, since a reverse order may give a different answer.

Beginning with regular polygons, you attempted to tile an area. Some shapes completely filled an area without overlap or gaps. If exactly one shape was used to fill an area, it was called a tessellation.

Having explored the congruence transformation, you studied dilation. Dilation results in a similar image. You reviewed solving triangles using indirect measurement.

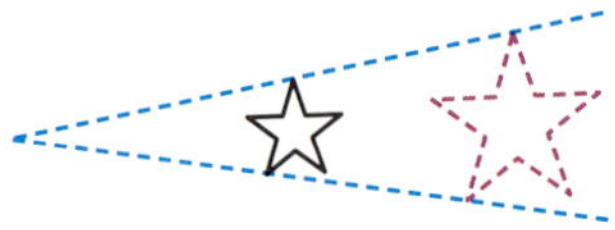

Dilation

VOCABULARY

congruence transformation
congruent
dilation
enlargement
enlargement factor
flip
image
line of reflection
line of symmetry
line of symmetry
mapping notation
order
pre-image
product of mappings
reflection
rotation
scale ratio
similar
size transformation
slide
symmetry
tessellation
tiling pattern
translation
turn

CHAPTER REVIEW

1. The mapping for a translation is given by $(x, y) \rightarrow (x + 2, y - 1)$. Find the vertices of the image $\triangle KPR$.

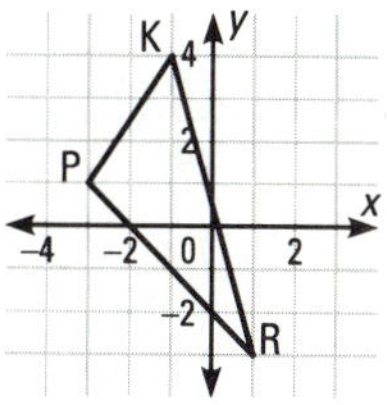

$\triangle RYQ$ has vertices $R(-4, 1)$, $Q(4, 2)$, $Y(0, 4)$. Find the vertices of $\triangle R'Y'Q'$ if $\triangle RYQ$ is reflected in the

2. x-axis.
3. y-axis.
4. Which points are invariant? (Note: Invariant means does not change.)

$\triangle PQR$ has vertices $P(-3, 4)$, $Q(-1, 2)$, and $R(-4, 0)$.

5. Find the vertices of the image of $\triangle PQR$ for a 90° clockwise rotation. Use the origin as the rotation center.
6. Compare the slopes of PQ and $P'Q'$; PR and $P'R'$; QR and $Q'R'$.
7. Compare the senses of the vertices of the original figure and image.

For each figure, write the number of lines of symmetry. Change the figure so that it has only one line of symmetry.

8. 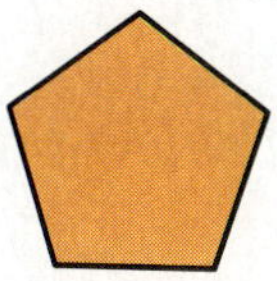

9. 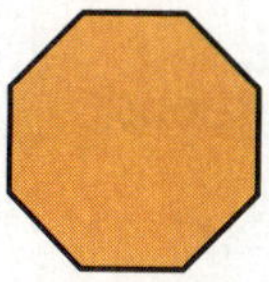

Write the order of rotational symmetry for each figure.

10. 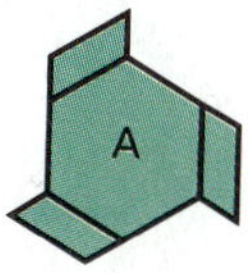

11. 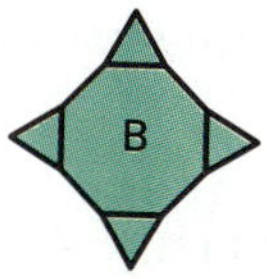

12. Find the image of $\triangle FGH$ given by $G(1, 5)$, $H(4, 3)$, $F(5, 6)$ if it is translated by the mapping $(x, y) \rightarrow (x - 2, y - 3)$ and then reflected in the x-axis.
13. Explain why regular hexagons tessellate.
14. Apply the dilation $(x, y) \rightarrow (2x, 2y)$ to find the image. What do you notice?

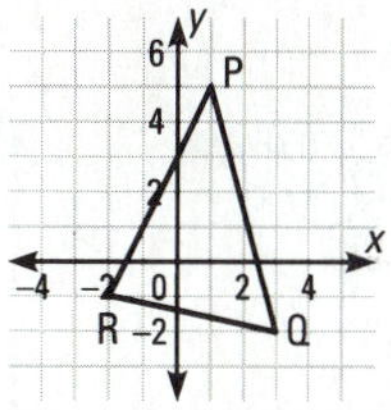

15. Adele's shadow is 4 m. She is 1.8 m tall, and her distance to the flagpole is 12 m. Find the height of the pole.

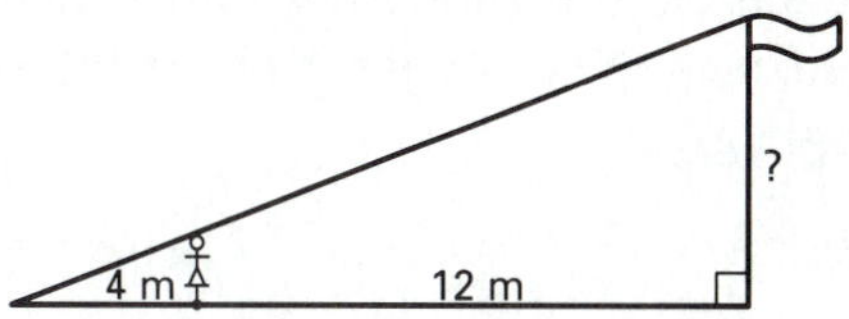

CHAPTER 14 TEST

The mapping for a translation is given by $(x, y) \to (x - 3, y + 5)$.

1. Find the coordinates of the image of $\triangle DEF$ given by $D(4, 0)$, $E(1, 2)$, and $F(3, 5)$.

Reflect figure *ABCDE* in

2. the x-axis **3.** the y-axis

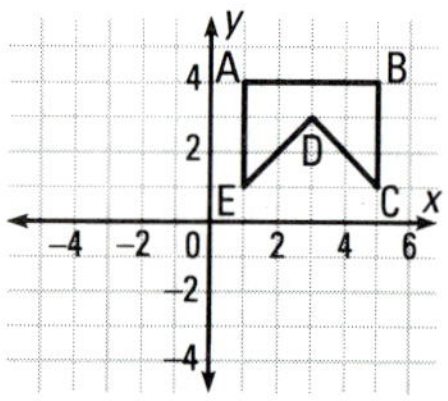

Apply the mapping $(x, y) \to (3x, 3y)$, to $\triangle DEF$.

4. Find the image of $\triangle DEF$.

5. When this transformation is applied what changes? What is invariant?

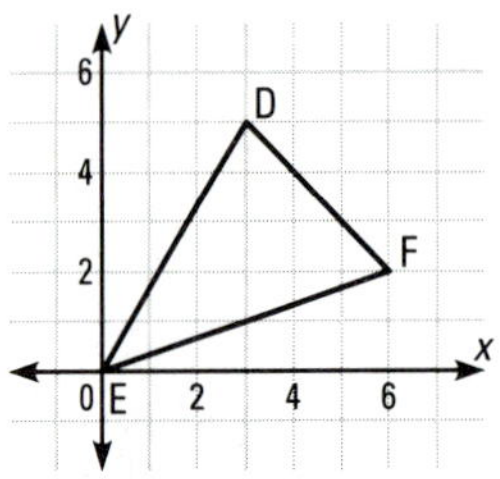

Copy each of the following and draw the rotation image about center O for the rotation angle given.

6. **7.** **8.**

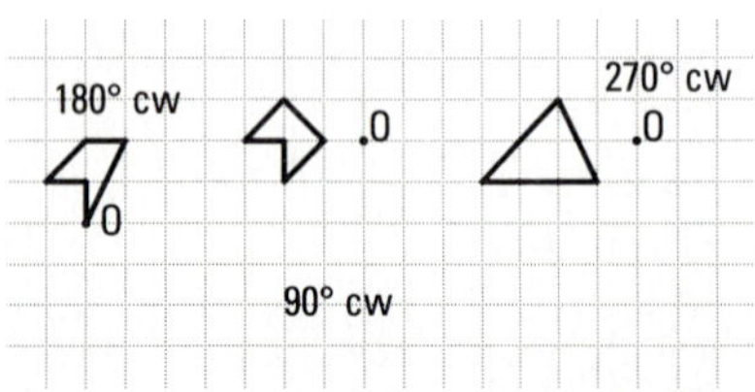

Draw a quadrilateral with the following number of lines of symmetry.

9. 4 **10.** 2 **11.** 0

What is the order of rotational symmetry for each figure?

12.

13.

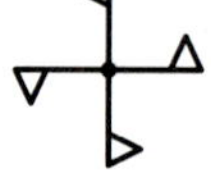

14.

Apply each transformation to the figure shown. Name the vertices of the image figure.

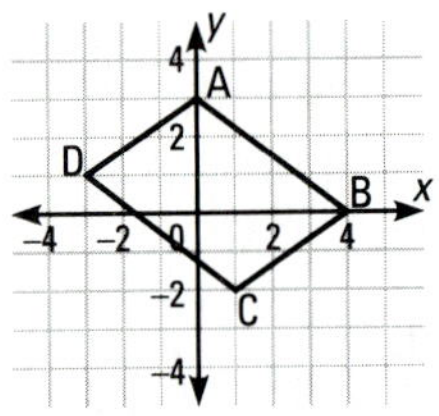

15. $(x, y) \to (-x, y)$ **16.** $(x, y) \to (x, -y)$

17. $(x, y) \to (x + 4, y - 3)$

18. $(x, y) \to (x + 1, y + 5)$, then reflected in the x-axis.

Tell whether the following regular polygons will or will not tessellate.

19. square **20.** equilateral triangle

21. pentagon **22.** hexagon

Which figures appear to be

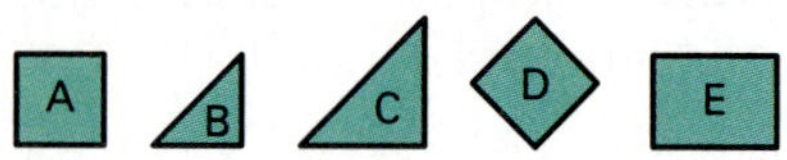

23. congruent?

24. similar?

CUMULATIVE REVIEW CHAPTERS 11–14

Refer to the graph below for Exercises 1–6.

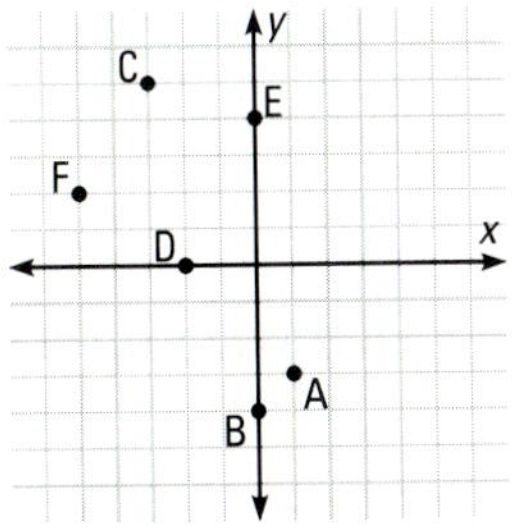

Give the coordinates of each point as an ordered pair.

1. C

2. E

3. D

Name the point on the graph that corresponds to the ordered pair.

4. $(1, -3)$

5. $(-5, 2)$

6. $(0, -4)$

Rewrite the equation in the slope-intercept form. Then name the slope and y–intercept of the line defined by each equation.

7. $5x + y = 11$

8. $2x - y = 6$

A jar contains 5 red marbles, 8 green marbles, and 2 white marbles. What is the probability that:

9. you will pick a white marble?

10. you will not pick a red marble?

11. you will pick a green marble and then a red marble (if the first marble is not replaced)?

Find the mean, median, range, and mode for each set of data.

12. 55, 59, 47, 44, 39

13. 87, 72, 91, 75, 87, 92

The mapping for a translation is given by $(x, y) \rightarrow (x + 2, y - 3)$.

14. Find the vertices of the image of $\triangle ABC$ given $A(-1, 2)$, $B(3, -2)$, $C(-3, -1)$.

REFERENCE SECTION

This Reference Section contains much useful information that will help you solve problems throughout this course. Being able to solve problems is a very important skill. However, being able to find the information needed to solve the problems is of equal importance.

Refer to the Reference Section Table of Contents or Reference Section Index to find where needed information is located.

REFERENCE SECTION TABLE OF CONTENTS

General Information

Mathematics Information

ADVERTISEMENTS

Housing

Apartment for rent
$750/month, 2 bedrooms, all appliances. Heat/electric $50/month. Optional garage $90/mo. Available immediately.
Call 1-708-555-1580.

Apartment for rent
All appliances included, heat/electric/parking included. 2 bedrooms $900/month.
Call 1-815-555-1234.

Apartment for rent
$400/month, 1 bedroom. Gas heat and electric $75/month. Parking (outdoor) included. Available immediately. Call 1-312-555-9664.

Apartment for rent
Lovely 1 bedroom. Lake view. Garage optional. Gym membership optional. Only $550/month.
Call 1-312-555-9876.

Mail Orders

School T-Shirts
Send us your school logo and we will print it on a T-shirt. Allow 6 weeks for delivery. Sizes S, M, L, XL

1236 **$24.95**

Compact Disk Holder
Holds up to 24 compact disks in a neat, orderly way. Can be personalized—specify name.

Plain 3672 **$12.95**
Personalized 3674 $1.50 per letter + $12.95

Shipping and Handling Charges

Total Order	S/H Charges
0–$20.00	$2.95
$20.01–$40.00	$4.55
$40.01–$60.00	$6.00
$60.01–$80.00	$7.75
$80.01–$99.99	$9.00
$100 and over	$9.99

Key Chains
Your favorite baseball team's logo on a key chain. Specify team. Allow 4 weeks for delivery

7235 **$6.95**

Your own set of baseball cards.
796 in all. State year desired. Yes! You can pre-order next year's cards now!

5254 Next year's set	**$29.95**
5243 This year's set	**$27.95**
5241 Last year's set	**$33.95**

Specify Year!

Insurance cost
Special/per order $1.99

PA residents add 6% sales tax.

Overnight/RUSH handling, add just $2.98

BUILDINGS

United States Skyscrapers

City	Building	Stories	Height ft	Height m
Chicago	Sears Tower	110	1454	443
New York	World Trade Center	110	1377	419
New York	Empire State	102	1250	381
Chicago	John Hancock Center	100	1127	343
Houston	Texas	75	1002	305
Los Angeles	United California Bank	62	858	261
San Francisco	Transamerica Pyramid	61	853	260
Pittsburgh	U. S. Steel Headquarters	64	841	256
Boston	John Hancock Tower	60	790	241
Minneapolis	IDS Tower	57	775	236

FOOD AND NUTRITION

Source for all Nutritional Information: Food and Nutrition Board, National Academy of Sciences National Research Council, Washington, DC

Recommended Daily Dietary Allowances

	Age (yr)	Weight (kg)	Weight (lb)	Height (cm)	Height (in)	Energy (kcal)	Protein (g)
Infants	0.0–0.5	6	13	60	24	kg × 115	kg × 2.2
	0.5–1.0	9	20	71	28	kg × 105	kg × 2.0
Children	1–3	13	29	90	35	1300	23
	4–6	20	44	112	44	1700	30
	7–10	28	62	132	52	2400	34
Boys	11–14	45	99	157	62	2700	45
	15–18	66	145	176	69	2800	56
Girls	11–14	46	101	157	62	2200	46
	15–18	55	120	163	64	2100	46
Men	19–22	70	154	177	70	2900	56
	23–50	70	154	178	70	2700	56
	51–75	70	154	178	70	2400	56
	76+	70	154	178	70	2050	56
Women	19–22	55	120	163	64	2100	44
	23–50	55	120	163	64	2000	44
	51–75	55	120	163	64	1800	44
	76+	55	120	163	64	1600	44

U.S. Recommended Daily Dietary Allowances for Teenagers (15–18)

	Protein (g)	Calcium (mg)	Iron (mg)	Vitamins A (IU)	B1 (mg)	B2 (mg)	Niacin (mg)	C (mg)	D (IU)
Males	56	1200	18	5000	1.4	1.6	18	50	400
Females	46	1200	18	5000	1.2	1.4	16	60	200

Calorie Count Of Selected Foods

Food	Approximate Amount	Food Energy (kcal)
Dairy Products, Fats		
Butter or Margarine	$\frac{1}{2}$ cup	815
Butter or Margarine	1 pat	35
Buttermilk	1 cup	100
Cheese:		
American	1 oz	105
Cottage	1 cup	235
Cream	1 oz	100
Parmesan	1 oz	130
Cream (half and half)	1 tbsp	20
Cream, sour	1 tbsp	25
Fats, vegetable	1 tbsp	110
Milk, skim	1 cup	85
Milk, whole	1 cup	150
Oils, salad or cooking	1 tbsp	120
Sherbet	1 cup	270
Yogurt (whole milk)	1 cup	140
Breads, Cereals, Desserts, Pastas, Snacks		
Bagel (egg or water)	1	165
Biscuit (2″ diam.)	1	105
Breads:		
Rye	1 slice	60
White	1 slice	70
Whole wheat	1 slice	65
Breakfast Cereals:		
Corn flakes	1 cup	95
Oatmeal	1 cup	130
Puffed rice	1 cup	60
Puffed wheat	1 cup	55
Shredded wheat	1 biscuit	90
Crackers, graham	4	110
Crackers, saltine	4	50
Gelatine dessert	1 cup	140
Honey	1 tbsp	65
Jams and preserves	1 tbsp	55
Macaroni, plain	1 cup	190
Macaroni w/cheese	1 cup	430
Muffin	1	120
Noodles (enriched)	1 cup	200
Pizza, cheese	1 med. slice	145
Popcorn (popped plain)	1 cup	25
Pretzel, sticks	10	10
Roll (hot dog or hamburger)	1	120
Spaghetti, plain	1 cup	190
Spaghetti w/meat balls and tomato sauce	1 cup	330
Sugar, granulated	1 tbsp	45
Syrup (table blends)	1 tbsp	60
Fruits and Juices		
Apple, raw ($2\frac{3}{4}$″ diam.)	1	80
Apple juice, unsweetened	1 cup	120
Apricots, canned in syrup	1 cup	220
dried, uncooked	1 cup	340
dried, cooked unsweetened	1 cup	100
Banana, raw	1	100
Blackberries, raw	1 cup	85
Blueberries, raw	1 cup	90

Food	Approximate Amount	Food Energy (kcal)
Cantaloupe, raw	$\frac{1}{2}$	80
Cherries, water packed	1 cup	105
Fruit cocktail, sweetened	1 cup	195
Grapefruit, raw		
white	$\frac{1}{2}$	45
canned in syrup	1 cup	180
Grapefruit juice		
sweetened	1 cup	135
unsweetened	1 cup	100
Grape juice, unsweetened	1 cup	165
Grapes, Thompson seedless	10 grapes	35
Lemon, raw	1	20
Lemonade	1 cup	105
Orange, raw ($2\frac{5}{8}''$ diam.)	1	65
Orange juice, unsweetened	1 cup	120
Peach, raw	1 medium	120
Peaches, canned in syrup	1 cup	200
Pear, raw ($2\frac{1}{2}''$ diam.)	1	100
Pears, canned in syrup	1	195
Pineapple, raw diced	1 cup	80
Pineapple, canned in syrup	1 cup	190
Pineapple juice, unsweetened	1 cup	140
Plum, raw	1	30
Prunes, cooked unsweetened	1 cup	255
Prune juice	1 cup	195
Raisins ($\frac{1}{2}$ oz package)	1 pkg	40
Raspberries, red raw	1 cup	70
Strawberries, raw	1 cup	55
Watermelon, wedge $4'' \times 8''$	1	110
Meat, Eggs, Poultry, Fish		
Bacon	2 slices	85
Beef, ground broiled	3 oz	235
Beef, roast (lean only)	3 oz	165
Beef, lean broiled steak	3 oz	220
Chicken, fried breast wo/bone	3 oz	170
Chili con carne w/beans	1 cup	340
Corned beef (canned)	3 oz	185
Eggs, scrambled in butter	1 large	95
Egg, boiled	1 large	80
Fish and seafood:		
Fish sticks	3 oz	150
Haddock, breaded fried	3 oz	140
Salmon, canned	3 oz	120
Tuna, canned in oil	3 oz	170
Ham, baked	3 oz	245
Ham, boiled	1 oz	65
Hot dog (frankfurter)	2 oz	170
Lamb, leg roasted	3 oz	235
Liver, beef fried	3 oz	195
Pork, chop broiled	2.7 oz	305
Pork, roast	3 oz	310
Pork, sausage links	2	120
Veal, cutlet	3 oz	185
Vegetables and Nuts		
Almonds, chopped	1 cup	775
Beans, green fresh cuts	1 cup	30
Beans, baby lima	1 cup	210
Beans, navy	1 cup	225
Beets, diced or sliced	1 cup	55
Broccoli, frozen chopped	1 cup	50

Food	Approximate Amount	Food Energy (kcal)
Cabbage, cooked	1 cup	30
Cabbage, raw	1 cup	15
Carrot, raw (7″ long)	1	30
Cauliflower, raw chopped	1 cup	31
Celery, diced raw	1 cup	20
Collards, from leaves	1 cup	65
Corn, canned	1 cup	175
Corn, sweet	1 ear	70
Cucumber, raw	1	20
Lettuce, iceberg chopped	1 cup	5
Mushrooms, raw sliced	1 cup	20
Onions, raw chopped	1 cup	65
Parsnips, diced	1 cup	100
Peanut butter	1 tbsp	95
Peanuts, roasted w/oil	1 cup	840
Peas, green canned	1 cup	150
Pecan halves	1 cup	810
Pepper, green raw	1	15
Pickle, sweet gherkin	1 small	20
Potato, baked	1 large	145
Potato, boiled	1 medium	105
Potato, french fried	10 pieces	135
Potato, mashed w/milk	1 cup	135
Rice, white instant	1 cup	180
Sauerkraut, canned	1 cup	40
Spinach, fresh	1 cup	40
Sweet potato, baked	1	160
Tomato, raw	1 medium	25
Tomato juice, canned	1 cup	45
Turnip greens, fresh	1 cup	30
Turnips, diced	1 cup	35
Walnuts, black chopped	1 cup	785
Miscellaneous Items		
Barbecue sauce	1 cup	230
Bean with pork soup	1 cup	170
Beef broth	1 cup	30
Bouillon cube	1	5
Club soda, unsweetened	12 oz	0
Cream of chicken soup	1 cup	180
Cream of mushroom soup	1 cup	215
Ginger ale	12 oz	115
Salad dressings:		
Blue cheese	1 tbsp	75
French	1 tbsp	65
Italian	1 tbsp	85
Mayonnaise	1 tbsp	170
Soda, fruit-flavored	12 oz	170
Tartar sauce	1 tbsp	75
Tomato soup w/milk	1 cup	175
Tomato soup w/water	1 cup	90
Vegetarian soup w/water	1 cup	80
White sauce	1 cup	408

Nutrient Value of Selected Food

Food	Portion	Calories (kcal)	Carbohydrates (gram)	Protein (gram)	Fats (gram)	Saturated Fatty Acids (gram)
Apple, raw	1 medium	80	20	tr	tr	0
Bacon, broiled	2 slices	85	tr	4	8	2.5
Beans, green frozen	1 cup	35	8	2	tr	0
Beef, roast	3 oz	165	0	2.5	7	2.8
Bran (40% bran)	1 cup	105	28	1	0	0
Bread:						
White	1 slice	70	13	2	1	0.2
Whole wheat	1 slice	65	14	3	1	0.1
Broccoli, cooked	1 cup	40	7	5	tr	0
Butter	1 pat	35	tr	tr	4	2.5
Carrots, raw	1 medium	30	7	1	tr	0
Cheese, processed	1 oz	82	2	5	6	3.8
Chicken, broiled						
$\frac{1}{2}$ broiler boned	6.2 oz	240	0	42	7	2.2
Coffee, unsweetened	6 oz	2	0	tr	tr	0
Corn flakes w/sugar	1 cup	95	21	2	tr	0
Eggs, scrambled						
in butter	1 large	95	1	6	7	2.8
Ham, boiled	3 oz	245	0	18	19	6.8
Jams & preserves	1 tbsp	55	14	tr	tr	0
Lamb chop	3.1 oz	360	0	18	32	14.8
Lettuce, chopped	1 cup	5	2	tr	tr	0
Macaroni w/cheese	1 cup	430	40	17	22	8.9
Margarine	1 pat	35	tr	tr	4	0.7
Milk:						
Whole	1 cup	150	11	8	8	5.1
Nonfat, skim	1 cup	85	12	8	tr	0.3
Orange	1 medium	65	16	1	tr	0
Juice, frozen	1 cup	120	29	2	tr	0
Pancakes, plain	1 medium	60	9	2	2	0.5
Peanut butter	1 tbsp	95	3	4	8	1.5
Pork chop, broiled	2.7 oz	305	0	19	25	8.9
Potatoes:						
Baked, peeled	1 large	145	33	4	tr	0
Mashed w/milk	1 cup	135	27	4	2	0.7
Raisins	1 cup	420	112	4	tr	0
Rice, instant	1 cup	180	40	4	tr	tr
Spaghetti, plain	1 cup	260	37	0	0	0
Sugar, granulated	1 tbsp	45	12	0	0	0
Syrup, table blend	1 tbsp	60	15	0	0	0
Tea, unsweetened	6 oz	2	tr	tr	tr	0
Watermelon	1 4″ × 8″ wedge	110	27	2	1	0

OCEANS AND RIVERS

Oceans of the World

Ocean	Area (mi^2)	Average Depth (ft)	Volume (mi^3)
Arctic	5,440,000	3,954	4,300,000
North Pacific	32,225,000	12,658	77,200,000
South Pacific	25,298,000	12,766	61,150,000
North Atlantic	18,059,000	10,778	36,850,000
South Atlantic	14,426,000	13,423	36,660,000
Indian	31,507,000	14,006	83,870,000
Antarctic	12,451,000	12,238	28,860,000

Ocean Trenches	Depth (m)
Mid America Trench	−6669
Ryukyu Trench	−7507
New Hebrides Trench	−7570
Peru Chile Trench	−8064
Aleutian Trench	−8100
Palau Trench	−8138
Japan Trench	−8412

River Lengths

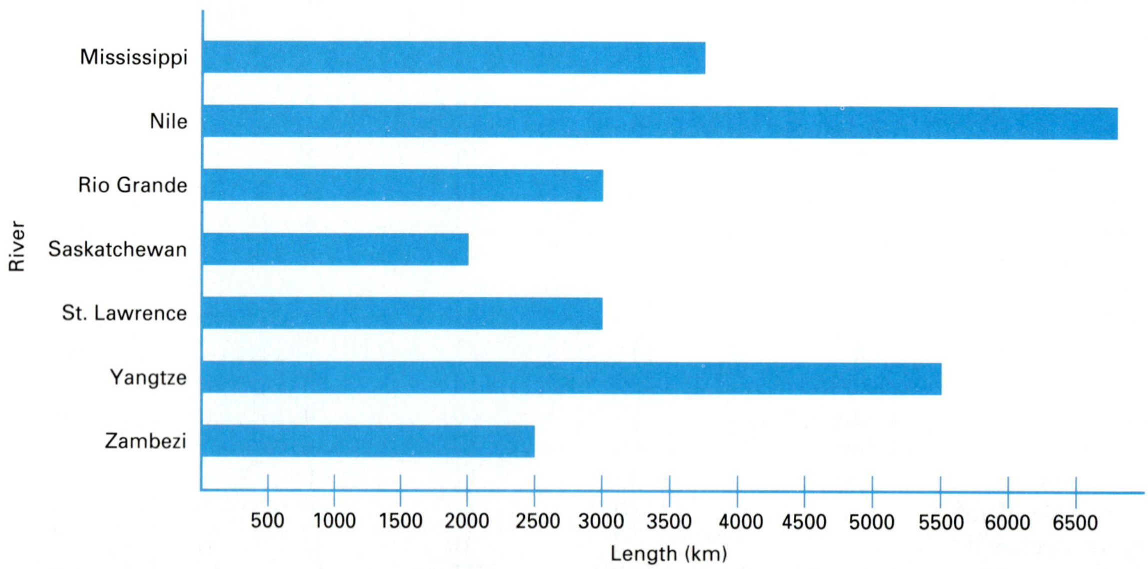

OPINION POLLS

Public Opinions

Poll 1: Television Preferences

	Men	Women	Teenagers	Total
Comedy	14	17	12	43
Drama	18	20	12	50
Game Shows	16	19	15	50
Information/News	22	23	18	63
Music Videos	1	2	27	30
Sports	29	19	16	64

Poll 2: Voter Preference

Council President		Council Secretary		Council Treasurer	
Adams	73	Jacobs	46	Rorke	53
Behrand	104	Kuntz	47	Sweeney	52
Chin	95	Lincoln	53	Tolski	35
Dayton	28	Martinez	54	Urbandale	60

POPULATION

In a recent year, the following tables show the population in the United States by age, to the nearest hundred thousand.

United States Population

Age	Male	Female
0–9	18,400,000	17,500,000
10–19	17,900,000	17,100,000
20–29	20,900,000	20,800,000
30–39	20,000,000	20,100,000
40–49	13,700,000	14,300,000
50–59	10,600,000	11,500,000
60–69	9,600,000	11,200,000
70–79	5,600,000	8,000,000
80+	2,000,000	3,600,000

Population by State

State	1900	1800
Connecticut	908,420	619,722
Delaware	184,735	59,096
Maine	694,466	96,540
Maryland	1,188,044	319,728
New Hampshire	411,588	183,858
North Carolina	1,893,810	393,751
Rhode Island	428,556	68,825
South Carolina	1,340,316	249,073
Vermont	343,641	85,425

SALES REPORTS

Buy It Here Shop

Month	Sales
Jan.	$17,500
Feb.	$15,000
Mar.	$13,000
Apr.	$12,000
May	$12,500
June	$14,000
July	$14,500
Aug.	$16,000
Sept.	$17,000
Oct.	$19,000
Nov.	$20,500
Dec.	$23,000

Super Delux Hamburgers Sold

April–May–June														
55	127	76	23	17	84	20	50	49	60	30	65	66	44	57
99	75	42	110	45	87	66	54	107	51	100	28	59	66	
82	77	63	39	52	95	23	81	46	79	34	99	61	98	73
118	22	89	36	101	55	63	61	53	28	84	49	66	52	45
117	31	45	111	85	71	54	36	74	52	80	76	67	94	72
11	58	69	122	70	88	73	48	100	98	89	73	108	14	
53	65	17												

"You Do It Shop" Yearly Sales

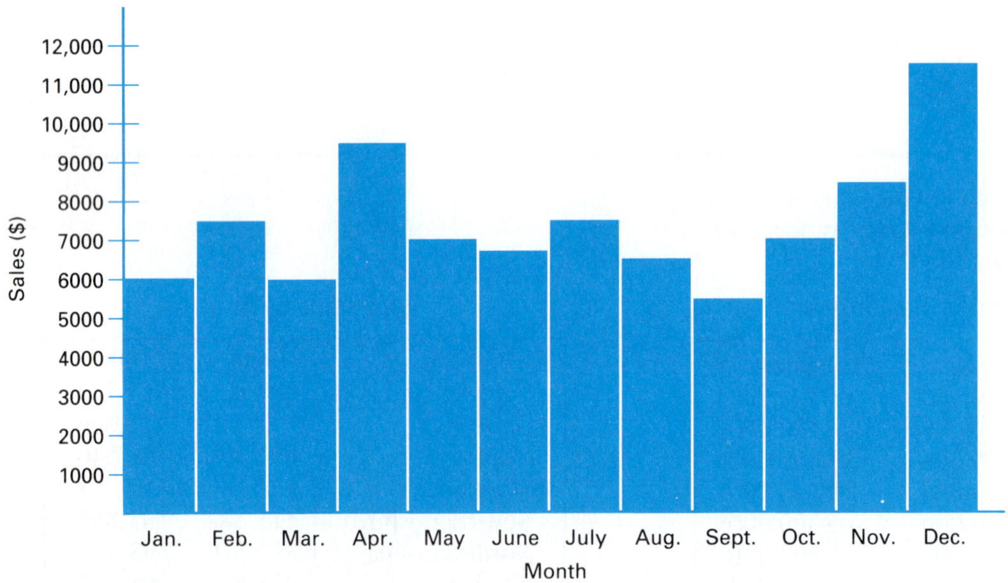

SOLAR SYSTEM

Planet	Ave. Distance From Sun (km)	Equatorial Diameter (km)
Mercury	58,000,000	4,880
Venus	108,000,000	12,100
Earth	150,000,000	12,756
Mars	228,000,000	6,780
Jupiter	778,000,000	142,800
Saturn	1,427,000,000	120,000
Uranus	2,870,000,000	50,800
Neptune	4,497,000,000	48,600
Pluto	5,900,000,000	2,200

Planet	Length of Day (in Earth Hours)	Length of Year (in Earth Years— 365.26 days = 1 year)
Mercury	1406.4	0.2409
Venus	5832	0.6152
Earth	23.934	1.00
Mars	24.623	1.88
Jupiter	9.842	11.86
Saturn	10.23	29.46
Uranus	23.0	84.01
Neptune	22.0	164.8
Pluto	153.0	248.4

Earth's Crust

Earth's Crust is made up of the main 8 elements:

Oxygen	47%	Calcium	4%
Silicon	28%	Magnesium	2%
Aluminum	8%	Sodium	3%
Iron	5%	Potassium	3%

SPORTS

Records

Fishing Records—With Rod and Reel

Type Fish	Weight
Fresh-Water Bass	60 pounds, 8 ounces
Fresh-Water Salmon	97 pounds, 4 ounces
Fresh-Water Trout	65 pounds, 0 ounces
Salt-Water Tuna	1496 pounds, 0 ounces
Salt-Water Shark	2664 pounds, 0 ounces

World Speed Skating Records—1988

Distance	Men's Time		Women's Time	
(Meters)	Minute(s)	Seconds	Minute(s)	Seconds
500	0	36.45	0	39.10
1000	1	12.58	1	17.65
1500	1	52.50	2	00.68
3000	3	59.27	4	11.94
5000	6	44.63	7	14.13
10,000	13	48.20	—	—

World Swimming Records

Distance	Men's Time		Women's Time	
(Meters)	Minute(s)	Seconds	Minute(s)	Seconds
FREESTYLE:				
50	0	22.18	0	24.98
100	0	48.42	0	54.73
200	1	47.44	1	57.75
400	3	47.38	4	5.45
800	7	50.64	8	17.12
1500	14	54.76	15	52.10
BACKSTROKE:				
100	0	54.91	1	0.59
200	1	58.14	2	9.91
BREASTSTROKE:				
100	1	1.65	1	7.91
200	2	13.34	2	27.40

Baseball Statistics

Name	Games Played	At Bats	Runs Scored	Hits	Runs Batted In	Stolen Bases	Base on Balls	Strike Outs
Barton	18	45	5	9	6	2	6	2
Ciosek	20	30	2	6	2	0	3	4
Dylan	15	40	6	12	10	3	2	5
Evers	14	28	1	8	6	1	1	2
Fells	22	50	9	18	15	7	2	2
Garcia	21	42	7	15	12	2	8	3
Hennison	16	23	8	12	12	0	1	7
Jamison	18	40	4	6	2	1	2	5
Kent	22	47	8	12	10	3	5	7
Lymon	15	30	7	8	9	2	8	2
Moreno	19	45	3	20	12	0	6	2
Nahrod	20	48	6	21	9	2	3	1

Miscellaneous Information

Type	Diameter (cm)	Average Weight (g)
Baseball	7.6	145
Basketball	24.0	596
Croquet ball	8.6	340
Field hockey ball	7.6	160
Golf ball	4.3	46
Handball	4.8	65
Soccer ball	22.0	425
Softball, large	13.0	279
Softball, small	9.8	187
Table tennis ball	3.7	2
Tennis ball	6.5	57
Volleyball	21.9	256

TIME ZONES

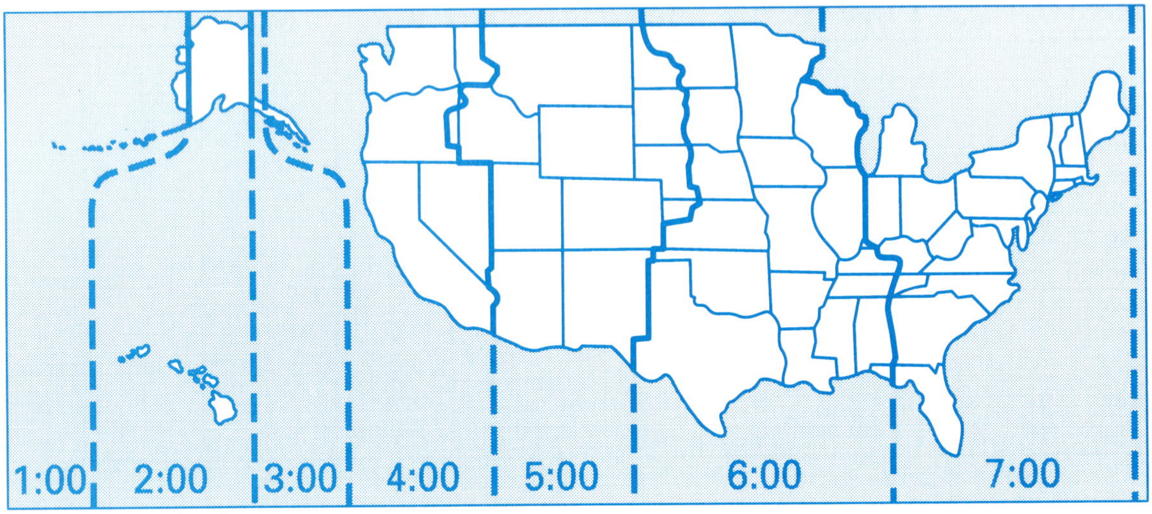

WEATHER

Wind Chill Factor Chart

Speed (km/h)	Actual thermometer reading (°C)												
	10	5	0	−5	−10	−15	−20	−25	−30	−35	−40	−45	−50
8	9	3	−3	−8	−13	−18	−23	−28	−33	−38	−44	−50	−57
16	4	−2	−9	−15	−21	−26	−33	−38	−44	−50	−57	−64	−72
24	2	−6	−13	−20	−26	−33	−40	−45	−51	−58	−65	−73	−81
32	0	−8	−16	−22	−30	−37	−44	−49	−55	−63	−71	−79	−88
40	−1	−9	−18	−25	−32	−40	−47	−53	−59	−67	−76	−83	−93
48	−2	−11	−19	−27	−34	−41	−50	−55	−62	−70	−78	−87	−97
56	−3	−12	−20	−28	−36	−42	−52	−57	−64	−72	−81	−89	−99

Highest Temperatures Recorded by Continent

Continent	Degrees Fahrenheit	Degrees Celsius
Africa (World)	136	58
North America (U.S.)	134	57
Asia	129	54
Australia	128	53
Europe	122	50
South America	120	49
Antarctica	59	15

Greatest Rainfalls in the World

Time	Inches	Centimeters
1 minute	1.23	3.1
20 minutes	8.1	20.5
42 minutes	12.0	30.5
12 hours	53.0	135.0
24 hours	74.0	188.0
5 days	152.0	386.0
1 month	366.0	929.6
12 months	1042.0	2647.0

Selected Great Snowfalls

Time	Location	Inches	Centimeters
19 hours	Bessans, France	68.0	173.0
24 hours	Colorado	76.0	192.5
1 month	California	390.0	991.0
1 storm	California	189.0	480.0
1 storm	Alaska	175.0	445.5
1 season	Washington	1122.0	2850.0
1 season	Alaska	974.5	2475.0
1 season	Canada	964.0	2446.5

Number of Tropical Storms & Hurricanes—1886–1987

	All Tropical Storms	Hurricanes
Jan.–Apr.	3	1
May	14	3
June	55	23
July	63	33
August	203	144
September	288	182
October	179	89
November	40	21
December	6	3

Height and Weight Tables

Men Height Ft	In.	Small frame	Medium frame	Large frame	Women Height Ft	In.	Small frame	Medium frame	Large frame
5	2	128–134	131–141	138–150	4	10	102–111	109–121	118–131
5	3	130–136	133–143	140–153	4	11	103–113	111–123	120–134
5	4	132–138	135–145	142–156	5	0	104–115	113–126	122–137
5	5	134–140	137–148	144–160	5	1	106–118	115–129	125–140
5	6	136–142	139–151	146–164	5	2	108–121	118–132	128–143
5	7	138–145	142–154	149–168	5	3	111–124	121–135	131–147
5	8	140–148	145–157	152–172	5	4	114–127	124–138	134–152
5	9	142–151	148–160	155–176	5	5	117–130	127–141	137–155

USING CALCULATORS

Calculators help everyone complete arithmetic computations quickly and accurately. However, there are certain skills that everyone needs when using calculators.

Know your calculator. Not all calculators are the same. As you read this section or try exercises, especailly when using specific keys on your calculator for the first time, use paper and pencil to find the answer first, then use your calculator. If you get the correct answer, you know you used the proper keys and pressed the correct sequence of keys. Later in this section is a discussion of the various kinds of calculators available.

Estimate. Whenever you use a calculator, you should have an idea of what the answer should be. For instance, you are to add 71.658 and 110.24. You get an answer of 82.682. You should estimate that your answer should be about 180, not 80. Try the exercise again. Is your new answer closer to your estimate?

Traditional Arithmetic Calculators

Many of the less expensive calculators are the traditional arithmetic calculators. However, even these calculators can be very useful when completing many different kinds of arithmetic computations. You need to learn to use your calculator efficiently. To do this, start by reading the manual that comes with your calculator. Then try the examples in the following material.

Use your calculator to complete each of the following exercises. See if you get the indicated answer.

a. 61 + 73 + 95 + 67 = 296

$$\begin{array}{r} 61 \\ 73 \\ 95 \\ +\ 67 \\ \hline 296 \end{array}$$

b.

$$\begin{array}{r} 50{,}871 \\ -\ 18{,}916 \\ \hline 31{,}955 \end{array}$$

c.

$$\begin{array}{r} 670 \\ \times\ 25 \\ \hline 16{,}750 \end{array}$$

d.

$$\begin{array}{r} 18 \\ 27\overline{)486} \end{array}$$

If you did not get the correct answer, try the exercise again. If you still cannot get the correct answer, read the manual that goes with your calculator, or ask your teacher for help.

Now try this: Press the keys in the order given, and then answer the questions.

[3] [+] [4] [=] [=] [=]

Does your calculator do chain addition?
Which number is added repeatedly?
Does your calculator do chain calculations for subtraction? for multiplication? for division?

Many calculators also have other keys such as

[%], [+/−], [√].

Use your manual and experiment to find how to correctly use each special key on your calculator.

Many calculators also have memory keys, which might look like this:

[CM], [RM], [M−], [M+].

In general, [CM] clears the memory, [RM] recalls whatever is in the memory, and [M−] and [M+] subtract from or add to whatever is in the memory. Again, study your manual and practice using these keys.

Scientific Calculators

It is now possible to purchase inexpensive scientific calculators. These calculators follow the correct order of operations. For instance, compare these two examples.

scientific calculator

[8] [+] [2] [×] [3] [=]

[14.]

arithmetic calculator

[8] [+] [2] [×] [3] [=]

[30.]

Scientific calculators also automatically change very large or very small numbers into exponential notation.

scientific calculator

[8] [6] [x^y] [5] [=]

4704270176.

or 4.704270176 [9]

arithmetic calculator

[8] [6] [×] [=] [=] [=] [=]

47.042701 E

Notice how the [x^y] key is used: the base is entered first, then the [x^y] key, then the power to which the base is to be raised. Notice that the arithmetic calculator overflows—that is, could not print the whole answer—so it indicated an error with an "E."

Below is an example of how each calculator handles very small numbers.

scientific calculator

[.] [0] [5] [x^y] [5] [=]

3.125 [-07]

arithmetic calculator

[.] [0] [5] [×] [=] [=] [=] [=]

0.

The scientific calculator tells you the answer is 3.125×10^{-7} or 0.0000003125. The arithmetic calculator underflows and gives zero as the answer.

To enter very large or very small numbers on a scientific calculator, use scientific notation.

number	scientific notation
475,000,000,000	4.7×10^{8}
0.00000000045	4.5×10^{-10}

key sequence

[4] [.] [7] [5] [EXP] [8]

[4] [.] [5] [EXP] [1] [0] [+/−]

Scientific calculators can accomplish many complex functions. However, it does take practice to learn to use them properly. Take time to read your manual and practice such uses.

Graphing Calculators

Another useful calculator is the graphing calculator. This type of calculator will automatically display the graph of an important function, a linear equation, or even a system of linear equations. You can determine what part of a graph you want to see by choosing the appropriate scale. Different scales, while producing different-looking graphs, will give the same solutions. For instance, here are the graphs of two equations that are done using two different scales.

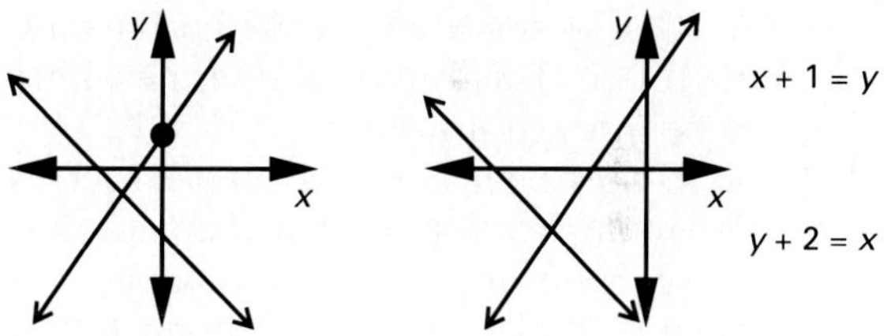

Scale A Scale B

Most graphing calculators also have a "trace point" which is indicated by a flashing dot or cursor. You can move the trace point to find the coordinates of the solution for a system of equations.

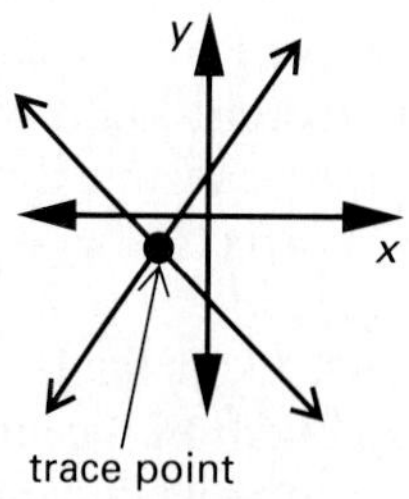

However, remember, this solution is an approximation. You must check your results!

Some graphing calculators can be used when you are dealing with statistics. For instance, you can enter a set of data and have the calculator display the data in many different ways, such as a histogram, a broken-line graph, or as individual-point plots.

USING COMPUTERS

Throughout this textbook, you will find references to the use of computers. The goal of this course is to alert you to the possible uses of computers in solving problems. A computer, like paper and pencils, rulers, and calculators, is a tool—it is to be used to help you learn to solve problems.

BASIC Programming Language

In order for the computer to understand what you want it to do, you must be able to communicate clearly with the computer. To do this, many different computer languages are used. One of the more common languages is BASIC. There are many different versions of BASIC, depending upon the computer systems being used.

As you look at various BASIC computer programs, you should note the following:

- Computers require input (information put into the computer) and a program to produce output (information the computer prints or puts out in some manner).
- BASIC programs are numbered using multiples of 10. This is done so that new lines can be inserted if needed.
- BASIC uses some special symbols, such as * for multiplication, / for division, and ^ for exponents or powers of a number.
- Computers, like some calculators, follow the algebraic order of operations.
- The computer follows the program exactly.

Have you ever heard the phrase "Garbage in, garbage out" and wondered what it meant? In computer terms, it means that if the program is not correct, the output may or may not be correct.

Just like calculators, computers vary in how they function. For instance, on some computers you use ** when you want to raise a number to a power. On others you use ^. The following general information will work for most computers. However, once again you must read the manual or ask someone for help if you have trouble with the computer.

- When typing in a computer program, it is essential to proofread what you have input. A comma, period, and semicolon all mean different things to a computer. If your program does not work, one of the first things to do is to "LIST" the program and check your typing.
- To get the computer to print out information, the word "PRINT" must be found in the program. If quotation marks are around a phrase or letter, it will be printed. If there are no quotation marks, the value of that phrase or letter will be printed. On some systems, if the phrase or letter has not been defined, you will get an error message.
- To put information into a program, you can use a "LET" statement, such as "LET X = 2," or an "INPUT" statement, such as "INPUT S." Another method of inputting data, is the use of the "READ" and "DATA" statements.
- You must understand the logic that is used in a computer program. For instance, if a program line says "LET X = X + 2," that tells the program to add 2 to whatever value X had and then use that (X + 2) as the new value for X.
- When working with inequalities, you need to know how to enter symbols such as $<$, $>$, $\leq$, or $\geq$. Many computers use these symbols:

<	is less than
>	is greater than
<=	is less than or equal to
>=	is greater than or equal to
<>	is not equal to

- Computers, like calculators, also use powerful functions that are stored in their memories. Some useful ones are for absolute value which is (in BASIC) ABS(X) and square root which is SQR(X). Notice that X needs to be specified in each case.

A Final Note

The goal of this information is to furnish some helpful tips that may make your use of computers easier. Remember to read the manual and ask for help. Do not be afraid to ask questions—they show that you want to learn how to use computers more effectively.

FORMULAS/USEFUL EQUATIONS

Geometric

l = length w = width
b = base h = height
s = side r = radius
d = diameter H = slant height
B = area of the base of a three-dimensional figure

Perimeter: *P*

Rectangle: $P = 2(l + w)$ or $2l + 2w$
Square: $P = 4s$
Parallelogram: $P = 2(s_1 + s_2)$ or $2s_1 + 2s_2$
Rhombus: $P = 4s$
Regular polygon with n sides: $P = ns$
Triangle: $P = s_1 + s_2 + s_3$
Circle (Circumference): $C = \pi d$ or $2\pi r$

Area: *A*

Rectangle: $A = lw$
Parallelogram: $A = lw$
Trapezoid: $A = \frac{1}{2}h(b_1 + b_2)$
Triangle: $A = \frac{1}{2}bh$
Circle: $A = \pi r^2$ or $\frac{1}{4}\pi d^2$
Square: $A = s^2$
Rhombus: $A = lw$

Surface Area: *SA*

Rectangular Solid: $SA = 2(lw + lh + wh)$
Cube: $SA = 3s^2$
Cylinder: $SA = 2\pi r(r + h)$ or $2\pi r^2 + 2\pi rh$
Sphere: $SA = 4\pi r^2$
Regular Pyramid: $SA = \frac{1}{2}PH + B$
Cone: $SA = \pi rH + \pi r^2$

Volume: *V*

Rectangular Solid: $V = lwh$ or Bh
Cube: $V = s^3$
Cylinder: $V = \pi r^2 h$ or Bh
Sphere: $V = \frac{4}{3}\pi r^3$
Regular Pyramid: $V = \frac{1}{3}Bh$
Cone: $V = \frac{1}{3}\pi r^2 h$

Pythagorean Theorem

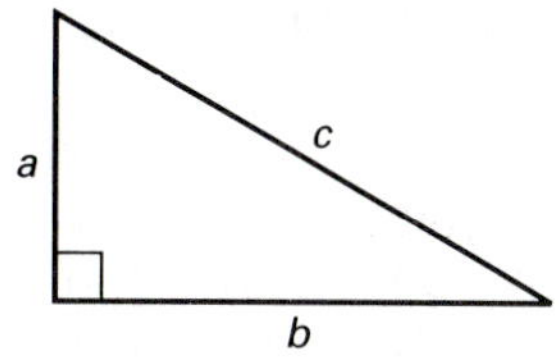

In a right triangle, as shown, $a^2 + b^2 = c^2$ or $c = \sqrt{a^2 + b^2}$

Other Formulas/Useful Equations

Temperature

C: degrees Celsius	F: degrees Fahrenheit
$C = \frac{5}{9}(F - 32)$	$F = \frac{9}{5}C + 32$

Permutations/Combinations

P: Permutation C: Combination
n: total number of items r: number of items taken at a time
$n!$: $(n)(n - 1)(n - 2) \ldots (2)(1)$

$_nP_r = \frac{n!}{(n - r)!}$ $_nC_r = \frac{n!}{(n - r)!r!}$

Distance

d: distance	r: rate	t: time
$d = rt$	$r = \frac{d}{t}$	$t = \frac{d}{r}$

Interest

I: interest A: total amount
p: principal r: rate
t: time in years

$I = prt$ $A = p + prt$ or $p + I$

LAWS OF EXPONENTS

$a^m \cdot a^n = a^{m+n}$ $(a^m)^n = a^{m \times n}$ $\frac{a^m}{a^n} = a^{m-n}$

PRIME NUMBERS

2	3	5	7	11	13	17	19
23	29	31	37	41	43	47	53
59	61	67	71	73	79	83	89
97	101	103	107	109	113	127	131
133	137	139	149	151	157	161	163
167	173	179	181	191	193	197	199
203	211	217	223	227	229	233	239
241	251	257	263	269	271	277	281
283	293	307	311	313	317	331	337
347	349	353	359	367	373	379	383
389	397	401	409	419	421	431	433
439	443	449	457	461	463	467	479
487	491	499					

DIVISIBILITY RULES

A number is divisible by	if
2	it is even (last digit is 0, 2, 4, 6, 8).
3	the sum of the digits is divisible by 3.
4	the last two digits are divisible by 4.
5	the last digit is 5 or 0.
6	it is divisible by 2 **and** 3.
8	the last three digits are divisible by 8.
9	if the sum of the digits is divisble by 9.
10	if the last digit is 0.

PROBLEM-SOLVING PLAN

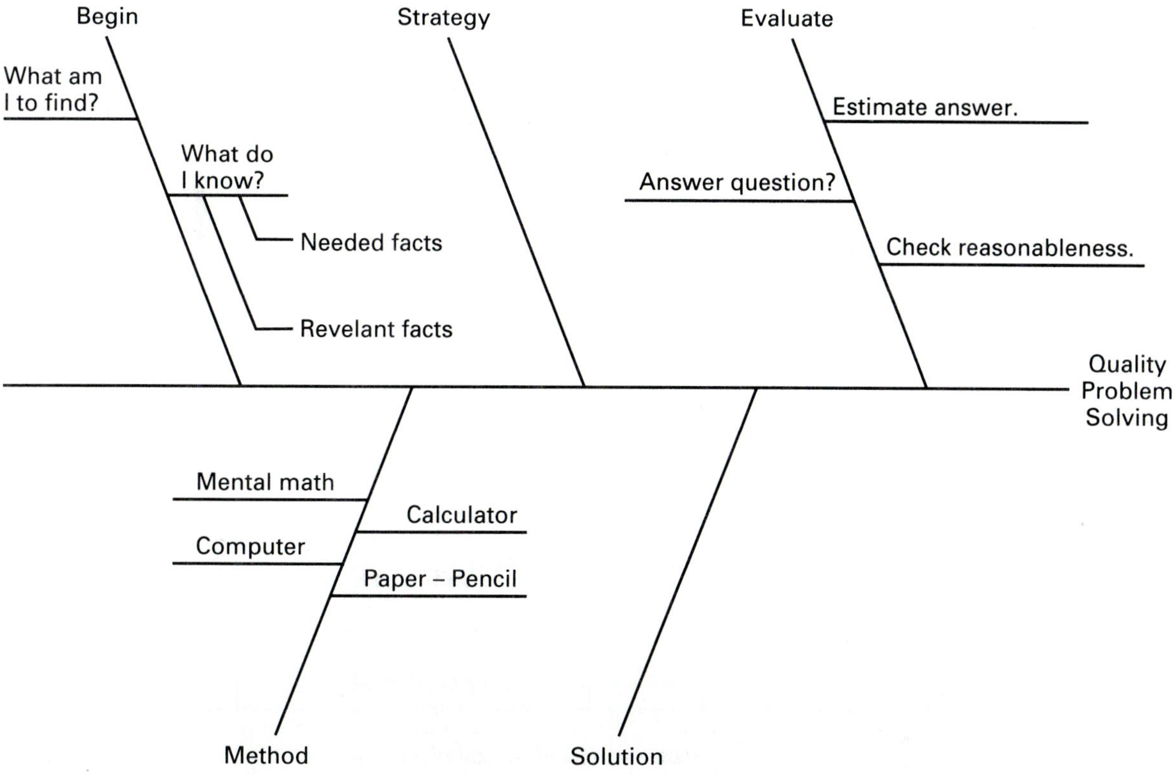

ESTIMATION

Round to estimate.

$$\begin{array}{rcr} 37,856 & \rightarrow & 38,000 \\ -15,200 & \rightarrow & \underline{15,000} \\ & & 23,000 \end{array} \qquad \begin{array}{rcr} 159 & \rightarrow & 200 \\ 278 & \rightarrow & 300 \\ 506 & \rightarrow & 500 \\ +314 & \rightarrow & \underline{300} \\ & & 1300 \end{array} \qquad \begin{array}{rcr} 567 & \rightarrow & 600 \\ \underline{\times 19} & \rightarrow & \underline{20} \\ & & 12,000 \end{array}$$

Use compatible (easy to use) numbers.

$$63\overline{)46,850} \rightarrow \begin{array}{r} 800 \\ 60\overline{)48,000} \end{array} \qquad 38\overline{)156,000} \rightarrow \begin{array}{r} 4,000 \\ 40\overline{)160,000} \end{array}$$

Remember: Estimate **before** you solve a problem.
Then estimate **after** you solve a problem to check that your answer is reasonable.

PLACE VALUE

billion	hundred–million	ten–million	million	hundred–thousand	ten–thousand	thousand	hundred	ten	one
10^9	10^8	10^7	10^6	10^5	10^4	10^3	10^2	10^1	1
1,000,000,000	100,000,000	10,000,000	1,000,000	100,000	10,000	1,000	100	10	1

SYMBOLS

$=$	is equal to
$\neq$	is not equal to
$\cong$	is congruent to
$\sim$	is similar to
$\approx$	is approximately equal to
$\parallel$	is parallel to
$\perp$	is perpendicular to
$>$	is greater than
$<$	is less than
$\geq$	is greater than or equal to
$\leq$	is less than or equal to
( )	parentheses: "Do this operation first."
3^2	exponent (2), base (3)

%	percent
$0.\overline{3}$	repeating decimal
π	pi: $\approx \frac{22}{7}$ or 3.14
$\sqrt{\ }$	square root
$\lvert x \rvert$	absolute value of x
$\overleftrightarrow{AB}$	line AB
$\overline{AB}$	line segment AB
$\overrightarrow{AB}$	ray AB
$\angle ABC$	angle ABC
$\llcorner$	right angle
$^\circ$	degrees
$A \rightarrow A'$	point A maps into A'
$A(1,-2)$	point A coordinates: $x = 1$, $y = -2$

MEASUREMENT

Metric Measurement

Length

1 centimeter (cm)	= 10 millimeters (mm)
10 centimeters or 100 millimeters	= 1 decimeter (dm)
10 decimeters or 100 centimeters	= 1 meter (m)
1000 meters	= 1 kilometer (km)

Area

100 square millimeters (mm^2)	= 1 square centimeter (cm^2)
10,000 square centimeters	= 1 square meter (m^2)
100 square meters	= 1 are (a)
10,000 square meters	= 1 hectare (ha)

Volume

1000 cubic millimeters (mm^3)	= 1 cubic centimeter (cm^3)
1000 cubic centimeters	= 1 cubic decimeter (dm^3)
1,000,000 cubic centimeters	= 1 cubic meter (m^3)

Capacity

1000 milliliters	= 1 liter (L)
1000 liters	= 1 kiloliter (kL)

Mass

1000 milligrams (mg)	= 1 gram (g)
1000 grams	= 1 kilogram (kg)
1000 kilograms	= 1 metric ton (t)

Temperature

0°C	= freezing point of water
37°C	= normal body temperature
100°C	= boiling point of water

U. S. Customary Measurement

Length

12 inches (in.) = 1 foot (ft)
3 feet or
36 inches = 1 yard (yd)
1760 yards or
5280 feet = 1 mile (mi)
6076 feet = 1 nautical mile

Area

144 square inches (in^2) = 1 square foot (ft^2)
9 square feet = 1 square yard (yd^2)
4840 square yards = 1 acre (A)

Volume

1728 cubic inches (in^3) = 1 cubic foot (ft^3)
27 cubic inches = 1 cubic yard (yd^3)

Capacity

8 fluid ounces (fl. oz) = 1 cup
2 cups = 1 pint (pt)
2 pints = 1 quart (qt)
4 quarts = 1 gallon (gal)

Weight

16 ounces (oz) = 1 pound (lb)
2000 pounds = 1 ton (T)

Temperature

32°F = freezing point of water
98.6°F = normal body temperature
212°F = boiling point of water

Metric/Customary Comparisons

5 centimeters is about the same length as 2 inches.
1 meter is slightly longer than 1 yard.
5 kilometers is about the same length as 3 miles.

RANDOM TABLES

Random Digits

12714	04647	38236	51890	90523
51145	20096	26373	76878	48959
68924	45167	70801	32592	30341
84807	91356	42792	23653	01054
20390	73491	51849	85877	66602
96231	52804	07132	43586	79014
62323	14015	65407	89957	69887
07073	68962	15518	24129	43304
74956	84538	27926	01178	53960
20434	33225	41090	18769	86175
59781	09656	27437	34802	15986
90238	07112	16403	55829	34674
23154	65732	76468	59718	90098
19771	58054	54222	95231	21280
86690	76394	81303	10764	87439
13333	97599	36300	67765	40813
61072	49227	71646	92484	00451
98442	05216	63266	39948	93558
46182	88944	78176	17720	35663
72171	64017	44895	10085	88319

Last-Four Digits of Random Telephone Numbers

−2778	−6594	−8000	−5925	−1124	−4443
−1676	−2336	−0361	−1988	−2978	−5001
−8722	−3635	−3243	−6714	−5080	−9912
−4300	−5982	−9812	−7806	−6415	−2347
−5918	−9130	−7372	−4549	−5671	−6680
−3758	−0306	−1467	−0428	−7739	−6340
−0015	−9946	−0117	−4978	−5481	−8840
−1057	−4104	−1233	−3022	−9899	−6211
−2825	−2565	−5573	−9682	−6663	−5720
−5636	−2294	−3394	−6514	−8171	−7952
−4015	−1400	−1606	−5709	−7002	−7239
−3327	−2118	−4750	−8995	−1846	−9753
−8444	−4181	−8299	−2381	−0150	−4997
−6232	−3323	−0864	−7671	−3467	−2888
−7536	−9573	−8082	−9965	−6678	−5504
−0289	−2041	−8319	−5259	−3934	−1356
−5169	−0554	−9031	−7144	−6907	−9221
−0616	−7586	−7479	−1584	−9398	−4205
−8648	−3145	−4691	−0783	−2437	−9455
−4838	−1793	−6199	−5851	−6066	−3526

TABLES—SQUARES AND SQUARE ROOTS

n	n^2	$\sqrt{n}$	$\sqrt{10n}$	n	n^2	$\sqrt{n}$	$\sqrt{10n}$
1.0	1.00	1.000	3.162	**5.5**	30.25	2.345	7.416
1.1	1.21	1.049	3.317	**5.6**	31.36	2.366	7.483
1.2	1.44	1.095	3.464	**5.7**	32.49	2.387	7.550
1.3	1.69	1.140	3.606	**5.8**	33.64	2.408	7.616
1.4	1.96	1.183	3.742	**5.9**	34.81	2.429	7.681
1.5	2.25	1.225	3.873	**6.0**	36.00	2.449	7.746
1.6	2.56	1.265	4.000	**6.1**	37.21	2.470	7.810
1.7	2.89	1.304	4.123	**6.2**	38.44	2.490	7.874
1.8	3.24	1.342	4.243	**6.3**	39.69	2.510	7.937
1.9	3.61	1.378	4.359	**6.4**	40.96	2.530	8.000
2.0	4.00	1.414	4.472	**6.5**	42.25	2.550	8.062
2.1	4.41	1.449	4.583	**6.6**	43.56	2.569	8.124
2.2	4.84	1.483	4.690	**6.7**	44.89	2.588	8.185
2.3	5.29	1.517	4.796	**6.8**	46.24	2.608	8.246
2.4	5.76	1.549	4.899	**6.9**	47.61	2.627	8.307
2.5	6.25	1.581	5.000	**7.0**	49.00	2.646	8.367
2.6	6.76	1.612	5.099	**7.1**	50.41	2.665	8.426
2.7	7.29	1.643	5.196	**7.2**	51.84	2.683	8.485
2.8	7.84	1.673	5.292	**7.3**	53.29	2.702	8.544
2.9	8.41	1.703	5.385	**7.4**	54.76	2.720	8.602
3.0	9.00	1.732	5.477	**7.5**	56.25	2.739	8.660
3.1	9.61	1.761	5.568	**7.6**	57.76	2.757	8.718
3.2	10.24	1.789	5.657	**7.7**	59.29	2.775	8.775
3.3	10.89	1.817	5.745	**7.8**	60.84	2.793	8.832
3.4	11.56	1.844	5.831	**7.9**	62.41	2.811	8.888
3.5	12.25	1.871	5.916	**8.0**	64.00	2.828	8.944
3.6	12.96	1.897	6.000	**8.1**	65.61	2.846	9.000
3.7	13.69	1.924	6.083	**8.2**	67.24	2.864	9.055
3.8	14.44	1.949	6.164	**8.3**	68.89	2.881	9.110
3.9	15.21	1.975	6.245	**8.4**	70.56	2.898	9.165
4.0	16.00	2.000	6.325	**8.5**	72.25	2.915	9.220
4.1	16.81	2.025	6.403	**8.6**	73.96	2.933	9.274
4.2	17.64	2.049	6.481	**8.7**	75.69	2.950	9.327
4.3	18.49	2.074	6.557	**8.8**	77.44	2.966	9.381
4.4	19.36	2.098	6.633	**8.9**	79.21	2.983	9.434
4.5	20.25	2.121	6.708	**9.0**	81.00	3.000	9.487
4.6	21.16	2.145	6.782	**9.1**	82.81	3.017	9.539
4.7	22.09	2.168	6.856	**9.2**	84.64	3.033	9.592
4.8	23.04	2.191	6.928	**9.3**	86.49	3.050	9.644
4.9	24.01	2.214	7.000	**9.4**	88.36	3.066	9.695
5.0	25.00	2.236	7.071	**9.5**	90.25	3.082	9.747
5.1	26.01	2.258	7.141	**9.6**	92.16	3.098	9.798
5.2	27.04	2.280	7.211	**9.7**	94.09	3.114	9.849
5.3	28.09	2.302	7.280	**9.8**	96.04	3.130	9.899
5.4	29.16	2.324	7.348	**9.9**	98.01	3.146	9.950
5.5	30.25	2.345	7.416	**10**	100.00	3.162	10.000

CUBES AND CUBE ROOTS

n	n^3	$\sqrt[3]{n}$	$\sqrt[3]{10n}$	$\sqrt[3]{100n}$	n	n^3	$\sqrt[3]{n}$	$\sqrt[3]{10n}$	$\sqrt[3]{100n}$
1.0	1.000	1.000	2.154	4.642	**5.5**	166.375	1.765	3.803	8.193
1.1	1.331	1.032	2.224	4.791	**5.6**	175.616	1.776	3.826	8.243
1.2	1.728	1.063	2.289	4.932	**5.7**	185.193	1.786	3.849	8.291
1.3	2.197	1.091	2.351	5.066	**5.8**	195.112	1.797	3.871	8.340
1.4	2.744	1.119	2.410	5.192	**5.9**	205.379	1.807	3.893	8.387
1.5	3.375	1.145	2.466	5.313	**6.0**	216.000	1.817	3.915	8.434
1.6	4.096	1.170	2.520	5.429	**6.1**	226.981	1.827	3.936	8.481
1.7	4.913	1.193	2.571	5.540	**6.2**	238.328	1.837	3.958	8.527
1.8	5.832	1.216	2.621	5.646	**6.3**	250.047	1.847	3.979	8.573
1.9	6.859	1.239	2.668	5.749	**6.4**	262.144	1.857	4.000	8.618
2.0	8.000	1.260	2.714	5.848	**6.5**	274.625	1.866	4.021	8.662
2.1	9.261	1.281	2.759	5.944	**6.6**	287.496	1.876	4.041	8.707
2.2	10.648	1.301	2.802	6.037	**6.7**	300.763	1.885	4.062	8.750
2.3	12.167	1.320	2.844	6.127	**6.8**	314.432	1.895	4.082	8.794
2.4	13.824	1.339	2.884	6.214	**6.9**	328.509	1.904	4.102	8.837
2.5	15.625	1.357	2.924	6.300	**7.0**	343.000	1.913	4.121	8.879
2.6	17.576	1.375	2.962	6.383	**7.1**	357.911	1.922	4.141	8.921
2.7	19.683	1.392	3.000	6.463	**7.2**	373.248	1.931	4.160	8.963
2.8	21.952	1.409	3.037	6.542	**7.3**	389.017	1.940	4.179	9.004
2.9	24.389	1.426	3.072	6.619	**7.4**	405.224	1.949	4.198	9.045
3.0	27.000	1.442	3.107	6.694	**7.5**	421.875	1.957	4.217	9.086
3.1	29.791	1.458	3.141	6.768	**7.6**	438.976	1.966	4.236	9.126
3.2	32.768	1.474	3.175	6.840	**7.7**	456.533	1.975	4.254	9.166
3.3	35.937	1.489	3.208	6.910	**7.8**	474.552	1.983	4.273	9.205
3.4	39.304	1.504	3.240	6.980	**7.9**	493.039	1.992	4.291	9.244
3.5	42.875	1.518	3.271	7.047	**8.0**	512.000	2.000	4.309	9.283
3.6	46.656	1.533	3.302	7.114	**8.1**	531.441	2.008	4.327	9.322
3.7	50.653	1.547	3.332	7.179	**8.2**	551.368	2.017	4.344	9.360
3.8	54.872	1.560	3.362	7.243	**8.3**	571.787	2.025	4.362	9.398
3.9	59.319	1.574	3.391	7.306	**8.4**	592.704	2.033	4.380	9.435
4.0	64.000	1.587	3.420	7.368	**8.5**	614.125	2.041	4.397	9.473
4.1	68.921	1.601	3.448	7.429	**8.6**	636.056	2.049	4.414	9.510
4.2	74.088	1.613	3.476	7.489	**8.7**	658.503	2.057	4.431	9.546
4.3	79.507	1.626	3.503	7.548	**8.8**	681.472	2.065	4.448	9.583
4.4	85.184	1.639	3.530	7.606	**8.9**	704.969	2.072	4.465	9.619
4.5	91.125	1.651	3.557	7.663	**9.0**	729.000	2.080	4.481	9.655
4.6	97.336	1.663	3.583	7.719	**9.1**	753.571	2.088	4.498	9.691
4.7	103.823	1.675	3.609	7.775	**9.2**	778.688	2.095	4.514	9.726
4.8	110.592	1.687	3.634	7.830	**9.3**	804.357	2.103	4.531	9.761
4.9	117.649	1.698	3.659	7.884	**9.4**	830.584	2.110	4.547	9.796
5.0	125.000	1.710	3.684	7.937	**9.5**	857.375	2.118	4.563	9.830
5.1	132.651	1.721	3.708	7.990	**9.6**	884.736	2.125	4.579	9.865
5.2	140.608	1.732	3.733	8.041	**9.7**	912.673	2.133	4.595	9.899
5.3	148.877	1.744	3.756	8.093	**9.8**	941.192	2.140	4.610	9.933
5.4	157.464	1.754	3.780	8.143	**9.9**	970.299	2.147	4.626	9.967
5.5	166.375	1.765	3.803	8.193	**10**	1000.000	2.154	4.642	10.000

BIBLIOGRAPHY

Albers, Donald J. ed., *et. al. More Mathematical People.* Washington, DC: The Mathematical Association of America, 1990.

Bell, E. T. *The Last Problem.* Washington, DC: The Mathematical Association of America, 1990.

———. *Mathematics: Queen and Servant of Science.* Washington, DC: The Mathematical Association of America, 1987.

Devlin, Keven. *Mathematics: The New Golden Age.* Washington, DC: The Mathematical Association of America, 1990.

Eves, Howard. *Great Moments in Mathematics After 1650.* Washington, DC: The Mathematical Association of America, 1982.

———. *Great Moments in Mathematics Before 1650.* Washington, DC: The Mathematical Association of America, 1982.

Gardner, Martin. *Mathematical Magic Show.* Washington, DC: The Mathematical Association of America, 1990.

———. *Riddles of the Sphinx and Other Mathematical Puzzle Tales.* Washington, DC: The Mathematical Association of America, 1990, 1987.

Halmos, Paul R. *I Want to Be A Mathematician.* Washington, DC: The Mathematical Association of America, 1987.

Hollingdale, Stuart. *Makers of Mathematics.* Washington, DC: The Mathematical Association of America, 1989.

Honsberger, Ross. *More Mathematical Morsels.* Washington, DC: The Mathematical Association of America, 1990.

———. *Mathematical Morsels.* Washington, DC: The Mathematical Association of America, 1979.

REFERENCE SECTION INDEX

SELECTED ANSWERS

CHAPTER 1 PATTERNS AND PROPERTIES OF NUMBERS

SECTION 1–1, page 6

Exercises 1. C.J. **3.** Numerical ranking allows you to quickly find the hottest city; alphabetical order allows you to quickly find the temperature of a city. **7.** PB-10806 **9.** 4 hr 45 min

SECTION 1–2, page 10

Exercises 1. D **3.** B **5.** C **7.** 2116 **9.** 1,100,001 **11.** 7010 m **13.** $<$ **15.** $<$ **17.** $>$ **19.** Tens; 70 **21.** Ten thousands; 70,000 **23.** Millions; 7,000,000 **25.** T **27.** Twenty-seven dollars **29.** Three hundred forty-eight dollars **31.** 1,392,000 km **33.** Hundreds **35.** tens.

SECTION 1–3, page 14

Exercises 5. C **7.** A **9.** 8,500 L **11.** 1,000 **13.** 3,600; 400; 1,800; 2,000 **15.** 60,000; 150,000; 200,000; 100,000

SECTION 1–4, page 19

Exercises 3. 4^2 **5.** 3^3 **7.** 1; 2 **9.** 3; 7 **11.** 2; 4; 6 **13.** 5; 10; 15 **15.** 7; 14; 21 **17.** 9; 18; 27 **19.** 5^3 **21.** 4^4 **23.** 1×100; 2×50; 4×25; 5×20; 10×10; {1, 2, 4, 5, 10, 20, 25, 50, 100} **25.** 1×45; 3×15; 5×9; {1, 3, 5, 9, 15, 45} **27.** No **29.** Yes **31.** No **33.** No **35.** 16 **37.** 1 **39.** 27 **41.** 1000

SECTION 1–5, page 22

Exercises 1. P **3.** C **5.** P **7.** C **9.** C **11.** 3 **13.** 4 **15.** 6 **17.** $2^2 \times 3$ **19.** 3×5^2 **21.** 2×7^2 **23.** 30 **25.** 200 **27.** 630 **29.** 2; 2 **31.** 5; 5 **33.** 7; 7 **35.** 15, 21, 35, or 105 **37.** 11, 13 **39.** 29, 31 **41.** 53, 59, 61 **43.** 83, 89, 91, 97 **45.** 234, 243, 324, 342, 432, 423 **47.** D **49.** A

SECTION 1–6, page 27

Exercises 1. 1, 2, 4 **3.** 1 **5.** 1, 3, 9 **7.** 1, 2, 3, 4, 6, and 12 **9.** 2 **11.** 12 **13.** 9 **15.** 10 **17.** 12 **19.** 10 **21.** 9, 18, 27 **23.** 19, 38, 57 **25.** 7, 14, 21 **27.** 30th **29.** GFC: 1; LCM: 60 **31.** GFC: 4; LCM: 120 **33.** GFC: 4; LCM: 240 **35.** 4

SECTION 1–7, page 30

Exercises 1. 17 **3.** 6 **5.** 2 **7.** The parentheses alters the order of operations. **9.** 16 **11.** 4 **13.** 1 **15.** 2 **17.** 12 **19.** 3 **21.** $>$ **23.** $<$ **25.** $>$ **27.** $36 \div (4 + 2)$ **29.** $8 + 45 \div (9 - 6)$ **31.** $\frac{104}{8} = 13$ **33.** $\frac{4}{4} = 1$

SECTION 1–8, page 35

Exercises 1. Associative **3.** Distributive **5.** $24 **7.** No **9.** False **11.** False

CHAPTER 1 REVIEW, page 37

1. 37,800 **3.** 99605 **5.** $99 **7.** You missed the inn. **9.** 4200 **11.** {1, 2, 3, 6, 7, 14, 21, 42} **13.** $2^2 \times 3 \times 5$ **15.** Prime **17.** Prime **19.** GCF = 2×7, or 14; LCM = $2^2 \times 3 \times 7$ or 84 **21.** 6900 **23.** 3600 **25.** 78

CHAPTER 2 WORKING WITH INTEGERS

SECTION 2–1, page 42

Exercises 1. +6 or 6 **3.** −14 **5.** +9 or 9 up **7.** 3 m left **9.** 12 m left **11.** 0 (no move) **13.** 8840 m **15.** −390 m **17.** 2 **19.** 36 **21.** 5 **23.** 8 **25.** 5 **27.** $<$ **29.** $>$ **31.** $>$ **37.** 24 in.

SECTION 2–2, page 46

Exercises 1. 9 **3.** −5 **5.** −8 **7.** positive **9.** The sign of the number farther from zero, or zero **11.** −3 **13.** −8 **15.** 0 **17.** −7 **19.** 1 **21.** 1 **23.** 0 **25.** 9 **29.** Both scores are the same, −1. **31.** +3°

SECTION 2–3, page 50

Exercises 1. 0 **3.** Their sum is 0. **5.** −7 + −6 **7.** −3 + 12 **9.** 3 **11.** 4 **13.** −2 **15.** −6 **17.** −124°F **19.** 13 − −3, which equals 16 is greater. **21.** −5 **23.** −4 **25.** 4 − −3 = 7 **27.** −15 − −4 = −11 **29.** 3 **31.** 5 **33.** Molten lava; 1730 − 480 = 1250° C. **35.** 2-yard gain

SECTION 2–4, page 54

Exercises 1. 4, 0, 4, −8 **3.** −12 **5.** They're the same. **7.** A positive integer **9.** −4 **11.** 12 **13.** −12 **15.** −24 **17.** −72 **19.** −120 **21.** −240 **23.** −35 gallons

SECTION 2–5, page 58

Exercises 1. 5 **3.** −5 **5.** A positive integer **7.** A positive integer **9.** −6 **11.** 3 **13.** 0 **15.** −1 **17.** −1 **19.** −7780 m **21.** 1,780,000,000 km **25.** No; some parts could be too shallow for diving.

SECTION 2–6, page 62

Exercises 1. (a) Work inside the parentheses first. (b) Divide left to right. (c) Divide. **3.** Divide **5.** Multiply **7.** −27 **9.** 36 **11.** −42 **13.** −40; **15.** −2; **17.** 6; **19.** −25 (−24 + −14) = 950; **21.** (−144 + −48 + 96)/(−12) = 8; **23.** (−8) (−2) + (−24/6) = 12; **25.** C = 60, D = 1; **27.** C = 48, D = 1; **29.** A not equal to 0, B not equal to 0

SECTION 2–7, page 66

Exercises 1. −28°C **3.** −55°C **5.** 24 km/h **7.** 8 g **9.** 3 to 11 pounds **11.** 288,698 people

CHAPTER 2 REVIEW, page 68

Exercises 1. Integers are the set of whole numbers and their opposites. **3.** 12 **5.** $<$ **7.** $<$ **9.** Positive, a negative integer times a negative integer equals a positive integer. **11.** -14 **13.** 0 **15.** -7 **17.** -17 **19.** -12 **21.** 55 **23.** -5 **25.** -105 **27.** 4 **29.** 15 **31.** -5 **33.** 4 **35.** -6 points **37.** $(-16 + 3) - -2 = -11$ **39.** $(12 + -1) + -11 = 0$ **41.** b

CHAPTER 3 FRACTIONS AND DECIMALS

SECTION 3–1, page 79

Exercises 1. 0.5 **3.** 0.385 **5.** 0.036 **7.** 2.5 **9.** 2.865 **11.** $\frac{197}{1000}$ **13.** $\frac{23}{1000}$ **15.** $\frac{3}{10}$ **17.** 0.47 is not shaded. **19.** tenths **21.** hundredths **23.** hundreds **25.** $\frac{1}{4}$; 0.25 pound **27.** $\frac{7}{8}$; 0.875 cup

SECTION 3–2, page 83

Exercises 1. b **3.** a **5.** $\frac{1}{2}$ **7.** $\frac{3}{4}$ **9.** $\frac{1}{6}$ **11.** $<$ **13.** $>$ **15.** $>$ **17.** $>$ **19.** $>$ **21.** 5 **23.** 4 **25.** 2 **27.** $\frac{1}{7}, \frac{2}{7}, \frac{3}{7}, \frac{4}{7}, \frac{5}{7}, \frac{6}{7}$

SECTION 3–3, page 87

Exercises 1. $>$ **3.** $>$ **5.** $<$ **7.** $>$ **9.** $<$ **11.** $>$ **13.** 4.76 **15.** 5.1 **17.** 0 **19.** 0 **21.** 66,900 **23.** $0.38

SECTION 3–4, page 89

Exercises 1. a **3.** b **5.** b **7.** a **9.** 10.5 **11.** 25.8 **13.** −3.36 **15.** 5.45 **17.** 7.16 **19.** Estimate: $14; actual amount: $13.53 **21.** Estimate: $17; actual amount: $18.15 **23.** $.62 **25.** 7.5 **27.** −0.87 **29.** 9.974 **31.** $93.05 **33.** Answers will vary. **35.** 100-meter freestyle Alfred H., Jean S., Galaoia Z. **37.** Yes

SECTION 3–5, page 94

Exercises 1. b **3.** a **5.** $\frac{1}{2}$ **7.** 1 **9.** $\frac{1}{5}$ **11.** $\frac{11}{12}$ **13.** $1\frac{1}{2}$ **15.** $\frac{1}{12}$ **17.** Cassandra, $\frac{5}{12}$ hour longer **19.** $\frac{3}{8}$ **21.** $-\frac{9}{10}$ **23.** $-\frac{7}{8}$ **25.** $-1\frac{1}{20}$ **27.** $\frac{5}{8}$ **29.** $1\frac{1}{2}$ yards **31.** $\frac{3}{4}$ hour **33.** $1\frac{7}{10}$ **35.** $-4\frac{1}{4}$ **37.** 0

SECTION 3–6, page 98

Exercises 1. b **3.** a **5.** b **7.** a **9.** $5.52 **11.** $9.72 **13.** $11.92 **15.** 13,650 g **17.** 0.7 **19.** 8.6 **21.** 0.3 **23.** 0.09 **25.** Ready toothpaste at 125 ml for $1.29 is the better buy.

SECTION 3–7, page 102

Exercises 1. $-\frac{3}{8} - \frac{3}{8}$ **3.** 24 students **5.** $\frac{1}{3}$ **7.** $24 **9.** Distributive Property **11.** $-\frac{1}{15}$ **13.** $\frac{7}{12}$ **15.** $-\frac{2}{5}$ **17.** Lost $375 in value **19.** not enough information **21.** $\frac{1}{16}$ **23.** $\frac{1}{4}$ **25.** See TE **27.** 33

SECTION 3–8, page 106

Exercises 1. 1.25 **3.** No reciprocal; error message **5.** $10\frac{1}{2}$ times **7.** $\frac{35}{32}$ or $1\frac{3}{32}$ **9.** $\frac{8}{75}$ **11.** -12 **13.** $\frac{1}{18}$ **15.** $1\frac{1}{3}$ block per minute **17.** 80 customers **19.** 20 classes

CHAPTER 3 REVIEW, page 109

1. $\frac{5}{11}$ **3.** $\frac{3}{7}$ **5.** 0.125 **7.** $<$ **9.** $>$ **11.** $<$ **13.** -2 **15.** $\frac{8}{9}$ **17.** $\frac{11}{24}$ **19.** $\frac{5}{6}$ **21.** -2 **23.** -7.79 **25.** -21.08 **27.** 0.5 **29.** (a) $24.95 (b) $3.00 **31.** b **33.** $23.33

CHAPTER 4 DATA ANALYSIS

SECTION 4–1, page 115

Exercises 1. Newspaper; the local paper has a section on rock. **3.** School survey. **5.** J: Hangin' On **7.** H: Downtown Train **11.** H; J **13.** Orange **15.** Example: Take a census. **17.** Biased; most teens with summer jobs would be at work, and could not answer a telephone survey. **19.** Probably not biased

SECTION 4–2, page 120

Exercises 1. 14 **3.** 7 **5.** 13 **7.** 60 **9.** No **11.** 13 **13.** Science Fiction **15.** 15.0–15.9 and 16.0–16.9 **17.** 63 **19.** Estimate near 15.5 seconds; estimate falls in the 15.0–15.9 second interval.

SECTION 4–3, page 124

Exercises 1. 3 **3.** 3 **5.** 35 **7.** 19 cm **9.** 22 **11.** About 55 or 56 **13.** Quick visual result; high, low and median values can be seen easily. **15.** 66

SECTION 4–4, page 128

Exercises 1. Pictograph **3.** In order: 12; 16; 6; 10 **5.** Horizontal bar graph **7.** 10 **9.** Strong showing comparisons; weak on detail and accuracy. **11.** Day 6; about $1500

SECTION 4–5, page 131

Exercises 1. About the same **3.** Nile: $3\frac{1}{2}$ times as long 1000 km **9.** Yangtze; 1750 km **11.** 3750 km (to nearest 25 km) **13.** Nile, Mississippi, Yangtze **17.** 20 students **19.** Walk; other **21.** 380 **25.** About 250 **29.** More than 2 times; almost 4 times

SECTION 4–6, page 135

Exercises 9. How does the amount of thiamine compare to the amount of iron? How does the amount of riboflavin compare to the amount of niacinamide? How does the amount of iron compare to the amount of niacinamide? **11.** Stem and leaf plot; ranges can be seen more easily. **19.** It starts large and gets smaller toward noon, and then bigger toward evening.

SECTION 4–7, page 139

Exercises 1. They record the same data; the graphs use a different scale. **3.** Sample answer: \$8000. The profit growth will continue at the same rate. **5.** 8 times **7.** By doubling the dimensions, the volume appears 8 times larger. The increase looks greater. **9.** The bottle increases in area as it increases in height, giving the impression that production is rapidly increasing.

SECTION 4–8, page 144

Exercises 1. Probably not; nearest \$100 **5.** Pictograph **7.** More exact than pictograph; less eye-catching than pictograph **9.** $\frac{2}{5}, \frac{5}{12}, \frac{5}{12}, \frac{5}{13}$ **11.** About 50 **13.** About 80-85

CHAPTER 4 REVIEW, page 147

Exercises 1. Take a census. Advantage: everyone's opinion counts; disadvantage: too time consuming. Sample the population. Advantage: quick way of getting a general idea as to how everyone feels; disadvantage: not everyone's opinion counts. **5.** 8:00–9:00 **7.** 8°C

CHAPTER 5 WORKING WITH VARIABLES

SECTION 5–1, page 154

Exercises 1. b **3.** a **5.** $x + 12$ **7.** $14 - y$ **9.** $3(2) - 2(3)$; $6 - 6$; 0 **11.** $4(2)(3) = 24$ **13.** $\frac{3+2}{3-2} = \frac{5}{1} = 5$ **15.** 29 **17.** 24 **19.** 15 **21.** 23 **23.** Two more wins than the Concords and Argonauts

SECTION 5–2, page 158

Exercises 1. (a) Same variables; (b) Different coefficients; (c) Same variables **3.** B **5.** M **7.** B **9.** L **11.** U **13.** 11 **15.** 1 **17.** $4a + 4b$ **19.** $3a + 3b + 2c$ **21.** $4k + 6m$ **23.** $3a + 4b + 3c$ **25.** No monomials; binomials: 17, 18, 20, 21, 24; trinomials: 19, 22, 23 **27.** 3 **29.** 3 **31.** 2, 2, a, a **33.** -1, 3, x, x, y **35.** 2, 2, x, y, z

SECTION 5–3, page 162

Exercises 1. d **3.** e **5.** b **7.** $5x - y$ **9.** $3x + 2x + -4y + 2y = 5x - 2y$ **11.** $4x^2$ **13.** $-6x^2$ **15.** $-x^2 - 2x + 5$ **17.** $-2a - 4b$ **19.** 3 **21.** $5y + 4$ **23.** $-x - y$ **25.** $-2m$ **29.** $3a + 5b$ **31.** $7x - 4y$

SECTION 5–4, page 166

Exercises 1. 3 and -5 are multiplied together. **3.** 6 **5.** 5 **7.** 5 **9.** $(a^3)(a^2) = a^5$ Counting is the same as adding; when multiplying we can add the exponents. **11.** -15 **13.** -8 **15.** -75 **17.** $-20ab$ **19.** $6x^4$ **21.** $-18a^2b^2y^3$ **23.** $\frac{1}{8}$ **25.** $3am + 2an$ **27.** $-4y + 10$ **29.** $18ab - 12ac$ **31.** $(6xy)(6xy) = 36x^2y^2$ **33.** $(2x)(5x - 2) = 10x^2 - 4x$ **35.** 576 sq units

SECTION 5–5, page 170

Exercises 1. Distribute the 3. **3.** Distribute the -1. **5.** Distribute the 3 and -2. **7.** d **9.** a **11.** b **13.** $23 - m$ **15.** $7y - 63$ **17.** $-11y + 4$ **19.** 22 **21.** -12 **23.** 2.5 cents **25.** 312.5 cents **29.** 1 cm × 2 cm = 2 sq cm; 2 cm × 3 cm = 6 sq cm; 4 cm × 5 cm = 20 sq cm

SECTION 5–6, page 174

Exercises 1. False **3.** True **5.** $2a$ **7.** -7 **9.** $5t$ **11.** $6x^2$ **13.** $3y$ **15.** $-6n$ **17.** 8 **19.** $-3xy$ **21.** $-6m + 3$; $-6m$; 3 **23.** $2y - 8$ **25.** $4x^2 - 2y^2$ **27.** 6; 2; 24 **29.** $-4x^2 + x + 3$ **31.** $3x^2 + 2x - 5$ **33.** $\frac{m}{s}$

SECTION 5–7, page 179

Exercises 1. $(2)(3)(x)(x)(y)$ **3.** $(2)(2)(2)(x)(y)(y)$ **5.** $(2)(2)(a)(a)(b)$ **7.** $(-1)(3)(3)(a)(x)(x)(b)$ **9.** $2m$ **11.** e **13.** a **15.** d **17.** $x(5x - 3)$ **19.** $b(4a - b)$ **21.** $4(y^2 - 4)$ **23.** $mn(36 - 25mn)$ **25.** (a) 90; (b) 90; (c) (b) is the factored form of (a); (d) b

SECTION 5–8, page 182

Exercises 1. \$668 profit to the nearest dollar **3.** A: \$1189.95 profit; B: \$943.85 profit **5.** \$17.50; 17.5 ÷ 100 = \$0.18 **7.** The average cost per copy decreases as the number of copies increases. **9.** Langleys' holds 234 cu ft more water. **11.** 0.202 **13.** A race or contest in which disadvantages or advantages are imposed on contestants, so that each may have an equal chance of winning

SECTION 5–9, page 187

Exercises 1. d **3.** a **5.** f **7.** 4 more than n **9.** t tripled **11.** $7x - 4$ **13.** $-x^2 - 4y^2$ **15.** $-6x^2y$ **17.** $-2a$ **19.** $4x^2$ **21.** $15x^3 - 18x^2$ **23.** $2(12) = 24$

CHAPTER 5 REVIEW, page 189

Exercises 1. $(3) + 8 = 11$ **3.** $(3) + (7) = 10$ **5.** $3(7) - 5 = 21 - 5 = 16$ **7.** 3; 1 **9.** -1; 3 **11.** $-4n$, $2n$ **13.** B **15.** M **17.** $4m + 4n$ **19.** $8a^2$ **21.** $2y^5$ **23.** $-4m + 8$ **25.** $-4g + 8$ **27.** $-9x + 18$ **29.** $n + 3$ **31.** $-4b + 8c$ **33.** $-4x$ **35.** C = 100 **37.** $(3kn) - (3k)(-4n)$

CHAPTER 6 SOLVING EQUATIONS AND INEQUALITIES

SECTION 6–1, page 196

Exercises 1. g **3.** b **5.** h **7.** c **9.** $x + 4 = 12$ **11.** $4 + x = 16$ **13.** $16 = x - 4$ **15.** $25 = x + 5$ **17.** $2d - 12 = 25$ **19.** $T - 3 = 83$ **21.** $(6 + n) + 5 = 18$ **23.** $n + 2n + 3 = 213$.

SECTION 6–2, page 200

Exercises 1. $x + 7 = 10$ **3.** $m + 7 = 13$ **5.** $s - 5 = 0$ **7.** $4n = 32$ **9.** $4z = 12$ **11.** Subtract 5. **13.** Multiply by 2.

15. Multiply by $\frac{1}{3}$. **17.** No **19.** No **21.** Yes **23.** Yes **25.** Yes **27.** Yes **29.** Yes **31.** Yes

SECTION 6-3, page 204

Exercises 1. Add -8 or subtract. **3.** Add 1.5. **5.** Add -10. **7.** Subtract 5. **9.** Subtract 5. **11.** Add 10. **13.** $y + 0 = 9$; $y = 9$ **15.** Correct **17.** Incorrect; $m = 21$ **19.** $y = 10$ **21.** $z = 8$ **23.** $s = 8$ **25.** $r = -3$ **27.** $m = 4.8$ **29.** $x = 16$ **31.** $x = -16$ **33.** $p + 3 = 20$, $p = 17$; Joe answered 17 problems correctly. **35.** $8 = 2 + n$; $n = 6$; there are 6 letters in the name. Examples are Festag, Stazik.

SECTION 6-4, page 208

Exercises 1. $\frac{1}{4}$ **3.** $\frac{1}{2}$ **5.** $\frac{1}{6}$ **7.** $-\frac{1}{3}$ **9.** 3 **11.** Incorrect; $m = 36$ **13.** Incorrect; $x = 18$ **15.** $a = 6$ **17.** $k = 7$ **19.** $x = 6$ **21.** $k = 1.6$ **23.** $m = 16$ **25.** Its average speed is $9\frac{1}{5}$ mi/h or 9.2 mph. **27.** Owen's speed is 9.7 meters per second or 9.7 m/s. **29.** The average speed of the bus was 52 mi/h or 52 mph.

SECTION 6-5, page 212

Exercises 1. $(L)\ x = 1$, $(I)\ x = 0$, $(M)\ x = -1$, $(S)\ x = -2$, $(E)\ x = 2$ **3.** c **5.** e **7.** C **9.** $n = 3$ **11.** $x = 0$ **13.** $x = 8$ **15.** $x = -\frac{1}{2}$ **17.** E **19.** G **21.** A **23.** C **25.** 12 degrees latitude southward, 16 degrees longitude eastward

SECTION 6-6, page 216

Exercises 1. Subtract 4; simplify. **3.** Multiply by 5; add 1. **5.** Add 8. **7.** Subtract 12. **9.** a **11.** b **13.** Subtract 3; simplify. **15.** Add 12; simplify. **17.** $x = 10$ **19.** $y = 19$ **21.** $y = 18$ **23.** $x = 3.1$ **25.** $x = 4$ **27.** $x = 2$ **29.** $l = 4$ **31.** $l = 35$ **33.** Solve $2n + 3 = 27$, $n = 12$. **35.** $h = -3$; the octopus has 3 hearts. **37.** $x = 2$; the word has 2 letters.

SECTION 6-7, page 220

Exercises 1. d **3.** a **5.** c **7.** Subtract $5x$ (or subtract $9x$). **9.** $m = 2$ **11.** $k = 9$ **13.** $y = -2$ **15.** $n = 9$; nine nations competed in 1896. **17.** $x = 5$; there are 5 Great Lakes. **19.** $y = 5$; five events are in the pentathlon. **21.** $c = \frac{1}{9}$; $\frac{1}{9}$ of an iceberg shows above the water. **23.** $y = 2$ **25.** $n = 5$

SECTION 6-8, page 224

Exercises 1. $2y - 6$ **3.** $-2x + 12$ **5.** $9 - 6y$ **7.** $\frac{x}{6}$ **9.** $12y + 3 = 9$ **11.** $3x - 2x = 12$ **13.** $5x + 10 = 15$ **15.** $10 = 2y + 6$ **17.** $y = \frac{7}{2}$ **19.** $x = -6$ **21.** $n = 6$ **23.** $m = \frac{1}{2}$ **25.** $y = \frac{1}{12}$ **27.** $n = 30$ **29.** $m = 10$ **31.** $x = 2$ **33.** $x = 10$ **35.** $y = -6$ **37.** 95°F

SECTION 6-9, page 228

Exercises 1. b **3.** d **5.** When the inequality has $>$ or $<$ **7.** $x \leqslant -5$ **9.** $x \leqslant -2$ **11.** $x > -3$ **21.** $x > 14$ **23.** $-10 < x < 0$ **25.** $-10 \leqslant x \leqslant -5$

SECTION 6-10, page 233

Exercises 1. Add -5 to both sides; inequality sign stays the same. **3.** Multiply both sides by 2; inequality sign stays the same. **5.** Add 8 to both sides; inequality sign stays the same. **7.** b **9.** a **11.** $n < 9$ **13.** $t \leq 11$ **15.** $m < 4$ **17.** $x < 6$ **19.** $y \leq 11$ **21.** $2k < 6$; $k < 3$ **23.** $-3x + 5 > 11$; $x < -2$

CHAPTER 6 REVIEW, page 234

Exercises 1. $3n = 12$ **3.** $\frac{n}{5} - 3 = 11$ **5.** M **7.** A **9.** $K = 13$ **11.** $n = 19$ **13.** $x = 18$ **15.** $d + 15 = 28$; the number of days is 13. **17.** $x = 21$ **19.** $y = -8$ **21.** $t = 12$ **23.** $x = 7$ **25.** $n = 4$ **27.** $x = 11$ **29.** $x = 9.3$ **31.** $x = 8$; there are 8 horses. **33.** $n = 2$ **35.** $x = -25$ **37.** $x = 3$ **39.** $r = 3$ **41.** $k = \frac{11}{5}$ **47.** $n > 5$ **49.** $x \geq -33$

CHAPTER 7 PROBLEM SOLVING

SECTION 7-1, page 242

Exercises 1. Virja = \$550; Yannak = \$450 **3.** \$3276 **5.** 1435 m **7.** 14,592

SECTION 7-2, page 247

Exercises 1. A **3.** D **5.** 73 days **7.** $\frac{6}{41}$

SECTION 7-3, page 250

Exercises 1. 10 pennies, 1 nickel, 1 dime, 1 quarter; *or* 5 pennies, 7 nickels, 1 dime **3.** $5.24x$ **5.** $2x - 1$ **7.** 220 **9.** Yes **11.** 6 **13.** 6

SECTION 7-4, page 254

Exercises 1. False **3.** True **5.** 10,000 **7.** 48 **9.** Sometimes true **13.** 18 square mm

SECTION 7-5, page 259

Exercises 1. 5 **3.** 1, 2, 3, 4, 6, 7, 8, 9 **5.** 14; 16 **7.** 26; 26 **9.** $n = 5$ **11.** 2.16 **13.** 6924

SECTION 7-6, page 262

Exercises 1. I **3.** II **5.** 1) a = amount of blocks for two days; b = amount of blocks on Monday; c = amount of blocks on Tuesday; 11 blocks. 2) a = how long the trip took; b = the distance of the path; c = the speed he went up hill; 5 seconds. 3) a = money in account; b = amount deposited; c = amount withdrawn; \$250 in her account. 4) a = amount of money needed; b = number of cards; c = price per card; 50 cents. **7.** 7.5 miles **9.** 688 m **11.** \$160 **13.** \$40 **15.** \$40

SECTION 7-7, page 266

Exercises 1. 36 **5.** 24 ways **7.** If Sharon had them all right then Barbara would have missed them all. Thus if Barbara had them all right then Sharon would have missed them all. Zelda must have had them all right. Sharon would have missed 1, and Barbara missed 2. **9.** 8 squares

SECTION 7–8, page 270

Exercises 1. 7 cubes; 10 cubes **3.** 19 **5.** 16 **7.** $a_n = n(n + 4)$ where n is a member of the set {1, 2, 3, . . . }

CHAPTER 7 REVIEW, page 272

1. Cost of party **3.** Total of 12 people at the party **5.** 20 posts **7.** Sometimes true **9.** 20 possible choices **11.** Car 2 **13.** 96 **15.** $a_1 = 3$; $a_n = a_{n-1} + 1$ where $n = \{2, 3, 4, \ldots\}$ **17.** 12 units

CHAPTER 8 RATIOS, PROPORTIONS, AND PERCENTS

SECTION 8–1, page 280

Exercises 1. 3:5 **3.** 8:3 **5.** 4:3 **9.** Bs to As **11.** 40:67 **13.** M: 23:45; B: 22:39; J: 25:42 **15.** 5:6 **17.** 1:2 **19.** 35 **21.** 2:35

SECTION 8–2, page 284

Exercises 1. 6:8 = 12:16 = 15:20 **3.** 12:8 **5.** $\frac{6}{5}$ **7.** 1:2 **9.** 2:7 **11.** 2:3 **13.** 2:1 **15.** 10:1 **17.** 2:3 **19.** 4:6 = 2:3 **21.** 12:16 = 3:4 **23.** 4 **25.** 9 **27.** 34 **29.** 343 **31.** 24 **33.** 6

SECTION 8–3, page 288

Exercises 1. 1.62 cents **3.** 28.71 cents **5.** $1.69 **7.** .59 cent **9.** 0.03 cent **11.** 3.4 cents **13.** 4.0 cents **15.** 28 cents **17.** 26 cents **19.** B **21.** C **23.** $0.88 **25.** $0.86

SECTION 8–4, page 292

Exercises 1. Inches per hour **3.** Pounds per number of minutes **5.** 20 minutes **7.** 48 blocks **9.** $3.25 **11.** 66.3 cents each **13.** B **15.** $\frac{AC}{DF} = \frac{BC}{EF}$; 5 **17.** If A items cost B cents and you want to find how many items you can buy for D cents, you write the proportion $\frac{A}{B} = \frac{x}{D}$ and solve by cross multiplication. You then divide by B and get $x = \frac{(A)(D)}{B}$. **19.** $x = 36$ **21.** She forgot to change 1 hour to 60 minutes. She got the number of problems done in 45 minutes.

SECTION 8–5, page 296

Exercises 1. $\frac{43}{100}$; .43; 43% **3.** $\frac{19}{100}$; .19; 19% **5.** 25% **7.** 10% **9.** 800% **11.** 73% **13.** 9% **15.** 12.5% **17.** 70% **19.** 10% **21.** 40% **23.** 66.7% **25.** 81.8% **27.** 80% **29.** 62.5%

SECTION 8–6, page 300

Exercises 1. 28%; 70%; $\frac{7}{10}$ is greater. **3.** 80%; 60%; $\frac{4}{5}$ is greater. **5.** 69.3%; 69.5%; $\frac{66}{95}$ is greater. **7.** 17.9%; 17.1%; $\frac{28}{156}$ is greater. **9.** 16% **11.** 10% **13.** 24% **15.** 15.3% **17.** 40%; 60%; white **19.** Second test: 66.7% vs 75% **21.** (a) 17.6% (b) 82.4% (c) 100% **23.** Rice: 52.4% vs 49%

SECTION 8–7, page 304

Exercises 1. d **3.** a **5.** 1 m **7.** 30 cars **9.** 1.5 g **11.** 12 oz **13.** 4.5 km **15.** 16 people **17.** 105 **19.** 52.6 **21.** 360 **23.** 19 **25.** 4 **27.** 1 cent **29.** 3 **31.** $51.30; $290.70

SECTION 8–8, page 308

Exercises 1. Part **3.** Whole **5.** $\frac{1}{100} = \frac{12}{x}$ **7.** $\frac{6}{100} = \frac{21}{x}$ **9.** 1600 **11.** 280 **13.** 30 **15.** 32 **17.** 400 **19.** 500 **21.** 53.3 **23.** 80 **25.** 2500 **27.** 450 **29.** 12.9 ft **31.** 819.7 lb **33.** $11 **35.** 42,000

SECTION 8–9, page 312

Exercises 1. Part **3.** Part **5.** Part **7.** What percent of 80 is 60? **9.** 60 is 33% of what number? **11.** 25% **13.** 48 **15.** 60 **17.** 25% **19.** 77.7% **21.** $31.86 **23.** (a) 14.4 (b) 17.28 (c) 106.25 **25.** 101.5 lb

CHAPTER 8 REVIEW, page 314

1. 25:28 **3.** 25:33:28 **5.** No **7.** No **9.** 20 **11.** 3 **13.** 72 **15.** .88 cent; .89 cent; 3 dozen **17.** 12 hours **19.** 37.5% **21.** 5.4% **23.** 75%; 70%; $15 **25.** 45 **27.** 567 **29.** 50 **31.** $20.00 **33.** $29,200

CHAPTER 9 EXPONENTS AND MEASUREMENT

SECTION 9–1, page 322

Exercises 1. 100,000 **3.** 16 **5.** m^{12} **7.** $8y^3$ **9.** $\frac{m^4}{n^4}$ **11.** When dividing powers with like bases, subtract the exponents. **13.** $(3^2)(3^3) = 3^{2+3} = 3^5 = 243$; $(9)(27) = 243$; $\frac{5^3}{5^2} = 5^{3-2} = 5^1 = 5$; $\frac{125}{25} = 5$; $(4^3)^2 = 4^{3 \times 2} = 4^6 = 4096$; $(64)^2 = 4096$ **15.** Positive **17.** Positive **19.** Negative **21.** −24 **23.** 144 **25.** $<$ **27.** $=$ **29.** Division; $k = 1$ **31.** Division; $x = 4$

SECTION 9–2, page 326

Exercises 1. 27, 9, 3, 1, $\frac{1}{3}$, $\frac{1}{9}$ or $\frac{1}{3^2}$ **3.** 0.1 **5.** 10^{-3} **7.** 0.00001 **9.** 0.0000001 **11.** −3 **13.** 3.8 **15.** $9.25 \times 10^{-3} = 0.00925$ **17.** $4.51 \times 10^6 = 4{,}510{,}000$ **19.** 3.21 − E04 **21.** 6.96 + E07 **23.** 1.0×10^{-6}; 3.6×10^9

SECTION 9–3, page 331

Exercises 1. $1\frac{2}{3}$ ft **3.** 1320 ft **5.** 5.7 cm **7.** Inches **9.** Feet **11.** Feet **13.** 7, 9, 11, and 12: centimeters; 8: millimeters; 10: kilometers

SECTION 9–4, page 334

Exercises 1. 7 ft 6 in. **3. a.** 20 cm; **b.** 16 cm; **c.** 16 cm; **d.** 24 cm; **e.** 12 cm **5. a. 8 in.; b.** $10\frac{1}{2}$ in.; **b** has the greater perimeter.

SECTION 9–5, page 338

Exercises 1. A **3.** B **5.** 6 m^2 **7.** 9 m^2 **15.** 314 m^2 **17.** 12.56 yd^2 **19.** 2038.1 m^2 **21.** 346.2 m^2 **23.** 9

SECTION 9-6, page 342

Exercises 1. 12 **3.** 6 m^2 **5.** $y^2 - 2y$ **7.** $6m^2 - 2m$ **9.** $-3n^2 + 12$ **11.** $x^2 + 5x + 4x + 20 = x^2 + 9x + 20$ **13.** They are the same because either bracket can be distributed. **15.** $(x + 5), (x + 5)$ **17.** $6(y - 3)$ **19.** $x^2 + 5x - 24$ **21.** $x^2 - 4y^2$ **23.** $25a^2 - 4b^2$ **25.** They are the difference of two squares.

SECTION 9-7, page 346

Exercises 1. a. 72 cm^3; b. 128 cm^3; c. 125 cm^3 **3.** 2 g **5.** 12 kg **7.** 1 kL **9. a.** 13.8 m^3; **b.** 82 cm^3 **11.** 20 cm; 11.25 cm **13.** Ounces or milliliters **15.** Pounds or kilograms

SECTION 9-8, page 350

Exercises 1. SA = $6a^2$ **3.** *a.* V = $24x^2$ cu units; SA = $40x + 12x^2$ sq units **5.** V = $12c^3 + 12c^2d$ cu units; SA = $38c^2 + 14cd$ sq units **7.** 142 cm^2 **9.** 96.5 cm^2 **11.** l = 1; w = 2; h = 3; V = 6; SA = 22; and l = 2; w = 2; h = 2; V = 8; SA = 24.

SECTION 9-9, page 355

Exercises 1. Volume of ingredients (cups); width of pie tin (cm or inches) **3.** Volume of water (milliliters or cups); capacity of container (milliliters or cups) **5.** Height of hoop (meters or feet) **7.** About 45 seconds **9.** About 20 minutes **11.** 20°–25° C; 68°–77° F **13.** 11.6 days **15.** 1:00 PM

CHAPTER 9 REVIEW, page 356

1. 3^8 **3.** y^2 **5.** −64 **7.** 7.2 + E07 **9.** $1\frac{1}{2}$ **11.** Inches, centimeters **13.** Inches, millimeters **15.** $7x + 9y$ **17.** 144 **19.** 9 **21.** $4x^2 - 4x + 1$ **23.** 77.1 m^3 **25.** A **27.** SA = 52 sq units; V = 24 cu units **29.** Length and area

CHAPTER 10 GEOMETRY

SECTION 10-1, page 364

Exercises 5. *AB, KL, FG, UV* **7.** An acute angle has a measure greater than 0° but less than 90°. **9.** An obtuse angle has a measure greater than 90° but less than 180°. **17.** Two angles are adjacent if they have the same vertex, a common side, and do not share any interior points. **19.** Two angles are complementary if the sum of their measures is 90°.

SECTION 10-2, page 368

Exercises 1. Measures of the angles are the same. **3.** Vertical angles have the same measure. **5.** The base angles have the same measure. **7.** The angles all measure 60°. **9.** The angles of an equilateral triangle all measure 60°. **11.** $y = 30°$ **13.** $x = 121°$ **15.** $m = 60°$, $y = 60°$, $k = 60°$ **17.** $x = 50°$

SECTION 10-3, page 372

Exercises 1. Isosceles **3.** Scalene **5.** Equiangular **7.** Right **9.** True; one obtuse angle and two acute angles of different measures **11.** False; an equilateral triangle has 3 equal angles that measure 60°. **13.** Rectangle **15.** Hexagon **17.** 40 cm^2 **19.** 18 cm^2

SECTION 10-4, page 376

Exercises 1. 9 **3.** 64 **5.** 4 **7.** 10 **9.** 2.65 **11.** 4.90 **13.** $x = 11.66$ **15.** $w = 15$ **17.** 80 m **19.** 80 m; 45 m

SECTION 10-5, page 380

Exercises 1. Rational **3.** Rational **5.** Rational **7.** Rational **9.** Irrational **11.** 3.14159; through the millionths place **13.** $6 < \sqrt{40} < 7$ **15.** $-9 < \sqrt{75} < -8$ **17.** All real numbers less than 3 **19.** The members of both sets are less than 3. Set A contains all numbers less than 3, whereas set B contains only integers less than 3. **23.** No; $\frac{2}{3} - (\frac{1}{2} - \frac{3}{4}) \neq (\frac{2}{3} - \frac{1}{2}) - \frac{3}{4}$

SECTION 10-6, page 384

Exercises 1. 7 cm; 10 cm **3.** Similar figures: *A, C, G, H,* and *K; B, D, E,* and *J; F* and *I.* Congruent figures: *A* and *K; D, E,* and *J; G* and *H; F* and *I.* **5.** $\frac{AB}{DE} = \frac{EF}{BC}; \frac{CA}{FD}$ **7.** $d = 50$ m **9.** You assume that the jet rises at a constant rate throughout its climb.

SECTION 10-7, page 388

Exercises 1. $C \approx 31.4$ in. **3.** You walk 343.52 m around the track. **5.** They are all 120°. **7.** $\frac{72°}{360°} = 20\%$ **9.** About 28%

SECTION 10-8, page 392

Exercises 1. Regular polygon is both equilateral and equiangular. All the faces of a regular polyhedron are congruent regular polygons. **3.** V = 595 cm^3 **5.** Container C holds the most; C, D, A, B **7.** V ≈ 618,894 cm^3 **11.** V ≈ 73.58 m^3

CHAPTER 10 REVIEW, page 394

1. True **3.** False **5.** True **7.** $x = 93°$ **9.** Right triangle **11.** $x = 3$ cm **13.** Irrational **15.** Rational **19.** C = 50.24 cm; A = 200.96 cm^2 **21.** Cylinder

CHAPTER 11 RELATIONS AND FUNCTIONS

SECTION 11-1, page 402

Exercises 1. (2,2) **3.** (2,−3) **5.** (5,−7) **7.** (−5,5) **9.** *A* **11.** *J* **13.** *Q* **15.** *R* **17.** *H* **19.** *W* (0,6); *L* (0,2); *N* (0,−6) They all lie on the *y*-axis. **21.** *U, H, V, J, D* **23.** The *x*-coordinates indicate different directions along the *x*-axis. The *y*-coordinates indicate different directions along the *y*-axis. **35.** Square **37.** M

SECTION 11-2, page 406

Exercises 1. 1, 2, 3 **3.** R = {0, 1, 2, 3} **5.** D = {1, 2, 3, 4} **9.** The second number is the square of the first plus 1. **11.** 17 **13.** 255 **15.** 32 **17.** A relation rule between two numbers such that one more than twice the domain gives the range **19.** A relation rule between two numbers such that twice the sum of the domain and one gives the range **21.** Relation rule: is sister to; (Cassandra, David), (Lina, Januk)

SECTION 11-3, page 410

Exercises 1. D = {0, 10, 15, 30, 50, 60, 90, 130, 150, 180} **3.** 6 **7.** (May,−500); (June,250); (July,500); (August,750); (September,100) **9.** D,H **13.** R = {30, $30\frac{1}{4}$, $30\frac{1}{2}$, $30\frac{3}{4}$, 31}

SECTION 11-4, page 414

Exercises 1. 1; −1; 3; −3; 5 **3.** 10; 1; 10 **5.** −7; −5; −3 **11.** (−1,−4); (0,3); (1,2); (2,−1); (3,0) **17.** (5,5); (6,6); (−5,−5); $y = x$ **19.** (2,6); (3,7); (4,8); $y = x + 4$

SECTION 11-5, page 418

Exercises 1. Function **3.** Not a function; all x-coordinates are the same. **5.** Function **7.** Function **9.** Function **11.** No; (2,2), (2,−2) or (0,4), (0,−4) **13.** (−2,−7); (−1,−5); (0,−3); (1,−1); (2,1); function **15.** (−2,4); (−1,1); (0,0); (1,1); (2,4); function **17.** Yes; the domain is given and each value is used only once. **19.** Yes; each person has only one birthday. **21.** Yes; each student has only one average.

CHAPTER 11 REVIEW, page 420

1. B (5,4); C (4,0); H (−1,4); I (−4,0); L (−4,−3) **3.** H, J **5.** 3, 4, 5, 6 **7.** R = {3, 4, 5, 6} **15.** (2,3) **19.** Function **21.** Function

CHAPTER 12 GRAPHING RELATIONS AND FUNCTIONS

SECTION 12-1, page 429

Exercises 1. 1 **3.** 3 **5.** 1 **7.** −1 **9.** 5 **11.** A, C, E; B, D, F **13.** 1, 3, 5 **15.** (t,T) = (0,0), (1,2.5), (2,5)

SECTION 12-2, page 432

Exercises 1. $\frac{1}{4}$ **3.** 11 **5.** Slope = $\frac{4}{-2} = -2$ **7.** Rise: −3; run: 2; slope: $\frac{-3}{2}$ **9.** No slope **11.** Slope at zero **13.** Points are on the same line. **15.** Not collinear

SECTION 12-3, page 436

Exercises 1. b, a, c, f, d, e **3.** $\frac{3}{5}$ **5.** $-\frac{1}{6}$ **7.** $y = -3x$; −3 **9.** $y = -2x - 1$; −2 **11.** Neither **13.** Parallel **15.** Perpendicular **17.** Parallel

SECTION 12-4, page 440

Exercises 1. Slope: 2; y-intercept; (0,0) **3.** Slope: −3; y-intercept: (0,1) **5.** Slope: $-\frac{2}{3}$; y-intercept: (0,0) **23.** (−2,0) **25.** (−7,0) **27.** $(-\frac{1}{2}, -2)$ **29.** (2,−3)

SECTION 12-5, page 444

Exercises 1. (2,3) **3.** No **5.** Point of intersection: (1,1) **7.** Point of intersection: (1,−1) **9.** Point of intersection: (−3,4) **11.** $y = -2x + 3$; $y = \frac{3}{4}x + 3$ **13.** $y = -x + 1$; $y = -x + 1$ **15.** 11: (0,3); 12: No solution; lines are parallel; 13: Infinite number of solutions **17.** Point of intersection: (−1,2) **19.** Point of intersection: (2,6) **21.** 5 apples and 9 oranges **23.** 2;8

CHAPTER 12 REVIEW, page 446

1. Yes **3.** Yes **5.** $y = -2x + 6$ **7.** $y = \frac{1}{3}x$ **9.** $-\frac{3}{2}$ **11.** 2 **19.** c, b, a, d **21.** $-\frac{4}{5}$ **23.** $\frac{3}{2}$ **25.** Parallel **27.** Neither **29.** Slope: 0; y-intercept: (0,6) **31.** Slope: $\frac{1}{2}$; y-intercept: (0,−3) **41.** (−5,0) **43.** (−4,0) **47.** $x + y = 0$; $x = y - 4$

CHAPTER 13 PROBABILITY AND STATISTICS

SECTION 13-1, page 454

Exercises 1. 1 **5.** 70% **7.** 50; 50 **9.** The probability did not match the actual occurrence exactly. Also, there is a wider variation at certain times but the more times the coin is tossed, the closer the actual results become to the actual mathematical probability. **11.** $\frac{4}{9}$ **13.** 0 **15.** $\frac{3}{10}$ **17.** $\frac{17}{25}$ **19.** $\frac{1}{13}$ **21.** $\frac{1}{26}$ **23.** $\frac{5}{13}$ **25.** The diagram lets you see the progression as each coin is tossed. The beginning of each tree represents the 2 possibilities when the nickel is tossed. The second column shows the 2 possibilities when the dime is tossed. The last column indicates the 2 possibilities when the penny is tossed, with each column taking in the possibilities of the previous coin tossed. **27.** Probability of 1 head and 2 tails is $\frac{3}{8}$.

SECTION 13-2, page 459

Exercises 1. 10,000 **3.** 20 **5.** 358,800 **7.** 143,640 **9.** 144 **11.** $\frac{1}{100}$

SECTION 13-3, page 462

Exercises 1. 60 **3.** 840 **5.** 56 **7.** 6 **9.** 210 **11.** 5040 **13.** 30,240 **15.** 180 **17.** 720

SECTION 13-4, page 466

Exercises 1. 15 **3.** 220 **5.** 210 **7.** 15 **9.** 2100 **11.** 11,760 **13.** 435 **15.** No, it should be called a permutation lock because the order of the numbers makes a difference.

SECTION 13-5, page 470

Exercises 1. $\frac{1}{12}$ **3.** $\frac{1}{4}$ **5.** $\frac{1}{4}$ **7.** $\frac{1}{4}$ **9.** $\frac{1}{8}$ **11.** $\frac{2}{15}$ **13.** $\frac{9}{35}$ **15.** $\frac{1}{20}$ **17.** $\frac{1}{495}$ **19.** $\frac{3}{25}$ **21.** $\frac{7}{60}$

SECTION 13-6, page 474

Exercises 1. 478 **3.** $\frac{239}{500}$ **5.** 86 **7.** $\frac{11}{76}$ **13.** $\frac{11}{36}$ **15.** 388

SECTION 13-7, page 478

Exercises 1. 8 **3.** Since the number appears the same number of times, there is no mode. **5.** 39.9, 40, 38 **7.** 7.6, 7.7, 5.7 **9.** 12.6; 12; 8 and 18 **13.** It is not representative of most of the set of data. **17.** 40–50 **19.** 40–50 **23.** No; the ranges are not small enough.

CHAPTER 13 REVIEW, page 480

1. $\frac{1}{4}$ **3.** $\frac{1}{2}$ **5.** The probabilities will usually be different but the more times you do it, the closer you will get to the theoretical probability. **7.** 60 **9.** 30 **11.** 10,080 **13.** 165 **15.** $\frac{1}{5}$ **17.** $\frac{7}{75}$ **19.** 516 **21.** 162.2; 179.5; 85 **25.** The total number of points in 4 meets since this is based on the mean and accounts for extreme differences.

CHAPTER 14 TRANSFORMATIONAL GEOMETRY

SECTION 14-1, page 488

Exercises 1. Slide 5 right and 2 up **3.** Flip and a slide. **7.** $P(-5,-2) \rightarrow P'(4,3)$; $Q(-4,-2) \rightarrow Q'(5,3)$; $R(-4,-4) \rightarrow R'(5,1)$; $S(-6,-4) \rightarrow S'(3,1)$; $T(6,-3) \rightarrow T'(3,2)$ **9.** Preimage plus 5 **11.** $S'(1,5)$; $T'(-2,2)$; $U'(1,-1)$; $V'(4,2)$ **13.** $H'(3,1)$; $J'(6,1)$; $K'(6,-1)$; $L'(4,-1)$; $M'(4,-4)$; $N'(3,-4)$ **15.** $A'(-2,4)$; $B'(-6,2)$; $C'(-1,-1)$ **17.** $AC = \sqrt{26}$; $A'C' = \sqrt{26}$ **19.** $BC = \sqrt{34}$, $B'C' = \sqrt{34}$ **21.** $A'(2,1)$; $B'(6,3)$; $C'(4,7)$; $D'(0,5)$ **23.** In a translation, the figure and its image will: 1) have the same slopes of corresponding sides. 2) have the same lengths of corresponding sides. 3) have same angle measures of corresponding angles.

SECTION 14-2, page 492

Exercises 1. What is red, striped and crawls? **3.** The answer in 1 is a red, striped crawler. **5.** 2:30 **11.** U **17.** $(-3,5)$; $(3,5)$ **19.** y-axis **21.** The figure that is formed by the image and the original figure is divided in half by the y-axis. **23.** They are all equal. **25.** They are opposites. **29.** The order of the variables changes from clockwise to counterclockwise. The shape of the figure does not change.

SECTION 14-3, page 496

Exercises 1. R **3.** P **5.** 90° clockwise **7.** 180° clockwise **13.** Square **17.** They are equal. **19.** They change.

SECTION 14-4, page 500

Exercises 1. 1 **3.** 3 **5.** 2 **7.** 1 **11.** 5 **13.** 8 **15.** No lines of symmetry; yes, rotational symmetry **17.** 1 line of symmetry **19.** No lines of symmetry; yes, rotational symmetry **21.** 4

SECTION 14-5, page 504

Exercises 5. Slide or translation **9.** $A'(-5,1)$; $B'(-4,3)$; $C'(-2,0)$ **11.** $P'(-4,-4)$; $O'(-2,0)$; $R'(-3,7)$; $P''(4,-4)$, $O''(2,0)$, $R''(3,7)$ **13.** They are equal. **15.** $P'(4,-6)$; $O'(2,-2)$; $R'(3,5)$; $P''(4,-4)$; $O''(2,0)$; $R''(3,7)$ **17.** Changes are the order of vertices and slopes of corresponding line segments. Measures of corresponding sides and angles do not change.

SECTION 14-6, page 508

Exercises 7. 5 basic shapes. **9.** Rhombus **11.** Rectangle **13.** Reflectional and rotational

SECTION 14-7, page 512

Exercises 1. $A \rightarrow A'$; $B \rightarrow B'$; $C \rightarrow C'$; $D \rightarrow D'$ **3.** $\frac{A'B'}{AB} = 2:1$; $\frac{B'C'}{BC} = 2:1$; $\frac{C'D'}{CD} = 2:1$; $\frac{D'A'}{DA} = 2:1$ **9.** Illegible **11.** Clockwise **13.** EF is not congruent to $E'F'$; EG is not congruent to $E'G'$; FG is not congruent to $F'G'$. Conclusion: no, corresponding sides are not equal. **15.** Angle measures and order of vertices remain unchanged; lengths of line segments change. **17.** Corresponding angles of at least two angles of two triangles are congruent. **19.** $\frac{AB}{ED} = \frac{BC}{DC}$; $\frac{d}{25} = \frac{40}{20}$; $\frac{d}{25} = \frac{2}{1}$; $d = 50$ meters The width of the river is 50 meters. **21.** $(-2,-4) \rightarrow (1,2)$; $(6,-2) \rightarrow (-3,1)$; $(2,-4) \rightarrow (1,-2)$; $(x,y) \rightarrow (\frac{x}{-2}, \frac{y}{-2})$

CHAPTER 14 REVIEW, page 515

1. $K'(1,3)$; $P'(-1,0)$; $R'(3,-4)$ **3.** $R'(4,1)$; $Q'(-4,2)$; $Y'(0,4)$ **5.** $P'(4,3)$; $Q'(2,1)$. $R'(0,4)$ **7.** The order of the vertices is the same. **9.** 8 lines of symmetry **11.** 4 **13.** Each interior angle = 120°. Sum at one point = 360°, which fills the space about a point. **15.** $\frac{1.8}{4} = \frac{x}{4} + 12$; $x = 7.2$

GLOSSARY

absolute value The distance of any number, x, from zero on the number line. Represented by $|x|$.

acute angle An angle measuring between 0° and 90°.

acute triangle A triangle having three acute angles.

addition property of equality For any numbers a and b, if $a = b$, then $a + c = b + c$. $3 = 3$, so $3 + 6 = 6 + 3$.

addition property of inequalities For any numbers a and b, if $a > b$, then $a + c > b + c$. $6 > -3$, so $6 + 4 > -3 + 4$.

additive identity The number which, when added to a given number, results in that given number. For real numbers, the additive identity is zero.

additive identity property The sum of a number and the additive identity (0) is the original number. $5 + 0 = 5$

additive inverse A number which, when added to a given number, yields a sum of zero.

additive inverse property The sum of a number and its additive inverse is zero. $5 + -5 = 0$

algebraic expression An expression involving at least one variable.

angle A plane figure formed by two rays having a common endpoint.

arc A part of a circle.

area The number of square units needed to cover a surface.

arithmetic sequence A sequence of numbers in which the difference between each two consecutive terms is the same.

associative property If three or more numbers are added or multiplied, the numbers can be regrouped without changing the result. $4 + (6 + 5) = (4 + 6) + 5$; $4 \times (6 \times 5) = (4 \times 6) \times 5$

average The sum of a set of numbers divided by the number of numbers in that set.

axes Two perpendicular lines used for reference in a coordinate plane.

---

bar graph A means of displaying statistical information in which horizontal or vertical bars are used to compare quantities.

base (of a power) The repeating factor in a power. In 2^3, the base is 2; see also *power*.

BASIC A computer language. BASIC is an acronym for Beginner's All-purpose Symbolic Instruction Code.

binomial A polynomial having two unlike terms.

broken-line graph A means of displaying statistical information by connecting graphs of ordered pairs to show changes in quantities.

---

capacity The amount of liquid or dry ingredients a container can hold.

Cartesian plane See *coordinate system.*

Celsius The metric temperature scale.

center of a circle The point from which all points on the circle are equidistant.

central angle An angle in the plane of a circle having its vertex at the center of the circle.

chord A line segment joining two distinct points on a circle.

circle The set of points on a plane that are all the same distance from a given point.

circle graph A means of displaying statistical information in which parts of a circle are used to compare parts of a whole.

circumference The distance around a circle.

class interval A statistical method for grouping data in a frequency chart.

closed If an operation is performed on any two numbers in a given set and the result is also a member of the set, then the set is closed, with respect to that operation (closure property).

coefficient A number by which a variable or group of variables is multiplied. In $7ab^2$, 7 is the coefficient.

collinear Three or more points lying in a straight line.

combination An arrangement of members of a set in which numbers do not repeat but in which the order does not matter.

common denominator A denominator that is a multiple of all original denominators. For example, in the problem $\frac{1}{15} + \frac{1}{45}$, common denominators would include 45, 90, 135, and so on. A common denominator is the same as the common multiple of the numbers, which are denominators.

common factor A number that is a factor of two or more numbers.

commutative property If two numbers are added or multiplied, the operations can be done in any order. $4 \times 5 = 5 \times 4$; $5 + 4 = 4 + 5$

compass An instrument used to draw circles and arcs and to transfer measurements.

composite number A whole number greater than one having more than two factors.

cone A three-dimensional figure having a circular base and one vertex.

congruent Figures having exactly the same shape and size.

conjecture A guess or conclusion based on assumed or known knowledge but without proof.

constant A monomial that does not contain a variable.

coordinate of a point A number associated with a point on a number line or an ordered pair of numbers associated with a point on a grid.

coordinate system Two perpendicular number lines forming a grid.

cross products For $\frac{a}{b} = \frac{c}{d}$, the cross products are ad and bc. In a proportion, cross products are equal.

cube a right prism having six congruent square faces.

cylinder A three-dimensional figure having two parallel and congruent circular bases.

---

data Statistical information, most often gathered in numerical form.

decagon A ten-sided polygon.

decimal form A number written in the decimal system (base 10) using the digits 0, 1, 2, 3, 4, 5, 6, 7, 8, 9.

decimal numeral system The number system based on powers of ten, including the use of zero.

degree A common unit of measurement for angles or temperature.

degree of a monomial The sum of the exponents of the variables of the monomial.

degree of a polynomial The greatest degree of any monomial within the polynomial.

denominator In the fraction $\frac{2}{3}$, the number 3 is the denominator. The denominator tells the number of equal parts into which a whole has been divided.

density The idea that between any two real numbers there is at least one other real number.

dependent events A pair of events where the outcome of the second event depends upon the outcome of the first event.

diagonal A line segment joining two non-adjacent vertices of a polygon.

diameter A line segment passing through the center of a circle and having endpoints on the circle.

dilation A transformation producing a similar, but not necessarily congruent, figure.

distributive property A property of numbers illustrated by $3 \times (5 + 4) = (3 \times 5) + (3 \times 4)$; $3 \times (5 - 4) = (3 \times 5) - (3 \times 4)$.

divisible A number is divisible by another number if the remainder is zero or if the second number is a factor of the first number.

division property of equality For any numbers a, b, and c, where $c \neq 0$, if $a = b$, then $\frac{a}{c} = \frac{b}{c}$.

division property of inequalities For any numbers a, b, and c, where $c \neq 0$, if $a > b$ and $c > 0$, $\frac{a}{c} > \frac{b}{c}$; if $a > b$ and $c < 0$, $\frac{a}{c} < \frac{b}{c}$. $12 > 9$, so $\frac{12}{-3} > \frac{9}{3}$; $12 > 9$, so $\frac{12}{-3} < \frac{9}{-3}$.

domain The set of first elements in the ordered pairs of a relation.

---

enlargement A figure similar to a given figure but larger.

equation A mathematical sentence containing an equal sign (=).

equidistant All points being the same distance from a given point.

equilateral triangle A triangle having three congruent sides.

equivalent fractions Fractions that name the same number. For example, $\frac{2}{4} \times \frac{3}{3} = \frac{6}{12}$ means $\frac{2}{4}$ and $\frac{6}{12}$ are equivalent fractions.

estimate An approximation of an answer or measurement.

evaluate To find the value of an algebraic expression.

even number Any number divisible by two.

event A result of a probability investigation.

expanded form A decimal number is written as sums of products of powers of tens. Expanded form shows the place value of each digit. $421 = 4 \times 100 + 2 \times 10 + 1$

experimental probability The probability, expressed as a fraction $\frac{m}{n}$, where m is the number of times a desired result actually occurred in n tries.

exponent A number showing how many times the base is used as a factor. For 2^3, the exponent is 3.

exponential form Number written using exponents. $2 \times 2 \times 2 = 2^3$

extrapolation Extending and estimating data based on given information.

extremes In a proportion, the first and fourth terms.

---

face value A digit that tells how many ones, tens, hundreds, and so on are in the number.

factor Any number multiplied by another number to produce a product.

factoring Finding two or more factors of a number or a polynomial.

Fahrenheit The customary temperature scale in the United States.

flip A transformation resulting in the one-to-one mapping of a figure when reflected over a line.

flow chart An outline of steps needed to solve a problem.

formula A specific equation giving a rule for relationships between quantities.

frequency chart A method of organizing statistical information.

function A relation in which each member of the domain is paired with one, and only one, member of the range.

Fundamental Counting Principal This principal states that if there are a ways of doing one task and b ways of doing a second task, then there are ab different ways in which these two tasks can be completed.

---

generalize To take known information or patterns and make a general rule.

gram The basic metric unit of mass.

greatest common factor The largest integer that is a factor of two or more numbers.

---

heptagon A polygon having seven sides.

hexagon A polygon having six sides.

histogram A graph, similar to a bar graph, used to display statistical information.

hypotenuse The side opposite the right angle in a right triangle.

---

identity See *additive identity* and *multiplicative identity.*

image If A is mapped onto A′, then A′ is the image of A.

improper fraction A fraction in which the numerator is greater than the denominator.

independent events A pair of events where the outcome of the second event does not depend on the outcome of the first event.

inductive reasoning Logical reasoning where a generalization is made from a limited number of specific examples.

inequality A mathematical sentence involving the symbols $<$, $>$, $\leq$, or $\geq$.

integers The set of whole numbers and their opposites.

interpolation Interpreting and estimating data between given values.

inverse See *additive inverse* and *multiplicative inverse.*

inverse operations Operations that undo each other, such as multiplication and division.

irrational numbers Real numbers that cannot be expressed in the form $\frac{a}{b}$, where a and b are integers.

isosceles triangle A triangle having at least two congruent sides.

---

language form A number written in words. 421 = four hundred twenty-one.

least common multiple The smallest number that is a multiple of two or more numbers.

legend A list providing explanations of symbols used in a map or pictograph.

legs In a right triangle, the two sides that are not the hypotenuse.

like fractions Fractions having the same denominator. For example, $\frac{5}{8}$ and $\frac{1}{8}$ are like fractions.

like terms Mathematical expressions containing the same variables. $9a^2b$ and $15a^2b$

line graph See *broken-line graph.*

line of symmetry A line separating a plane figure into congruent figures, so that when one part is reflected over that line it matches the other part exactly.

line segment The set of points containing two endpoints and all points between them.

linear equation An equation for which the graph is a straight line.

linear measurement The distance between two points, generally expressed in customary or metric units.

liter Basic unit of capacity in the metric system.

lowest terms A fraction in which the numerator and denominator do not have a common factor other than one.

---

mapping notation A rule assigning each member of a set (or figure) to exactly one number of a second set (or figure). Generally shown as $2 \rightarrow 4$ or $A \rightarrow A'$.

mass The amount of matter an object contains.

matrix An ordered table or listing of numerical data.

mean See *average.*

means In a proportion, the second and third terms.

measures of central technology Numerical values giving an overall picture of a set of data, generally including the mean, the median, and the mode.

median In a list of data ordered from least to greatest or greatest to least, the middle number or the average of the middle two numbers.

meter The basic unit of length in the metric system.

metric system A measurement system using meter, gram, and liter as the basic units.

mixed number A number that is the sum of a whole number and a common fraction, such as $1\frac{1}{2}$.

mode In a list of data, the number or item occuring most frequently.

monomial A mathematical expression that is either a constant, a variable, or a product of numbers and variables.

multiple The product of any number and another whole number.

multiplication property of equality For any numbers a and b, if $a = b$, then $ac = bc$. $6 = 6$, so $6 \times 6 = 6 \times 3$.

multiplication property of inequalities For any numbers a, b, and c with $c > 0$, if $a > b$, then $ac > bc$. If $c < 0$ and $a > b$, then $ac < bc$. So, $-2 > -3$, so $-2 \times 2 > -3 \times 2$, but $-2 \times -6 < -3 \times -6$.

multiplicative identity The number which, when multiplied by a given number, results in that given number. For real numbers, the multiplicative identity is one.

multiplicative identity property The product of a number and the multiplicative identity (1) is the original number. $35 \times 1 = 35$

multiplicative inverse A number which, when multiplied with a given number, yields a product of one. The reciprocal of a number.

---

natural number One of the numbers 1, 2, 3, 4, . . . also called counting numbers.

negative number A number with a point on the left side (negative side) of zero on a number line.

number line A line on which points are identified by numbers. The number is called the coordinate of the point. On a number line, the points are evenly spaced, and the magnitude of the number increases as you move from left to right.

numeral A name or symbol used to represent a number.

numerator In the fraction $\frac{2}{3}$, the number 2 is the numerator.

---

obtuse angle Any angle measuring between 90° and 180°.

obtuse triangle A triangle having one obtuse angle.

octagon A polygon having eight sides.

odd numbers Any number not divisible by two.

opposites Two integers having a sum of zero. -27 and 27

ordered pair Two numbers named in a specific order.

origin The point of intersection of the x-axis and y-axis in a coordinate plane.

outcome A possible result of a probability event.

---

parallel lines Lines lying in the same plane that do not intersect.

parallelogram A quadrilateral having both pairs of opposite sides parallel.

pentagon A polygon having five sides.

percent A hundredth part of a whole. $19\% = \frac{19}{100}$

perimeter The distance around a plane figure.

permutation An ordered arrangement of numbers in which members are not repeated.

perpendicular lines Intersecting lines that form right angles.

pi (π) The ratio of the circumference of a circle to its diameter.

pictograph A means of displaying statistical information that presents data in a pictorial manner.

place value Value of a digit depending on its position in a number.

plane A flat surface having no boundaries.

point A specific location in space.

polygon A simple, closed plane figure formed by three or more line segments.

polyhedron A three-dimensional shape formed by four or more plane surfaces.

power A number that can be expressed using an exponent. 2^3

prime factorization A composite number expressed as a product of prime numbers.

prime number A whole number greater than one having exactly two distinct factors, one and the number itself.

prism A three-dimensional shape having two bases that are parallel and congruent.

probability The chance or possibility of a specified event happening.

proper fraction A fraction in which the numerator is less than the denominator.

proportion Two equivalent ratios set up to make a mathematical sentence.

protractor An instrument for measuring angles.

pyramid A three-dimensional figure having three or more triangular faces and a polygonal base.

Pythagorean Theorem In a right triangle, the sum of the squares of the length of the legs is equal to the square of the length of the hypotenuse. $a^2 + b^2 = c^2$

---

quadrant One of the four regions formed by the axes of the coordinate plane.

quadrilateral A polygon having four sides.

---

radius A line segment having one endpoint at the center of a circle and the other endpoint on the circle.

range (relation) The set of second elements in the ordered pairs of a relation.

range (statistics) The difference between the greatest and least numbers in a set of numerical data.

ranking An ordered list of numbers.

rate A ratio of two measurements having different units.

ratio A comparison of two numbers, represented in one of the following ways: 2 to 5, 2 out of 5, 2:5, or $\frac{2}{5}$.

rational number A number in the form of an $\frac{\text{integer}}{\text{integer}}$, where the denominator does not equal zero. All integers are rational numbers, for example, $2 = \frac{2}{1}$.

ray A part of a line having one endpoint and extending without end in one direction.

real numbers The set of irrational and rational numbers together.

reciprocals Two rational numbers whose product is one. $\frac{2}{3}$ and $\frac{3}{2}$, $-\frac{1}{2}$ and $-\frac{2}{1}$

rectangle A parallelogram having four right angles.

reflexive property of equality For any number a, $a = a$.

regular polygon A polygon having all sides congruent and all angles congruent.

relation A set of ordered pairs.

repeating decimal A decimal in which a digit or group of digits repeats. A bar above the group of repeating decimals is used to express the repeating decimal, such as $\frac{2}{3} = 0.\overline{6}$.

rhombus A parallelogram having four congruent sides.

right angle An angle having a measure of 90°.

right triangle A triangle having a right angle.

Roman numerals A numeration system used by the Romans that utilized letters, such as I, V, X, L, C, and M.

---

sample space The set of all possible outcomes for a situation.

scale factor The numeric ratio used to produce an enlarged or reduced drawing of a picture or an object.

scalene triangle A triangle having no congruent sides.

scientific notation A means of expressing a number as a product of a number greater than or equal to one and less than ten and a power of ten.

similar figures Figures having the same shape.

simplify To combine like terms and/or to replace an expression with most common, or simplest, value.

slope The ratio of the change in y to the change in x, or the ratio of the change in the rise to the run.

slope-intercept form A linear equation in the form $y = mx + b$, where m is the slope of the graph of the equation and b is the y-intercept.

software Computer programs.

solution A replacement set for a variable that makes a mathematical sentence true.

solution set The set of numbers from the replacement set that makes an open sentence true.

sphere A three-dimensional figure consisting of points that are the same distance from the center.

square (geometric) A parallelogram having all congruent sides and all right angles.

square (numeric) The product of a number and itself.

square root One of two equal factors of a number. Since $(-5)(-5) = 25$ and $(5)(5) = 25$, $\sqrt{25} = 5$ or -5. The principal square root of 25 is 5.

statistics The field of mathematics involving the collection, analysis, and presentation of data.

stem-and-leaf display A means of organizing data in which certain digits are used as stems, and the remaining digits are used as leaves.

substitute To replace a variable with another variable, expression, or constant.

surface area The sum of the areas of all the faces of a three-dimensional figure.

symmetric property of equality For all numbers a and b, if $a = b$, then $b = a$.

symmetry A figure has symmetry if it is its own image under reflection in a line.

system of linear equations Two or more linear equations in the same two variables.

---

tally chart A systematic method of recording amounts of numerical data by utilizing tally marks (|).

term A number, variable, or product or quotient in a mathematical sentence or expression.

terminating decimal A decimal in which the only repeating digit is 0. For example, $\frac{1}{2} = 0.5$.

theoretical probability A probability determined without performing an experiment.

transitive property of equality For all numbers a, b, and c, if $a = b$ and $b = c$, then $a = c$.

trapezoid A quadrilateral having one, and only one, pair of parallel sides.

triangle A polygon having three sides.

trinomial An equation having three unlike terms.

---

variable Placeholders in mathematical expressions or sentences.

vertex The common endpoint of two rays that form an angle.

vertical angles Congruent angles formed by two intersecting lines.

volume The number of cubic units needed to fill a space.

---

whole number The set of natural numbers and zero. 0, 1, 2, 3, . . .

***x*-axis** The horizontal line of the pair of perpendicular lines that form a coordinate plane.

***x*-coordinate** The first number in an ordered pair.

***x*-intercept** The value of x in a linear equation when $y = 0$.

***y*-axis** The vertical line of the pair of perpendicular number lines that form a coordinate plane.

***y*-coordinate** The second number in an ordered pair.

***y*-intercept** The value of y in a linear equation when x is 0.

INDEX